DMU 0596661 01 0

AF689254

Applications of Plant Metabolic Engineering

Applications of Plant Metabolic Engineering

Edited by

R. Verpoorte

Leiden University, The Netherlands

A. W. Alfermann

Heinrich-Heine-Universität, Düsseldorf, Germany

and

T. S. Johnson

Reliance Life Sciences Pvt. Ltd, Navi Mumbai, India

Springer

A C.I.P. Catalogue record for this book is available from the Library of Congress.

ISBN 978-1-4020-6030-4 (HB)
ISBN 978-1-4020-6031-1 (e-book)

Published by Springer,
P.O. Box 17, 3300 AA Dordrecht, The Netherlands.

www.springer.com

DE MONTFORT UNIVERSITY LIBRARY:
B/code
Fund: 16 Date: 7/7/08
Sequence:
Class: 660.63
Suffix: APP

Printed on acid-free paper

All Rights Reserved
© 2007 Springer
No part of this work may be reproduced, stored in a retrieval system, or transmitted in any form or by any means, electronic, mechanical, photocopying, microfilming, recording or otherwise, without written permission from the Publisher, with the exception of any material supplied specifically for the purpose of being entered and executed on a computer system, for exclusive use by the purchaser of the work.

CONTENTS

CONTRIBUTORS

Asaph Aharoni Weizmann Institute of Science, P.O. Box 26, Rehovot 76100, Israel

V.A. Bapat Plant Cell Culture Technology Section, Nuclear Agriculture and Biotechnology Division, Bhabha Atomic Research Centre, Trombay, Mumbai 400 085, India

Faith C. Belanger The Biotechnology Center for Agriculture & the Environment, School of Environmental and Biological Science, Rutgers, The State University of New Jersey, New Brunswick, NJ 08903, USA Phone 732-932-8165x304, Fax 732-932-6535

Esha Bhattacharya NFCL (Nagarjuna Fertilizers Corporations Ltd). Panjagutta, Hyderabad 500 082, India

Robert G Birch Botany Department, School of Integrative Biology, The University of Queensland, Brisbane 4072 Australia

Harro J. Bouwmeester Plant Research International, P.O. Box 16, 6700 AA Wageningen, The Netherlands /harro.bouwmeester@wur.nl

Ian B. Cole Chemistry, I.K. Barber School of Arts & Sciences, University of British Columbia Okanagan, Kelowna, British Columbia, Canada, V1V 1V7

Howard G. Damude DuPont Experimental Station, Wilmington, DE 19880 USA

Henry Daniell Pegasus Professor & Trustee Chair, University of Central Florida, 4000 Central Florida Blvd, Dept. Molecular Biology & Microbiology Biomolecular Science, Bldg # 20, Room 336, Orlando FL 32816-2364, USA./Tel: 407-823-0952, Fax: 407-823-0956 /daniell@mail.ucf.edu

T.R. Ganapathi Plant Cell Culture Technology Section, Nuclear Agriculture and Biotechnology Division, Bhabha Atomic Research Centre, Trombay, Mumbai 400 085, India

Suvi T Häkkinen VTT Technical Research Centre of Finland, P.O. Box 1000, FI – 02044 VTT (Espoo), Finland

Takashi Hashimoto Graduate School of Biological Sciences, Nara Institute of Science and Technology, 8916-5 Takayama, Ikoma, Nara 630-0192, Japan /hasimoto@bs.naist.ac.jp

Daphna Havkin-Frenkel The Biotechnology Center for Agriculture & the Environment, School of Environmental and Biological Science, Rutgers, The State University of New Jersey, New Brunswick, NJ 08903, USA Phone 732-932-8165x304, Fax 732-932-6535

Koji Inai Graduate School of Biological Sciences, Nara Institute of Science and Technology, 8916-5 Takayama, Ikoma, Nara 630-0192, Japan

Anthony J. Kinney DuPont Experimental Station, Wilmington, DE 19880 USA

Argelia Lorence Arkansas Biosciences Institute, Arkansas State University, P.O. Box 639, State University, AR 72467

Joost Lücker University of British Columbia, Faculty of Land and Food Systems, Wine Research Centre, 216-2205 East Mall, Vancouver, B.C., V6T 1Z4, Canada

Erin K. Marasco Dept. Biochemistry, Molecular Biology and Biophysics, University of Minnesota, St. Paul, MN 55108, USA

Susan J. Murch Chemistry, I.K. Barber School of Arts & Sciences, University of British Columbia Okanagan, Kelowna, British Columbia, Canada, V1V 1V7

Craig L Nessler Virginia Agricultural Experiment Station, Virginia Tech, 104 Hutcheson Hall, Blacksburg, VA 24061 /cnessler@vt.edu

Masaaki Noji Graduate School of Pharmaceutical Sciences, Chiba University, Yayoi-cho 1-33, Inage-ku, Chiba 263-8522, Japan

Kirsi-Marja Oksman-Caldentey VTT Technical Research Centre of Finland, P.O. Box 1000, FI – 02044 VTT (Espoo), Finland /kirsi-marja.oksman@vtt.fi

Waraporn Putalun Department of Pharmacognosy, Graduate School of Pharmaceutical Sciences, Kyushu University, 3-1-1 Maidashi, Higashi-ku, Fukuoka Japan 812-8582

M.V. Rajam Department of Genetics, University of Delhi – South campus, Benito Juarez Road, New De

Heiko Rischer VTT Technical Research Centre of Finland, P.O. Box 1000, FI – 02044 VTT (Espoo), Finland

Tracey Ruhlman Dept. of Molecular Biology & Microbiology, University of Central Florida, Biomolecular Science, Building #20, Room 336, Orlando, FL 32816-2364, USA

Kazuki Saito Graduate School of Pharmaceutical Sciences, Chiba University, Yayoi-cho 1-33, Inage-ku, Chiba 263-8522, Japan, RIKEN Plant Science Center, 1-7-22 Suehiro-cho, Tsurumi-ku, Yokohama 230-0045, Japan /ksaito@faculty.chiba-u.jp
Fumihiko Sato Department of Plant Gene and Totipotency, Graduate School of Biostudies, Kyoto University, Kyoto, 606-8502, Japan

Praveen K. Saxena Department of Plant Agriculture, University of Guelph, Guelph, Ontario, Canada, N1G 2W1

Claudia Schmidt-Dannert Dept. Biochemistry, Molecular Biology and Biophysics, University of Minnesota, 1479 Gortner Avenue St. Paul, MN 55108, USA./Tel: 1-612-625-5782 /schmi232@umn.edu

Yukihiro Shoyama Faculty Pharmaceutical Science, Nagasaki International University, 2825-7 Hansutenbosu-cho, Sasebo, Nagasaki 859-3298, Japan

L. Srinivas Plant Cell Culture Technology Section, Nuclear Agriculture and Biotechnology Division, Bhabha Atomic Research Centre, Trombay, Mumbai 400 085, India

G.B. Sunil Kumar Plant Cell Culture Technology Section, Nuclear Agriculture and Biotechnology Division, Bhabha Atomic Research Centre, Trombay, Mumbai 400 085, India

Hiroyuki Tanaka Department of Pharmacognosy, Graduate School of Pharmaceutical Sciences, Kyushu University, 3-1-1 Maidashi, Higashi-ku, Fukuoka Japan 812-8582

R. Verpoorte Division of Pharmacognosy, section Metabolomics, IBL, Leiden University, PO Box 9502, 2300RA Leiden, The Netherlands/VERPOORT @LACDR.LeidenUniv.NL

INTRODUCTION

R. VERPOORTE
Division of Pharmacognosy, section Metabolomics, IBL, Leiden University
PO Box 9502, 2300RA Leiden, The Netherlands,
Email: VERPOORT@LACDR.LeidenUniv.NL

In the past years the interest in plant secondary has increased rapidly. Three major reasons can be mentioned for this. First because plants are a major source for the production of medicines and the development of novel medicines; second because plants contain health promoting secondary metabolites, third because of the interest in the resistance of plants against pests and diseases in which the secondary metabolism plays a crucial role. Seven years ago we edited a book (Verpoorte and Alfermann 2000) on the engineering of plant secondary metabolism. A general overview was given of plant secondary metabolism, and the strategies one could envisage for engineering plant secondary metabolite pathways. Furthermore, a number of examples were presented describing the state-of-the-art of engineering plant secondary metabolism. Now we have again compiled a series of papers on the engineering of plant metabolism.

Obviously in the past period quite a few applications have been reported. Some of them were successful, others were less successful and the unsuccessful ones we will probably never hear of. Reasons for failure are often basic biological problems: the regeneration of transgenic plants from transformed cells, and the stability of transformed cell lines or transformed plants. The toolkit for transformation and overexpressing genes has improved and consequently the number of successful transformations increased. However, the major difficulties concern the fact that the biosynthetic pathways involved proved to be much more complicated than originally thought. Engineering a single step may result in an increase of the immediate product but not necessarily in an increase of the final product of the pathway. As we discussed in the previous book, problems of pathway architecture, interaction between various pathways in the total metabolic network, enzyme complexes, compartmentation, feedback inhibition, and regulation all play an important role. It means that unraveling pathways on all levels should have the highest priority. Eventually this might enable us to design efficient approaches to pathway engineering.

PATHWAY ELUCIDATION

The key to genetic engineering is the detailed knowledge of the pathways of interest. The step-by-step approach for elucidation of pathways remains an important, though elaborate, tool in biosynthetic studies. Retrobiosynthetic studies and labeling experiments have shown to be excellent tools to confirm pathways on the level of intermediates (e.g. Eisenreich et al., 2004). Once the intermediates are known, one has to identify the enzymes involved. However, the isolation of enzymes catalyzing the individual steps of a pathway is hampered by, among others, low levels of the enzyme, instability of the enzyme, and problems in obtaining the substrate for measuring activity. Consequently many of the secondary metabolite pathways still have quite a few black boxes, for which paper chemistry has proposed intermediates, but for which no actual experimental evidence exists.

To elucidate pathways various molecular biological approaches have been advocated. Many are based on making "mutants" by knocking out genes (transposon tagging, RNAi, etc.). However, the problem is the identification of the steps which have been blocked in a mutant. Plants in which an essential biosynthetic gene for the flower pigments is affected are immediately observed by eye. In a split second one can screen hundreds of plants for the flower color. However, in case of a colorless metabolite in roots or leaves elaborate analytical methods are needed to identify a mutant. This explains why the flavonoids/anthocyanin biosynthesis is one of best known biosynthetic pathways (Springob et al., 2003).

FUNCTIONAL GENOMICS

Because of the problems in pathway mapping, functional genomics was thought to be a way to elucidate secondary metabolite pathways on all levels from genes to products. Functional genomics aims at determining the function of genes. Transcriptomic data, proteomic data, metabolomic data and physiological functions are all matched through biostatistical methods and bioinformatics. In case of organisms with a known genome sequence such an approach may be successful. But lack of sequence data is a major constraint in studying secondary metabolism in non-model plants.

Proteomics is not the panaceae to solve these problems, as only a small percentage of all proteins will be observed. Particularly low abundance proteins will not be observed (Jacobs et al., 2000, 2005; Chen and Harmon 2006). Secondary metabolism often only represents a small part of the total metabolism, e.g. the energy needed for the biosynthesis of alkaloids was found to be less than 1% of the total metabolism in the development of *Cinchona* seedlings (Aerts et al., 1990, 1991). The enzymes involved may be below the level of detection. For example in proteomics of *Catharanthus roseus* cell cultures some 100 proteins were found to be induced when alkaloid biosynthesis was turned on. Only two of these are known indole alkaloid biosynthesis enzymes (Jacobs et al., 2000, 2005). About 60 had homology with peptide sequences from primary metabolite genes from other plants, whereas the peptides of about 40 proteins did not match with any known sequence. To identify the genes encoding these proteins and determine their function would be quite difficult and elaborate.

Goossens and co-workers (Goossens et al., 2003; Oksman-Caldentey et al., 2004; Rischer et al. 2006) developed a cDNA-amplified fragment-length polymorphism method that in combination with targeted metabolomics can be used to identify genes involved in certain pathways. Indeed it was shown that in this way a number of genes involved with the induction of alkaloid biosynthesis in *Catharanthus roseus* can be identified, though many of them are primary metabolism related genes, and not directly involved in the pathway. Genes with sequences not matching any known genes are candidates for structural genes of species specific pathways, but it requires extensive further studies to identify the precise role.

Metabolomics, the latest of the – omics family, aims at the qualitative and quantitative analysis of all metabolites in an organism (Fiehn, 2001; Rochfort, 2005; Ryan and Robards, 2006). Metabolomics can be considered as the chemical characterization of a phenotype, and is thus an important tool in functional genomics. It can be used to measure the levels of compounds under different conditions. By correlating these data with proteomic and transcriptomic data one may get information about genes involved in the regulation of pathways and the structural genes involved.

The integration of all the – omics data and physiological data, i.e. taking a holistic view at the organism at all levels without a starting hypothesis, is a novel approach to biological research now known as systems biology. Also for plants this approach is now recognized as a very promising way to study for example plant interaction with insects or microorganisms (Oksman-Caldentey et al., 2004; Sweetlove and Fernie, 2005; Verpoorte et al., 2005).

Even though the various tools of functional genomics can be helpful in identifying genes involved in secondary metabolite pathways, none of them is capable of identifying all intermediates, proteins or genes involved in a pathway. Besides problems of low concentrations, the major problem is that in a living system, the changes in levels of transcripts, activity of enzymes and level of metabolites have different dynamics. The final result of an induction at gene level is only observed many hours or days later, if one even at all can speak about a final result in a dynamic system.

COMPARTMENTATION

The compartmentation of secondary metabolite biosynthetic pathways has received much attention in the past years. Several reviews on this topic have been published (e.g. Kutchan, 2005; Yazaki, 2005). If we take *Catharanthus roseus* as an example it has been shown that both intra- and intercellular compartmentation do play an important role. The early terpenoid precursors from the MEP-terpenoid pathway and geraniol-10-hydroxylase are made in different cells (internal phloem parenchyma) than the other important precursor tryptamine (epidermis). The last step of the biosynthesis of the terpenoid precursor secologanin occurs also in these epidermis cells. Strictosidine synthase present in the vacuoles of these cells catalyzes the condensation of tryptamine and secologanin to yield the intermediate strictosidine, which is the starting point for several different pathways leading to different types of terpenoid indole alkaloid skeletons (for a review see van der Heijden et al. 2004). The branch leading to vindoline is present in other specialized cells (ideoblasts and laticifers), thus requiring intercellular transport of strictosidine or a later product from the vindoline branch.

Concerning the intracellular compartmentation, it is known that plastids are the source of the terpenoid precursors and tryptophan. Decarboxylation of tryptophan occurs in the cytosol, whereas strictosidine is produced in the vacuole from the precursors secologanin and tryptamine in the vacuole. Further steps are again outside the vacuole. The required glucosidase, for example, is localized in the ER (Geerlings et al, 2000), whereas a crucial step in the vindoline biosynthesis occurs in green chloroplasts (for a review see van der Heijden et al., 2004; Kutchan, 2005). This has implications for engineering alkaloid production in the native host of the pathway. One needs to express the gene in the correct compartment and the correct type of cell, otherwise no or little effect is achieved. But even more important, it means that the flux through a pathway is not only controlled by structural genes catalyzing a chemical reaction, but also by transport from the site of production of a precursor to the site of the next enzyme.

TRANSPORT

Because of the different compartments involved in biosynthetic pathways, the intermediates need to be reallocated to the proper compartment. Reallocation is a complex phenomenon in plants and plant cells. Diffusion is always involved in the reallocation of compounds. Affinity for lipid membranes (lipophylic properties of a compound) and intra- and extracellular fluids (hydrophilic properties of a compound) are important factors for diffusion driven transport through membranes (Blom et al. 1991). On top of that active transport through membranes may occur through e.g. a proton antiport mechanism or ABC-type of transporters (such as proteins belonging to the PDR, MRP and MDR families). For example from measuring transport of alkaloids and iridoids into isolated *C. roseus* vacuoles, we concluded that bidirectional transport occurs through different type of transporters (MRD out and ABC and MRP proteins in) with quite different rates for the different *C. roseus* alkaloids and secologanin (Roytrakul, 2004; Roytrakul and Verpoorte, 2007). In other cell organelles and the cell membranes similar processes might occur. Furthermore, conjugation of compounds with e.g. glutathione under the influence of glutathione transferases and peroxidases may play a role in the vacuolar transport of certain compounds (Dean and Devarenne, 1997; Grotewold 2004; Yazaki 2005). Transport is thus extremely complex as besides diffusion driven transport, different types of active transport are involved, with different directions and for each single compound a different selectivity. Biosynthetic rates might thus very well be controlled on the level of transport.

Besides transport also storage is an important aspect of secondary metabolite production. Vacuoles are storage organelles, but import of the products is required. For example overexpression of the terpenoid indole alkaloid pathway genes encoding tryptophan decarboxylase and strictosidine synthase in tobacco cells in combination with feeding of the precursor secologanin did not result in any storage of the products. Instead the products were excreted into the medium, which is opposite to the situation in *C. roseus* cells where the alkaloids are stored in the vacuole (Hallard et al., 1997).

In this introduction I will not try to give a complete overview of all aspects of compartmentation, transport and storage. I only want to conclude that the green factory in many aspects is very similar to an industrial factory, (e.g. a factory assembling cars). Both require energy for the production process, transport from the sites of the production of building blocks to the site where these are assembled to yield the final product and a storage site for the stock of the final product. It might thus be possible to apply technical engineering strategies to plan plant metabolic engineering.

TARGETS FOR METABOLIC ENGINEERING

Metabolic engineering is possible, but what are the targets? Why should one like to alter the metabolism of plants?

The following goals can be mentioned:

- Improved quality for producer (farmer)
  - Improved yield
  - Improved resistance against pests and diseases
  - Improved traits for cultivation and harvesting
- Improved quality for processing (industry)
  - Storage of food
  - Suppress level of unwanted products (e.g. toxic compounds) or improve quality of product (e.g. starch, lignins)
  - Higher level of specialty chemicals, e.g. for medicines
  - Fiber quality
  - Biofuel viscosity, stability
- Novel compounds for drug development (industry)
- Improved quality for consumer
  - Taste of food
  - Color of food or flowers
  - Increased level of health improving compounds
  - Lower level of undesired compounds

Looking at this list of possibilities it is clear that the applications concern changes in primary metabolism or in secondary metabolism. It also implies that choices have to be made, e.g. does one go for yields or quality (Morris and Sands 2006; Singh et al. 2006). Secondary metabolism is per definition species specific, it serves the producing organism to survive in its ecosystem. In plants it is, among others, involved in defense against pests and diseases, and in attracting pollinators. Furthermore taste, flavor and color of our food are related to secondary metabolism. Also various health effects of food are connected with secondary metabolites. The defense compounds are of different character, some are constitutively expressed (phytoanticipins), others are only biosynthesized after wounding or in infection (phytoalexins) (Zhao et al. 2005). That means that the regulation of secondary metabolism in part is developmentally regulated, in part is dependent of external (stress) signals.

Starting from ubiquitous primary metabolites as precursors the number of steps in secondary metabolite pathways differs considerably. The biosynthesis of the phytoalexin resveratrol from ubiquitous primary metabolites consists of only a single step, catalyzed by one single enzyme, encoded by one single gene (Hain and Grimmig, 2000). Whereas the biosynthesis of an indole alkaloid like vinblastine, includes at least 30 different steps, at least three different cells types and four different cellular compartments, and consequently also is regulated by transport systems (van der Heijden et al. 2004; Pasquali et al., 2006). Because secondary metabolism is species-specific, the knowledge about most pathways is limited, and very few pathways in plants have been fully elucidated to all levels of intermediates, enzymes and genes.

STRATEGIES

For developing a strategy for metabolic engineering of plant secondary metabolism, one has to keep all the above mentioned aspects in mind. There is a clear difference in approach for increasing or decreasing the flux through a pathway.

Decreasing a flux could for example be of interest in case of undesired (toxic) compounds, or to cut off certain pathways that compete with the pathway of interest. Also catabolic pathways might be of interest to cut, in order to increase the level of a desired compound. To decrease a flux, the level of the protein of interest can be decreased by an antisense or RNAi approach or by overexpresssing an antibody of the selected enzyme of the target pathway. As long as not any vital pathway is knocked out, this approach should be easy with a good chance of success.

To increase the level of a compound, one needs to know the pathway into much detail to be able to select targets for engineering. This should result in the identification of possible sites for modification, e.g. overcoming limiting steps. As mentioned above, only a few genes of plant secondary metabolite pathways are known. Engineering long pathways thus requires extensive studies to elucidate the pathway. One may also consider the use of microbial genes to achieve certain reactions in plants for which the encoding plant genes are not known yet. The production of salicylate in plants by overexpression of microbial genes is such an example (Verberne et al. 2000).

One of the problems with pathway engineering is that many genes need to be transformed into the plant. Instead one may also consider the overexpression of regulatory genes. This may result in the induction of a series of genes of a secondary metabolite pathway (Grotewold et al. 1998). An example is the engineering of the signal transduction pathway leading to the induction of a pathway of interest, e.g. overexpression of a transcription factor (Gantet and Memelink, 2002; Memelink et al. 2001; Memelink 2005a, 2005b). This approach does not necessarily result in an increased level of the desired end products as it was found for alkaloids in *C. roseus*, other flux limiting steps may remain (Memelink et al. 2001).

Another possibility is to modify a constitutive pathway into an inducible one by introducing inducible promoters. This has the advantage that one can separate growth and secondary metabolite production, and avoiding a competition between the two processes for the limited energy and precursor pools in the plant cells. Highest production of secondary metabolites in a bioreactor can be achieved in a fed-batch-type of process in which growth and product formation are in different phases of the process (Verpoorte and ten Hoopen, 2006; Zhao and Verpoorte 2007).

NEW COMPOUNDS

To produce new compounds for a plant, one can add extra steps to an already existing route, or introduce a new enzyme for the plant that catalyzes an early step in a route. Examples are the production of alkaloids in the hairy roots of an iridoid producing plant *Weigela* "Stryriaca" by overexpression of tryptophan decarboxylase and strictosidine synthase (Hallard, 2000). Another example is the overexpression of a terpene synthase (see Luecker et al., Chapter 9; Aharoni et al., 2005; Dudareva et al., 2006), resulting in the formation of a novel molecular skeleton for the host plant. This molecule may serve as a scaffold for further reactions catalyzed by enzymes in the plant to yield a novel product for that plant, and maybe even a totally novel compound. Such a recombinatorial biochemistry approach aiming at production of novel compounds is of interest for developing novel leads for drug development (Julsing et al. 2006). It might also be of interest for increasing the resistance of the plant against pest and diseases. However, from the point of view of safety, it might need quite some work to proof that the new compound(s) is(are) not toxic for the consumer.

Producing a known compound in another plant might be of interest for several reasons. The most obvious is the introduction of health promoting compounds in food plants (Yonekur-Sakakibara and Saito, 2006). Golden rice is an excellent example of extra nutrional value that can be created by metabolic engineering to increase the vitamine A level (Al-Babili and Beyer, 2005). Flavonoid production is a target for metabolic engineering in plants, with the aim to increase antioxidant levels in food (Forkmann and Martens, 2001; Schijlen et al. 2004).

An other reason could be that the target plant has better properties for producing the compound than others. The *Atropa belladonna* plant producing scopolamine is such an example (Hashimoto et al. 1993).

In studies on metabolic engineering one often sees only small increases in fluxes if compared to wild type plants or plant cells. Apparently the complex metabolic networks do not allow major reshuffling of carbon fluxes. Therefore rebuilding pathways in the simplest possible cells is seen as a possibility to reach high productivity. Synthetic biology aims at the construction of such production cells in which all unnecessary genetic material has been cut to leave a cell rigged for growth. In such cells the desired pathways could be introduced. Obviously such cells will be of microbial origin. The possibility of using microorganisms for the production of plant products has been shown by several studies in the past years, e.g. terpenoids (Chemler et al., 2006; Kutchan, 1989; Martin et al., 2001; Ro et al., 2006; Szczebara et al., 2003; Withers and Keasling, 2007) and alkaloids (Kutchan, 1989; Geerlings et al., 2001). The total biosynthesis of hydrocortisone in yeast is an excellent example of the potential of this approach for production of medicines (Szczebara et al. 2003).

CONSTRAINTS

In principle there are infinite possibilities for applying metabolic engineering, but they are limited by:

- public acceptance
- safety issues
- lack of knowledge of biosynthetic pathways
- regeneration of genetically transformed plants
- viability of plants with altered metabolism

The major area for application on the short term is the production of medicinal compounds, either known or new, in plants or plant cell or tissue cultures. At present the commercial application of genetically modified secondary metabolism is in altered flower colors.

PERSPECTIVES

To end this introduction I want to mention the great perspectives of plant metabolic engineering. In comparing the present book with that of 7 years ago we can see that some dreams already have become true, but much more can be expected to become true in the coming years. Above I dealt with some of the results and quite extensively with the difficulties, as we need to identify the bottlenecks to further open the way for applications. Predicting the future is difficult, there will always be unexpected breakthroughs that will change the complete picture. But at least for certain technologies it is easy to predict that the methods, the tools, will become easier and faster. For example for sequencing genomes it is easy to predict that in 10 years it will be a minor project to sequence the genome of a plant, which means that for the functional genomics at least the problem of non-model plants will be solved. In the mean time the number of plant genes with a known function will also grow steadily, and consequently it will become easier to identify the function of genes in other plants, and thus to elucidate pathways. Also metabolomics will further develop into fluxomics, instead of taking a "picture" at one single time point, one will make a "film" of the metabolic fluxes, in microorganisms this sort of studies are already in progress, for plants it will come in due time. Fluxomics will be of great use to identify the bottlenecks in a pathway that needs to be addressed to be able to increase the flux.

With such improved tools, the great chemical potential of plants for making complex chemical structures will be further developed. The advances made with recombinatorial biochemistry in microorganisms and metagenomics (Chemler et al., 2006; Ordovas and Moose, 2006) show us that we may expect a lot from exploring the metagenome of all plants for useful genes to make novel complex compounds with interesting biological activities.

Considering the various goals, one may expect that transgenic plant cell cultures producing medicines might be easily accepted by all people now opposing transgenic plants. In fact plant cell cultures might be an excellent platform for metabolic engineering using different modules containing sets of genes that could be used for recombinatorial biochemistry. Transgenic plants producing desired fine chemicals or with improved traits for non-food applications are other important areas where we may expect interesting applications. However, for food use it may take a longer time as in that case safety studies will be required, but on the longer term the perspectives seems bright for plants with increased levels of health promoting constituents.

Engineering the green cell factory is in its infancy, but it looks like a healthy baby with a bright future ahead!

REFERENCES

Aerts R.J., van der Leer T., van der Heijden R. and Verpoorte R. (1990) Developmental regulation of alkaloid production in *Cinchona* seedlings. J. Plant Physiol. 136(1990)86–91.

Aerts R.J., Snoeijer W., Aerts-Teerlink O., van der Meijden E. and Verpoorte R. (1991) Control and biological implications of alkaloid synthesis in *Cinchona* seedlings. Phytochemistry 30, 3571–3577.

Aharoni A., Jongsma M.A. and Bouwmeester H.J. (2005) Volatile science? Metabolic engineering of terpenoids in plants. Trends Plant Sci. 10, 594–602.

Al-Babili S., and Beyer P. (2005) Golden Rice – five years on the road – five years to go? Trends Plant Sci. 10, 565–573.

Blom T.J.M., Sierra M., van Vliet T.B., Franke-van Dijk M.E.I., de Koning P., van Iren F., Verpoorte R. en Libbenga K.R. (1991) The transport and accumulation of the alkaloid ajmalicine and the bioconversion of ajmalicine into serpentine in isolated vacuoles of *Catharanthus roseus* (L.) G. Don. Planta 183, 170–177.

Chemler J.A.., Yan Y.J., Koffas M.A.G.(2006) Biosynthesis of isoprenoids, polyunsaturated fatty acids and flavonoids in *Saccharomyces cerevisiae*. Microb. Cell Fact. 5, 20.

Chen S.X., and Harmon A.C. (2006) Advances in plant proteomics. Proteomics 6, 5504–5516.

Dean J.V.,and Devarenne T.P. (1997) Peroxidase-mediated conjugation of glutathione to unsaturated phenylpropanoids. Evidence against glutathione S-transferase involvement. Physiol. Plant. 99, 271–278.

Dudareva N., Negre F., Nagegowda D.A., and Orlova I. (2006) Plant volatiles: Recent advances and future perspectives. Crit. Rev. Plant Sci. 25, 417–440.

Eisenreich W., Bacher A., Arigoni D., et al. (2004) Biosynthesis of isoprenoids via the non-mevalonate pathway Cell. Mol. Life Sci. 61, 1401–1426.

Fiehn O., (2001) Combining genomics, metabolome analysis, and biochemical modelling to understand metabolic networks. Comp. Funct. Genom., 155–168.

Forkmann G, Martens S: Metabolic engineering and applications of flavonoids. Curr Opin Biotechnol 2001, 12: 155–160.

Gantet P. and Memelink J. (2002) Transcription factors: tools to engineer the production of pharmacologically active plant metabolites. Trends Pharm. Sci. 23, 563–569.

Geerlings A., Martinez-Lozano Ibanez M., Memelink J., van der HeijdenR. and Verpoorte R. (2000) The strictosidine ß-D-glucosidase gene from *Catharanthus roseus* is regulated coordinately with other terpenoid-indole alkaloid biosynthetic genes and the encoded enzyme is located in the endoplasmic reticulum. JBC 275, 3051–3056.

Geerlings A., Redondo F.J., Contin A., et al. (2001) Biotransformation of tryptamine and secologanin into plant terpenoid indole alkaloids by transgenic yeast. Appl. Microbiol. Biotechnol. 56, 420–424.

Goossens A., Hakkinen S.T. Laakso I. et al. (2003) A functional genomics approach toward the understanding of secondary metabolism in plant cells. Proc. Natl. Acad. Sci USA 100, 8595–8600.

Grotewold E, Chamberlin M, Snook M, Siame B, Butler L, Swenson J, Maddock S, St Clair G, Bowen B: Engineering secondary metabolism in maize cells ectopic expression of transcription factors. Plant Cell 1998, 10: 721–740.

Grotewold E (2004) The challenges of moving chemicals within and out of cells: insights into the transport of plant natural products. Planta 219, 906–909.

Hain R, and Grimmig B. (2000) In: Metabolic engineering of plant secondary metabolism. Verpoorte R. and Alfermann A.W. (Eds). Kluwer Academic Publishers, Dordrecht.

Hallard D.A.C. (2000) Transgenic plant cells for the production of indole alkaloids. PhD Thesis, Leiden 2000.

Hallard D., van der Heijden R., Verpoorte R. et al. (1997) Suspension cultured transgenic cells of *Nicotiana tabacum* expressing tryptophan decarboxylase and strictosidine synthase cDNAs from *Catharanthus roseus* produce strictosidine upon feeding of secologanin. Plant Cell Rep. 17, 50–54.

Hashimoto T., Yun D.-J., and Yamada Y. (1993) Production of tropane alkaloids in genetically engineered root cultures. Phytochemistry 32, 713–718.

Jacobs D.I., van der Heijden R. and Verpoorte R. (2000) Proteomics in plant biotechnology and secondary metabolism research. Phytochem. Anal. 11, 277–287.

Jacobs D.I., Gaspari M., van der Greef J., et al. (2005) Proteome analysis of *Catharanthus roseus* cultured cells for the identification of proteins involved in alkaloid biosynthesis and finding of novel sequences. Planta 221, 690–704.

Julsing M.K., Koulman A., Woerdenbag H.J., Quax W.J., and, Kayser O. (2006) Combinatorial biosynthesis of medicinal plant secondary metabolites Biomol. Engin. 23, 265–279.

Kutchan T.M. (1989) Expression of enzymatically active cloned strictosidine synthase from the higher plant *Rauvolfia serpentina* in *Escherichia coli*. FEBS Lett. 257: 127–130.

Kutchan T. (2005) A role for intra- and intercellular translocation in natural products biosynthesis. Curr. Opin. Plant Biol. 8, 292–300.

Ordovas J.M., and Mooser V. (2006) Metagenomics: the role of the microbiome in cardiovascular diseases. Curr. Opin. Lipidol. 17, 157–161.

Pleiss J. (2006) The promise of synthetic biology. Appl. Microbiol. Biotechnol. 73, 735–739.

Martin V.J.J., Yoshikuni Y. and Keasling J.D. (2001) The in-vivo synthesis of plant sesequiterpenes by *Escherichia coli*. Biotechnol. Bioengin. 75, 497–503.

Memelink J. (2005a) Tailoring the plant metabolome without a loose stitch. Trends Plant Sci. 10,305–307.

Memelink J (2005b) The use of genetics to dissect plant secondary pathways. Curr. Opin. Plant Biol. 8, 230–235.

Memelink J., Verpoorte R. and Kijne J.W. (2001) ORCAnization of jasmonate-responsive gene expression in alkaloid metabolism. Trends Plant Sci. 6, 212–219.

Morris C.E. and sands D.C. (2006) The breeder's dilemma – yield or nutrition. Nature Biotechnol. 24, 1078–1080.

Oksman-Caldentey K.M., Inze D., and Oresic M. (2004) Connecting genes to metabolites by a systems biology approach. Proc. Natl. Acad. Sci USA 101, 9949–9950.

Pasquali G., Porto D.D., Fett-Neto A.G. (2006) Metabolic engineering of cell cultures versus whole plant complexity in production of bioactive monoterpene indole alkaloids: Recent progress related to old dilemma. J. Biosci. Bioengin. 101, 287–296.

Rischer H., Oresic M., Seppanen-Laakso T., et al. (2006) Gene-to-metabolite networks for terpenoid indole alkaloid biosynthesis in *Catharanthus roseus* cells. Proc. Natl. Acad. Sci USA 103, 5614–5619.

Ro D.-K., Paradise E.M., Ouellet M., et al. Production of the antimalaria drug precursor artemisinic acid in engineered yeast. Nature 440, 940–943.

Rochfort S. (2005) Metabolomics reviewed: A new "Omics" platform technology for systems biology and implications for natural products research. J. Nat. Prod. 68, 1813–1820.

Roytrakul S. (2004) Transport of alkaloids and its precursors through the vacuolar membrane of *Catharanthus roseus*. PhD-Thesis.

Roytrakul S. and Verpoorte R. (2007) Role of vacuolar transporter proteins in plant secondary metabolism: *Catharanthus roseus* cell culture. Phytochem. Rev. in press.

Ryan D. and, Robards K. (2006) Analytical chemistry considerations in plant metabolomics. Sep. Purif. Rev. 35, 319–356.

Schijlen E.G.W., de Vos C.H.R., van Tunen A.J., et al. (2004) Modification of flavonoid biosynthesis in crop plants. Phytochemistry 65, 2631–2648.

Singh O.V., Ghai S., Paul D., and Jain R.K. (2006) genetically modified crops: success, safety assessment and public concern. Appl. Microbiol. Biotechnol. 71, 598–607.

Springob K., Nakajima J., Yamazaki M, et al. (2003) Recent advances in the biosynthesis and accumulation of anthocyanins. Nat. Prod. Rep. 20, 288–303.

Sweetlove L.J. and, Fernie A.R. (2005) Regulation of metabolic networks: understanding metabolic complexity in the systems biology era. New Phytol. 168, 9–23.

Szczebara F.M., Chandelier C., Villeret C., et al. (2003) Total biosynthesis of hydrocortisone from a simple carbon source in yeast. Nat. Biotechnol. 21, 143–149.

van der Heijden R., Jacobs D.I., Snoeijer W., Hallard D. and Verpoorte R. (2004) *Catharanthus roseus* alkaloids: Pharmacognosy and biotechnology. Curr. Med. Chem. 11, 1241–1253.

Verberne M., Verpoorte R., Bol J., et al. (2000) Overproduction of salicylic acid in plants by bacterial transgenes enhances pathogen resistance. Nature Biotechnol. 18, 779–783.

Verpoorte R. and Alfermann A.W. (2000) Metabolic engineering of plant secondary metabolism. Kluwer Academic Publishers, Dordrecht.

Verpoorte R., Choi Y.H., and Kim H.K. (2005) Ethnopharmacology and systems biology: A perfect holistic match. J. Ethnopharmacol. 100, 53–56.

Verpoorte R., and ten Hoopen H.J.G. (2006) Plant Cell Biotechnology. In Basic Biotechnology. C. Ratledge and B. Kristiansen (Eds). Cambridge University Press, Cambridge, pp 549–577.

Withers S.T. and Keasling J.D. (2007) Biosynthesis and engineering of isoprenoid small molecules. Appl. Microbiol. Biotechnol. 73, 980–990.

Yazaki K. (2005) Transporters of secondary metabolites. Curr. Opin. Plant Biol. 8, 301–307.

Yonekura-Sakakibara K. and Saito K. (2006) Review: genetically modified plants for the promotion of human health. Biotechnol. Lett. 28, 1983–1991.

Zhao J., L.C. Davis L.C., Xiaoyan Tang, and Verpoorte R. (2005) Elicitor Signal Transduction Leading to Production of Plant Secondary Metabolites. Biotechnol. Adv. 23, 283–333.

Zhao J., and Verpoorte R. (2007) Scaleup Production of Indole Alkaloids by *Catharanthus roseus* Cell Cultures in Bioreactor: From Biochemical Processing to Metabolic Engineering. Phytochem. Rev. in press.

CHAPTER 1

BIOSYNTHESIS OF PLANT NATURAL PRODUCTS AND CHARACTERIZATION OF PLANT BIOSYNTHETIC PATHWAYS IN RECOMBINANT MICROORGANISMS

ERIN K. MARASCO AND CLAUDIA SCHMIDT-DANNERT[§]

Dept. Biochemistry, Molecular Biology and Biophysics, University of Minnesota, St. Paul, MN 55108, USA

Abstract: Plant natural products are important medicinal and flavor and fragrance compounds. Many of these metabolites have complex structures that cannot be produced economically through total chemical synthesis. The production of natural products for consumer use or as scaffolds for more complex molecules has to rely upon extraction from plant materials or the development of engineered production hosts. Limitations in the development of engineered biosynthetic pathways stem from incomplete knowledge of plant biosynthetic pathways. There is a strong dependence on recombinant microorganisms to elucidate plant biosynthetic pathways and characterize individual biosynthetic enzymes. As more pathways are characterized, the emphasis on microorganisms will shift from single enzyme studies to the construction of short biosynthetic pathways for the targeted overproduction of compounds

Keywords: recombinant; microbial pathway engineering; plant natural products; microbial gene expression

1. INTRODUCTION

Humans have been exploiting plant natural products for centuries as medicines, dyes and to improve food quality. The two largest industries that rely on plant extracts are the pharmaceutical and consumers products industry; jointly they possess a combined market value in 2002 of US $31.5 billion in the USA alone (Raskin et al., 2002; Papnikolaw 1998). Almost one quarter of all prescribed pharmaceuticals

[§] To whom the correspondence should be addressed: Claudia Schmidt-Dannert, Dept. Biochemistry, Molecular Biology and Biophysics, University of Minnesota, 1479 Gortner Avenue St. Paul, MN 55108, USA. E-mail: schmi232@umn.edu Phone: 1-612-625-5782

R. Verpoorte et al. (eds.), Applications of Plant Metabolic Engineering, 1–43.
© 2007 *Springer.*

contain compounds that are directly or indirectly (through semi-synthesis) derived from plants (Cragg et al., 1997; Oksman-Caldentey et al., 2004). Industries such as food, beverage, pharmaceuticals, nutraceuticals, soaps and detergents, cosmetics and toiletries all rely on aroma additives derived from plants.

Many promising plant phytochemicals with prospects in the development of new pharmaceuticals have such complex chemical structures that total chemical synthesis is economically prohibitive. Thus, production of these compounds or precursor scaffolds (which can be further modified synthetically) has to rely on extraction from plant materials or the development of engineered production hosts for their targeted overproduction. Aroma and flavor compounds, on the other hand, typically have less complex chemical structures allowing their production via synthetic routes. However, consumers' concerns about health and environmental impact of synthetic chemicals and the production thereof has created a market for "natural" flavors and aromas (Vanderhaegen et al., 2003). Such "natural" ingredients can either be directly derived from plant materials or generated by enzymatic activities or fermentation, which includes the use of engineered hosts, according to the US Food and Drug Administration (Food and Drug Administration, 2001).

To develop consumer goods based on plant derived natural products it is important to understand metabolite biosynthesis on a molecular level. Knowledge of biosynthetic enzymes and their corresponding genes, as well as the complex underlying regulation of metabolic pathways in plants, would enable overproduction of plant phytochemicals of value in engineered plant or microbial cells. Moreover, designed biosynthetic pathways comprised of biosynthetic enzymes expressed at modified levels and/or enzymes with altered activities can create new structural diversity, encompassing structures not found in nature. However, genes of secondary metabolic plant pathways are not clustered as in microorganisms and the tremendous duplication of genes encoding certain biosynthetic enzyme classes creates significant challenges in the characterization of biosynthetic pathways despite the improved availability of plant genomic information. Genome sequences and EST data from plant model organisms such as *Arabidopsis thaliana*, *Medicago truncutata* and *Clarkia breweri* reveal that the most prolific enzyme families (cytochrome P450s, glycosyltransferases, and methyltransferases), known to be involved in the biosynthesis of natural products, have undergone extensive gene duplication (Kliebenstein et al., 2001; Pichersky et al., 2006; D'auria and Gershenzon 2005), which severely limits homology-based predictions of gene functions as well as comparative genomics approaches to decipher biosynthetic pathways (Frick and Kutchan, 1999). Hence, often tedious and time consuming genetic/molecular biology or metabolic/biochemistry approaches have been used to study plant pathways (Rohloff and Bones, 2005). More recent approaches utilize genomic, proteomic and metabolomic data (Sumner et al., 2003; Trethewey, 2004; Oksman-Caldentey and Saito, 2005; Fridman et al., 2005; Jacobs et al., 2005; Verpoorte and Memelink, 2002; Ounaroon et al., 2003) to study metabolic networks of plants. As part of these efforts and with the availability of EST and genome data, an increasing number of plant genes have been cloned and expressed in microbial hosts for biochemical annotation of function (Aharoni et al., 2000; Achnine et al., 2005).

In this chapter we will review how recombinant microorganisms have been used for the elucidation of plant biosynthetic pathways as well as production of plant natural products. Although recombinant microorganisms have until now mostly been used for the characterization of catalytic activities of newly discovered biosynthetic enzymes, a number of recent examples demonstrate the feasibility of installing multi-step plant pathways in engineered microbial cells for the production of plant natural products. We will use these examples to illustrate the potential of engineered microbial hosts for the synthesis of diverse phytochemicals. Futhermore, we show how this approach forms a platform to further diversify the spectrum of plant natural products through techniques such as manipulation of enzyme functions and precursor supply, and combinatorial biosynthesis. We have organized this chapter according to natural product classes and their main industrial application – flavor and aroma or medicinal compounds. It should be noted though, that some natural product classes (e.g. terpenes, apocarotenoids, flavonoids) have compounds with applications in both consumer markets and will be discussed in this review under the market of which the majority of its members fall.

2. FLAVOR AND AROMA COMPOUNDS

Plant derived volatile aroma compounds are one of the best studied groups of secondary metabolites with over 1000 volatiles identified from plants (Pichersky et al., 2006). Volatiles serve important functions in plants acting as signaling molecules, phytoalexins, and attractants for pollinators (for reviews of plant derived volatile aroma compounds see (Pichersky and Gershenzon, 2002; Dudareva et al., 2004; Dudareva and Pichersky, 2000). Aroma synthesizing plants are used as flavorings, preservatives and herbal remedies and the production of flavor and fragrance compounds has many important industrial applications (Goff and Klee, 2006). Applications include: addition or replenishment of flavors and fragrances to processed foods with extended shelf lives, metabolic engineering of more fragrant or colorful flowers and fruits and increasing vitamin or pigment content of foods.

Biotechnological production of aroma compounds can be carried out through plant cell and tissue culture, biocatalysis and biotransformations by microorganisms or *de novo* synthesis (reviewed in (Krings and Berger, 1998; Vandamme, 2002)). Currently more is known about microbial production of aroma and flavor compounds than plant biosynthesis (Marasco and Schmidt-Dannert, 2003). Poor characterization of secondary metabolite pathways in plants is due in part to the minute quantities of aroma compounds found in plants that are often difficult to isolate. Conventional breeding is impeded by the complexity of the biochemical pathways involved in the biosynthesis of aroma and flavor formation, and the sensitivity to environmental conditions. Furthermore, there is a fundamental lack of established methodology to characterize flavor chemistry and organoleptic properties (Lewinsohn et al., 2001). An understanding of aroma secondary metabolite pathways in plants on a molecular level similar to the level of understanding researchers have achieved for microbial systems will provide access to the genes and enzymes involved allowing for improvements and manipulations of the pathways.

Aroma compounds are divided into three major classes of compounds: phenylpropanoid (benzenoid) derivatives, fatty acid derivatives and isoprenoid derivatives. Plant medicinal natural products also derive from phenylpropanoid and isoprenoid pathways. Unique to aroma biosynthesis are fatty acid derivatives. In contrast, many medicinal phytochemicals are alkaloid derivatives, and few aroma compounds are generated from alkaloid precursors.

2.1. Fatty acid derived flavor and aroma compounds

Important fruit flavors with low odor thresholds are derived from unsaturated fatty acids during ripening and described as fruity, coconut-like, buttery, sweet or nutty aromas. β-oxidative cleavage and decarboxylations of fatty acids form compounds such as lactones and volatile esters. Flavor formation is initiated when free fatty acids are released from lipids by lipases and then specific double bonds are peroxidized by lipoxygenases to produce hydroperoxides. For example, in tomatoes, the cleavage by 13-hydroperoxide from linoleic or linolenic acids produces "grassy" or "green" notes from hexanal and *cis*-3-hexenal respectively. These aldehydes can be further reduced by an alcohol dehydrogenase to form hexanol and *cis*-3-hexanol (Vick and Zimmerman, 1986) (Figure 1). The availability of appropriate unsaturated fatty acids is an important factor affecting fruit flavor and aroma development because they provide the aliphatic esters, alcohols, acids and carbonyls that contribute to fruit flavors. Modifications of the unsaturated fatty acid pools can direct the formation of specific products of value.

Esters are the main component of apple (*Malus domestica)*, pear *(Pyrus communis)*, and banana (*Musa sapientum)* flavors. Esterification is the final step in volatile ester formation and results from trans-acylation by an acyl-CoA moiety to the alcohol and is carried out by alcohol acyltransferase (AAT) enzymes. (Other volatile esters that derive from methylsalicylate, most likely from the phenylpropanoid pathway, will be discussed in a later section). Several AAT genes (i.e. from melons, apples, bananas and strawberries) have been cloned and expressed in *E. coli* (Aharoni et al., 2000; Chau et al., 2004a; Yahyaoui et al., 2002; Souleyre et al., 2005; Beekwilder et al., 2004; D'auria et al., 2002; Shalit et al., 2003). Recombinant systems allow for different combinations of alcohols and acyl-CoAs to be studied to identify the specificity and flexibility of AAT activity. DNA microarrays were used to identify a strawberry AAT (SAAT) involved in the formation of over 100 fruity esters in strawberries. The SAAT enzyme is a member of the BAHD family of acetyltransferases whose members have diverse roles in secondary metabolism. Aharoni et al. (2000) expressed the recombinant protein in *E. coli* and found it had wide substrate specificity with a preference for medium chain length alcohols and that volatile ester formation depended on the availability of acyl-CoA molecules and alcohol substrates (Aharoni et al., 2000).

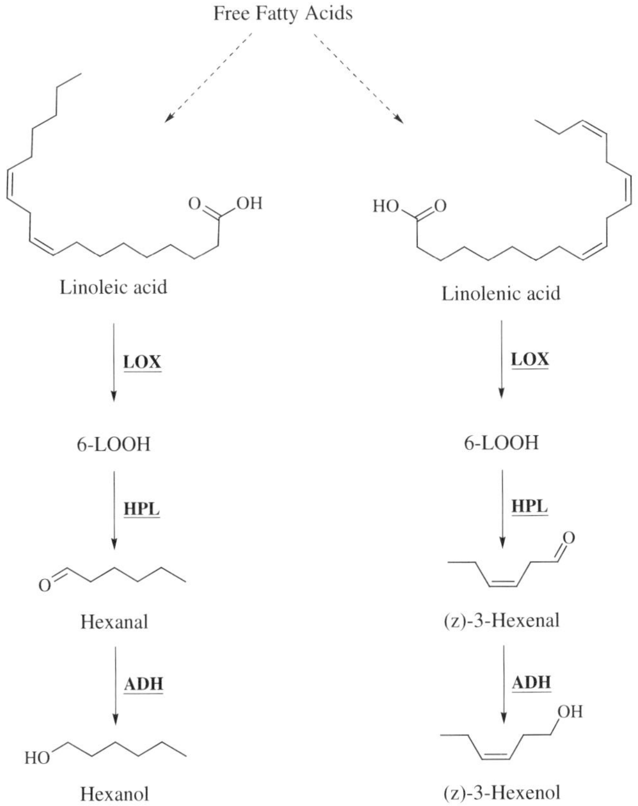

Figure 1. Example biosynthetic pathway for the formation of fatty acid derived volatile compounds (C5 and C6) starting from linoleic and linolenic acids. Enzyme abbreviations are as follows: LOX, lipoxygenase; HPL, hydroperoxide lyase; ADH, alcohol dehydrogenase

Beekwilder et al. (2004) compared wild strawberry (*Fragaria vesca*) and banana (*Musa sapientum*) AATs to the cultivated strawberry *Fragaria x ananassa*. They found substrate preference could not be determined by sequence similarity and that recombinant substrate preferences were not reflected in the fruit volatile profiles. This suggests that the volatile ester profile of a fruit species may be governed by the supply of precursors. A similar study created transgenic *P. hybrida* with the rose AAT and observed higher levels of benzyl acetate and phenylacetate demonstrating the possibility of directing the production of pathway volatiles (Guterman et al., 2006). Metabolic engineering of precursors within plants can create new ester profiles or conversely the AAT enzymes can be introduced into new precursor backgrounds to create novel esters. Directed biotransformation within engineered microbial cells is also a mechanism for creating new aroma products. Advantages to microbial production would be fewer volatiles that mask the scents, stricter control over protein expression levels and less modification of compounds to non-volatile forms through glycosylation reactions (Dudareva and Negre, 2005).

2.2. Volatile benzenoids

Most phenylpropanoid and benzenoid compounds derive from the shikimate pathway via phenylalanine. The first committed step in phenylpropanoid and benzenoid formation is catalyzed by L-phenylalanine ammonium lyase (PAL), which is the deamination of phenylalanine to *trans*-cinnamic acid. *Trans*-cinnamic acid is the branchpoint between two pathways, one forming the volatile benzenoid compounds and the other medicinal compounds such as stilbenes and flavonoids discussed later.

Benzoic and salicylic (2-hydroxyl benzoic) acids are synthesized from *trans*-cinnamic acids (Boatright et al., 2004a) and are precursors for the synthesis of important floral scents such as methyl benzoate and methyl salicylate. *Trans*-cinnamic acid loses 2 carbon units in route to becoming volatile. Metabolic flux models suggest several mechanisms for the loss of the two carbon units (Boatright et al., 2004b). One mechanism is a CoA-dependent β-oxidation similar to the β-oxidation of fatty acids. The *trans*-cinnamic acid is activated to the cinnamoyl-CoA ester and then hydrated to form 3-hydroxy-3-phenylpropionyl-CoA. The hydroxyl group of the intermediate is oxidized to a ketone and the β-keto thioester is subsequently cleaved to form benzoyl-CoA. The alternative CoA-independent, non-β-oxidative pathway involves the hydration of the free cinnamic acid to 3-hydroxy-3-phenylpropionic acid. A reverse aldol reaction forms the benzaldehyde which is then oxidized to benzoic acid. Finally, a combination pathway has been described that is CoA-dependent and non-β-oxidative (Figure 2). Enzymes and genes involved in the early biosynthesis of benzenoids have not been decribed on a molecular level, analyses has been restricted to analyses of pathway intermediates and flux analysis (Boatright et al., 2004b). However, several methyltransferases that convert benzenoid compounds into the corresponding carboxyl-methyl esters have been identified. Also a benzoyl-CoA:benzyoyl:benzyl alcohol benzoyl transferase which forms benzylbenzoate has been identified (D'auria et al., 2002).

Methyl esters of benzenoids are responsible for the scents of many flowering plants and find applications as flavor and aroma compounds. For example, methyl jasmonate is used in the perfume industry; methyl salicylate found in leaves and flowers of wintergreen is a flavor ingredient in many types of candy, food and medicine; methyl cinnamate is found in basil; methyl benzoate is a major component of ylang-ylang oil and in the aroma and flavor of tropical fruits. Six benzenoid *O*-methyltransferase enzymes have been identified and heterologously expressed (reviewed in (Effmert et al., 2005)). All six enzymes belong to the same SAM dependent class of carboxyl O-methyltransferases, but recognize different benzenoid substrates which has been established by functional characterization of enzymes expressed in *E. coli*. The first methyltranferase identified was from *Clarkia breweri* and shown to involved in floral scent production (Ross et al., 1999).

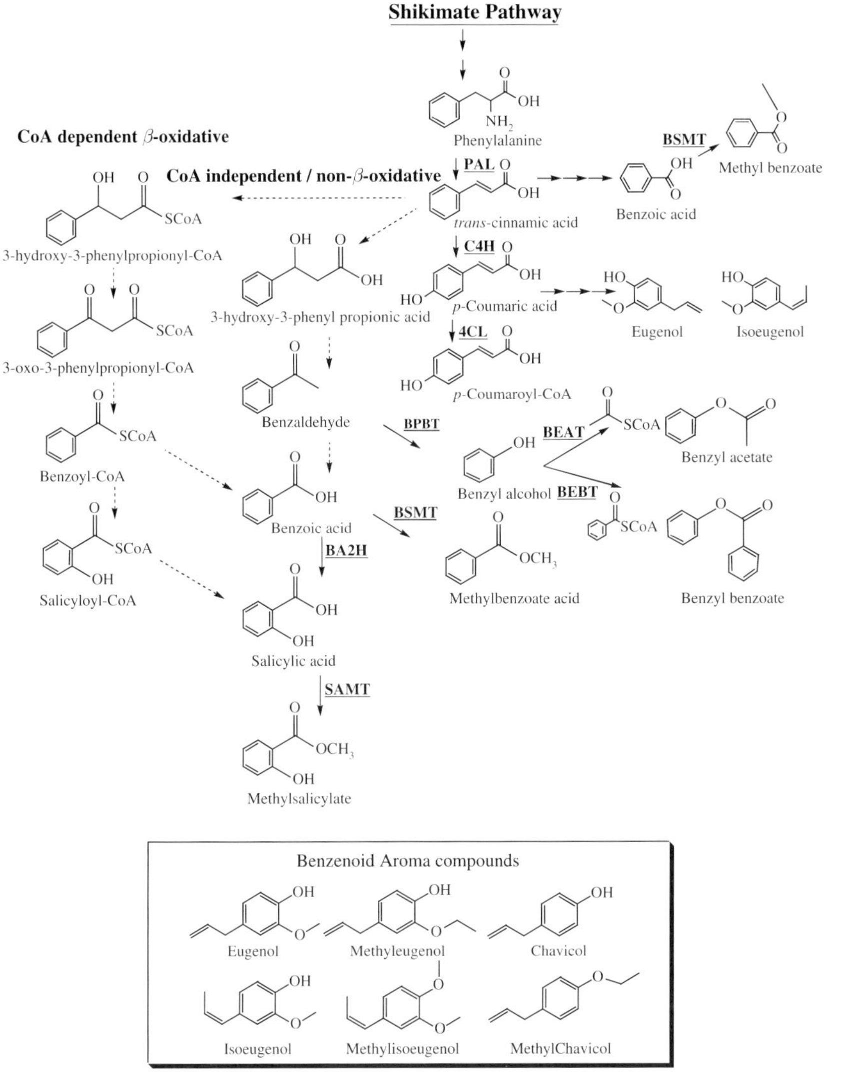

Figure 2. Proposed biosynthetic pathway of benzenoid compounds starting from the amino acid phenylalanine. Adapted from Boatwright et al. 2004. Cloned genes are bolded and underlined. Enzyme abbreviations are as follows: PAL, phenylalanine ammonium lyase; C4H, P450 cinnamate-4-hydroxylase; 4CL, 4-coumaroyl-CoA-ligase; BSMT, *S*-adenosyl-l-Met:benzoic acid carboxyl methyltransferase; BPBT, benzoyl-CoA:benzyl alcohol/phenylethanol benzoyltransferase; BEAT, acetyl-CoA:benzyl alcohol acetyltransferase; BEBT, benzoyl-CoA:benzyl alcohol benzoyl transferase; BA2H, benzoic acid 2-hydroxylase; SAMT, salicylic acid carboxyl methyltransferase

This enzyme was highly specific for the conversion of salycilate to its corresponding methylester, which was also observed for an enzyme identified from snapdragons (*Antirrhinum majus*) (Negre et al., 2002). In contrast, another methyltransferase isolated from *Stephanotis floribunda* forms methyl benzoate but accepts several different benzoic and cinnamic acid derivatives (Pott et al., 2002). Other identified enzymes methylate benzoate (Murfitt et al., 2000); jasmonate (Seo et al., 2001); benzoate and salicylate (Chen et al., 2003),(Negre et al., 2003) or several benzoic and cinnamic acid derivatives (Pott et al., 2004).

2.3. Monoterpenes

Monoterpenes are the main component in the essential oils frequently used in flavorings, fragrances, and pharmaceuticals (Aggarwal et al., 2002; Croteau, 1998; Loza-Tavera, 1999; Mahmoud and Croteau, 2002; Pattnaik et al., 1997; Hohl, 1996) and are synthesized in the plastids of plants and are responsible for many of the odors associated with them (Colby et al., 1993). These compounds along with sesqui-, diterpenoids and carotenoids discussed later in the section on medicinal natural products, belong to the diverse class of terpenoid (or isoprenoid) natural products synthesized by plants and microorganisms.

2.3.1. Biosynthesis of terpenoid compounds

Over 30,000 terpenoid structures have been described from different terpenoid classes (Broun and Somerville, 2001), making terpenoids the largest group of natural products. These compounds are synthesized by successive condensations of C_5 isoprene units, isopentyl diphosphate (IDP) and dimethylallyl disphosphate (DMADP). Condensations are catalyzed by various chain length specific prenyltransferases in mainly head-to-tail reactions (Yan et al., 2001). A different class of enzymes is responsible for catalyzing the head-to-head condensation seen with carotenoids and sterols. Terpenoid classes are defined by the number of condensed isoprene units resulting in chains with 10, 15, 20, 30 or 40 carbons: In the case of volatile monoterpenoids, two isoprene units make up a ten carbon isoprenoid backbone (C_{10}) (Figure 3).

Additional modifications to the isoprenoid backbones including cyclizations, desaturations and oxidations produces the structural diversity in the terpenoid class (Dewick, 2002; Kuzuyama and Seto, 2003). Terpene synthases (also referred to as cylases, TPSs) catalyze cyclization of the linear isoprene diphosphate precursors geranyldiphosphate (GDP, C_{10}), farnesyldiphosphate (FDP, C_{15}), and geranylgeranyldiphosphate (GGDP, C_{20}) into mono-, sesqui-, and diterpenes respectively (Davis and Croteau, 2000; Bohlmann et al., 1998a). TPSs catalyze cyclization by first catalyzing removal of the diphosphate to form a highly reactive carbocation and then controlling carbocation migration as it steps through the backbone (involving hydride shifts, cyclizations, rearrangements and reprotonations) until final carbocation quenching and product release occurs. Terpene synthases all have a similar active site scaffold (Lesburg et al., 1997; Starks et al., 1997a), but different types of cyclases are categorized based on the isoprene chain length specificity, the way that they generate the carbocation, folding patterns and the quenching mechanism. The number of structures increases dramatically with increasing backbone length and conjugation. For example, 1,000 monoterpenoid structures have been reported from C10 GDP, whereas there are 7,000 known sesquiterpene structures from C15 FDP (Bohlmann et al., 1998a). In the case of monoterpenes, several different structural types can be distuingished: acyclic (e.g. myrcene, linalool), monocyclic (*d*-limonene) or bicyclic(e.g. α- and β-pinene, 3-carene) monoterpenoids. Additional oxidation and reduction, and isomerization reactions result in more complex monoterpenes such as geraniol, menthol, and camphor.

Sequence information is not predictive of the terpenoid products from TPSs and as a result many plant terpene synthases have been cloned, expressed in heterologous hosts and functionally characterized (Pichersky and Gershenzon, 2002; Dudareva et al., 2004; Starks et al., 1997b; Faldt et al., 2003; Bohlmann et al., 2000; Phillips et al., 2003; Chen et al., 2004; Martin et al., 2004). Monoterpene synthase from species of *Pinaceae*, *Lamiaceae*, *Rutaceae*, *Myrtaceae*, *Asteraceae*, and other plants have been cloned and functionally expressed in *E. coli* (Bohlmann et al., 1999; Yuba et al., 1996; Bohlmann et al., 1997; Williams et al., 1998; Wise et al., 1998; Jia et al., 1999). *Arabidopsis thaliana* alone has 32 putative cyclase genes identified in its genome, including 6 proven monoterpene synthases and 2 sesquiterpene synthases (Aubourg et al., 2002; Tholl et al., 2005) which are responsible for some of the 102 volatiles identified in *A. thaliana* (Rohloff and Bones, 2005). Ongoing functional genomics projects including functional analysis of recombinant enzymes aim at understanding the AtTPS families (Chen et al., 2003; Faldt et al., 2003; Bohlmann et al., 2000; Chen et al., 2004; Aharoni et al., 2003).

2.3.2. Engineered terpenoid biosynthesis

With a cyclase gene identified it is possible to engineer recombinant microbial cells for terpenoid synthesis (Carter et al., 2003; Huang et al., 1998; Huang et al., 2001; Martin et al., 2003) from endogenous IDP and DMADP precursors. These precursors are produced through two different pathways: the mevalonate (MVA) or mevalonate independent pathway (DXP) (Rodriguez-Concepcion and Boronat, 2002). The MVA pathway is predominantly in eukaryotes, archaea, and some eubacteria. The DXP pathway is used by the majority of eubacteria, including *E. coli* and green algae (Boucher and Doolittle, 2000). Plants, however, use both pathways to produce sesqui- and triterpenes via the MVA pathway in the cytosol and mono-, diterpenes and carotenoids via the DXP pathway in the chloroplasts.

In *E. coli*, there have been many studies aimed at increasing IDP and DMAPP pools by diverting the precursors away from competing endogenous pathways and directing flux towards the engineered pathway (Estevez et al., 2001; Farmer and Liao, 2000; Kajiwara et al., 1997; Kim and Keasling, 2001; Lee and Schmidt-Dannert, 2002; Matthews and Wurtzel, 2000). Significant challenges to this engineering approach include the complex regulation of the DXP pathway, achieving significant expression of plant enzymes in the host and toxicity of some DXP pathway intermediates when present in high quantities. The engineering of the

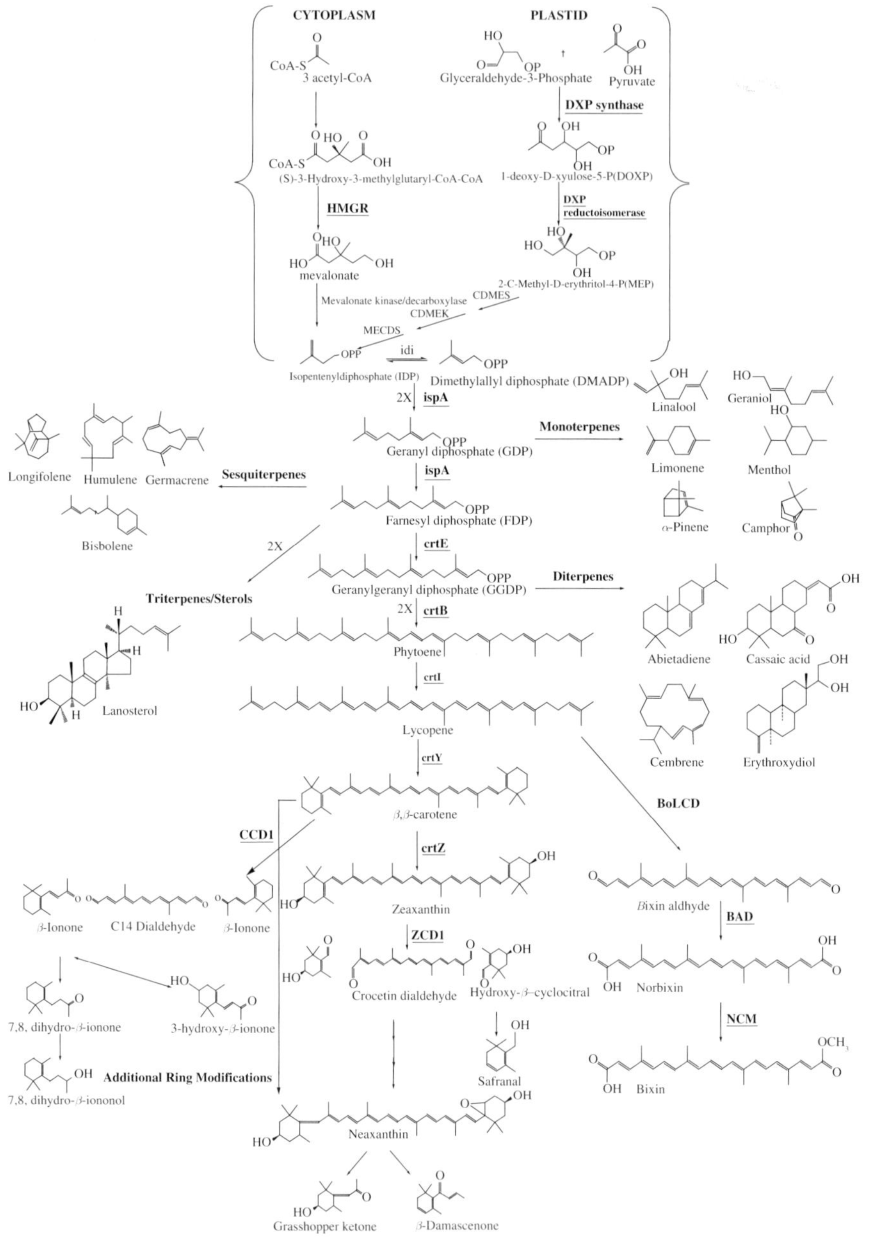

Figure 3. Biosynthetic pathways of isoprenoid derived compounds. Cloned genes are in bold and underlined. Enzyme abbreviations are as follows: HMGR, HMG-CoA reductase; Idi, IDP isomerase; Dxs, 1-deoxy-d-xylulose 5-phosphate synthase; IspA, FDP synthase; CrtE, GGDP synthase; CrtB, phytoene synthase; CrtI, phytoene desaturase; CrtY, lycopene cyclase; CrtZ, carotene hydroxylase; CCD, carotenoid cleavage dioxygenase; ZCD1, zeaxanthin cleavage enzyme; BoLCD, lycopene cleavage dioxygenase; BAD, bixin aldehyde dehydrogenase; NCM, norbixin carboxyl methyltransferase

eukaryotic MVA pathway into *E. coli* as an alternative pathway for the production of terpenoids was one successful attempt at alleviating the regulation and toxicity issues (Martin et al., 2003). Along with precursor engineering, stoichiometric flux balance analysis has been used, further increasing terpenoid production levels in recombinant cells (Alper et al., 2005a; Alper et al., 2005b). In addition to ensuring sufficient pools of IDP and DMADP in *E. coli*, production of terpenoids other than sesquiterpenoids (from endogenous FDP in *E. coli*) requires expression of heterologous prenyltransferases for synthesis of respective prenyldiphosphate precursors such as GDP (C_{10}) or GGDP (C_{20}) for mono-, diterpenoid and carotenoids production.

Synthesis of 10 out of 17 monoterpenoid compounds of lemon oil, widely used in the beverage and cosmetic industry, has been achieved in *E. coli* by cloning and expressing four *Citrus* monoterpene synthases (Lucker et al., 2002). These cyclases mainly synthesized (+)-limonene, (–)-β-pinene, and γ-terpinene, the major monoterpenpoids of lemon oil. cDNAs for various limonene synthases have also been cloned from *Mentha* species (Colby et al., 1993); *Abies grandis* (Bohlmann et al., 1997); *Perilla frutescens* (Yuba et al., 1996); and *Schizonepeta tenuifolia* (Maruyama et al., 2001).

The formation of monoterpenes from several species of mint plants has been fully elucidated (Croteau et al., 2005). Menthol is a key monoterpene in peppermint (*Mentha piperita)* and provides its characteristic organoleptic cooling sensation. Biosynthesis of menthol involves cyclization of GDP by limonene synthase to limonene followed by hydroxylation by a P450 limonene-3-hydroxylase to *trans*-isopiperitenol (Lange and Croteau, 1999; Lupien et al., 1999; Alonso et al., 1992). *Trans*-isopipertitenol is then converted through the activity of a dehydrogenase, an isomerase and three reductases to (–)-menthol (Ringer et al., 2003; Ringer et al., 2005; Davis et al., 2005). In spearmint *(Mentha spicata)*, a homologous monoterpenoid hydroxylation reaction occurs but with differerent regiospecificity (C6) resulting in the formation of *trans*-carveol (Lupien et al., 1999; Haudenschild et al., 2000), which is converted to (–)-carvone by a dehydrogenase that is homologous to the peppermint enzyme (Ringer et al., 2005) (Figure 4). A single amino acid substitution has been found to convert the limonene-6-hydroxylase from spearmint into a limonene-3-hydroxylase (Schalk and Croteau, 2000).

The simpler spearmint pathway was installed in *E. coli* in a stepwise manner to develop a monoterpene production system (Carter et al., 2003). Overexpression of GDP synthase, limonene synthase, P450 hydroxylase and dehydrogenase in *E. coli* led to the secretion of 5 mg/L limonene into the culture medium. However, none of the targeted product carvone was detected, although feeding studies of recombinant *E. coli* with limonene demonstrated that the two last steps of the pathway were functional. Limited availability of isoprenoid precursors and the excretion of pathway intermediate limonene into the medium were identified as the main bottlenecks for efficient carvone production in *E. coli*. These problems could be overcome by increasing a) the isoprenoid precursor pool using the above described engineering strategies and b) activity of the two downstream pathway enzymes. It is well-known that high P450 activity is very difficult to reconstitute

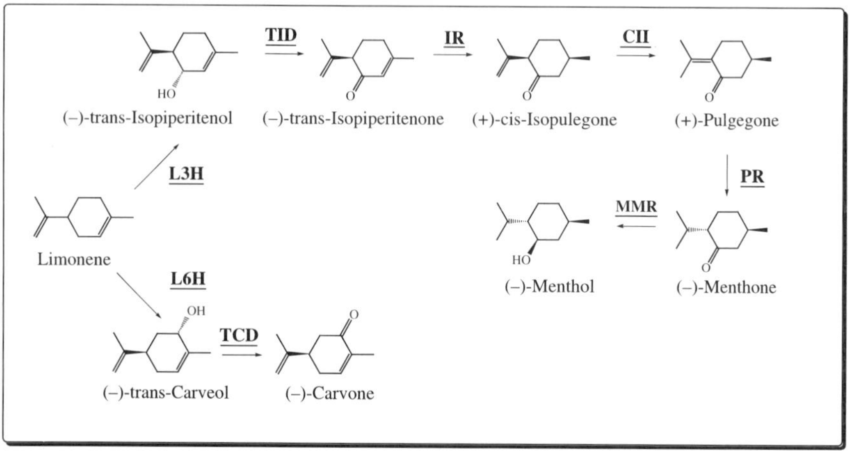

Figure 4. Menthol biosynthetic pathway from the monoterpene limonene. Cloned genes are bolded and underlined. Enzyme abbreviations are as follows: L3H, P450 limonene-3-hydroxylase; L6H, P450 limonene-6-hydroxylase; TID, (−)-trans-isopiperitenol dehydrogenase; IR, (−)-isopiperitenone reductase; CII, (+)-cis-isopulegone isomerase; PR, (+)-pulegone reductase; MMR, (−)-menthone:(−)-menthol reductase; TCD, (−)-trans-carveol dehydrogenase

in *E. coli* as these enzymes require a compatible NADPH reductase system not present in *E. coli*. Moving this pathway into yeast as a recombinant host, which is capable of functional P450 expression, may alleviate some of the problems seen in *E. coli*.

Alternatively, recombinant *E. coli* or yeast expressing the last two enzymes of the pathway may be used to convert limonene to carvone in a biotransformation reaction. Many limonene biotransformation processes utilizing bacteria, fungi, yeasts and plants have been described (Duetz et al., 2003). For example, a thermostable limonene hydratase expressed in *E. coli* conferred the ability to grow on limonene to the cells (Savithiry et al., 1997). *Rhodococcus globerulus* PWD8 was found to convert (+)-limonene into *trans*-carveol and carvone (Duetz et al., 2001; De Carvalho and Da Fonseca, 2003). *R. erythropolis* DCL14 is a strain able to grow on limonene as sole carbon source and contains several NAD, dichlorophenolindophenol (DCPIP)- and NADP-dependent carveol dehydrogenases that convert carveol into carvone (De Carvalho and Da Fonseca, 2002). There is considerable interest in the development of efficient limonene biotransformation reactions for the production of enantiomerically pure compounds such as α-terpineol, perillyl alcohol, carveol, carvone, and menthol because chemical synthesis results in complex product mixtures that require purification and thus causes high bulk compound prices (US$30–60/kg) (Duetz et al., 2003). Limonene and derivatized compounds are important building pharmaceutical building blocks. For example, the monoterpene (–)-borneol is used as the starting material in one synthesis of the diterpene anticancer drug Taxol (Holton et al., 1994).

3. MEDICINAL NATURAL PRODUCTS

Over 1000 phytochemicals have been employed to treat human ailments and diseases such as cancer, heart disease and microbial infections. Sixty percent of anticancer or anti-infectives are natural products or analogs (Tietze et al., 2003). Major classes of compounds that produce medicinally valuable and/or biologically active products include isoprenoids, flavonoids, and alkaloids (Figure 5). Of these classes, alkaloids

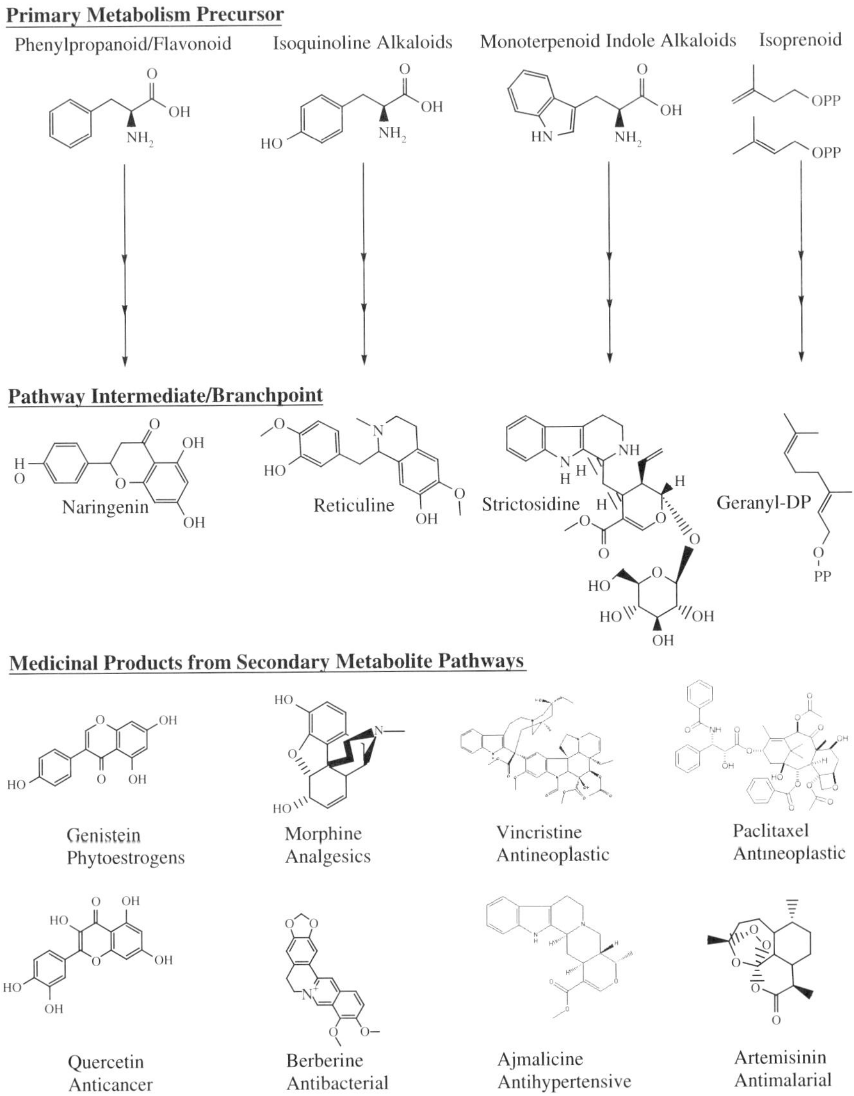

Figure 5. Overview of secondary metabolite pathways involved in medicinal natural product biosynthesis

are the most abundant compounds. However, the complexity of alkaloid biosynthetic pathways is such that there are still many enzymes and genes unknown. The flavonoid and isoprenoid pathways are better characterized and therefore will be more intensely discussed in this chapter.

3.1. Alkaloids

Alkaloids are a structurally diverse class of nitrogen containing compounds with over 12,000 structures elucidated from plants (Verpoorte and Memelink, 2002; Wink, 2003, 1999). Well known alkaloid compounds include purine alkaloids (caffeine and theobromine), tropane alkaloids (cocaine and scopolamine), benzylisoquinoline alkaloids (berberine and morphine) and monoterpenoid indole alkaloids (vinblastine, ajmaline) Alkaloids are classified based on their primary metabolite: purine alkaloids are produced from adenine or guanine, tropane alkaloids (TA) are produced from ornithine, isoquinoline alkaloids (IQA) are synthesized from tyrosine, and monoterpenoid indole alkaloids (MIA) derive from tryptophan. The central metabolite is converted into a branch point intermediate: xanthosine (purine alkaloids), homospermidine (TA's), norcoclaurine (IQA's) and strictosidine (MIA's) (Figures 5). These central intermediates are tailored by acetylation, hydroxylation, glycosylation, and methylation reactions into a diverse array of over 12,000 alkaloid structures produced by different plant species.

3.1.1. Purine alkaloids

Caffeine (1,3,7-trimethylxanthine) is a purine alkaloid found in high concentrations in coffee and tea. It is a product of nucleic acid catabolism and synthesized through a series of three *S*-adenosyl-L-methionine (SAM) dependent methylation steps carried out by *N*-methyltransferases. The elucidation of the caffeine biosynthetic pathway is a good example of using recombinant microbial cells to characterize a pathway and then using that information to alter the content of the metabolites in plants. The pathway has been characterized from tea (*Camellia sinesis)* and coffee *(Coffea arabica)* (Kato et al., 1996; Ashihara and Suzuki, 2004; Ashihara et al., 1997). Briefly, xanthosine is converted to 7-methylxanthosine by a 7-methylxanthosine synthase. The 7-methylxanthosine intermediate is converted to 7-methylxanthine, followed by two successive additions of methyl groups to form theobromine and caffeine. In tea plants, the final two steps are carried out by a bifunctional enzyme, caffeine synthase (Ashihara and Crozier, 1999). In coffee plants, theobromine synthase catalyzes 3-*N*-methylation of the purine ring of mono-methylxanthine forming theobromine (3,7-trimethylxanthine), then theobromine is methylated to caffeine by caffeine synthase (Figure 6).

All three enzymes have been cloned and recombinantly expressed in *E. coli.* 7-methylxanthosine synthase (known as either CmXRS1 or CaXMT1) was found to be specific for xanthosine (Mizuno et al., 2003). Theobromine synthase (CTS1, CTS2, and CaMXMT1) also has a narrow substrate range (Ogawa et al., 2001). Caffeine synthases (CS or CaDXMT) cloned from tea leaves and coffee plants have

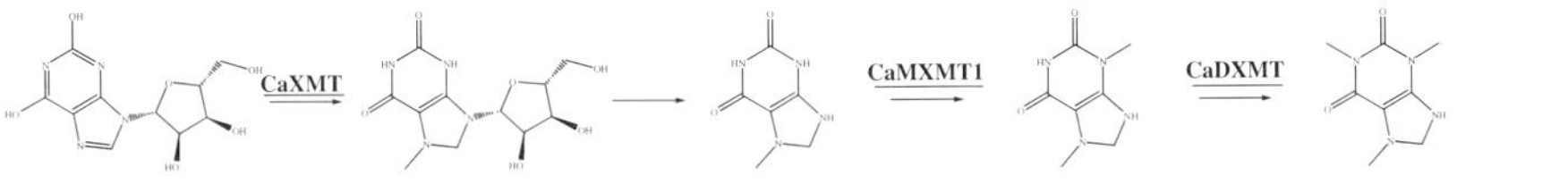

Xanthosine 7-methylxanthosine 7-methylxanthine Theobromine (3,7-dimethylxanthine) Caffeine (1,3,7-dimethylxanthine)

Figure 6. Overview of the caffeine biosynthetic pathway from *Coffea arabica*. Enzyme abbreviations are as follows: CaXMT, 7-methylxanthosine synthase; CaMXMT1, theobromine synthase; CaDXMT, caffeine synthase. All the enzymes have been cloned and functionally expressed in *E. coli*

a broad substrate specificity (Uefuji et al., 2003; Kato et al., 1999). The recombinant enzyme catalyzes 3-*N*-methylation and 1-*N*-methylation of the purine ring of mono and di-methylxanthines. The coffee caffeine synthase had higher relative activity towards theobromine than the tea caffeine synthase. Accumulation of purine alkaloids is dependent on *N*-methyltransferase substrate specificity and the rate of caffeine biosynthesis also appears to be regulated by *N*-methyltransferase activity.

Uefuji et al. (2003) combined all three recombinant methytransferases and crude *E. coli* extract to produce caffeine *in vitro* (Uefuji et al., 2003). The characterized caffeine biosynthetic pathway was then manipulated by RNA interference of the theobromine synthase (CaMXMT1) reducing caffeine levels in the coffee bean 30 to 50% of the control (Ogita et al., 2004). The same pathway was expressed in tobacco (*Nicotiana tabacum)* producing caffeine at levels of up to 5 ug/g of fresh dry weight of leaves was achieved in leaves. Caffeine production in transgenic plants may confer self defense to plants in the future.

3.1.2. Benzylisoquinoline alkaloids

Benzylisoquinoline alkaloids (BIQA) are a large diverse class of compounds used as analgesics (morphine, codeine), antimicrobials (berberine, sanguinarine), and muscle relaxants (papaverine and (+)-tubocurarine). All of the >2,500 benzylisoquinoline alkaloids pass through the central intermediate (*S*)-norcoclaurine (Facchini and Chappell, 1992). Biosynthesis of BIA's is initiated by the condensation of 4-dihydroxyphenylethylamine (dopamine) and 4-hydroxyphenylacetaldehyde (4-HPAA) catalyzed by (*S*)-norcoclaurine synthase (NCS). A recombinant NCS has recently been functionally expressed in *E. coli* (Samanani and Facchini, 2002; Samanani et al., 2004). *(S)*-norcoclaurine is modified by three related methyltransferases, a 6-*O*, 4-*O* and a *N*-methyltransferase (Seo et al., 2001, Frenzel and Zenk, 1990; Morishige et al., 2000; Morishige et al., 2002) in addition to a P450 monooxygenase, to form reticuline (Loeffler and Zenk, 1990). Reticuline is the branchpoint between the berberine and morphine pathways and (*R*)- or (*S*)-reticuline are potential substrates for the manufacture of various compounds that have shown antimalarial or anticancer activities (Angerhofer et al., 1999). Although all six enzymes to form reticuline have been cloned and heterologously expressed independently, there are no reports of microbial production of reticuline. A fermentative process may out-produce the transgenic reticuline accumulating poppy plants created by Allen et al. (2004) (Allen et al., 2004).

The antimicrobial berberine is synthesized through 13 different enzymatic reactions. Excluding the final oxidase, all the enzymes necessary for the conversion of norclaurine to berberine have been cloned (Seo et al., 2001; Facchini and Chappell, 1992; Morishige et al., 2000; Dittrich and Kutchan, 1991; Chou and Kutchan, 1998; Ikezawa et al., 2003; Takeshita et al., 1995; Pauli and Kutchan, 1998). Once the final enzyme is cloned, it will be feasible to produce berberine in alternative organisms. A heterologous host such as *Saccharomyces cerevisiae* could be used in the production of berberine and provide BIQA compounds for further modifications.

The biosynthetic pathway to morphine in the Oriental poppy *Papaver somniferum* (opium poppy) is an even longer synthesis pathway consisting of 17 reactions (Kutchan, 1996; Kutchan, 1995). Many of the enzymatic steps that have been identified and functionally expressed in insect cell culture (*S. frugiperda* Sf9 cells) or in *E. coli* (Gerardy and Zenk, 1993a; Grothe et al., 2001; Gerardy and Zenk, 1993b; Frick and Kutchan, 1999; Pauli and Kutchan, 1998; Grothe et al., 2001; Dai et al., 2006; Grothe et al., 2001; Unterlinner et al., 1999). However, biosynthesis in heterologous hosts is not yet feasible because there are still a number of biosynthetic steps in the morphinan pathway for which the enzymes have not been identified or cloned. An alternative approach is microbial biotransformations of morphine into valuable derivatives (Rathbone and Bruce, 2002). A stable reusable recombinant morphine/codeine biotransformation using *Pseudomonas* enzymes has been reported (Boonstra et al., 2001).

3.1.3. Monoterpenoid indole alkaloids

Fifteen MIA compounds are currently used in pharmacological and therapeutic applications including the antineoplastic drugs vinblastine and vincristine, the toxin strychnine, the vasodilator yohimbine, antihypertensive ajmalicine and antiarrhythmic ajmaline. Despite their widespread clinical uses, MIA biosynthesis is the least well-characterized pathway on a molecular level of all plant natural product classes. MIAs have amazingly complex synthesis pathways. For example, the formation of vindoline involves over 30 enzymes and has 35 intermediates (Van Der Heijden et al., 2004). Deciphering their intricate biosynthetic pathways would enable overproduction of intermediates in metabolically engineered plant or microbial cells. These intermediates would then act as scaffolds for further chemical modifications to the desired drug targets. A recent patent takes this approach for the production of the intermediates strictosidine and its aglycone. Previously cloned strictosidine synthase and glucosidase enzymes (Kutchan et al., 1988; Geerlings et al., 2000; Gerasimenko et al., 2002) were introduced into yeast and fed precursors tryptamine and secologanin (Geerlings 1998). Recombinant yeast over expressing the strictosidine synthase and strictosidine glucosidase catalyze the formation of the MIA intermediate cathenamine (Geerlings et al., 2001).

Of MIA compound pathways, the pharmacological important alkaloids vindoline (a key precursor to the cytostatic alkaloid vinblastine found in the Madagascar periwinkle (*Catharanthus roseus*) and the antiarrhythymic alkaloid ajmaline (isolated from the medicinal plants *Rauvolfia serpentine* and *Tetraphylla vomitoria*) have been characterized the most. Elucidation of ajmaline is more complete than vindoline because six of the ten enzymes involved in ajmaline synthesis have been identified and researchers are interested in complete reconstitution of the pathway in heterologous hosts (Ruppert et al., 2005; Dogru et al., 2000; Warzecha et al., 2000; Mattern-Dogru et al., 2002; Von Schumann et al., 2002; Bayer et al., 2004).

Vinorine synthase is responsible for the formation of the first ajmalan-type alkaloid structure and was functionally expressed in *E. coli* as part of the larger aim of overexpressing the entire ajmaline pathway in engineered microbial cells (Bayer et al., 2004). This acetyltransferase belongs to the benzylalcohol acetyl-, anthocyanin-*O*-hydroxy-cinnamoyl-, anthranilate-*N*-hydroxy-cinnamoyl/benzoyl-, deacetylvindoline acetyltransferase (BAHD) superfamily of acyltransferases. Other members of this family play a role in biosynthesis of a number of plant natural products and have been biochemically characterized in many different pathways: from alkaloid pathways, deacetylvindoline 4-*O*-acetyltransferase (DAT) is involved in vindoline biosynthesis and salutaridinol 7-*O*-acetyltransferase (SALAT) in morphine biosynthesis (Laflamme et al., 2001; Grothe et al., 2001), in terpenoid pathways, taxadiene-5α-ol-*O* acetyltransferase (TAT) and 10-deacetylbaccatin III-10-*O*-acetyl transferase (DBAT) acetylate and benzyoylate Taxol precursors respectively (Walker et al., 2000), in benzenoid pathways BEBT, BAMT, SAMT, BEAT acyltranferases are important in formation of aroma compounds (D'auria et al., 2002; Ross et al., 1999; Murfitt et al., 2000; Dudareva et al., 2000), and anthocyanin biosynthetic pathways (Fujiwara et al., 1997; Suzuki et al., 2000; Yonekura-Sakakibara et al., 2000; Fujiwara et al., 1998a; Fujiwara et al., 1998b). Members of this family have been identified in disparate pathways based on conserved amino acid sequences (Bayer et al., 2004).Very recently, the crystal structure of vinorine synthase was solved representing the first structure of a member of the BAHD superfamily (Ma et al., 2005). This structure will help in understanding the mechanism and activity of BAHD members in the different pathways (Table 1).

3.2. Sesqui- and diterpenes

Sesquiterpenes arise from the C_{15} precursor FDP and form important flavor and fragrances (e.g. nootkatone and patchouli alcohol), antimicrobial phytoalexins (e.g. capsidiol) and medicinal compounds (e.g. artemisinin). Terpene cyclases are responsible for producing most sesquiterpene structures, and plants frequently have multiple cyclase genes (Trapp and Croteau, 2001) (for example, basil has nine cyclases that form monoterpenes and sesquiterpenes (Lijima et al., 2004)) making unraveling terpenoid biosynthesis in plants a difficult task. Sesquiterpene synthases have been characterized from a number of plant species including tobacco (*Nicotiana tabacum*; (Facchini and Chappell, 1992)), tomato (*Lycopersicon esculentum*; (Colby et al., 1998)), *Hyoscyamus muticus*; (Back and Chappell, 1995), *Arabidopsis* (*Arabidopsis thaliana*; (Chen et al., 2003), maize (*Zea mays*; (Schnee et al., 2002; Degenhardt and Gershenzon, 2000)) and grand fir

Table 1. Benzylalcohol acetyl-, anthocyanin-*O*-hydroxy-cinnamoyl-, anthranilate-,*N*-hydroxy-cinnamoyl/benzoyl-, deacetylvindoline acetyltransferase (BAHD) superfamily of acyltransferases (BAHD) superfamily representatives involved in secondary metabolism

Enzyme	Proper Name	Activity	Plant	Ref.
VS	vinorine synthase	vindoline biosynthesis	*R. serpentine*	(Bayer et al., 2004)
DAT	deacetylvindoline 4-*O*-acetyltransferase	vindoline biosynthesis: vindoline	*C. roseus*	(Laflamme et al., 2001)
SALAT	salutaridinol 7-*O*-acetyltransferase	morphine biosynthesis: thebaine	*P. somniferum*	(Grothe et al., 2001)
TAT	taxa-4(20),11(12)-dien-5 alpha-ol-*O*-acetyl transferase	acetylate Taxol precursor: Taxa-4(20),11(12)-dien-5a-yl acetate	*T. cuspidata*	(Walker and Croteau 2000a)
DBAT	10-deacetylbaccatin III-10-*O*-acetyl transferase	Taxol precursor:Baccatin III	*T. cuspidata*	(Walker and Croteau 2000b)
BEBT	benzoyl-CoA:benzyl alcohol benzoyl transferase	benzenoid biosynthesis: benzyl benzoate	*C. breweri*	(D'auria et al., 2002)
BPBT	benzoyl-CoA:benzyl alcohol/phenylethanol benzoyltransferase	benzenoid biosynthesis: benzyl benzoate/phenethyl benzoate	*P. hybrida*	(Boatright et al., 2004b)
BAMT	S-adenosyl-L-Met:benzoic acid carboxyl methyltransferase	benzenoid biosynthesis		(Murfitt et al., 2000)
SAMT	salicylic acid carboxyl methyltransferase	benzenoid biosynthesis	*C. breweri*;	(Ross et al., 1999)
			A. belladonna;	(Fukami et al., 2002)
			A. majus	(Negre et al., 2002)
			S. floribunda	(Pott et al., 2002)
BEAT	acetyl-CoA:benzyl alcohol acetyltransferase	benzenoid biosynthesis: benzyl acetate	*C. breweri*	(Dudareva and Pichersky 2000)
SAAT	strawberry alchohol acetyltransferase	Medium chain aliphatic and benzyl esters	*F. x ananassa*	(Aharoni et al., 2000)
RhAAT	*R. hybrida* alcohol acetyltransferase	geranyl acetate and other volatile esters	*Rosa* hybrid	(Shalit et al., 2003)

(*Abies grandis*; (Bohlmann et al., 1998b)). Most cyclases are readily expressed in *E. coli* for functional characterization of their product spectrum. The majority of cyclases produce one major cyclization product, but some produce many products. Two cyclases identified from grand fir are the best examples of the complex multi-step, multi-product reactions catalyzed by a single cyclases: δ-selinene and γ-humulene cyclases produce 34 and 52 different products respectively (Steele et al., 1998).

Structural features that determine substrate specificity and the diverse product profile of cyclases is currently under intense investigation as it would allow the direct synthesis of valuable compounds. Domain swapping experiments between *epi*-aristolochene synthase from *Nicotiana tabacum* and the vetispiradiene synthase of *Hyoscyamus muticus* resulted in chimera that could produce varied product profiles (Back and Chappell, 1995). A more systematic approach of identifying active site residues that confer product plasticity was more effective (Yoshikuni et al., 2006). Using the promiscuous γ-humulene synthase novel mutant terpene synthases that catalyzed the synthesis of one or a few different sesquiterpenes were created.

The next step beyond engineering individual plant cyclase genes in a microbial host for biochemical studies and improved activity is the biosynthesis of plant terpenoids by engineered microbial cells. Flavor and aroma compounds are significantly less complex than medicinal structures, but medicinal terpenoids could be synthesized by extending the native *E. coli* FDP pathway further with heterologous terpenoid biosynthetic genes. Two successful examples of researchers taking this approach can be seen in the biosynthesis of amorph-4,11-diene and taxadiene, the undecorated terpenoid scaffolds of the antimalarial drug artemisinin and the antitumor drug Taxol, respectively.

3.2.1. Artemisinin

A highly publicized sesquiterpene lactone endoperoxide from sweet wormwood (*Artemisia annua*), artemisinin is being investigated because of its effectiveness in treating all known strains of malaria. Metabolite production in plants via extraction of artemisinin from plant sources or engineering plants for enhanced metabolite production is one option for treatment of malaria in developing countries. Alternatively, fermentation processes using engineered microorganisms to artemisinin or its precursor sesquiterpenoid scaffolds could yield a low cost method of producing an inexpensive and effective malaria drug. At this stage, it is impossible to generalize which production platform, engineered plants or microorganisms is superior.

Artemisinin is biosynthesized from farnesyl diphosphate through the undecorated, hydrocarbon sesquiterpenoid intermediate amorph-4,11-diene. This reaction is catalyzed by amorph-4,11-diene synthase, which has been cloned and characterized in *E. coli* (Mercke et al., 2000; Picaud et al., 2005). To date, the subsequent enzymatic steps of the pathway are unknown although intermediates are being isolated from *A. annua* leaves and gland secretory cells (Bertea et al., 2005). Intermediates include different oxygenated compounds, artemisic alcohol, -aldehyde and – acid and then the final endoperoxide lactone artemisinin (Figure 7).

Amorph-4,11-diene is the target of heterologous production strategies because the complete pathway to artemisinin is still undefined. Significant levels of amorph-4,11-diene were recently produced by combining the *S. cerevisae* mevalonate pathway and the amorpha-4,11-diene synthase in *E. coli* (Martin et al., 2003). The production levels of around 100 mg of amorpha-4,11-diene per liter of *E. coli* culture were almost two orders of magnitude higher than those reported for other terpenoids in recombinant microbial hosts. One reason for this higher yield was the use of the yeast mevalonate dependent pathway. The MVA pathway was not

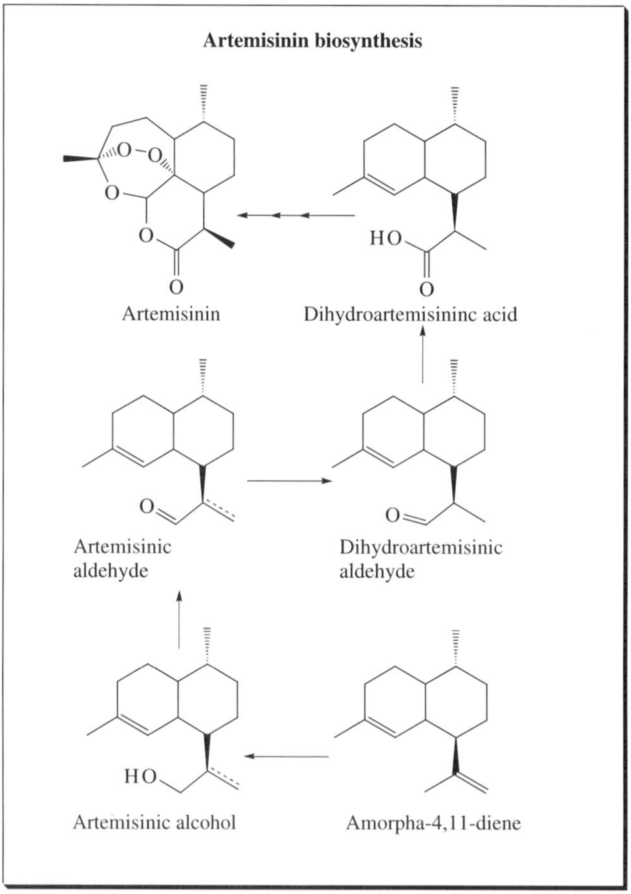

Figure 7. Proposed artemisinin biosynthetic pathway from amorpha-4,11-diene based on intermediates isolated from plants. The intermediate scaffold amorpha-4,11-diene can be produced in engineered microorganisms

under strict control like *E. coli's* endogenous DXP isoprenoid pathway and as such is able to provide sufficient FDP for subsequent cyclization by amorpha-4,11-diene synthase. Another boost to production came from using a codon-optimized version of the synthase gene. Previous studies of plant sesquiterpenoid production showed that poor expression of plant sesquiterpenoid synthases in *E. coli* limited terpenoid production (Vincent and Martin, 2001).

3.2.2. Taxol

Taxol (paclitaxel) is a diterpene that is a very important chemotherapeutic agent used to treat ovarian, breast, lung, neck, bladder, skin and cervical cancers (Jennewein and Croteau, 2001). Most of the Taxol and its derivatives used in cancer chemotherapies and scientific research have been isolated from the bark of the yew (*Taxus brevifolia)* or semi-synthesized from Taxol precursors baccatin III and 10-deacetylbaccatin III which were isolated from natural sources. In order to preserve supplies and increase production, alternative approaches to obtaining Taxol are under intense investigation. Elucidation of the biosynthesis and rate-limiting metabolic steps of Taxol and its derivatives may lead to the development of higher yielding production strategies using recombinant plant cell cultures or microbial cells.

The isoprenoid precursor geranylgeranyl diphosphate (GGDP) is cyclized to taxa-4(5), 11(12)-diene in the first committed step in the biosynthesis of Taxol. This reaction is catalyzed by taxadiene synthase, an enzyme first cloned and characterized a decade ago (Wildung and Croteau, 1996). The taxane skeleton is then further functionalized by a series of eight oxidative modifications catalyzed by cytochrome P450 enzymes, three CoA dependent acylations and side chain additions catalyzed by CoA-thioester dependent transferases (Walker and Croteau, 2001) (Figure 8). A total of nineteen enzymatic reactions are required to produce Taxol.

Enzymes involved in six of the eight total cytochrome P450 hydroxylation reactions (Schoendorf et al., 2001; Jennewein et al., 2004a; Jennewein et al., 2003; Jennewein et al., 2001; Chau and Croteau, 2004; Chau et al., 2004b) have been cloned and expressed in *E. coli*. Also cloned and functionally expressed in *E. coli* were two acetyl transferases, two benzoyltransferases, a phenylpropanoid side chain CoA-transferase (Walker and Croteau, 2000c, 2000d; Walker et al., 2002a; Walker et al., 2002b) and a phenylalanine aminomutase involved in Taxol side chain biosynthesis (Walker et al., 2004). Engineering Taxol or key intermediate biosynthesis in *E. coli* requires that the other enzymes be cloned and functionally expressed in *E. coli*. At present, only key intermediate scaffolds (which can be used for further synthetic modification to Taxol or Taxol derivatives) can be produced in heterologous hosts.

In order to heterologously produce diterpenes,the GGDP synthase from *Taxus canadensis* was overexpressed in *E. coli* (Hefner et al., 1998). Taxadiene was produced in cells transformed with both the GGDP synthase and taxadiene synthase co-overexpressed with two DXP pathway enzymes, DXP synthase and IDP isomerase (Huang et al., 2001). This recombinant four step pathway led to

Taxol Biosynthesis

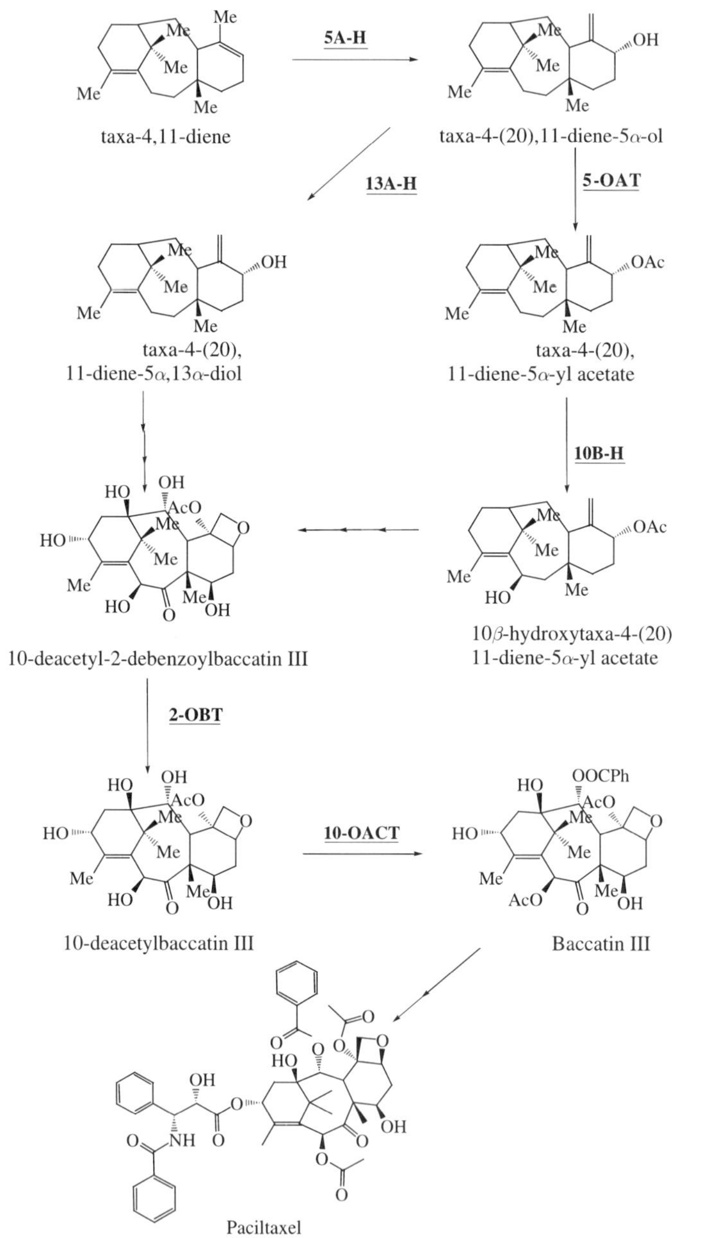

the production of 1.3 mg/L of taxadiene, the unmodified terpenoid scaffold for Taxol. Modifications to the taxadiene synthase were necessary to obtain expression of soluble protein in *E. coli.* In order to produce more complex intermediates or biologically active Taxol molecules in a microbial host, additional biosynthetic enzymes need to be cloned and functionally active.

One difficult aspect of Taxol biosynthetic pathway engineering in *E. coli* is the large number of P450 reactions. Reconstitution of cytochrome P450 monooxygenase activity in the chosen microbial host is a serious technical hurdle because P450 monooxygenases require a complementary NADPH:cytochrome P450 reductase for efficient electron transport from co-factor to monooxygenase. There are no P450 monooxygenase genes found in *E. coli*, but yeasts such as *S. cerevisiae* do have endogenous microsomal P450's and reductases. Heterologous *S. cerevisiae* WAT11 cells (Pompon et al., 1996) overexpress a reductase from *Arabidopisis* and have been used to express active P450's in cell lysates (Schoendorf et al., 2001; Jennewein et al., 2004a; Chau and Croteau, 2004; Jennewein et al., 2004b). More efficient reductase activity was observed with a homologous reductase from *Taxus.* It was used in *S. cerevisiae* for transgenic redox coupling with the Taxol P450 10 β-hydroxylase (Jennewein et al., 2005).

The ability of *S. cerevisiae* to functionally express P450 enzymes led to an attempt to engineer a five-step biosynthetic pathway leading to the intermediate taxa-4(20), 11(12)-dien-5α-acetoxy-10β-ol. This pathway requires functional co-expression of GGDP synthase, taxadiene synthase, three P450's (taxadiene 5α, 10β and 13α-hydroxylase, two acetyl-transferases (taxadiene 5-*O* and 10-*O*-acetyl-transferases) and a 5-*O*-benzoyl transferase (Dejong et al., 2006). Five of the enzymes (GGPP synthase, taxadiene synthase, taxadiene 5α and 10β-hydroxylase and 5-*O*-acetyl transferase) were instated in *S. cerevisiae*, but the 5α-hydroxylase had only limited expression leading to taxadiene production. This roadblock prevented flux further through the pathway. Engineering efficient redox-coupling between P450 enzymes and NADPH and high P450 expression levels are hurdles that must be overcome to achieve functional Taxol biosynthesis in yeast cells. However, the production of the intermediate taxadiene from microbial hosts *E. coli* and *S. cerevisiae* is already 100-fold higher than engineered *Arabidopsis* (Besumbes, 2004).

3.2.3. Carotenoids

Carotenoids are structurally diverse higher terpenes that are widespread in plants and microorganisms. The isoprenoid-based pigments have a variety of functions in

Figure 8. Proposed Taxol biosynthetic pathway starting from the precursor taxa-4,11-diene. Taxa-4, 11-diene can be expressed in engineered microbial cells as a scaffold. Cloned genes are bolded and underlined. Enzyme abbreviations are as follows: 5A-H, taxadiene-5α-hydroxylase; 13A-H, 13α-hydroxylase; 5-OAT, taxadien-5α-ol acetyl transferase; 10B-H, 10β-hydroxylase; 2-OBT, 2α-*O*-benzoyltransferase; 10-OACT, 10-deacetylbaccatin III-10-*O*-acetyl transferase

microbial, plant and animal systems: such as in photosynthesis, as antioxidants, in coloration and as Vitamin A precursors (Lee and Schmidt-Dannert, 2002; Becker et al., 2003; Fraser and Bramley, 2004). They are also of great industrial interest and used as colorants and antioxidants in feed, food, nutraceuticals, and pharmecueticals (Fraser and Bramley, 2004; Hirschberg, 2001). Few carotenoids can be chemically synthesized or isolated from their natural sources (Ernst, 2002; Valla et al., 2004). In the past few decades, there has been major interest in the identification of microbial and plant carotenoid biosynthetic genes and engineering microorganisms for the production of diverse carotenoid structures (Lee and Schmidt-Dannert, 2002; Umeno et al., 2005). Major pathways have been identified and installed in non-carotenogenic recombinant microorganisms leading to the production of carotenoids with C_{40} (derived from GGDP) or C_{30} (derived from FDP) backbones. Most known carotenoid enzymes of microbial or plant origin can function cooperatively when installed in a host allowing for the recombinant biosynthesis of diverse cyclic and acyclic structures in *E. coli* (Lee et al., 2003; Mijts et al., 2004; Mijts et al., 2005; Sandmann, 2002). Carotenoids are especially amenable to genetic engineering because of the color screening, complementation of both microbial and plant enzymes and plasticity of the biosynthetic enzymes. Combinatorial and directed evolution strategies have produced an even wider array of carotenoid structures synthesized by engineered pathways (Umeno et al., 2005).

The first committed step in carotenoid biosynthesis is the head-to-head condensation of two GGDP molecules to produce phytoene (Cunningham and Gantt, 1998). Non-carotenogenic hosts like *E. coli* do not produce the isoprenoid precursor GGDP and need the geranyl geranyl diphosphate synthase (CrtE) to add an isoprenoid unit to the C15 intermediate FDP. Two GGDP molecules are condensed by the activities of the phytoene synthase (CrtB). The phytoene intermediate is subjected to four desaturation reactions catalyzed by phytoene desaturase (CrtI) to form the first colored carotenoid, lycopene. This core pathway can be further modified by cyclization, glucosylation, and diverse oxygenation reactions using characterized carotenoid enzymes to form acyclic, cyclic carotenoids and xanthophylls in *E. coli* (Lee et al., 2003) (Figure 3).

The ease of combinatorial assembly of genes from different origins into functional pathways within *E. coli* and the ease of screening carotenoid products has made them good target compounds for directed evolution and other scaffold diversification strategies (Schmidt-Dannert et al., 2000). Individual biosynthetic genes within an assembled pathway can be subjected to *in vitro* evolution to create enzyme mutants with new catalytic activities (Schmidt-Dannert, 2000). The chromatic properties of carotenoids allow for novel structures to identified if the structural modifications alter the chromophore (Lee et al., 2004). Both the phytoene desaturase and the lycopene cyclase (responsible for the formation of β-carotene) have been the subject of directed evolution studies. A phytoene desaturase variant that carries out addition desaturations to a lycopene backbone was isolated and forms 3,4-didehydrolycopene and the fully conjugated 3',4', 3,4-tetradehyrolycopene (Schmidt-Dannert et al., 2000). A mutant cyclase was added to this recombinant pathway and formed the monocyclic torulene. When additional microbial carotenoid modifying enzymes were added to the evolved pathways, a large number of different structures were produced in *E. coli* (Lee et al., 2003). Other directed evolution and combinatorial biosynthesis studies have generated carotenoids with unusual length backbones (C_{35}, C_{45}, C_{50}) (Umeno and Arnold, 2004; Umeno et al., 2002). The discovery of novel enzymes such as the carotenoid oxygenase from *Staphylococcus aureus* can expand the array of novel C_{30} and C_{40} compound through combinatorial biosynthesis (Mijts et al., 2005).

Overall yields of carotenoids from recombinant *E. coli* are low (approximately 1 mg/g dry cell weight, corresponding to 5–10 mg/L of culture). There are three major improvements that need to be made for the development of high producing fermentation processes. The first is the optimization of the isoprenoid precursor pool. A great deal of metabolic engineering of carotenoid pathways has focused on this aspect including the overexpression of isoprenoid genes such as *dxr* (1-deoxy-d-xylulose-5-phosphate reductoisomerase), *dxs* (1-deoxy-d-xylulose-5-phosphate synthase), *idi* (IDP isomerase) and *ispA* (FDP synthase), either individually or in combination (Jones et al., 2000; Matthews and Wurtzel, 2000; Farmer and Liao, 2000; Kim and Keasling, 2001). An alternative strategy is to redirect the flux from pyruvate to glyceraldehye-3-phosphate by overexpressing *pps* (phosphoenolpyruvate) or *pck* (phosphoenolpyruvate carboxykinase) or by inactivation of *pyk* (pyruvate kinase). The second improvement needed is balancing the expression of carotenogenic genes for efficient transformation of precursors into products. Balancing the rate limiting enzymes has been shown to increase flux through the carotenoid pathway (Farmer and Liao, 2000; Kim and Keasling, 2001). Flux control through the carotenoid pathway can be achieved by modulating enzyme expression levels to accumulate different ratios of intermediates (Smolke et al., 2001). The third improvement is the need for sufficient storage space for the lipophilic carotenoids. Additional membrane storage may be engineered into *E. coli* or a more suitable heterologous host with high storage capabilities such as yeasts or photosynthetic bacteria.

3.2.4. *Apocarotenoids*

The oxidative cleavage of carotenoids produces apocarotenoids, which constitute a structurally diverse and widespread compound class in nature. The diversity in apocarotenoid structures comes from the more than 600 known carotenoid structures, variations in cleavage sites, and subsequent oxidative modifications and glycosylations of the cleavage products (Winterhalter and Rouseff, 2002). Some apocarotenoids are very potent aroma compounds (e.g. β-ionone) and important pigments (e.g. crocin and bixin) of high industrial value. Other apocarotenoids have important biological activities as signaling molecules, hormone (e.g. absisic acid) or vitamin A (e.g. retinal) precursors (Bouvier et al., 2005).

The formation of apocarotenoids can be either through non-specific mechanisms such as photo-oxidation or lipoxygenase co-oxidation (Wache et al., 2003) or through specific enzymatic activity. In the last decade, a class of enzymes responsible for the specific oxidative cleavage of carotenoids has been identified in plants, animals (retinal formation) and more recently, in microorganisms. Collectively these enzymes are known as carotenoid cleavage dioxygenases (CCDs), although they are subdivided into distinct classes depending on the site of double bond cleavage of the conjugated isoprenoid backbone of carotenoids [reviewed in (Bouvier et al., 2005)]. Here, we will discuss CCDs cloned and characterized from plants.

9-*cis*-epoxy dioxygenases (NCEDs), the first CCDs described, are involved in the rate limiting steps of abscisic acid formation in plants (Schwartz et al., 1997). Xanthophylls with a 9-*cis* conformation are cleaved at the 11,12 double bond to form xanthoxin which is then converted into the plant growth regulator abscisic acid. Several NCEDs from tomato (Burbidge et al., 1999), avocado (Chernys and Zeevaart, 2000), cowpca (Iuchi et al., 2000), bean (Qin and Zeevaart, 1999) and *Arabidopsis* (Iuchi et al., 2001) have been identified and functionally characterized in *E. coli*.

Many aroma and flavor compounds derive from a second class of CCDs. These enzymes cleave cyclic carotenoids at the 9,10 and 9',10' double bond forming a C_{14} dialdehyde and two volatile C_{13} cyclohexone derivatives (e.g. β-ionone, dihydroactinidiolide, damascenol and β-cyclocitral). Enzymes that cleave β-carotene symmetrically have been described from a number of different sources (Schwartz et al., 2001; Simkin et al., 2004a; Cao et al., 2005; Simkin et al., 2004b). The resulting ionone structure can rearrange to a number of different molecules producing the aromas associated with fruits and plants (Leffingwell, 2003). The substrate specificity of these enzymes is not stringent; for example, CCD from tomato was shown to cleave at the 9,10 and 9'10' positions of β-carotene, zeaxanthin, lutein, violaxanthin and neoxanthin which have different ionone ring modifications.

Two examples of important industrial apocarotenoid products include the formation of the pigment bixin and the spice saffron. A lycopene-specific 5,6 (5',6')-cleavage dioxygenase (BoLCD) from *Bixa orellana* is responsible for the formation of bixin dialdehyde and a C7 cleavage product MHO (6-methyl-5-hepten-2-one). Expression of the lycopene CCD and two additional enzymes bixin aldehyde dehydrogenase and norbixin carboxyl methyltransferase in *E. coli* with a lycopene accumulating background resulted in bixin biosynthesis (Bouvier et al., 2003a). Saffron, the most expensive spice in the world (US$1000–2000/kg), is comprised of water soluble norcarotenoid glycosides and found in the dry stigma of *Crocus sativus*. It is composed of three major apocarotenoids; crocins which are responsible for the pigmentation, safranal which is responsible for the aroma and picrocrocins which are the bitter tasting glucosides of safranal. The majority of the color derives from crocetin esters which are created by the cleavage of zeaxanthin (Tarantilis et al., 1995). This cleavage reaction is the committed step in the formation of these apocarotenoid and is catalyzed by a zeaxanthin-specific 7,8 (7',8')-cleavage dioxygenase (*CsZCD*). This enzyme from *Crocus sativus* was characterized and found to form crocetin dialdehyde and 3-hydroxy-β-cyclocitral *in vitro* (Bouvier et al., 2003b). One cleavage product, 3-hydroxy-β-cyclocitral is a critical intermediate in the formation of safranal (Bouvier et al., 2003b). The other product crocetin, can be recognized by a GTase enzyme and converted to glycosides. The enzyme (UGTCs2) involved in glyucosylation reactions was functionally expressed in bacterial systems and the recombinant enzyme was shown to recognize and glucosylate crocetin, crocetin b-D-gentibiosyl ester and crocetin β-D-glucosyl (Moraga et al., 2004). The heterologous expression of both the *Cs*ZCD and the GTase with a carotenoid biosynthesis pathway would lead to a competitive alternative to natural crocin production by production in hosts such as *E. coli* or yeast.

3.2.5. *Flavonoids*

Flavonoids are one of the best studied natural product classes because they exhibit a multitude of important functions in plants (Harborne and Williams, 2000; Dixon et al., 2002a). Functions of flavonoids include coloration, defenses against herbivory and pathogens, and signalling. Flavonoids act medicinally by scavenging free radicals, inhibiting enzymes, and having anti-inflammatory and cytotoxic antitumor activities (Harborne and Williams, 1998; Dixon and Ferreira, 2002).

Flavonoids are characterized by a cyclic phenylchromane ring structure derived from one phenylpropene unit (such as 4-coumaric-, caffeic-, or ferulic acid) and three acetate units from malonyl-CoA. Functionalization of this scaffold leads to three different compound classes: (i) flavanols, anthocyanins, condensed tannins, (ii) isoflavonoids, (iii) stilbenes (Winkel-Shirley, 2001).

Flavonoid pathway elucidation originally focused on anthocyanins, the brightly colored flavonoid structures that attract pollinators and seed dispersers to plants. Identification of genes and regulatory mechanisms involved in anthocyanin biosynthesis was greatly aided by the visible, screenable colored phenotypes (Forkmann and Martens, 2001). In recent years, many flavonoid genes have been have been cloned from diverse plant species and heterologously expressed in microbial hosts for enzymatic characterization. This has led flavonoids, as a compound class, to have the most detailed pathways maps in terms of genes, enzymes, and pathway intermediates [reviewed in (Dixon et al., 2002b; Winkel-Shirley, 2001)].

Biosynthesis of flavonoids begins with a four step reaction series (Figure 9). The first reaction is the deamination of L-phenylalanine by phenylammonia lyase (PAL) which forms cinnamic acid. Cinnamic acid is hydroxylated by a P450 cinnamate-4-hydroxylase (C4H) to coumaric acid. The carboxyl group of coumaric acid is then activated with coenzyme A by coumaroyl-CoA-ligase (4CL) to form 4-coumaroyl-CoA.

4-coumaroyl-CoA can undergo the first committed step in flavonoid biosynthesis, the condensation with three malonyl-CoA molecules catalyzed by a unimodular type III polyketide synthases (Jez et al., 1999; Ferrer et al., 1999; Austin and Noel, 2003; Jez et al., 2001). Two different types of polyketide synthases can carry out this condensation: either a chalcone or stilbene synthase. Chalcone synthases (CHS) produces a naringenin chalcone (4,2',4',6'-tetrahydroxychalcone) through intramolecular C6—C1 Claisen type cyclization. Stilbene synthases (STS) are

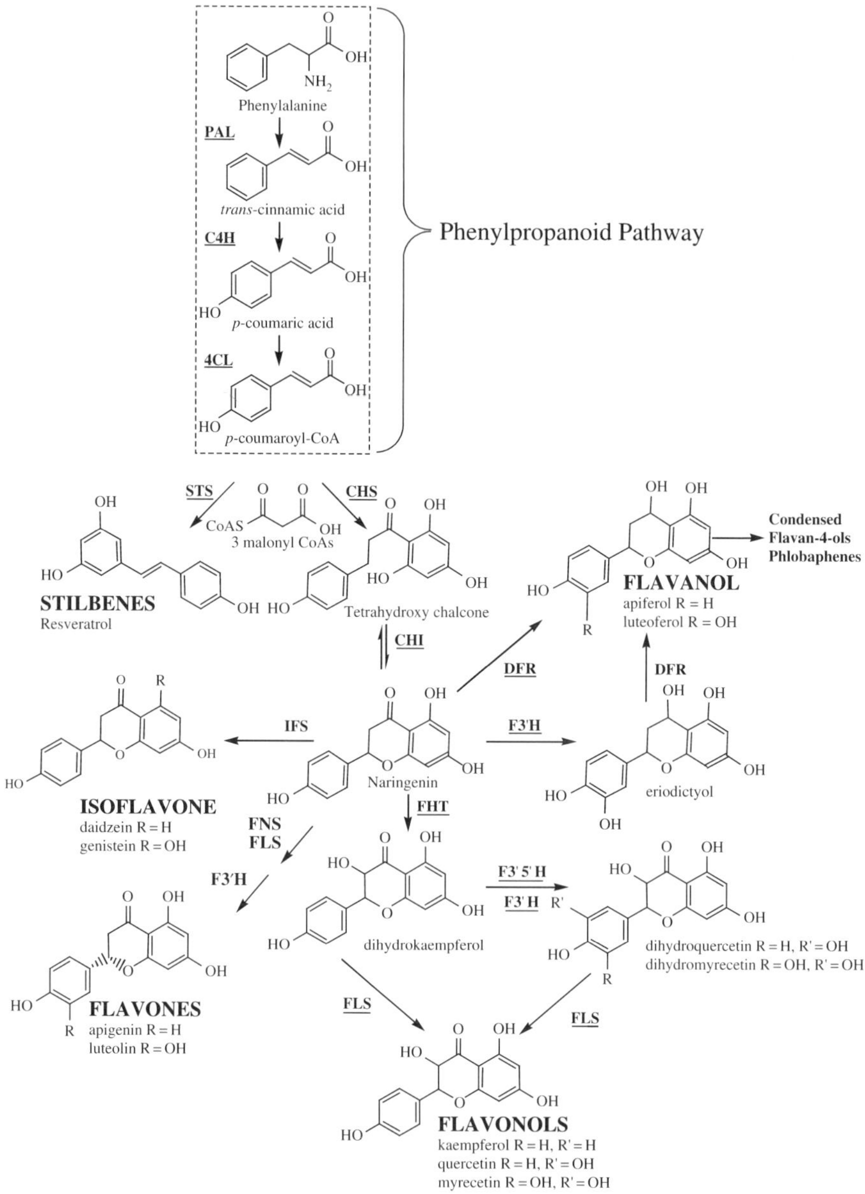

Figure 9. Biosynthetic pathways leading to select flavonoid-derived products in plants. Major biosynthetic routes for which enzymes have been characterized on a molecular level are shown. Common names for differently oxidized and substituted phenylcoumarane structures are capitalized and in bold. Cloned genes are in bold and underlined. Enzyme abbreviations are as follows: 4CL: coumaroyl-CoA-ligase; C4H, cinnamate-4-hydroxylase; CHI, chalcone isomerase; CHS, chalcone synthase; DFR, dihydroglavonol-4-reductase; F3'H, flavonoid 3'-hydroxylase; F3'5'H, flavonoid 3'5'-hydroxylase; FLS, flavonol synthase; IFS, isoflavone synthase; PAL, phenylammonia lyase; STS, stilbene synthase

homologous to the CHS, but the condensation mechanism involves an additional decarboxylation step and catalyses a different ring closure via an intramolecular C2—C7 aldol condensation (Austin et al., 2004). This results in two aromatic rings separated by two carbons instead of three.

From the chalcone product (2-*S*)-naringenin, a series of hydroxylations and oxidations catalyzed by cytochrome P450 monooxygenases and 2-oxo-glutarate dependent dioxygenases yield flavones, flavanones and flavonols. Further modifications such as glycosylations, methylations, acylations and prenylations convert the structures into brightly colored anthocyanins. Through a different pathway, naringenin is transformed into diverse isoflavonoid structures.The first reaction in isoflavonoid biosynthesis is a hydroxylation catalyzed by a cytochrome P450 enzyme 2-hydroxyisoflavone synthase (IFS). The resulting product is an unstable 2-hydroxyisoflavanone which is spontaneously dehydrated to form isoflavones daidzein and genistein. The diverse array of isoflavonoid structures found in legumes are the result of reductions, oxidations, methylations and glycosylations on the core isoflavone structure (Aoki et al., 2000).

Newly discovered flavonoid compounds frequently have important biological activities. Bioactive benzophenone derivatives from the St. John's wort family of plants (*Hypericum*) and stilbene derivatives such as resveratrol (trihydroxystilbene) are now wideley accepted to treat ailments and provide health benefits. Stilbenes have been linked to a number of beneficial medicinal effects such as anti-inflammatory activities, vasodilation and anticancer activities (Howitz et al., 2003; Fremont, 2000; Soleas et al., 1997). Other flavanols, isoflavan, and prenylated isoflavonoids have been described as having antibacterial, antifungal, or antiviral activity (Harborne and Williams, 1998; Dixon and Steele, 1999). Flavonoids therefore represent an important class of medicinally valuable compounds. The biosynthetic production of flavonoids in engineered microbial host would produce enantiomerically pure flavonoids in good yields from inexpensive precursors. There are several successful examples of reconstructing flavonoid pathways in microbial hosts.

The first successful reconstitution pathway was engineered into *E. coli* cells and produced low levels of naringenin and pinocembrin (Hwang et al., 2003). This pathway was a hybrid of enzymes harvested from yeast, bacterial, and plant genomes. The use of a bifunctional ammonia lyase on simple precursors tyrosine and phenylalanine resulted in the formation of two products (Hwang et al., 2003). An attempt was made to engineer a different pathway into *E. coli* using solely *Arabidopsis thaliana* genes for PAL, C4H, 4CL, and CHS (Watts et al., 2004). However, expression problems of the plant cytochrome P450 C4H resulted in a pathway block that could only be overcome by feeding exogenous 4-coumaric acid. To bypass this blockage in *E. coli*, a bacterial tyrosine ammonium lyase (TAL) from *Rhodobacter sphaeroides* was substituted for the *A. thaliana* PAL. The TAL used tyrosine as a starter unit and formed a shortened pathway. Production levels of naringenin were 250-fold greater (~21mg/L) than the plant, yeast, and bacterial hybrid pathway. Biotransformation experiments with this pathway found that feeding different phenylpropanoids directly to the cells resulted in modified flavonoid structures. The compound phloretin was synthesized from 3-(4-hydroxyphenyl)propionic acid.

The addition of chalcone isomerases (CHI) to naringenin and pinocembrin producing pathways extended the pathway to the corresponding (2*S*)-flavanones (Miyahisa et al., 2005). Flavones have also been produced in recombinant *E. coli* by expressing five plant specific flavone biosynthetic enzymes, a 4-coumaryl-CoA ligase, a chalcone synthase, a chalcone isomerase, flavone synthase and a 7-*O*-methyltransferase (Leonard et al., 2006). Through biotransformations of phenylpropanoid 4-coumaric acid, 0.415 mg/L of apigenin and 0.208 mg/L of the methylated product genkwanin were produced. Feeding studies with caffeic acid resulted in only minor product formation of luteolin.

Leonard et al. (2006) used the flavone synthase I from parsley because unlike the more common flavone synthase II, FSI is not a P450 monooxygenase (Martens et al., 2001). Flavonoid biosynthesis involved many P450 hydroxylations and oxygenation reactions and it is difficult to express eukaryotic P450's in the active form in *E. coli.* For that reason, *S. cerevisiae,* a more suitable host for functional P450 expression, is also being investigated for heterologous flavonoid production. Naringenin (7 mg/L) and pinocembrin (0.8 mg/L) were produced in *S. cerevisiae* by a pathway constructed with a PAL from *Rhodosporidium toruloides,* 4CL from *A. thaliana*, and the CHS from *Hypericum androsaemum* (Jiang et al., 2005). Several derailment products accumulated and the pathway flux was limited by the tyrosine precursor pool. More naringenin (up to 28 mg/L) was produced by another pathway in *S. cerevisiae* that utilized genes from different plant species (Yan et al., 2005). This pathway was extended further with two different types of flavone synthases, a soluble 2-oxoglutarate dependent dioxygenase and a membrane bound cytochrome P450, to produce the flavones chrysin, apigenin, and luteolin (Leonard et al., 2005).

Stilbene compounds have also been heterologously produced in engineered microbial hosts. Resveratrol was produced in *S. cerevisiae* using a hybrid poplar and grapevine pathway (Becker et al., 2003). The production levels (~1 ug/L) were considerably lower than those achieved by an *E. coli* engineered pathway (~100 mg/L) (Watts et al., 2006). The *E. coli* pathway consisted of a 4-coumaroyl ligase from *A. thaliana* and a stilbene synthase from *Arachis hypogaea.* Phenylpropanoids were fed to the recombinant cells and 4-coumaric acid was converted to resveratrol and caffeic acid to piceatannol; ferulic acid was not converted to isorhapontigenin however.

4. CONCLUSIONS

The major thesis of this chapter was to show how recombinant microbial systems have aided in plant biosynthetic pathway elucidation and represents a potential method for industrial production of plant natural products. This chapter examined both medicinal natural products and flavor/fragrance compounds from several different compound classes: fatty acid derivatives, alkaloids, isoprenoids and flavonoids. Much of the work in plant secondary metabolites overlap between medicinal and aroma pathway discovery, as can be seen clearly with the acyltransferase family of enzymes. Characterization of aroma biosynthesis is occurring at an increased pace due to new technologies to assess and analyze volatile compound production and constituents (Pichersky et al., 2006). These discoveries regarding major enzymatic steps in aroma biosynthesis will aid in the prediction of biosynthetic genes involved in the complex pathways that form medicinal compounds. Heterologous expression of enzymes in *E. coli* or *S. cerevisiae* has been used for single enzyme characterization as well as the reconstitution of short pathways. Enzymes discovered by genomic and proteomic methods need to be individually characterized to determine substrate specificity and the product profile (particularly evident in the terpenoid cyclase family). The assembly of pathways in microorganisms allows for the stepwise identification of intermediates, which is critical in understanding pathway blockages or regulation points. The expression of pathways in recombinant hosts has identified some difficulties in optimizing flux through pathways and other complex regulatory controls. Plant enzymes may need to be modified to work optimally in heterologous pathways. Despite the challenges ahead, we are optimistic that microbial fermentation of plant natural products will provide high amounts of compounds for further research, to act as scaffolds for semi-synthesis or as direct consumer products. A stunning example of the feasibility of reconstituting complex biosynthetic pathways in microorganisms is the engineering of a 13 gene heterologous pathway for hydrocortisone biosynthesis in yeast (Szczebara et al., 2003). Hybrid pathways containing genes from different plant and/or microbial sources can be built in microorganisms to produce both natural and designer secondary metabolites (such as the production of bixin in lycopene producing cells). Critical to the success of this approach is the continued effort of plant biologists in elucidating biosynthetic pathways. In addition, combined efforts between plant biologists, metabolic engineers and natural products biochemists will be necessary to capitalize on all the advancements made in the respective fields.

REFERENCES

Achnine L, Huhman DV, Farag MA, et al. (2005) Genomics-based selection and functional characterization of triterpene glycosyltransferases from the model legume *Medicago truncatula*. *Plant J* 41(6):875–887.

Aggarwal KK, Khanuja SPS, Ahmad A, et al. (2002) Antimicrobial activity profiles of the two enantiomers of limonene and carvone isolated from the oils of *Mentha spicata* and *Anethum sowa*. *Flav Fragr* 17(1):59–63.

Aharoni A, Keizer LCP, Bouwmeester HJ, et al. (2000) Identification of the SAAT gene involved in strawberry flavor biogenesis by use of DNA microarrays. *Plant Cell* 12(5):647–662.

Aharoni A, Giri AP, Deuerlein S, et al. (2003) Terpenoid metabolism in wild-type and transgenic *Arabidopsis* plants. *Plant Cell* 15(12):2866–2884.

Allen RS, Millgate aG, Chitty JA, et al. (2004) RNAi-mediated replacement of morphine with the nonnarcotic alkaloid reticuline in opium poppy. *Nat Biotechnol* 22(12):1559–1566.

Alonso WR, Rajaonarivony JI, Gershenzon J, et al. (1992) Purification of 4S-limonene synthase, a monoterpene cyclase from the glandular trichomes of peppermint (*Mentha x piperita*) and spearmint (*Mentha spicata*). *J Biol Chem* 267(11):7582–7587.

Alper H, Jin YS, Moxley JF, et al. (2005a) Identifying gene targets for the metabolic engineering of lycopene biosynthesis in *Escherichia coli*. *Metab Eng* 7(3):155–164.

Alper H, Miyaoku K, Stephanopoulos G (2005b) Construction of lycopene-overproducing *E. coli* strains by combining systematic and combinatorial gene knockout targets. *Nat Biotechnol* 23(5):612–616.

Angerhofer CK, Guinaudeau H, Wongpanich V, et al. (1999) Antiplasmodial and cytotoxic activity of natural bisbenzylisoquinoline alkaloids. *J Nat Prod* 62(1):59–66.

Aoki T, Akashi T, Ayabe S (2000) Flavonoids of leguminous plants: Structure, biological activity, and biosynthesis. *J Plant Res* 113(1112):475–488.

Ashihara H, Crozier A (1999) Biosynthesis and catabolism of caffeine in low-caffeine-containing species of *Coffea*. *J Agricult Food Chem* 47(8):3425–3431.

Ashihara H, Gillies FM, Crozier A (1997) Metabolism of caffeine and related purine alkaloids in leaves of tea (*Camellia sinensis L*). *Plant Cell Physiol* 38(4):413–419.

Ashihara H, Suzuki T (2004) Distribution and biosynthesis of caffeine in plants. *Front Biosci* 9:1864–1876.

Aubourg S, Lecharny A, Bohlmann J (2002) Genomic analysis of the terpenoid synthase (AtTPS) gene family of *Arabidopsis thaliana*. *Mol Genet and Genom* 267(6):730–745.

Austin MB, Bowman ME, Ferrer J-L, et al. (2004) An aldol switch discovered in stilbene synthases mediates cyclization specificity of type III polyketide synthases. *Chem Biol* 11(9):1179–1194.

Austin MB, Noel aJP (2003) The chalcone synthase superfamily of type III polyketide synthases. *Nat Prod Rep* 20(1):79–110.

Back K, Chappell J (1995) Cloning and bacterial expression of a sesquiterpene cyclase from *Hyoscyamus muticus* and its molecular comparison to related terpene cyclases. *J Biol Chem* 270(13): 7375–7381.

Bayer A, Ma XY, Stöckigt J (2004) Acetyltransfer in natural product biosynthesis – functional cloning and molecular analysis of vinorine synthase. *Bioorg Med Chem* 12(10):2787–2795.

Becker JVW, Armstrong GO, Van Der Merwe MJ, et al. (2003) Metabolic engineering of *Saccharomyces cerevisiae* for the synthesis of the wine-related antioxidant resveratrol. *FEMS Yeast Research* 4(1):79–85.

Beekwilder J, Alvarez-Huerta M, Neef E, et al. (2004) Functional characterization of enzymes forming volatile esters from strawberry and banana. *Plant Physiol* 135(4):1865–1878.

Bertea CM, Freije JR, Van Der Woude H, et al. (2005) Identification of intermediates and enzymes involved in the early steps of artemisinin biosynthesis in *Artemisia annua*. *Planta Medica* 71(1):40–47.

Besumbes O (2004) Metabolic engineering of isoprenoid biosynthesis in *Arabidopsis* for the production of taxadiene, the first committed precursor of Taxol. *Biotechnol Bioeng* 88(2):168–175.

Boatright J, Negre F, Chen XL, et al. (2004a) Understanding in vivo benzenoid metabolism in petunia petal tissue. *Plant Physiol* 135(4):1993–2011.

Boatright J, Negre F, Chen XL, et al. (2004b) Understanding in vivo benzenoid metabolism in petunia petal tissue. *Plant Physiol* 135(4):1993–2011.

Bohlmann J, Crock J, Jetter R, et al. (1998b) Terpenoid-based defenses in conifers: cDNA cloning, characterization, and functional expression of wound-inducible (E)-alpha-bisabolene synthase from grand fir (*Abies grandis*). *Proc Natl Acad Sci USA* 95(12):6756–6761.

Bohlmann J, Martin D, Oldham NJ, et al. (2000) Terpenoid secondary metabolism in *Arabidopsis thaliana*: cDNA cloning, characterization, and functional expression of a myrcene/(E)-β-ocimene synthase. *Arch Biochem Biophys* 375(2):261–269.

Bohlmann J, Meyer-Gauen G, Croteau R (1998a) Plant terpenoid synthases: Molecular biology and phylogenetic analysis. *Proc Natl Acad Sci* 95(8):4126–4133.

Bohlmann J, Phillips M, Ramachandiran V, et al. (1999) cDNA cloning, characterization, and functional expression of four new monoterpene synthase members of the Tpsd gene family from grand fir (*Abies grandis*). *Arch Biochem Biophys* 368(2):232–243.

Bohlmann J, Steele CL, Croteau R (1997) Monoterpene synthases from Grand fir (*Abies grandis*) – cDNA isolation, characterization, and functional expression of myrcene synthase, (–)(4S)-limonene synthase, and (–)-(1S,5S)-pinene synthase. *J Biol Chem* 272(35):21784–21792.

Boonstra B, Rathbone DA, Bruce NC (2001) Engineering novel biocatalytic routes for production of semisynthetic opiate drugs. *Biomol Eng* 18(2):41–47.

Boucher Y, Doolittle WF (2000) The role of lateral gene transfer in the evolution of isoprenoid biosynthesis pathways. *Mol Microbiol* 37(4):703–716.

Bouvier F, Dogbo O, Camara B (2003a) Biosynthesis of the food and cosmetic plant pigment bixin (Annatto). *Science* 300(5628):2089–2091.

Bouvier F, Rahier A, Camara B (2005) Biogenesis, molecular regulation and function of plant isoprenoids. *Progr Lipid Res* 44(6):357–429.

Bouvier F, Suire C, Mutterer J, et al. (2003b) Oxidative remodeling of chromoplast carotenoids: Identification of the carotenoid dioxygenase CsCCD and CsZCD genes involved in *Crocus* secondary metabolite biogenesis. *Plant Cell* 15(1):47–62.

Broun P, Somerville C (2001) Progress in plant metabolic engineering. *Proc Natl Acad Sci USA* 98(16):8925–8927.

Burbidge A, Grieve TM, Jackson A, et al. (1999) Characterization of the ABA-deficient tomato mutant notabilis and its relationship with maizeVp14. *Plant J* 17(4):427–431.

Cao YR, Guo XL, Zhang Q, et al. (2005) Isolation and characterization of carotenoid cleavage dioxygenase gene in halophyte *Suaeda salsa*. *Plant Growth Regul* 46(1):61–67.

Carter OA, Peters RJ, Croteau R (2003) Monoterpene biosynthesis pathway construction in *Escherichia coli*. *Phytochemistry* 64(2):425–433.

Chau MD, Walker K, Long R, et al. (2004a) Regioselectivity of taxoid-O-acetyltransferases: heterologous expression and characterization of a new taxadien-5 alpha-ol-O-acetyltransferase. *Arch Biochem Biophys* 430(2):237–246.

Chau M, Croteau R (2004) Molecular cloning and characterization of a cytochrome P450 taxoid 2 alpha-hydroxylase involved in Taxol biosynthesis. *Arch Biochem Biophys* 427(1):48–57.

Chau M, Jennewein S, Walker K, et al. (2004b) Taxol biosynthesis: Molecular cloning and of a cytochrome P450 characterization taxoid 7 beta-hydroxylase. *Chem Biol* 11(5):663–672.

Chen F, Ro D-K, Petri J, et al. (2004) Characterization of a root-specific *Arabidopsis* terpene synthase responsible for the formation of the volatile monoterpene 1,8-cineole. *Plant Physiol* 135(4): 1956–1966.

Chen F, Tholl D, D'auria JC, et al. (2003) Biosynthesis and emission of terpenoid volatiles from *Arabidopsis* flowers. *Plant Cell* 15(2):481–494.

Chernys JT, Zeevaart JAD (2000) Characterization of the 9-cis-epoxycarotenoid dioxygenase gene family and the regulation of abscisic acid biosynthesis in avocado. *Plant Physiol* 124(1):343–354.

Chou W-M, Kutchan TM (1998) Enzymatic oxidations in the biosynthesis of complex alkaloids. *Plant J* 15(3):289–300.

Colby SM, Alonso WR, Katahira EJ, et al. (1993) 4S-limonene synthase from the oil glands of spearmint (*Mentha spicata*). cDNA isolation, characterization, and bacterial expression of the catalytically active monoterpene cyclase. *J Biol Chem* 268(31):23016–23024.

Colby SM, Crock J, Dowdle-Rizzo B, et al. (1998) Germacrene C synthase from *Lycopersicon esculentum cv*. VFNT Cherry tomato: cDNA isolation, characterization, and bacterial expression of the multiple product sesquiterpene cyclase. *Proc Natl Acad Sci USA* 95(5):2216–2221.

Cragg GM, Newman DJ, Snader KM (1997) Natural products in drug discovery and development. *J Nat Prod* 60(1):52–60.

Croteau RB (1998) Monoterpene biosynthesis: Prospects for pathway engineering. Abstracts of Papers of the American Chemical Society 216:U55.

Croteau RB, Davis EM, Ringer KL, et al. (2005) (–)-Menthol biosynthesis and molecular genetics. *Naturwissenschaften* 92(12):562–577.

Cunningham F, Gantt E (1998) Genes and enzymes of carotenoid biosynthesis in plants. *Annual Review of Plant Physiology and Plant Mol Biol* 49:557–583.

D'auria JC, Chen F, Pichersky E (2002) Characterization of an acyltransferase capable of synthesizing benzylbenzoate and other volatile esters in flowers and damaged leaves of *Clarkia breweri*. *Plant Physiol* 130(1):466–476.

D'auria JC, Gershenzon J (2005) The secondary metabolism of *Arabidopsis thaliana*: growing like a weed. *Curr Opin Plant Biol* 8(3):308–316.
Dai Y, Wang HZ, Li H, et al. (2006) Increased expression of MAP kinase kinase 7 causes deficiency in polar auxin transport and leads to plant architectural abnormality in *Arabidopsis*. *Plant Cell* 18(2):308–320.
Davis EM, Croteau R (2000) Cyclization enzymes in the biosynthesis of monoterpenes, sesquiterpenes, and diterpenes. In: Biosynthesis: Aromatic Polyketides, Isoprenoids, Alkaloids, vol. 209, pp 53–95.
Davis EM, Ringer KL, Mcconkey ME, et al. (2005) Monoterpene metabolism. cloning, expression, and characterization of menthone reductases from peppermint. *Plant Physiol* 137(3):873–881.
De Carvalho CCR, Da Fonseca MMR (2002) Maintenance of cell viability in the biotransformation of (–)-carveol with whole cells of *Rhodococcus erythropolis*. *J Mol Catal B: Enzymatic* 19–20(389–398.
De Carvalho C, Da Fonseca MMR (2003) Towards the bio-production of trans-carveol and carvone from limonene: induction after cell growth on limonene and toluene. *Tetrahedron-Asymmetry* 14(24):3925–3931.
Degenhardt JR, Gershenzon J (2000) Demonstration and characterization of (E)-nerolidol synthase from maize: a herbivore-inducible terpene synthase participating in (3E)-4,8-dimethyl-1,3,7-nonatriene biosynthesis. *Planta* 210(5):815–822.
Dejong JM, Liu YL, Bollon aP, et al. (2006) Genetic engineering of Taxol biosynthetic genes in *Saccharomyces cerevisiae*. *Biotechnol Bioeng* 93(2):212–224.
Dewick PM (2002) The biosynthesis of C-5-C-25 terpenoid compounds. *Nat Prod Rep* 19(2):181–222.
Dittrich H, Kutchan T (1991) Molecular cloning, expression, and induction of berberine bridge enzyme, an enzyme essential to the formation of benzophenanthridine alkaloids in the response of plants to pathogenic attack. *Proc Natl Acad Sci USA* 88(22):9969–9973.
Dixon RA, Achnine L, Kota P, et al. (2002a) The phenylpropanoid pathway and plant defence; a genomics perspective. *Mol Plant Pathol* 3(5):371–390.
Dixon RA, Achnine L, Kota P, et al. (2002b) The phenylpropanoid pathway and plant defence – a genomics perspective. *Mol Plant Pathol* 3(5):371–390.
Dixon RA, Ferreira D (2002) Genistein. *Phytochemistry* 60(3):205–211.
Dixon RA, Steele CL (1999) Flavonoids and isoflavonoids – a gold mine for metabolic engineering. *Trends Plant Sci* 4(10):394–400.
Dogru E, Warzecha H, Seibel F, et al. (2000) The gene encoding polyneuridine aldehyde esterase of monoterpenoid indole alkaloid biosynthesis in plants is an ortholog of the α/β hydrolase super family. *Eur J Biochem* 267(5):1397–1406.
Dudareva N, Murfitt LM, Mann CJ, et al. (2000) Developmental regulation of methyl benzoate biosynthesis and emission in snapdragon flowers. *Plant Cell* 12(6):949–961.
Dudareva N, Negre F (2005) Practical applications of research into the regulation of plant volatile emission. *Curr Opin Plant Biol* 8(1):113–118.
Dudareva N, Pichersky E (2000) Biochemical and molecular genetic aspects of floral scents. *Plant Physiol* 122(3):627–634.
Dudareva N, Pichersky E, Gershenzon J (2004) Biochemistry of plant volatiles. *Plant Physiol* 135(4):1893–1902.
Duetz WA, Bouwmeester H, Van Beilen JB, et al. (2003) Biotransformation of limonene by bacteria, fungi, yeasts, and plants. *Appl Microbiol Biotechnol* 61(4):269–277.
Duetz WA, Fjallman aHM, Ren S, et al. (2001) Biotransformation of D-limonene to (+) trans-carveol by toluene-grown *Rhodococcus opacus* PWD4 Cells. *Appl Environ Microbiol* 67(6):2829–2832.
Effmert U, Saschenbrecker S, Ross J, et al. (2005) Floral benzenoid carboxyl methyltransferases: From *in vitro* to *in planta* function. *Phytochemistry* 66(11):1211–1230.
Ernst H (2002) Recent advances in industrial carotenoid synthesis. *Pure Appl Chem* 74(11):2213–2226.
Estevez JM, Cantero A, Reindl A, et al. (2001) 1-Deoxy-D-xylulose-5-phosphate synthase, a limiting enzyme for plastidic isoprenoid biosynthesis in plants. *J Biol Chem* 276(25):22901–22909.
Facchini PJ, Chappell J (1992) Gene Family for an elicitor-induced sesquiterpene cyclase in tobacco. *Proc Natl Acad Sci USA* 89(22):11088–11092.

Faldt J, Arimura G-I, Gershenzon J, et al. (2003) Functional identification of AtTPS03 as (E)-β-ocimene synthase: A monoterpene synthase catalyzing jasmonate- and wound-induced volatile formation in *Arabidopsis thaliana*. *Planta* 216(5):745–751.

Farmer WR, Liao JC (2000) Improving lycopene production in *Escherichia coli* by engineering metabolic control. *Nat Biotechnol* 18(5):533–537.

Ferrer JL, Jez JM, Bowman ME, et al. (1999) Structure of chalcone synthase and the molecular basis of plant polyketide biosynthesis. *Nat Struct Mol Biol* 6(8):775–784.

Forkmann G, Martens S (2001) Metabolic engineering and applications of flavonoids. *Curr Opin Biotech* 12(2):155–160.

Fraser PD, Bramley PM (2004) The biosynthesis and nutritional uses of carotenoids. *Progr Lipid Res* 43(3):228–265.

Fremont L (2000) Biological effects of resveratrol. *Life Sci* 66(8):663–673.

Frenzel T, Zenk MH (1990) Purification and characterization of three isoforms of S-adenosyl-methionine: (R,S)-tetrahydrobenzylisoquinoline-N-methyltransferase from *Berberis koetineana* cell cultures. *Phytochemistry* 29(11):3491–3497.

Frick S, Kutchan TM (1999) Molecular cloning and functional expression of O–methyltransferases common to isoquinoline alkaloid and phenylpropanoid biosynthesis. *Plant J* 17(4):329–339.

Fridman E, Wang J, Iijima Y, et al. (2005) Metabolic, genomic, and biochemical analyses of glandular trichomes from the wild tomato species *Lycopersicon hirsutum* identify a key enzyme in the biosynthesis of methylketones. *Plant Cell* 17(4):1252–1267.

Fujiwara H, Tanaka Y, Fukui Y, et al. (1997) Anthocyanin 5-aromatic acyltransferase from *Gentiana triflora* – Purification, characterization and its role in anthocyanin biosynthesis. *Eur J Biochem* 249(1):45–51.

Fujiwara H, Tanaka Y, Fukui Y, et al. (1998b) Purification and characterization of anthocyanin 3–aromatic acyltransferase from *Perilla frutescens*. *Plant Sci* 137(1):87–94.

Fujiwara H, Tanaka Y, Yonekura-Sakakibara K, et al. (1998a) cDNA cloning, gene expression and subcellular localization of anthocyanin 5-aromatic acyltransferase from *Gentiana triflora*. *Plant J* 16(4):421–431.

Fukami H, Asakura T, Hirano H, et al. (2002) Salicylic acid carboxyl methyltransferase induced in hairy root cultures of *Atropa belladonna* after treatment with exogeneously added salicylic acid. *Plant Cell Physiol* 43(9):1054–1058.

Geerlings A, Ibanez M M-L, Memelink J, et al. (2000) Molecular cloning and analysis of strictosidine β-D-glucosidase, an enzyme in terpenoid indole alkaloid biosynthesis in *Catharanthus roseus*. *J Biol Chem* 275(5):3051–3056.

Geerlings A, Redondo FJ, Contin A, et al. (2001) Biotransformation of tryptamine and secologanin into plant terpenoid indole alkaloids by transgenic yeast. *Appl Microbiol Biotechnol* 56(3–4):420–424.

Geerlings A, Van Der Heijden R, Memelink J, et al. (1998) A method for producing terpenoid indole alkaloids. EP 98204116.2, 4 December 1998.

Gerardy R, Zenk MH (1993a) Purification and characterization of salutaridine: NADPH 7-oxidoreductase from *Papaver somniferum*. *Phytochemistry* 34(1):125–132.

Gerardy R, Zenk MH (1993b) Formation of salutaridine from (R)reticuline by a membrane-bound cytochrome P-450 enzyme from *Papaver somniferum*. *Phytochemistry* 32(1):79–86.

Gerasimenko I, Sheludko Y, Ma X, et al. (2002) Heterologous expression of a *Rauvolfia* cDNA encoding strictosidine glucosidase, a biosynthetic key to over 2000 monoterpenoid indole alkaloids. *Eur J Biochem* 269(8):2204–2213.

Goff SA, Klee HJ (2006) Plant volatile compounds: sensory cues for health and nutritional value? Science 311(5762):815–819.

Grothe T, Lenz R, Kutchan TM (2001) Molecular characterization of the salutaridinol 7-O-acetyltransferase involved in morphine biosynthesis in opium poppy *Papaver somniferum*. *J Biol Chem* 276(33):30717–30723.

Guterman I, Masci T, Chen XL, et al. (2006) Generation of phenylpropanoid pathway-derived volatiles in transgenic plants: Rose alcohol acetyltransferase produces phenylethyl acetate and benzyl acetate in petunia flowers. *Plant Mol Biol* 60(4):555–563.

Harborne JB, Williams CA (1998) Anthocyanins and other flavonoids. *Nat Prod Rep* 15(6):631–652.

Harborne JB, Williams CA (2000) Advances in flavonoid research since 1992. *Phytochemistry* 55(6):481–504.

Haudenschild C, Schalk M, Karp F, et al. (2000) Functional expression of regiospecific cytochrome P450 limonene hydroxylases from mint (*Mentha spp.*) in *Escherichia coli* and *Saccharomyces cerevisiae*. *Arch Biochem Biophys* 379(1):127–136.

Hefner J, Ketchum REB, Croteau R (1998) Cloning and functional expression of a cDNA encoding geranylgeranyl diphosphate synthase from *Taxus canadensis* and assessment of the role of this prenyltransferase in cells induced for Taxol production. *Arch Biochem Biophys* 360(1):62–74.

Hirschberg J (2001) Carotenoid biosynthesis in flowering plants. *Curr Opin Plant Biol* 4(3): 210–218.

Hohl RJ (1996) Monoterpenes as regulators of malignant cell proliferation. Adv Exp Med Biol 401(137–146).

Holton RA, Somoza C, Kim HB, et al. (1994) First total synthesis of Taxol: Functionalization of the B-ring. *J Am Chem Soc* 116(4):1597–1598.

Howitz KT, Bitterman KJ, Cohen HY, et al. (2003) Small molecule activators of sirtuins extend *Saccharomyces cerevisiae* lifespan. *Nature* 425(6954):191–196.

Huang EX, Huang QL, Wildung MR, et al. (1998) Overproduction, in *Escherichia coli*, of soluble taxadiene synthase, a key enzyme in the taxol biosynthetic pathway. *Protein Expr and Purif* 13(1):90–96.

Huang QL, Roessner CA, Croteau R, et al. (2001) Engineering *Escherichia coli* for the synthesis of taxadiene, a key intermediate in the biosynthesis of taxol. *Bioorg Med Chem* 9(9):2237–2242.

Hwang EI, Kaneko M, Ohnishi Y, et al. (2003) Production of plant specific flavanones by *Escherichia coli* containing an artificial gene cluster. *Appl Environ Microbiol* 69(5):2699–2706.

Iijima Y, Davidovich-Rikanati R, Fridman E, et al. (2004) The biochemical and molecular basis for the divergent patterns in the biosynthesis of terpenes and phenylpropenes in the peltate glands of three cultivars of basil. *Plant Physiol* 136(3):3724–3736.

Ikezawa N, Tanaka M, Nagayoshi M, et al. (2003) Molecular cloning and characterization of CYP719, a methylenedioxy bridge forming enzyme that belongs to a novel P450 family, from cultured *Coptis japonica* cells. *J Biol Chem* 278(40):38557–38565.

Iuchi S, Kobayashi M, Taji T, et al. (2001) Regulation of drought tolerance by gene manipulation of 9-cis-epoxycarotenoid dioxygenase, a key enzyme in abscisic acid biosynthesis in *Arabidopsis*. *Plant J* 27(4):325–333.

Iuchi S, Kobayashi M, Yamaguchi-Shinozaki K, et al. (2000) A stress-inducible gene for 9-cis-epoxycarotenoid dioxygenase involved in abscisic acid biosynthesis under water stress in drought-tolerant cowpea. *Plant Physiol* 123(2):553–562.

Jacobs DI, Gaspari M, Van Der Greef J, et al. (2005) Proteome analysis of the medicinal plant *Catharanthus* roseus. *Planta* 221(5):690–704.

Jennewein S, Croteau R (2001) Taxol: biosynthesis, molecular genetics, and biotechnological applications. *Appl Microbiol Biotechnol* 57(1–2):13–19.

Jennewein S, Long RM, Williams RM, et al. (2004a) Cytochrome P450 taxadiene 5 alpha-hydroxylase, a mechanistically unusual monooxygenase catalyzing the first oxygenation step of taxol biosynthesis. *Chem Biol* 11(3):379–387.

Jennewein S, Park H, Dejong JM, et al. (2005) Coexpression in yeast of *Taxus* cytochrome P450 reductase with cytochrorne P450 oxygenases involved in taxol biosynthesis. *Biotechnol Bioeng* 89(5):588–598.

Jennewein S, Rithner CD, Williams RM, et al. (2001) Taxol biosynthesis: Taxane 13alpha –hydroxylase is a cytochrome P450-dependent monooxygenase. *Proc Natl Acad Sci USA* 98(24):13595–13600.

Jennewein S, Rithner CD, Williams RM, et al. (2003) Taxoid metabolism: Taxoid 14 beta-hydroxylase is a cytochrome P450-dependent monooxygenase. *Arch Biochem Biophys* 413(2):262–270.

Jennewein S, Wildung MR, Chau M, et al. (2004b) Random sequencing of an induced *Taxus* cell cDNA library for identification of clones involved in Taxol biosynthesis. *Proc Natl Acad Sci USA* 101(24):9149–9154.

Jez JM, Ferrer JL, Bowman ME, et al. (2001) Structure and mechanism of chalcone synthase-like polyketide synthases. *J Ind Microbiol Biotechnol* 27(6):393–398.

Jez JM, Ferrer JL, Bowman M, et al. (1999) Structure of chalcone synthase and the molecular basis of plant polyketide biosynthesis. *FASEB J* 13(7):A1392–A1392.

Jia JW, Crock J, Lu S, et al. (1999) (3R)-linalool synthase from *Artemisia annua* L.: cDNA isolation, characterization, and wound induction. *Arch Biochem Biophys* 372(1):143–149.

Jiang HX, Wood KV, Morgan JA (2005) Metabolic engineering of the phenylpropanoid pathway in *Saccharomyces cerevisiae*. *Appl Environ Microbiol* 71(6):2962–2969.

Jones KL, Kim S-W, Keasling JD (2000) Low-copy plasmids can perform as well as or better than high-copy plasmids for metabolic engineering of bacteria. *Metab Eng* 2(4):328–338.

Kajiwara S, Fraser PD, Kondo K, et al. (1997) Expression of an exogenous isopentenyl diphosphate isomerase gene enhances isoprenoid biosynthesis in *Escherichia coli*. *Biochem J* 324:421–426.

Kato M, Kanehara T, Shimizu H, et al. (1996) Caffeine biosynthesis in young leaves of *Camellia sinensis*: In vitro studies on N-methyltransferase activity involved in the conversion of xanthosine to caffeine. *Physiol Plant* 98(3):629–636.

Kato M, Mizuno K, Fujimura T, et al. (1999) Purification and characterization of caffeine synthase from tea leaves. *Plant Physiol* 120(2):579–586.

Kim SW, Keasling JD (2001) Metabolic engineering of the nonmevalonate isopentenyl diphosphate synthesis pathway in *Escherichia coli* enhances lycopene production. *Biotechnol Bioeng* 72(4):408–415.

Kliebenstein DJ, Lambrix VM, Reichelt M, et al. (2001) Gene duplication in the diversification of secondary metabolism: Tandem 2-oxoglutarate-dependent dioxygenases control glucosinolate biosynthesis in *Arabidopsis*. *Plant Cell* 13(3):681–693.

Krings U, Berger RG (1998) Biotechnological production of flavours and fragrances. *Appl Microbiol Biotechnol* 49(1):1–8.

Kutchan TM (1995) Alkaloid biosynthesis – The basis for metabolic engineering of medicinal plants. *Plant Cell* 7(7):1059–1070.

Kutchan TM (1996) Heterologous expression of alkaloid biosynthetic genes – a review. *Gene* 179(1):73–81.

Kutchan TM, Hampp N, Lottspeich F, et al. (1988) The cDNA clone for strictosidine synthase from *Rauvolfia serpentina* DNA sequence determination and expression in *Escherichia coli*. *FEBS Lett* 237(1–2):40–44.

Kuzuyama T, Seto H (2003) Diversity of the biosynthesis of the isoprene units. *Nat Prod Rep* 20(2):171–183.

Laflamme P, St-Pierre B, De Luca V (2001) Molecular and biochemical analysis of a Madagascar Periwinkle root-specific minovincinine-19-hydroxy-O-acetyltransferase. *Plant Physiol* 125(1):189–198.

Lange BM, Croteau R (1999) Genetic engineering of essential oil production in mint. *Curr Opin Plant Biol* 2(2):139–144.

Lee PC, Mijts BN, Schmidt-Dannert C (2004) Investigation of factors influencing production of the monocyclic carotenoid torulene in metabolically engineered *Escherichia coli*. *Appl Microbiol Biotechnol* 65(5):538–546.

Lee PC, Momen AZR, Mijts BN, et al. (2003) Biosynthesis of structurally novel carotenoids in *Escherichia coli*. *Chem Biol* 10(5):453–462.

Lee PC, Schmidt-Dannert C (2002) Metabolic engineering towards biotechnological production of carotenoids in microorganisms. *Appl Microbiol Biotechnol* 60(1–2):1–11.

Leffingwell J 2003 Saffron. Available online. http://www.leffingwell.com/saffron.htm. Cited 8 January 2006.

Leonard E, Chemler J, Lim KH, et al. (2006) Expression of a soluble flavone synthase allows the biosynthesis of phytoestrogen derivatives in *Escherichia coli*. *Appl Microbiol Biotechnol* 70(1):85–91.

Leonard E, Yan YJ, Lim KH, et al. (2005) Investigation of two distinct flavone synthases for plant-specific flavone biosynthesis in *Saccharomyces cerevisiae*. *Appl Environ Microbiol* 71(12):8241–8248.

Lesburg CA, Zhai GZ, Cane DE, et al. (1997) Crystal structure of pentalenene synthase: Mechanistic insights on terpenoid cyclization reactions in biology. *Science* 277 (5333): 1820–1824.

Lewinsohn E, Schalechet F, Wilkinson J, et al. (2001) Enhanced levels of the aroma and flavor compound S-linalool by metabolic engineering of the terpenoid pathway in tomato fruits. *Plant Physiol* 127(3):1256–1265.

Loeffler S, Zenk MH (1990) The hydroxylation step in the biosynthetic pathway leading from norcoclaurine to reticuline. *Phytochemistry* 29(11):3499–3503.

Loza-Tavera H (1999) Monoterpenes in essential oils – Biosynthesis and properties. In: Chemicals via higher plant bioengineering, vol 464. pp 49–62.

Lucker J, El Tamer MK, Schwab W, et al. (2002) Monoterpene biosynthesis in lemon. cDNA isolation and functional analysis of four monoterpene synthases. *Euro J Biochem* 269(13):3160–3171.

Lupien S, Karp F, Wildung M, et al. (1999) Regiospecific cytochrome P450 limonene hydroxylases from mint (Mentha) species: cDNA isolation, characterization, and functional expression of (–)-4S-limonene-3- hydroxylase and (–)-4S-limonene-6-hydroxylase. *Arch Biochem Biophys* 368(1): 181–192.

Ma X, Koepke J, Panjikar S, et al. (2005) Crystal structure of vinorine synthase, the first representative of the BAHD superfamily. *J Biol Chem* 280(14):13576–13583.

Mahmoud SS, Croteau RB (2002) Strategies for transgenic manipulation of monoterpene biosynthesis in plants. *Trends Plant Sci* 7(8):366–373.

Marasco E, Schmidt-Dannert C (2003) Towards the biotechnological production of aroma and flavor compounds in engineered microorganisms. *Appl Biotechnol Food Sci Policy* 1(3):145–157.

Martens S, Forkmann G, Matern U, et al. (2001) Cloning of parsley flavone synthase I. *Phytochemistry* 58(1):43–46.

Martin DM, Faldt J, Bohlmann J (2004) Functional characterization of nine Norway Spruce TPS genes and evolution of gymnosperm terpene synthases of the TPS-d subfamily. *Plant Physiol* 135(4):1908–1927.

Martin VJJ, Pitera DJ, Withers ST, et al. (2003) Engineering a mevalonate pathway in *Escherichia coli* for production of terpenoids. *Nat Biotechnol* 21(7):796–802.

Maruyama T, Ito M, Kiuchi F, et al. (2001) Molecular cloning, functional expression and characterization of d-limonene synthase from *Schizonepeta tenuifolia*. *Biol Pharm Bull* 24:373–377.

Mattern-Dogru E, Ma X, Hartmann J, et al. (2002) Potential active-site residues in polyneuridine aldehyde esterase, a central enzyme of indole alkaloid biosynthesis, by modelling and site-directed mutagenesis. *Eur J Biochem* 269(12):2889–2896.

Matthews PD, Wurtzel ET (2000) Metabolic engineering of carotenoid accumulation in *Escherichia coli* by modulation of the isoprenoid precursor pool with expression of deoxyxylulose phosphate synthase. *Appl Microbiol Biotechnol* 53(4):396–400.

Mercke P, Bengtsson M, Bouwmeester H J, et al. (2000) Molecular cloning, expression, and characterization of amorpha-4,11-diene synthase, a key enzyme of artemisinin biosynthesis in *Artemisia annua L. Arch Biochem Biophys* 381(2):173–180.

Mijts BN, Lee PC, Schmidt-Dannert C (2004) Engineering carotenoid biosynthetic pathways. In: Methods in enzymology, Academic Press, vol 388, pp 315–329.

Mijts BN, Lee PC, Schmidt-Dannert C (2005) Identification of a carotenoid oxygenase synthesizing acyclic xanthophylls: Combinatorial biosynthesis and directed evolution. *Chem Biol* 12(4): 453–460.

Miyahisa I, Kaneko M, Funa N, et al. (2005) Efficient production of (2S)-flavanones by *Escherichia coli* containing an artificial biosynthetic gene cluster. *Appl Microbiol Biotechnol* 68(4):498–504.

Mizuno K, Kato M, Irino F, et al. (2003) The first committed step reaction of caffeine biosynthesis: 7-methylxanthosine synthase is closely homologous to caffeine synthases in coffee (*Coffea arabica L.*). FEBS Lett 547(1–3):56–60.

Moraga A, Nohales P, Perez J, et al. (2004) Glucosylation of the saffron apocarotenoid crocetin by a glucosyltransferase isolated from *Crocus sativus* stigmas. *Planta* 219(6):955–966.

Morishige T, Dubouzet E, Choi K-B, et al. (2002) Molecular cloning of columbamine O-methyltransferase from cultured *Coptis japonica* cells. *Euro J Biochem* 269(22):5659–5667.

Morishige T, Tsujita T, Yamada Y, et al. (2000) Molecular characterization of the S-adenosyl-L-methionine:3'-hydroxy-N-methylcoclaurine 4'-O-methyltransferase involved in isoquinoline alkaloid biosynthesis in *Coptis japonica*. *J Biol Chem* 275(30):23398–23405.

Murfitt LM, Kolosova N, Mann C J, et al. (2000) Purification and characterization of S-adenosyl-L-methionine: Benzoic acid carboxyl methyltransferase, the enzyme responsible for biosynthesis of the volatile ester methyl benzoate in flowers of *Antirrhinum majus*. *Arch Biochem Biophys* 382(1):145–151.

Negre F, Kish CM, Boatright J, et al. (2003) Regulation of methylbenzoate emission after pollination in snapdragon and petunia flowers. *Plant Cell* 15(12):2992–3006.

Negre F, Kolosova N, Knoll J, et al. (2002) Novel S-adenosyl-L-methionine: salicylic acid carboxyl methyltransferase, an enzyme responsible for biosynthesis of methyl salicylate and methyl benzoate, is not involved in floral scent production in snapdragon flowers. *Arch Biochem Biophys* 406(2): 261–270.

Ogawa M, Herai Y, Koizumi N, et al. (2001) 7-Methylxanthine methyltransferase of coffee plants – Gene isolation and enzymatic properties. *J Biol Chem* 276(11):8213–8218.

Ogita S, Uefuji H, Morimoto M, et al. (2004) Application of RNAi to confirm theobromine as the major intermediate for caffeine biosynthesis in coffee plants with potential for construction of decaffeinated varieties. *Plant Mol Biol* 54(6):931–941.

Oksman-Caldentey K-M, Inze D, Oresic M (2004) Connecting genes to metabolites by a systems biology approach. *Proc Natl Acad Sci USA* 101(27):9949–9950.

Oksman-Caldentey KM, Saito K (2005) Integrating genomics and metabolomics for engineering plant metabolic pathways. *Curr Opin Biotech* 16(2):174–179.

Ounaroon A, Decker G, Schmidt J, et al. (2003) (R,S)-Reticuline 7-O-methyltransferase and (R,S)-norcoclaurine 6-O-methyltransferase of *Papaver somniferum* – cDNA cloning and characterization of methyl transfer enzymes of alkaloid biosynthesis in opium poppy. *Plant J* 36(6):808–819.

Papnikolaw J (1998) Market brief. *Chemical Market Reporter* 253(20):20.

Pattnaik S, Subramanyam VR, Bapaji M, et al. (1997) Antibacterial and antifungal activity of aromatic constituents of essential oils. *Microbios* 89(358):39–46.

Pauli HH, Kutchan TM (1998) Molecular cloning and functional heterologous expression of two alleles encoding (S) N methylcoclaurine 3' hydroxylase (CYP80B1), a new methyl jasmonate inducible cytochrome P 450 dependent mono oxygenase of benzylisoquinoline alkaloid biosynthesis. *Plant J* 13(6):793–801.

Phillips MA, Wildung MR, Williams DC, et al. (2003) cDNA isolation, functional expression, and characterization of (+)-alpha-pinene synthase and (−)-alpha-pinene synthase from loblolly pine (*Pinus taeda*): Stereocontrol in pinene biosynthesis. *Arch Biochem Biophys* 411(2):267–276.

Picaud S, Brodelius M, Brodelius PE (2005) Expression, purification and characterization of recombinant (E)-beta-farnesene synthase from Artemisia annua. *Phytochemistry* 66(9):961–967.

Pichersky E, Gershenzon J (2002) The formation and function of plant volatiles: perfumes for pollinator attraction and defense. *Curr Opin Plant Biol* 5(3):237–243.

Pichersky E, Noel JP, Dudareva N (2006) Biosynthesis of plant volatiles: Nature's diversity and ingenuity. *Science* 311(5762):808–811.

Pompon D, Louerat B, Bronine A, et al. (1996) Yeast expression of animal and plant P450s in optimized redox environments. In: Cytochrome P450, Pt B, vol 272: pp 51–64.

Pott MB, Pichersky E, Piechulla B (2002) Evening specific oscillations of scent emission, SAMT enzyme activity, and SAMT mRNA in flowers of *Stephanotis floribunda*. *J Plant Physiol* 159(8):925–934.

Pott MB,Hippauf F,Saschenbrecker S, et al. (2004) Biochemical and structural characterization of benzenoid carboxyl methyltransferases involved in floral scent production in Stephanotis floribunda and Nicotiana suaveolens. *Plant Physiol* 135(4):1946–1955.

Qin X, Zeevaart JAD (1999) Inaugural Article: The 9-cis-epoxycarotenoid cleavage reaction is the key regulatory step of abscisic acid biosynthesis in water-stressed bean. *Proc Natl Acad Sci USA* 96(26):15354–15361.

Raskin I, Ribnicky DM, Komarnytsky S, et al. (2002) Plants and human health in the twenty-first century. *Trends Biotech* 20(12):522–531.

Rathbone DA, Bruce NC (2002) Microbial transformation of alkaloids. *Curr Opin Microbiol* 5(3):274–281.

Ringer KL, Davis EM, Croteau R (2005) Monoterpene metabolism. Cloning, expression, and characterization of (–)-isopiperitenol/(–)-carveol dehydrogenase of peppermint and spearmint. *Plant Physiol* 137(3):863–872.

Ringer KL, Mcconkey ME, Davis EM, et al. (2003) Monoterpene double-bond reductases of the (–)-menthol biosynthetic pathway: isolation and characterization of cDNAs encoding (–)-isopiperitenone reductase and (+)-pulegone reductase of peppermint. *Arch Biochem Biophys* 418(1):80–92.

Rodriguez-Concepcion M, Boronat A (2002) Elucidation of the methylerythritol phosphate pathway for isoprenoid biosynthesis in bacteria and plastids. A metabolic milestone achieved through genomics. *Plant Physiol* 130(3):1079–1089.

Rohloff J, Bones AM (2005) Volatile profiling of *Arabidopsis thaliana* – Putative olfactory compounds in plant communication. *Phytochemistry* 66(16):1941–1955.

Ross JR, Nam KH, D'auria JC, et al. (1999) S-adenosyl-methionine:salicylic acid carboxyl methyltransferase, an enzyme involved in floral scent production and plant defense, represents a new class of plant methyltransferases. *Arch Biochem Biophys* 367(1):9–16.

Ruppert M, Ma X, Stöckigt J (2005) Alkaloid biosynthesis in *Rauvolfia* – cDNA cloning of major enzymes of the ajmaline pathway. *Curr Org Chem* 9(15):1431–1444.

Samanani N, Facchini PJ (2002) Purification and characterization of norcoclaurine synthase. The first committed enzyme in benzylisoquinoline alkaloid biosynthesis in plants. *J Biol Chem* 277(37):33878–33883.

Samanani N, Liscombe DK, Facchini PJ (2004) Molecular cloning and characterization of norcoclaurine synthase, an enzyme catalyzing the first committed step in benzylisoquinoline alkaloid biosynthesis. *Plant J* 40(2):302–313.

Sandmann G (2002) Combinatorial biosynthesis of carotenoids in a heterologous host: A powerful approach for the biosynthesis of novel structures. *Chembiochem* 3(7):629–635.

Savithiry N, Cheong TK, Oriel P (1997) Production of alpha-terpineol from *Escherichia coli* cells expressing thermostable limonene hydratase. Appl Biochem Biotechnol Enzym Eng Biotechnol 63–65(213–220).

Schalk M, Croteau R (2000) A single amino acid substitution (F363I) converts the regiochemistry of the spearmint (–)-limonene hydroxylase from a C6-to a C3-hydroxylase. *Proc Natl Acad Sci USA* 97(22):11948–11953.

Schmidt-Dannert C (2000) Engineering novel carotenoids in microorganisms. *Curr Opin Biotechnol* 11(3):255–261.

Schmidt-Dannert C, Umeno D, Arnold FH (2000) Molecular breeding of carotenoid biosynthetic pathways. *Nat Biotechnol* 18(7):750–753.

Schnee C, Kollner TG, Gershenzon J, et al. (2002) The maize gene terpene synthase 1 encodes a sesquiterpene synthase catalyzing the formation of (E)-beta -farnesene, (E)-nerolidol, and (E,E)-farnesol after herbivore damage. *Plant Physiol* 130(4):2049–2060.

Schoendorf A, Rithner CD, Williams RM, et al. (2001) Molecular cloning of a cytochrome P450 taxane 10 beta-hydroxylase cDNA from *Taxus* and functional expression in yeast. *Proc Natl Acad Sci USA* 98(4):1501–1506.

Schwartz SH, Qin X, Zeevaart JAD (2001) Characterization of a novel carotenoid cleavage dioxygenase from plants. *J Biol Chem* 276(27):25208–25211.

Schwartz SH, Tan BC, Gage DA, et al. (1997) Specific oxidative cleavage of carotenoids by VP14 of maize. *Science* 276(5320):1872–1874.

Seo HS, Song JT, Cheong JJ, et al. (2001) Jasmonic acid carboxyl methyltransferase: A key enzyme for jasmonate-regulated plant responses. *Proc Natl Acad Sci USA* 98(8):4788–4793.

Shalit M, Guterman I, Volpin H, et al. (2003) Volatile ester formation in roses: identification of an acetyl-coenzyme. A geraniol/citronellol acetyltransferase in developing rose petals. *Plant Physiol* 131(4):1868–1876.

Simkin AJ, Underwood BA, Auldridge M, et al. (2004a) Circadian regulation of the PhCCD1 carotenoid cleavage dioxygenase controls emission of β-ionone, a fragrance volatile of *Petunia* flowers. *Plant Physiol* 136(3):3504–3514.

Simkin aJ, Schwartz SH, Auldridge M, et al. (2004b) The tomato carotenoid cleavage dioxygenase 1 genes contribute to the formation of the flavor volatiles β-ionone, pseudoionone, and geranylacetone. *Plant J* 40(6):882–892.

Smolke CD, Martin VJJ, Keasling JD (2001) Controlling the metabolic flux through the carotenoid pathway using directed mRNA processing and stabilization. *Metab Eng* 3(4):313–321.

Soleas GJ, Diamandis EP, Goldberg DM (1997) Resveratrol: A molecule whose time has come? and gone? Clin Biochem 30(2):91–113.

Souleyre EJF, Greenwood DR, Friel EN, et al. (2005) An alcohol acyl transferase from apple (cv. Royal Gala), MpAAT1, produces esters involved in apple fruit flavor. *FEBS Journal* 272(12): 3132–3144.

Starks CM, Back KW, Chappell J, et al. (1997a) Structural basis for cyclic terpene biosynthesis by tobacco 5-epi-aristolochene synthase. *Science* 277(5333):1815–1820.

Starks CM, Back K, Chappell J, et al. (1997b) Structural basis for cyclic terpene biosynthesis by tobacco 5-epi-aristolochene synthase. *Science* 277(5333):1815–1820.

Steele CL, Crock J, Bohlmann J, et al. (1998) Sesquiterpene synthases from Grand Fir (*Abies grandis*). Comparison of constitutive and wound-indued activities, and cDNA isolation, characterization, and bacterial expression of delta-selinene synthase and gamma-humulene synthase. *J Biol Chem* 273(4):2078–2089.

Sumner LW, Mendes P, Dixon RA (2003) Plant metabolomics: large-scale phytochemistry in the functional genomics era. *Phytochemistry* 62(6):817–836.

Suzuki K, Xue HM, Tanaka Y, et al. (2000) Flower color modifications of *Torenia hybrida* by cosuppression of anthocyanin biosynthesis genes. *Mol Breeding* 6(3):239–246.

Szczebara FM, Chandelier C, Villeret C, et al. (2003) Total biosynthesis of hydrocortisone from a simple carbon source in yeast. *Nat Biotechnol* 21(2):143–149.

Takeshita N, Fujiwara H, Mimura H, et al. (1995) Molecular cloning and characterization of S-adenosyl-L-methionine-scoulerine-9-O-methyltransferase from cultured cells of *Coptis japonica. Plant and Cell Physiol* 36(1):29–36.

Tarantilis PA, Tsoupras G, Polissiou M (1995) Determination of saffron (*Crocus sativus* L) components in crude plant extract using high-performance liquid-chromatography uv-visible photodiode-array detection mass spectrometry. *J Chromatogr A* 699(1–2):107–118.

Tholl D, Chen F, Petri J, et al. (2005) Two sesquiterpene synthases are responsible for the complex mixture of sesquiterpenes emitted from Arabidopsis flowers. *Plant J* 42(5): 757–771.

Tietze LF, Bell HP, Chandrasekhar S (2003) Natural product hybrids as new leads for drug discovery. *Ang Chem-Int Ed* 42(34):3996–4028.

Trapp SC, Croteau RB (2001) Genomic organization of plant terpene synthases and molecular evolutionary implications. *Genetics* 158(2):811–832.

Trethewey RN (2004) Metabolite profiling as an aid to metabolic engineering in plants. *Curr Opin Plant Biol* 7(2):196–201.

US Food and Drug Administration (2001) Title 21. Code of Federal Regulation Paragraph 101.22 ((a)(3) and (a)(1)).

Uefuji H, Ogita S, Yamaguchi Y, et al. (2003) Molecular cloning and functional characterization of three distinct N-methyltransferases involved in the caffeine biosynthetic pathway in coffee plants. *Plant Physiol* 132(1):372–380.

Umeno D, Arnold FH (2004) Evolution of a pathway to novel long-chain carotenoids. *J Bacteriol* 186(5):1531–1536.

Umeno D, Tobias AV, Arnold FH (2002) Evolution of the C30 carotenoid synthase CrtM for function in a C40 pathway. *J Bacteriol* 184(23):6690–6699.

Umeno D, Tobias AV, Arnold FH (2005) Diversifying carotenoid biosynthetic pathways by directed evolution. *Microbiol Mol Biol Rev* 69(1):51–78.

Unterlinner B, Lenz R, Kutchan TM (1999) Molecular cloning and functional expression of codeinone reductase: the penultimate enzyme in morphine biosynthesis in the opium poppy *Papaver somniferum*. *Plant J* 18(5):465–475.

Valla aR, Cartier DL, Labia R (2004) Chemistry of natural retinoids and carotenoids: Challenges for the future. *Curr Org Synth* 1(2):167–209.

Van Der Heijden R, Jacobs DI, Snoeijer W, et al. (2004) The *Catharanthus* alkaloids: Pharmacognosy and biotechnology. *Curr Med Chem* 11(5):607–628.

Vandamme EJ (2002) Bioflavours and fragrances via fermentation and biocatalysis. *J Chem Tech & Biotechnol* 77(12):1323–1332.

Vanderhaegen B, Neven H, Coghe S, et al. (2003) Bioflavoring and beer refermentation. *Appl Microbiol Biotechnol* 62(2–3):140–150.

Verpoorte R, Memelink J (2002) Engineering secondary metabolite production in plants. *Curr Opin Biotech* 13(2):181–187.

Vick BA, Zimmerman DC (1986) The lipoxygenase pathway. In: Stumpf PK, Mudd JB, Nes WD (eds) The Metabolism, Structure, and Function of Plant Lipids, Plenum Press, New York p 383–390.

Vincent JJ, Martin Y (2001) The in vivo synthesis of plant sesquiterpenes by *Escherichia coli. Biotechnol Bioeng* 75(5):497–503.

Von Schumann G, Gao S, Stöckigt J (2002) Vomilenine reductase – a novel enzyme catalyzing a crucial step in the biosynthesis of the therapeutically applied antiarrhythmic alkaloid ajmaline. *Bioorg Med Chem* 10(6):1913–1918.

Wache Y, Bosser-Deratuld A, Lhuguenot JC, et al. (2003) Effect of cis/trans isomerism of beta-carotene on the ratios of volatile compounds produced during oxidative degradation. *J Agricult Food Chem* 51(7):1984–1987.

Walker K, Croteau R (2000a) Taxol biosynthesis: Molecular cloning of a benzoyl-CoA : taxane 2 alpha-O-benzoyltransferase cDNA from *Taxus* and functional expression in *Escherichia coli. Proc Natl Acad Sci USA* 97(25):13591–13596.

Walker K, Croteau R (2000b) Molecular cloning of a 10-deacetylbaccatin III-10-O-acetyl transferase cDNA from Taxus and functional expression in *Escherichia coli. Proc Natl Acad Sci USA* 97(2):583–587.

Walker K, Croteau R (2000c) Molecular cloning of a 10-deacetylbaccatin III-10-O-acetyl transferase cDNA from *Taxus* and functional expression in *Escherichia coli. Proc Natl Acad Sci USA* 97(2):583–587.

Walker K, Croteau R (2000d) Taxol biosynthesis: Molecular cloning of a benzoyl- CoA:taxane 2alpha -O-benzoyltransferase cDNA from *Taxus* and functional expression in *Escherichia coli. Proc Natl Acad Sci USA* 97(25):13591–13596.

Walker K, Croteau R (2001) Taxol biosynthetic genes. *Phytochemistry* 58(1):1–7.

Walker K, Fujisaki S, Long R, et al. (2002a) Molecular cloning and heterologous expression of the C-13 phenylpropanoid side chain-CoA acyltransferase that functions in Taxol biosynthesis. *Proc Natl Acad Sci USA* 99(20):12715–12720.

Walker K, Klettke K, Akiyama T, et al. (2004) Cloning, heterologous expression, and characterization of a phenylalanine aminomutase involved in Taxol biosynthesis. *J Biol Chem* 279(52):53947–53954.

Walker K, Long R, Croteau R (2002b) The final acylation step in Taxol biosynthesis: Cloning of the taxoid C13-side-chain N-benzoyltransferase from *Taxus. Proc Natl Acad Sci USA* 99(14):9166–9171.

Walker K, Schoendorf A, Croteau R (2000) Molecular cloning of a taxa-4(20),11(12)-dien-5α-ol-O-acetyl transferase cDNA from *Taxus* and functional expression in *Escherichia coli. Arch Biochem Biophys* 374(2):371–380.

Warzecha H, Gerasimenko I, Kutchan TM, et al. (2000) Molecular cloning and functional bacterial expression of a plant glucosidase specifically involved in alkaloid biosynthesis. *Phytochemistry* 54(7):657–666.

Watts KT, Lee PC, Schmidt-Dannert C (2004) Exploring recombinant flavonoid biosynthesis in metabolically engineered *Escherichia coli. Chembiochem* 5(4):500–507.

Watts KT, Lee PC, Schmidt-Dannert C (2006) Biosynthesis of plant-specific stilbene polyketides in metabolically engineered *Escherichia coli. BMC Biotechnol* 6.

Wildung MR, Croteau R (1996) A cDNA clone for taxadiene synthase, the diterpene cyclase that catalyzes the committed step of taxol biosynthesis. *J Biol Chem* 271(16):9201–9204.

Williams DC, Mcgarvey DJ, Katahira EJ, et al. (1998) Truncation of limonene synthase preprotein provides a fully active 'Pseudomature' form of this monoterpene cyclase and reveals the function of the amino-terminal arginine pair. *Biochemistry* 37(35):12213–12220.

Wink M (1999) Introduction: biochemistry, role and biotechnology of secondary metabolites. Sheffield Academic Press, Sheffield.

Wink M (2003) Evolution of secondary metabolites from an ecological and molecular phylogenetic perspective. *Phytochemistry* 64(1):3–19.

Winkel-Shirley B (2001) Flavonoid biosynthesis. A colorful model for genetics, biochemistry, cell biology, and biotechnology. *Plant Physiol* 126(2):485–493.

Winterhalter B, Rouseff RL (2002) Carotenoid-derived aroma compounds: An introduction. In: Winterhalter B, Rouseff R L, (Eds) Carotenoid derived aroma compounds, vol 802. Amercian Chemical Society, Washington DC, pp 1–19.

Wise ML, Savage TJ, Katahira E, et al. (1998) Monoterpene synthases from common sage (*Salvia officinalis*) – cDNA isolation, characterization, and functional expression of (+)-sabinene synthase, 1,8-cineole synthase, and (+)-bornyl diphosphate synthase. *J Biol Chem* 273(24):14891–14899.

Yahyaoui FEL, Wongs-Aree C, Latche A, et al. (2002) Molecular and biochemical characteristics of a gene encoding an alcohol acyl-transferase involved in the generation of aroma volatile esters during melon ripening. *Eur J Biochem* 269(9):2359–2366.

Yan W, Jang G-F, Haeseleer F, et al. (2001) Cloning and characterization of a human β, β-carotene-15, 15'-dioxygenase that is highly expressed in the retinal pigment epithelium. *Genomics* 72(2):193–202.

Yan Y J, Kohli A, Koffas MAG (2005) Biosynthesis of natural flavanones in *Saccharomyces cerevisiae*. *Appl Environ Microbiol* 71(9):5610–5613.

Yonekura-Sakakibara K, Tanaka Y, Fukuchi-Mizutani M, et al. (2000) Molecular and biochemical characterization of a novel hydroxycinnamoyl-CoA: anthocyanin 3-O-glucoside-6'-O-acyltransferase from *Perilla frutescens*. *Plant Cell Physiol* 41(4):495–502.

Yoshikuni Y, Ferrin TE, Keasling JD (2006) Designed divergent evolution of enzyme function. *Nature* 440:1078–1082.

Yuba A, Yazaki K, Tabata M, et al. (1996) cDNA cloning, characterization, and functional expression of 4S-(-)-limonene synthase from *Perilla frutescens*. *Arch Biochem Biophys* 332(2):280–287.

CHAPTER 2

PLANT MOLECULAR FARMING: HOST SYSTEMS, TECHNOLOGY AND PRODUCTS

G.B. SUNIL KUMAR, T.R. GANAPATHI, L. SRINIVAS AND V.A. BAPAT
Plant Cell Culture Technology Section, Nuclear Agriculture and Biotechnology Division, Bhabha Atomic Research Centre, Trombay, Mumbai 400 085, India.

Abstract: Plants have been used for various medicines by mankind over centuries and one quarter of prescription drugs are still of plant origin. The advances made in genomics and proteomics resulted in the isolation of several useful genes. These have been expressed in various expression systems including plants. The plant based production of pharmaceutically and industrially useful recombinant proteins is referred as molecular farming. Plants have been successfully genetically engineered using different methods. Stable nuclear transformation by *Agrobacterium* or biolistic methods, plastid engineering, use of viral based transient expression, agroinfiltration and magnifection techniques. In this review, an overview of plant molecular farming, host systems employed, vaccinogen production in plants with an emphasis on hepatitis B vaccine, technology perspective as well as the commercialized products are described

Keywords: Antibodies, glycosylation, hepatitis B, host systems, molecular farming, recombinant proteins, vaccines

Abbreviations: CT: Cholera toxin, ER: Endoplasmic reticulum, GM-CSF: Granulocyte-macrophage colony-stimulating factor, HbcAg : Hepatitis B core antigen, HBsAg : Hepatitis B surface antigen, IgA: Immunoglobulin A, mAb: monoclonal antibody, LT-B: Heat labile enterotoxin B subunit, TSP: Total soluble protein, VEGF: Human vascular endothelial growth factor

1. INTRODUCTION

Plants have been used for medicinal purposes over centuries. One quarter of prescription drugs are still of plant origin (Fisher and Emans, 2000). Advances in genomics and proteomics resulted in identification of an exponential growing number of useful genes. It is unlikely that a single expression system could

R. Verpoorte et al. (eds.), Applications of Plant Metabolic Engineering, 45–77.
© 2007 *Springer.*

be utilized to express these genes. Each protein expression system has unique properties. Therapeutic proteins are often produced in mammalian cell culture system, as the proteins derived from these systems are similar to those of human origin in glycosylation and other biochemical properties. Although many of the therapeutic proteins have been expressed in this system, difficulties in scale-up and expense of the production limit their extensive use. The non-availability and/or high production costs prohibit the regular use of therapeutic proteins by low-income groups that constitute the majority of the population in the developing world. The daily average income of one billion people in the developing countries is less than one US dollar, which literally make these therapeutics unaffordable. Transgenic plants offer as an alternate system for safer and economical production of pharmaceutically important proteins. Molecular farming refers to the production of pharmaceutically and industrially useful recombinant proteins in plants (Franken et al., 1997). The market share of therapeutic proteins is expected to exceed one billion $ in 2008 and could reach $18.6 billion by 2013. By 2008, 8–10 human therapeutic proteins produced transgenically are expected to be released in the market and may reach 30 or more by 2013 (Twyman et al., 2005).

Plant molecular farming has its origin since the first higher plant was successfully transformed (Fraley et al., 1983). The first reporter gene (*uid a*) expressed in transgenic plants (Jefferson et al., 1987) is now a molecular farming product (Kusnadi et al., 1998; Witcher et al., 1998). First report of human antibody expression in transgenic plants was by During (1988) and expression of secretory antibodies by Hiatt et al. (1989), blood substitutes (Magnusen et al., 1998), human growth hormone fusion protein (Barta et al., 1986), interferon (De Zoeten et al., 1989), human serum albumin (Sijmons et al., 1990) and vaccines (Haq et al., 1995). The success of expression of these proteins paved the way to explore the plants as an attractive expression system for the production of biopharmaceuticals. In this article we outline the important types of plant expression systems employed and molecular farming as a technology and products will be discussed.

The technology of molecular farming involves the transfer of the desirable gene to an appropriate host system, optimization of the desirable pattern of gene expression, as well as the recovery of the recombinant protein and further pharmaceutical product development.

2. METHODS OF GENE TRANSFER TO PLANTS

The first step of the biopharmaceutical production in transgenic plants involves the genetic manipulation by employing different methods. Currently, five different methods are in use to transfer the foreign genes and express the recombinant proteins in plant systems.

Stable nuclear transformation: It is the most common method employed for the genetic transformation of plants. It has produced all of the transgenic traits available in the market today. It requires a method and regeneration system for the gene transfer into the plant cells either using *Agrobacterium tumefaciens* or biolistic method using a particle gun. In these methods foreign genes are stably incorporated into the host nuclear genome.

Plastid transformation: It provides an alternative method to produce recombinant proteins in plants with some unique advantages. This strategy adopts biolistic transformation technique to engineer the plastids. The gene integration involves homologous recombination. Each plant cell contains approximately 100 chloroplasts each with about 100 identical genomes, thus a single gene is represented 10,000 times with in one cell (Bendich, 1987). The copy number of the gene encoded in the two inverted repeat regions of higher plants can reach up to 20,000 copies resulting in levels of transgene expression as high as 46% of the total soluble protein (De Cosa et al., 2001). Absence of positional effects and gene silencing, ease of multi-gene engineering in a single transformation event, transgene containment via maternal inheritance in most of the crops and the lack of pleiotropic effects are the advantages of plastid engineering. Over 40 transgenes have been stably integrated and expressed via the tobacco chloroplast genome to confer important agronomic traits as well as to produce industrially valuable biomaterials and therapeutic proteins (Grevich and Daniell, 2005). However, one of the limitations would be protein stability as it changes even with refrigeration in fresh tissues like leaves. Tobacco appears to be the only species in which plastid transformation has been carried out routinely (Daniell et al., 2002). Extraction and purification need to be performed at very specific times following harvest to avoid recombinant protein degradation.

Transient expression using viral vectors: Plant viruses can replicate autonomously and efficiently to very high copy numbers. This property enables higher levels of recombinant protein production in a very short time. Transgene is expressed as a fusion protein with the coat protein of the plant virus. The plant viral expression system has been developed from RNA viruses like tobacco mosaic virus, cowpea mosaic virus, plum pox virus, potato virus X, alfalfa mosaic virus and tomato bushy stunt virus (Canizares et al., 2005). The recombinant virus infects the plant and then transiently expresses the target protein in plant tissue, which accumulates in the interstitial spaces. The interstitial fluid can then be collected by centrifugation under vacuum. This system is not suitable for protein needed in large quantities and products needs to be processed immediately as storage may cause its degradation.

Agroinfiltration: *Agrobacteria* carrying the expression vector are delivered into leaf tissue by vacuum infiltration. The transferred T-DNA is not integrated into the host chromosome but is present in the nucleus, where it is transcribed and this leads to transient expression of the gene of interest (Kapila et al., 1996). Advantages of this technique include multiple genes present in different populations of *Agrobacteria* can be simultaneously expressed, thus the assembly of complex multimeric proteins can be tested in planta (Vaquero et al., 1999). Transient expression is rapid and results in protein expression can be obtained in days facilitating the verification of the gene product for its function before moving onto large- scale production of transgenic plants. Different expression cassettes can be assessed in a short time (Fischer and Emans, 2000).

Magnifection: A recently developed technique by Icon Genetics known as magnifection in which *Agrobacterium* is used to deliver viral replicons (Gleba et al., 2005). The unique advantage of this technique is that different 5' modules carrying different organelle targeting signals, which can fuse in frame with the transgene after recombination and nuclear processing allow the easier testing of recombinant protein accumulation in different compartments of a cell with minimum cloning steps. This technique has the advantage of vector efficiency, efficient systemic DNA delivery of *Agrobacterium*, speed and expression level or yield of a plant RNA virus as well as posttranslational capabilities and low production costs of a plant. This process has been validated through expression of a number of antigens of microbial and viral origin and antibodies (Gleba et al., 2005). Hepatitis B core antigen (HBcAg) was expressed at levels of 7 % of total soluble protein (TSP) after seven days of post infection in tobacco leaves (Huang et al., 2006). Similarly, Plague antigens were expressed in tobacco at levels of 2mg/g F.W. of leaves and exhibited a protective efficacy to aerosol challenge when purified antigens were used to immunize guinea pigs subcutaneously (Santi et al., 2006). However, one of the limitations could be posttranslational modifications, specially the glycosylation pattern. Similar problems are associated with other alternative expression systems. The choice of method of gene transfer depends on the host systems and the type of recombinant protein that needs to be produced.

3. HOST SYSTEMS

The choice of host system should consider certain properties or characteristics of crop species for molecular farming. An optimal system would offer flexibility, ease of genetic manipulation, high protein accumulation, rapid scale-up, low production and handling cost, free of anti-nutritional factors, amenable breeding procedures and stability of transgene expression. No single crop species can have all these characteristics. However, depending on the value of the recombinant protein and its intended final use the crop species has to be selected (Delaney, 2002). The representative crop species, which have been used as the host systems for the expression of recombinant proteins, are described here and some of them are depicted in Fig. 1.

4. CROPS WITH LEAVES AS A MAJOR BIOMASS

4.1. Tobacco

Optimized tissue culture protocols and established genetic transformation methods makes tobacco an ideal candidate for molecular farming. The biopharmaceuticals produced in tobacco are listed in Table 1. Tobacco has several advantages for

Figure 1. Representative host systems for plant molecular farming. A-B: Fruits, C-D: *In vitro systems* (Plant cell cultures and Hairy roots), E: Microtubers, F: Seeds and G: Leafy biomass yielding crop

biopharmaceutical production, it is a prolific seed producer (3000 seeds/capsule), and availability of a wide range of germplasm, so the genetic background required for molecular farming could be customized to remove metabolites like nicotine that may affect end use. Tobacco is a non-food crop for mammals, thus the transgenic plants cannot enter food chain and the use of male sterile hybrids may allow containment. The crop is harvested before flowering, thereby preventing the production of pollen or seeds and it is grown from transplants that are produced in contained facilities, thus eliminating the release of seed that could remain dormant in the soil (Rymerson et al., 2002).

Biomass production and protein extraction protocols are established for tobacco (Woodleif et al., 1981). Under conditions of the high-density (100,000 plants/ha) field production and multiple harvests from re-growth following the first cutting, tobacco could produce fresh weight yields in excess of 50,000 kg/ha. Soluble protein levels in tobacco ranged between 2.3 and 2.8% of total biomass and concentrations as high as 9.2% are possible, when stems are excluded. Yields of extractable protein varied between 155 and 228 kg/ha (Woodleif et al., 1981). Controlled environmental studies conducted with transgenic tobacco plants expressing two chains of IgG1 antibody showed that antibody concentration in leaves varied with stage of development, growth temperature, light levels and correlated with total

Table 1. Production of biopharmaceuticals for human health care in transgenic tobacco plants

Bio-pharmaceutical	Use/therapy of	Expression levels	Reference
Human serum albumin	Blood substitute	0.025% TSP	Sijmons et al., 1990
Human Protein C	Anticoagulant	0.002% TSP	Cramer et al., 1996
Human granulocyte macrophage colony stimulating factor	Neutropenia	0.5% TSP in seeds	James et al., 2000
Human somatotropin	Hypo-pituitary dwarfism in children	7% TSP (chloroplasts)	Staub et al., 2000
Human erythropoietin	Regulation of erythrocyte mass/ Anemia	0.01%TSP	Matsumoto et al., 1995
Human epidermal growth factor	Wound repair and cell proliferation	–	Higo et al., 1993
Human interferon alpha	Hepatitis B and C, and hairy cell leukemia	–	Smirnov et al., 1990
Human interferon beta	Multiple sclerosis	–	Edelbaum et al., 1992
Human IL-2	Renal cell carcinoma	25-350 ng/g callus	Magnuson et al., 1998
Human IL-4	Immune and inflammatory responses modulation	1.1 μg/g callus	Magnuson et al., 1998
Human haemoglobin alpha, beta	Blood substitute	0.05% seed protein	Dieryck et al., 1997
Human homotrimeric collagen	Natural bio-material	0.1 μmg/g F.W.	Ruggiero et al., 2000
Angiotensin converting enzyme 1	Hypertension	100 μg/g F.W.	Hamamoto et al., 1993
Alpha Trichosanthin	AIDS	2% TSP	Kumagai et al., 1993
Glucocerebrosidase	Gauchers disease	1-10% TSP	Cramer et al., 1996
Human alpha lactalbumin	Nutraceutical	0.2% TSP	Takase and Hagiwara 1998
Human lactoferrin	Anti-microbial	0.3% TSP	Salmon et al., 1998

– Not reported

soluble protein levels (Stevens et al., 2000). This study demonstrates that total soluble protein can be used as a marker for agronomic studies aimed at enhancing recombinant protein yields.

4.2. Alfalfa

It is a perennial forage crop productive up to five years with a capacity for symbiotic nitrogen fixation. It can be propagated by stem cutting with a strong regenerative capacity, thus enables production of large clonal populations within a limited time period. The low secondary metabolite contents of its leaves makes it an ideal crop over tobacco for molecular farming. It has a large dry biomass yield per hectare.

The disadvantages of leafy crops are that the recombinant proteins are often unstable in dilute and aqueous environment, so the leaves must be dried or frozen or processed soon after harvest or harvested during night times to ensure lowest possible temperatures to extract the product. Presence of phenolic compounds released during protein extraction can interfere with down stream processing. Moreover, constitutive expression in the leaves could interfere with plant growth and development. Finally biosafety concern like potential exposure of herbivores to pharmaceuticals expressed in leaves and the leaching of these proteins into the environment need to be considered.

5. VEGETABLE AND FRUIT CROPS

5.1. Banana

Banana is an ideal host for the production of edible vaccines as it offers advantages like digestibility and palatability by the infants, availability through out the year in the tropics and subtropics where economical vaccines are required to immunize large segments of the population (Sunil Kumar et al., 2004). As many of the edible bananas does not set seeds and are vegetatively propagated through suckers, it is an ideal candidate for gene containment and there is no segregation of the transgene. The first report on hepatitis B vaccine production in banana fruits has been recently published (Sunil Kumar et al., 2005b).

Although expression levels of the antigen are low in banana fruits, certain strategies to be adopted include the use of a promoter of abundant pulp protein (Clendennen et al. 1998), or promoters of the proteins found in abundance in the ripe banana fruits (Peumans et al., 2002), codon optimization and use of banana UTRs (un-translated regions) may result in enhanced foreign gene expression in the fruits.

5.2. Potato

Potato (*Solanum tuberosum*) is one of the important food crops with high nutritional value and yield potential. It provides approximately half of the world's annual production of all roots and tubers, making it the largest non-cereal food crop. Potatoes offer several advantages for molecular farming, which include efficiency of genetic transformation by *Agrobacterium tumefaciens* with relatively short transformation and generation times, clonal propagation, availability of tissue specific promoters and capability of microtuber production, storage potential and the ability to feed raw potatoes to test animals.

Clonal propagation allows stable production of a transgenic plant lines. Putative potato transgenic plants can be induced to produce microtubers, thus they can be screened readily for tuber specific expression. Tubers are the storage organs of the potato and can be stored for a period of time before being consumed. Foreign proteins may be stable for long periods of time in tubers as long as the tuber does not sprout or become damaged. The benefits of being able to store tubers without any processing may make potatoes very desirable as a vaccine production system for those animals that do eat raw potatoes (Richter and Kipp, 1999).

5.3. Tomato

Tomato is the second most popular crop next to potato in the world. It is grown world wide and more palatable than potatoes. It has advantages like high biomass yield and use of green houses for containment and is a short duration crop. Different formulations like sauce, puree and freeze dried powder can be made from the transgenic tomato fruits expressing biopharmaceuticals. The Tomatoes were first used for the production of rabies vaccine candidates and to produce antibodies (Twyman et al., 2003).

6. SEED-BASED PRODUCTION SYSTEMS

The production of recombinant proteins in seeds is attractive for several reasons, which include their ability to accumulate storage proteins in a small and stable environment which is facilitated by their desiccated nature and the availability of molecular chaperones and disulfide isomerases in the developing seed ensures proper protein folding and the lack of proteases (especially in the endosperm tissue) (Muntz, 1998). The small size of most of the seeds allows the recombinant protein to accumulate in relatively higher concentration in a small compact biomass, even if the overall yield is lower than that of a high biomass leafy crop such as tobacco and the simple seed proteome consisting of fewer proteins may allow easier downstream processing. Similarly, low levels of many of the substances known to interfere with purification steps are present in seeds compared to the levels of phenolics and alkaloids present in tobacco leaves and oxalic acid present in alfalfa. The presence of oil bodies in certain seeds like safflower and rape seeds can be exploited to ease recombinant protein purification by oleosin fusion technology and seed specific expression may not interfere with vegetative growth of the plant and also limits contact with non-target organisms in the rhizosphere and those feed on leaves (Stoger et al., 2005).

6.1. Corn

It is one of the potential crops for recombinant protein production as seeds can be stored for years and stored proteins can retain its activity in storage. Seeds are particularly advantageous as recombinant proteins can be expressed at levels as high as 0.1% of total dry weight. Multimeric proteins can be properly processed and assembled. A wide range of proteins from 6000 daltons to 272, 000 daltons have been expressed in seeds. Corn is a very economical production system with cost of $2.5 to produce one bushel. There is a vast infrastructure for growing, harvesting and processing of corn seeds (Jilka, 1999). The high biomass yield, ease of genetic transformation and convenience of scale-up makes it an attractive system for recombinant protein production.

6.2. Cereals and legumes

Different crop species have been investigated for seed based production, these include cereals like rice, wheat and barley. Legumes like pea and soybean and oil seeds like rapeseed, and safflower have also been considered as potential crops for molecular farming.

Rice has several advantages such as, high grain yield, available genetic transformation protocols, capacity for rapid scale-up and availability of constitutive and endosperm specific promoters. It is a self-pollinating crop and has the advantage that it can be grown in green houses in the early stages of production. Barley is another potentially useful crop as the producer price is low and seed protein content is high. Wheat has been used rarely for the production of pharmaceutical proteins, but it has a low producer price compared with rice and is therefore potentially attractive.

Soybean has the advantage of high protein content in the seeds but the constraints like more time consuming transformation procedures and high oil content may limit its use. Pea has a similar grain yield and seed protein content to soybean. However, the producer price is about fifty per cent higher than soybean. Oil seeds offer a unique advantage as they facilitate targeting of the recombinant proteins to the oil bodies by oleosin fusion. The fusion protein can be recovered from oil bodies using a simple extraction procedure and the recombinant protein from its fusion partner by endo-protease digestion (Stoger et al., 2005).

7. HAIRY ROOTS

Agrobacterium rhizogenes causes hairy root disease in plants. These neoplastic roots are characterized by their high growth rate, genetic stability, biosynthetic capacity and growth in hormone free media. Hairy root cultures serve as model system for plant metabolism and physiology and as a technical alternative to plant cell suspension cultures for the production of therapeutics and speciality chemicals (Shanks and Morgan, 1999). A more recent advance in hairy root technology is to employ these cultures for the production of therapeutic proteins. Recovery of recombinant proteins from plant tissues poses a great challenge and contributes a major cost factor. Secretion based plant systems or *in vitro* cultivated hairy roots offer a novel approach that relies on the biosynthetic potential of the roots to synthesize and secrete recombinant protein into hydroponic medium continuously, referred as rhizosecretion (Borisjuk et al., 1999). This technology was validated by producing active forms of three recombinant proteins of different origin: bacterial xylanase, jellyfish green fluorescent protein and human secreted alkaline phosphatase in tobacco plants. The levels of secretion of recombinant protein were higher in hairy roots than in adventitious root cultures, which was a result of more active protein synthesis and/ or secretion in hairy root tissue compared to the adventitious roots as mRNA levels in both the tissues were similar (Gaume et al., 2003).

8. MOSS

Mosses are phylogenetically ancient organisms that have maintained a high genetic stability during their evolution. Greenovation, a Germany based company focused on a moss *Physcomitrella patens* as an expression host for biopharmaceutical production. Moss is unique among all multicellular plants analysed to date in exhibiting a very effective homologous recombination in its nuclear DNA. This allows targeted knock out of genes, which is a highly attractive tool for designing a production strain and it also facilitates the manipulation of glycosylation pathway as opposed to higher plants. They are grown under photoautotrophic conditions in a simple medium essentially consisting of water and minerals. Their cultivation in glass bioreactors is well established. The main phase of life cycle is haploid; by appropriate vegetative propagation diploid phase can be completely avoided. Engineering of haploid cells directly results in a desired phenotype (e.g., altered biochemical pathways) without the requirement for time consuming crossing steps. The excellent genetic stability of strains of *Physcomitrella patens* transgenic for recombinant human vascular endothelial growth factor (VEGF) were subcultured for several years without any selection pressure. The expression levels of VEGF remained constant and no alterations were detected in the isolated and sequenced transgene. Cryopreservation is also a well established storage technology for *Physcomitrella*. Viability of the strains from cryopreserved cultures was reported to be 100 per cent (Gorr and Jost, 2005).

Transformation of mosses can be achieved by simple polyethylene glycol mediated direct DNA transfer into protoplasts. Stably transformed plants can be obtained by the introduction of an expression cassette containing only linearized DNA, thus avoiding the introduction of viral or bacterial vector sequences into the genome of production strains. Optimization of the codon usage for the expression of human genes in *Physcomitrella* is not required. They perform extensive posttranslational modification of proteins, which include disulfide bridge formation and complex glycosylation. Optimization of gene expression and secretion in mosses was achieved by improved expression tools. Several expression and secretion regulating sequences such as promoters and signal peptides were isolated and characterized (Jost et al., 2005, Baur et al., 2005a). Further developments such as co-expression of a stabilizing protein (e.g. human serum albumin) exhibited great potential for further optimization (Baur et al., 2005b). For safe and flexible production of therapeutic proteins for human therapy, the humanization of plant protein glycosylation is essential. Transgenic strains of *Physcomitrella* were created which lacked α-(1,3)-fucose and β-(1,2)-xylose residues secreted a transiently expressed human VEGF in the same concentration as unmodified moss indicating that modification of glycosylation did not impair the secretory pathway of moss (Huether et al., 2005).

9. MICROALGAE

Recent success in the expression of recombinant proteins in microalga *Chlamydomonas reinhardtii* has shown the possibility of exploring novel expression systems. Lower eukaryotes such as microalgae can be employed to produce therapeutic proteins as they are capable of carrying out posttranslational modifications. This alga has beneficial attributes for the production of recombinant proteins. Nuclear, chloroplast and mitochondrial DNA can be transformed easily and relatively short period of time between the initial transformants and subsequent scale up, gametogenesis can be induced and genetic crosses can be carried out between haploid cells of opposite mating types, it can be grown either heterotrophically or phototrophically, wide variety of promoters can be used that are regulated by different factors, it can secrete glycosylated proteins and can be grown in cultures ranging from few ml to 500,000 litres in a contained and cost effective manner. These characteristics make *Chlamydomonas* an attractive system for the expression of human therapeutic proteins (Franklin and Mayfield, 2004).

10. OPTIMIZATION OF RECOMBINANT PROTEIN EXPRESSION IN TRANSGENIC PLANTS

The expression levels of recombinant proteins in plants are generally low. Recombinant proteins have been targeted to different organs or various sub-cellular locations to optimize recombinant protein yields.

11. EXPRESSION OF RECOMBINANT PROTEINS IN SEEDS

Both constitutive and tissue specific promoters have been used to express recombinant proteins in seeds. GM-CSF and the cytomegalovirus glycoprotein B were targeted to seeds using the rice glutelin promoter (Ganz et al., 1996; Tackaberry et al., 1999). Promoters of legumin 4 or unknown seed protein of *Vicia faba* were used to target ScFv antibodies to seed (Fiedler and Conrad 1995; Fiedler et al., 1997). The constitutive promoter CaMV 35S was used for seed expression of human hemoglobin (Dieryck et al., 1997). However, certain disadvantages like easy dispersal of seeds by wind etc. raises issues of containment.

12. ORGANELLE TARGETING

Transit peptides can target protein to various organelle compartments like ER, mitochondria, chloroplasts, vacuoles and the apoplast. Retention of proteins in ER by addition of C-terminal ER retention signal (KDEL or HDEL) increased recombinant protein accumulation levels. A single chain variable fragment of the antibodies (ScFv) expression in tobacco was enhanced 100 fold by incorporating KDEL sequence at the C-terminus (Schouten et al., 1996).

13. EXPRESSION IN CHLOROPLASTS

Plastids have an active homologous recombination system, which facilitates the precise, targeted integration of cloned DNA and its propagation through out the pool of plastid genomes present in an organelle. About 100 copies of genome are present in each plastid and 100 chloroplasts are present in each leaf cell. Several fold higher expression of recombinant proteins can be obtained by expressing them in plastids (Daniell et al., 2001). Staub et al. (2000) reported 300-fold increase in human somatotropin in chloroplast transformed tobacco plants compared to nuclear transformed plants.

14. APOPLAST TARGETING FOR SECRETION

Apoplast targeting for secretion of recombinant proteins yielded variable results as some proteins accumulated to high levels in the apoplast of plant cells such as *Aspergillus* phytase at 14% TSP (Pen et al., 1993), whereas some proteins accumulated in low levels such as erythropoietin at 0.0025% TSP (Matsumoto et al., 1995). The reasons for such variations are not yet clear.

15. CONSTRUCTION OF SYNTHETIC GENES WITH HOST PLANT PREFERRED CODON USAGE AND THE USE OF UNTRANSLATED REGIONS

The plant preferred codon usage enhances the recombinant protein expression as the tRNA pool for a given codon varies among species. Crystal protein genes of *Bacillus thuringiensis* were found to be poorly expressed in plants and the lower expression levels were attributed to the coding sequence itself. The GC content of bacterial genes is generally less than that of plants and they may contain AT sequences, which serve as polyadenylation signals of eukaryotic genes (Fischoff, 1992). Construction of synthetic genes with a plant preferred codon usage would enhance the expression levels (Sutton et al., 1992). A synthetic gene with a plant preferring codon, encoding LT-B was found to increase the LT-B expression levels in transgenic plants. (Lauterslager et al., 2001; Streatfeld et al., 2002). Mason et al. (1998) optimized the LT-B gene for plant mRNA processing by changing the AAT codon to GTG surrounding the translation initiation site, which resulted in 5–40 fold higher amount of LT-B in leaf tissues of potato, when compared to lines carrying the bacterial LT-B gene. Similarly, Richter et al. (2000) could obtain up to 16μg/g HBsAg in potato tubers with the use of 3' end of the soybean vegetative storage protein gene and 3' element from the potato pin II gene. Similarly, fusion of 5'untranslated leader sequence from alfalfa mosaic virus along with the RSV-F gene increased the transient expression level to 5.5 fold in apple leaf protoplasts. Further, adding the transcriptional enhancer from the pea plastocyanin gene increased the expression level to 7.7 fold (Sandhu et al., 1999). Specific sequences within mRNA itself, can affect its stability and translatability in plant cells, which include destabilizing motifs and cryptic splice sites (Newman et al., 1993; Taylor and Green, 1995). The expression levels of the target protein can be increased by constructing a synthetic gene considering the above mentioned criteria.

16. ROLE OF PROMOTERS

Cauliflower mosaic virus 35S promoter is extensively used for the expression of the recombinant proteins in plants. Though this promoter is active in many plant tissues and can give quite high levels of expression, it has certain limitations as it constitutively expresses the protein and the accumulation of the recombinant proteins to high levels may affect the yield or overall growth of the plant and is some times associated with co-suppression or gene silencing (Taylor, 1997). In addition, the 35S promoter is not functional in many mature tissues like fruits and fully expanded leaves, which reduces the full potential of biomass utilization. These problems could be alleviated by the use of inducible promoters, promoters with tight regulation of gene expression and tissue specific promoters. A post harvest inducible promoter termed the MeGA™, has been developed, it shows rapid and strong gene activation in response to mechanical stress like wounding or to a variety of defense elicitors (Cramer and Weissenborn, 1997). CropTech Scientists (USA) used this promoter as an effective one in driving high levels of inducible expression in all tissues of plants including fully expanded leaves. This technology offers advantages like minimizing the impact of environmental factors on protein yield and quality as the expression is induced in the post harvest conditions and the foreign protein expression will not affect the growth and development of the plant. The stability of the protein and its yield in fully active form can be maximized as all the recovered protein is newly synthesized (Cramer et al., 1999).

17. POSTTRANSLATIONAL MODIFICATIONS

Therapeutic proteins may require complex post translational modifications and/or assembly into multimeric forms for their biological activity. Plants seem to recognize and process correctly most of the signals within the mammalian polypeptides. Among various posttranslational modifications, glycosylation has been shown to play an important role in the physiological activities of mammalian glycoproteins (Kusnadi et al., 1997). Glycosylation may affect the biological activity, protein folding, stability and solubility properties of the recombinant proteins.

18. GLYCOSYLATION OF PLANT DERIVED RECOMBINANT ANTIBODIES

Many pharmaceuticals including antibodies require glycosylation for an increased *in vivo* activity and stability. The addition of N-glycans to several recombinant proteins increases their biological activity and half-life (Elliott et al., 2003). In plants, N-glycosylation starts in endoplasmic reticulum in a co-translational way, by the addition of an oligosaccharide precursor (Glc3 Man9 GlcNAC2) to the specific asparagine residue of the N-glycosylation recognition sequences (Asn-X-Ser/Thr) in the elongating polypeptide. The oligosaccharide undergoes several maturation steps involving removal or addition of sugar residues in the ER and the golgi complex. The plant glycosylation differs from mammalian N-glycosylation in the late golgi maturation steps, which results in the absence of α(1,6)-linked fucose and sialic acid and the presence of bisecting β(1,2)-xylose and core β(1,3)-fucose in the N-glycans (Gomord et al., 2004). The presence of IgE antibodies that are specific for plant N-glycan in the sera of human patients has been documented. Fay's group at CNRS, France characterized the specificity of glycan specific IgG4 in sera from allergic patients receiving grass pollen immunotherapy and demonstrated the specificity for both the α(1,3)-fucose and β(1,2)-xylose glycoepitopes (Gomord et al., 2005).

Few strategies were employed to modify the plant glycosylation pathway to produce the recombinant proteins with a desirable glycosylation pattern. These include the retention of glycosylated proteins bearing non-immunogenic glycans within the ER, inhibition of golgi glycosyltransferases and the expression of mammalian glycosyltransferases. The structural analysis of ER resident plant proteins revealed high mannose type N-glycans that are common to plants and mammals, and may not be immunogenic. Thus storage of recombinant proteins within the ER, upstream of the golgi complex where immunogenic glycoepitopes are added to the maturing plant N-glycans, may result in non-immunogenic glycoproteins. The addition of a C-terminal ER retention signal is sufficient for the retention of recombinant proteins in the plant ER (Gomord et al., 1997). This strategy was applied to retain human monoclonal antibody heavy chains retained in the ER of transgenic tobacco plants, where most glycans N-linked to this antibody were of high mannose type, but a fraction of about 10% of these glycans contained immunogenic glycoepitopes (Ko et al., 2003). The retention efficiency was further increased, when the C-terminal signal was fused to both heavy and light chains of a chimeric mouse-human antibody expressed in tobacco resulted in N-glycans exclusively of high mannose type (Sriraman et al., 2004). However, when compared with their mammalian homologues, antibodies produced using this strategy were less stable after injection in mice (Ko et al., 2003). This could be due to the degradation of antibodies with high mannose N-glycans after binding to the mannose receptor and endocytosis by macrophages (Wright and Morrison, 1998). In another study, a plant made antibody with complex N-glycans (Bardor et al., 2003a) has been shown to be as stable as its mammalian homologue in the blood stream of mice (Khoudi et al., 1999).

Inactivation of plant glycosyltransferases such as β(1,2)-xylosyltransferase and α(1,3)-fucosyltransferase will prevent glycoepitope formation on therapeutic proteins produced in transgenic plants. The inactivation of N-acetyl glucose aminyltransferase is sufficient to block the biosynthesis of complex N-glycans in *Arabidopsis* (von Schaewen et al., 1993). This glycosyltransferase has been cloned from several plants. However, the antisense expression of this in tobacco and potato plants could only decrease complex N-glycan biosynthesis in these plants (Wenderoth and von Schaewen, 2000). Recent results obtained in the moss *Physcomitrella patens* for inactivation of glycosyltransferases are encouraging. It is the only plant system which shows high frequency of homologous recombination which has been used to knockout xylosyl and fucosyl transferase gene leading to the disappearance of plant specific glycoepitopes without any effects on protein secretion (Koprivova et al., 2003). The analysis of several glycosyltransferases is providing the panel of specific signals that are sufficient for protein targeting within the different golgi subcompartments. These signals will help to target exogenous glycosyltransferases in the plant golgi apparatus for the required efficiency in the glycosylation pathway engineering (Gomord et al., 2004).

The other strategy includes the expression of mammalian glycosyltransferases in the plant golgi apparatus. The transfer of human β1,4-galactosyltransferase to tobacco modified the N-linked glycosylation pattern resulting in glycans with galactose residues at the terminal non-reducing ends. In addition, the absence of the dominant xylosidated and fucosylated type sugar chains confirms that the transformed cells can be used to produce glycoproteins without the highly immunogenic glycans (Palacpac et al., 1999). However, the tobacco derived antibodies exhibited complex mixture of N-glycans, some of them being partially humanized (Bakker et al., 2001). These strategies would be more efficient in plant systems such as alfalfa, where the N-glycosylation of antibodies is restricted to the predominant mature oligosaccharide chain with terminal N-acetylglucosamine residues. The glycans allow *in vitro* or *in vivo* remodeling into a human compatible N-glycosylation. Bardor et al. (2003b) have shown that *in vitro* galactosylation of alfalfa derived antibody using a 1,4-galactosyltransferase resulted in an efficient conversion of plant N-glycan into an oligosaccharide having homogenous galactosylation similar to that of murine antibody. These promising results obtained for the humanization of plant N-glycosylation may hopefully allow the development of plant systems producing biopharmaceuticals compatible for human therapy in the near future.

19. RECOVERY OF RECOMBINANT PROTEINS FROM PLANT SYSTEMS

Recombinant protein production in transgenic plants consists of two major components: one is the transgenic crop production and the other is the recovery of the recombinant protein. The cost breakdown for a highly purified therapeutic protein from transgenic corn shows that the transgenic raw material constitute only 5–10% of the total manufacturing cost, whereas extraction and purification make up as much as 90%. By contrast the purification cost of transgenic corn for the production of industrial protein amounts to only 35 to 40% of the total manufacturing costs (Nikolov and Hammes, 2002). Thus, bioprocessing of the pharmaceuticals produced in transgenic plants contribute to the most of the cost involved in its production. Various strategies/technologies were developed to reduce the recovery costs. Some of these include oleosin partitioning technology and the use of affinity tags.

20. OLEOSIN PARTITIONING TECHNOLOGY

Oil seeds store part of the energy needed for germination in organelles called oil bodies or oleosomes. Oil bodies are spherical structures consisting of an oil droplet surrounded by a half unit of phospholipid membrane. A unique class of proteins called oleosins surrounds the surface of these oil bodies. Oleosins are near ubiquitous seed proteins as they are present in all common oil seeds. 2–10% of total seed protein in various species comprises oleosin. The central domain of oleosin is crucial for its localization in oil bodies (van Rooijen and Moloney, 1995). It can tolerate fusion of foreign proteins to either N or C terminal end without the loss of oil body targeting efficiency. Oleosin fusions have been successfully used for targeting various recombinant proteins in the range of 7–55kDa. Oleosin fusion proteins are purified as less dense oil bodies from soluble contaminants by flotation centrifugation, enabling simple and rapid purification of recombinant proteins. A Canadian company, Sem-BioSys, employs this technology for the purification of recombinant proteins from oil seeds (Cramer et al., 1999).

21. AFFINITY TAG BASED SEPARATION

This technology employs development of fusion proteins, which are made by expressing the protein of interest and a protein or peptide that exhibits affinity for a specific ligand with a protease cleavage site in between the two. The fusion protein is first separated by affinity chromatography followed by removal of tag from recombinant protein by a protease treatment. The different types of affinity tag and ligand pairs include histidine residues-metal ions, maltose binding protein-amylose and *Staphylococcus* A protein -IgG. This technique was used in small- scale purification of human glucocerebrosidase -flag epitope fusion protein produced in tobacco (Cramer et al., 1996). The limitations of this strategy include the requirement of an additional step involving the removal of tag that contributes to higher down stream processing costs and the tags may alter folding or processing of the recombinant proteins.

22. PROTEIN TARGETING AND SECRETION

The recombinant proteins can be targeted to various compartments within the cell using different signal sequences or whole protein fusions. The desired protein purification is achieved by sub-cellular fractionation to enrich recombinant protein targeted organelles. Recombinant proteins are either destined for secretion or targeted to intracellular storage organelles. Organellar targeting may not allow the required post-translational modifications for a particular recombinant protein. The expression of human neuro peptide leu-enkephalin as an internal fusion protein between the N and C terminal ends of 2S albumin protein of *Arabidopsis in* the seeds of *Arabidopsis thaliana* and *Brassica napus* has been reported (Vandekerckhove et al., 1989). The protein based polymer GVGVP was used to purify the recombinant protein in one step without the use of chromatography resulting in minimized processing steps and costs. GVGVP is a protein-based polymer encoded by a synthetic gene, it exists as an extended molecule at low temperature and hydrophobically folds into β-spirals upon raising the temperature above the transition range. This polymer further aggregates by hydrophobic association to form twisted filaments. Daniell et al. (1996) used this method to purify insulin in a single step. Further, Decosa et al. (2001) used chaperonin protein to fold a foreign protein into cuboidal crystals allowing their purification in a single step by centrifugation. The added advantage of this method is protection of recombinant proteins from cellular proteases.

Recombinant proteins when secreted into the extra cellular media or periplasmic space by incorporating secretory signals as N-terminal fusions, enhances their stability and proper folding. During the course of secretion, the signal peptide is cleaved from the recombinant protein eliminating the requirement of an additional step of signal peptide removal. Sijmons et al. (1990) reported apoplast targeting of human serum albumin (HSA) in potato by fusing HSA gene with a tobacco pathogenesis related protein (PR-S signal sequence). Xylanase was targeted into apoplastic space of tobacco with a potato proteinase inhibitor II protein signal peptide fusion (Herbers et al., 1995). Incorporation of ER resident protein signal sequences allowed the secretion of recombinant proteins through the intercellular spaces into the guttation fluid in transgenic tobacco plants (Komarnytsky et al., 2000) and three heterologous proteins of diverse origin were secreted from the roots of transgenic tobacco plants using the same strategy (Borisjuk et al., 1999). Further, HBsAg was secreted into the spent medium by transformed tobacco cell suspension cultures, when a C-terminal ER retention signal sequence was fused to the 's' gene of HBsAg (Sunil Kumar et al., 2003).

23. RECOMBINANT PROTEINS EXPRESSED IN PLANT SYSTEMS

Recombinant proteins expressed in plants can be broadly categorized into four types: biopharmaceuticals for therapy, industrial proteins, monoclonal antibodies and antigens for vaccines (vaccinogens). Biopharmaceuticals have to be manufactured under stringent good manufacturing practices (cGMP) and the protein has to be of high purity (Horn et al., 2004). Therapeutic proteins expressed in plants include erythropoietin, -interferon, enkephalins, human serum albumin, glucocerebrosidase and granulocyte-macrophage colony stimulating factor. Applied Phytologics (USA) has engineered rice plants to produce human α-1-antitrypsin a potential therapeutic for cystic fibrosis, liver disease and hemorrhages. Glucocerebrosidase was produced in tobacco plants that have applications in the enzyme replacement therapy for Gaucher's disease. Hirudin is another important anti-coagulant that has been expressed in tobacco, oil seed rape and Ethiopian mustard. SemBioSys is commercially cultivating transgenic oil seed rape for the production of hirudin (Boothe et al., 1997).

24. PRODUCTION OF VACCINES IN RECOMBINANT PLANT SYSTEMS

Vaccination is the safest and cost effective strategy for combating infectious diseases. It greatly reduces mortality and morbidity rates. 33 million children born every year especially in the developing countries remain unvaccinated for vaccine preventable diseases (WHO, 2002). An ideal vaccine should have characteristics like efficacy as a single oral dose, heat stability, infant palatability and economical to achieve universal childhood vaccination on a global scale (Mitchell et al., 1993). It has been estimated that more than 30% of immunization injections are unsafe, due to the reuse of syringes. An alternative to parenteral administration of vaccines by syringe, oral delivery offers advantages in terms of safety and in immunity profiles.

Most of the infectious agents interact with the human mucosal surfaces of the nasal, oropharyngeal, respiratory, genitourinary and gastrointestinal tracts. Conventional vaccines involve parenteral immunization, which do not prevent these initial interactions. Moreover, many bacterial toxins bind to and interact with mucosal epithelial cells, in which significant damage to the host may occur before serum antibodies can play a role in protection. There is a need to focus on mucosal immunization as a means of inducing secretory IgA antibodies directed against specific pathogens of mucosal surfaces. These antibodies may block the attachment of invading pathogens at these sites. In addition, the existence of a common mucosal immune system permits immunization on one mucosal surface to induce secretion of antigen specific secretory IgA at distant mucosal sites. Mucosal immunization can be an effective means of inducing not only secretory IgA, but also systemic antibody and cell mediated immune responses (Freytag and Clements, 2004). Mucosally administered vaccines offer a number of potential advantages over parenterally delivered vaccines, these include the potential to confer mucosal as well as systemic immune responses. If compared with injectables the oral dosage forms have enhanced stability, increased shelf life, may not require cold storage (when delivered in edible plant tissues/seeds), elimination of needles and the need for trained personnel to deliver vaccines (Freytag and Clements, 2004).

Over the past fifteen years, several research groups worldwide have extensively investigated plants as bioreactors for vaccine production. Plant based vaccines have been produced against cholera, traveler's diarrhea, gastroenteritis, hepatitis B, rabies, autoimmune diabetes mellitus etc. Among these, extensive studies were carried out to optimize the hepatitis B vaccine production in recombinant plant systems.

25. HEPATITIS B VACCINE PRODUCTION IN PLANTS

Hepatitis B is one of the alarming diseases in the developing countries and affecting around two billion people worldwide. Of these, approximately 350 million are chronically infected. The late manifestation of the disease is hepatocellular carcinoma and cirrhosis of the liver. These conditions lead to around one million deaths annually, which are preventable by vaccination. The three major factors, which limit hepatitis B vaccination, include the cost of the vaccine and administration expenses, patient compliance is limited by the requirement of syringes and trained personnel and a significant proportion of individuals vaccinated is not protected by current vaccines (Streatfield, 2005). Thus there are opportunities for new hepatitis B vaccines, especially for oral formulation of this vaccine.

Hepatitis B surface antigen (HBsAg) has been expressed in various recombinant plant systems. Mason et al. (1992) reported for the first time the feasibility of HBsAg expression in plants. The transgenic plants accumulated the antigens at the levels of 66 ng/mg of total soluble protein (TSP). Recently, our group has demonstrated HBsAg accumulation in tobacco seeds at the levels of 6ng/g fresh weight (FW) (Sunil Kumar et al., 2006a). Richter et al. (2000) optimized the expression of HBsAg in transgenic' potato plants and noted the expression levels of 16 μg/g FW of tubers. Potato hairy roots were also employed to produce HBsAg at levels of 97.1 ng/g FW (Sunil Kumar et al., 2006b). Carrot cell suspension cultures were used to express HBsAg at levels of 25 ng/g FW (Imani et al., 2002). Kapusta et al., (1999) expressed HBsAg in lupin and lettuce at the levels of 150 ng/g FW of lupin callus. HBsAg was expressed with a plant signal peptide (signal peptide from soybean vegetative storage protein) resulted in a significant increase in the antigen accumulation at 226 ng/mg TSP (Sojikul et al., 2003). HBsAg was expressed in NT-1 cells of tobacco at the levels of 2 μg/g FW, when the antigen was fused with a C-terminal ER retention signal and the antigen was secreted into the spent medium (Sunil Kumar et al., 2003, 2005a). Peng et al. (2002) reported the HBsAg production in a macro alga (*Laminaria japonica*) at the levels of 2.497 μg/mg TSP. More recently we have reported the expression of HBsAg in transgenic banana fruits at the levels of 1 ng/g FW (Sunil Kumar et al., 2005b) and currently we are attempting to enhance the levels of expression in the fruits using different strategies. We have also transformed tomato variety megha to express HBsAg and the antigen accumulation in tomato fruits was confirmed (un published results).

Process parameters for extraction, stability and *in vitro* assembly of HBsAg derived from recombinant plant systems have been studied elaborately. Altering the sodium ascorbate concentration or buffer pH, the monoclonal antibody reactive epitopes of HBsAg expressed in plant systems was increased between 8 to 20 fold. Detergent concentration also influenced the antigen stability in plant cell lysates stored at 4 °C, under optimum concentrations stability was maintained for at least one month (Smith et al., 2002a). 2-Mercaptoethanol increased the antigenic activity of HBsAg up to four fold, an optimum temperature of 50 °C was found to increase the monoclonal antibody binding of HBsAg and in the presence of leupeptin, a proteinase inhibitor, higher antigenic activity was obtained (Dogan et al., 2000). HBsAg production in batch culture was optimized and the maximal HBsAg levels were obtained with soybean cell suspension cultures (20–22 mg/l). The monoclonal antibody reactive epitopes were substantially better in an early exponential phase (37%) in contrast to only 6% for stationary cultures (Smith et al., 2002b).

An effective vaccine against hepatitis B should contain faithfully reproduced conformational epitopes of HBsAg with extensively cross-linked proteins by disulfide bonding. HBsAg VLPs of appropriate size (22 nm) were demonstrated in purified tobacco leaf extracts (Mason et al., 1992). Buoyant density studies were carried out to compare the density of human serum derived HBsAg with that of HBsAg produced in transgenic tobacco plants (Mason et al., 1992), tobacco cells (Sunil Kumar et al., 2003) and banana plants (Sunil Kumar et al., 2005b). The plant derived HBsAg was found to accumulate intra-cellularly as tubular structures, with a complex size distribution and monoclonal antibody binding of 21–37% (Smith et al., 2003). The incorporation of C-terminal retention signal could improve the proportion of monoclonal antibody reactive HBsAg (49.3–67. 8%) in transgenic tobacco plants (Sunil Kumar et al., 2006a) and transgenic banana plants (Sunil Kumar et al., 2005b).

HBsAg derived from recombinant plant systems was assessed for its immunogenicity. HBsAg purified from tobacco leaves could elicit immune response in mice, which was qualitatively similar to that of yeast derived HBsAg and T-cells from the mice primed with tobacco derived HBsAg could be stimulated in *vitro* by the yeast derived HBsAg and by a synthetic peptide (Thanavala et al., 1995). Lettuce leaves expressing HBsAg were administered in two oral doses to three naïve human volunteers resulting in serum responses in excess of the benchmark of 10 mIU/ml in two of the volunteers with out any ill effects (Kapusta et al., 1999). The plant signal peptide fused HBsAg stimulated higher levels of serum IgG than native HBsAg, when injected into mice (Sojikul et al., 2003). Mice fed with three doses of potato tubers expressing HBsAg (5 μg HBsAg plus 10 μg cholera toxin for each dose) prompted a primary serum response peaking at approximately 70 mIU/ml three weeks after the last dose and subsequent boosting with yeast derived HBsAg injection resulted in an immediate high and prolonged antibody titer (Richter et al., 2000). Kong et al. (2001) demonstrated that mice injected with a sub-immunogenic dose of yeast derived HBsAg could be boosted with a high antibody titer, when they were subsequently fed three doses of 40 μg of HBsAg expressed in potato tuber plus 10 μg of cholera toxin. Immunogenicity studies were also carried out with or without middle protein sequences of HBsAg expressed in potato tubers using cholera toxin as an adjuvant. Higher antibody titers were obtained following a booster injection, when middle protein sequences were included with the surface antigen (Joung et al., 2004). These results suggest that inclusion of upstream sequences in addition to the major surface antigen may be useful in developing a more effective oral vaccine. Recently, Thanavala et al. (2005) demonstrated that 33 human volunteers who had been previously vaccinated against hepatitis B could show responses in excess of 10 mIU/ml antibody titers, when they were fed with two or three doses of raw potato tubers containing approximately one mg of HBsAg/dose. More than half of the individuals exhibited an increased antibody titer and tolerated the transgenic potato tubers similar to that of non-transgenic control tubers. A more practical hepatitis B vaccine could be developed by enhancing the expression levels, especially in edible plant tissues like transgenic banana fruits together with co-expression/co-administration of mucosal adjuvants.

26. MUCOSAL ADJUVANTS

The transition from innate to adaptive immunity is usually is dependent upon the co-administration of appropriate adjuvants. There are number of substances of bacterial origin, which have been investigated for their mucosal adjuvanticity. The three bacterial products with promising potential to function as mucosal adjuvants are the ADP-ribosylating enterotoxins [cholera toxin (CT) produced by various strains of *Vibrio cholerae* and the heat labile enterotoxin (LT) produced by enterotoxigenic E. *coli*, the synthetic oligodeoxynucleotides containing unmethylated CpG dinucleotides and monophosphoryl lipid A (MPL) (Freytag and Clements, 2004).

LT and CT elevate intracellular cyclic adenosyl monophosphate (cAMP) levels, which is in part related to their adjuvanticity function. MPL and CpG act through myeloid differentiation factor 88 (MyD88) dependent and independent pathways to elicit their effects on innate and adaptive immune responses. CT induces antigen specific $CD4^+Th2$ cells secreting cytokines that provide signals for the induction of antigen specific secretory IgA as well as serum IgG1, IgA and IgE responses in mouse models. In contrast to CT, LT induces mixed $CD4^+Th1$ and Th2 type cells with subsequent secretory IgA as well as serum IgG1, IgG2A and IgA responses. Thus LT seems to be a preferred mucosal adjuvant as it elicits a mixed cytokine response and lower levels of antigen specific IgE. LT and CT are both heteromultimeric proteins consisting of monomeric A subunit with ADP ribosylating activity and a pentameric B subunit which binds to the GM1 gangliosides. CT binds mostly to the ganglioside GM1 whereas LT binds not only GM1 but also other glycosphingolipids, glycoprotein receptors present in the intestine of rabbits and humans, polyglycosilceramides and paragloboside.

Although LT is a promising mucosal adjuvant, the toxicity of A-subunit limited its use both as an immunogen and adjuvant. Several groups reported the expression of LT-B in recombinant plant systems as a step towards developing vaccine for traveler's diarrhea (Haq et al., 1995, Mason et al., 1998, Lauterslager et al., 2001, Chikwamba et al., 2002, Streatfield et al., 2002). However, holotoxin (LT) is more effective both as an immunogen and adjuvant (Rappuoli et al., 1999). Holotoxoids, which are complete knockouts of enzymatic activity, have no toxicity *in vitro* or *in vivo*. One such mutant is LTK63, which is identical to wild type LT with the exception of active site wherein the bulky side chain of lysine 63 fills the active site (Van den Akker et al., 1997). LTK63 is an excellent mucosal adjuvant and its adjuvanticity has been demonstrated using a wide range of antigens including model antigens like ovalbumin and protective antigen from bacterial and viral pathogens. LTK63 is consistently a better immunogen than LT-B (Douce et al., 1998). Recently, Kang et al. (2004) reported the expression of LTK63 operon in tobacco chloroplasts. However, there are no reports on the expression of LTK63 in transgenic plants through nuclear transformation. Currently, our group is working towards production of LTK63 in banana and other recombinant plant systems.

Although, vaccine and adjuvant production in plants is in research and developmental stage, more progress has been made in the commercialization of industrial proteins and certain therapeutic proteins and monoclonal antibodies are close to commercialization.

27. COMMERCIALIZED PRODUCTS

27.1. Avidin

Avidin is a glycoprotein found in avian, reptilian and amphibian egg white and is used as a diagnostic reagent. The commercial source of avidin is chicken egg white, but it is relatively expensive due to the cost of maintaining live animals. Hood et al. (1997) reported the production of avidin in transgenic corn using a synthetic gene optimized for expression in corn. Expression levels of over 20% TSP has been achieved (Masarik et al., 2003). Processing methodologies to purify avidin from transgenic corn were developed (Kusnadi et al., 1998). This product is currently being sold by Sigma–Aldrich, USA.

28. β-GLUCURONIDASE

GUS (β-Glucuronidase) is a homotetrameric hydrolase that cleaves beta linked terminal glucuronic acid in mono and oligo-saccharides and phenols. GUS is widely used as a visual marker in transgenic plant research. It was first reported to be produced commercially in transgenic corn (Kusnadi et al., 1998, Witcher et al., 1998). Evangelista et al. (1998) carried out economic evaluation of the extraction and purification of corn derived GUS. Corn derived GUS is marketed by Sigma-Aldrich.

28.1. Trypsin

Trypsin is a protease used in a variety of commercial applications including the processing of some biopharmaceuticals with a significant market potential. Although it has been expressed in a variety of recombinant systems, none of these systems has been demonstrated to be commercially viable on a large scale. Expression of this protease at commercially viable scale in corn was possible by expressing this enzyme in an inactive zymogen form (Woodard et al., 2003). This will have considerable effect to eliminate animal source materials and reduce the fears of contamination of products by mammalian viruses and prions. Corn derived trypsin happens to be the first large-scale protein product from transgenic plant technology. This is also marketed by Sigma.

29. PRODUCTS NEAR TO COMMERCIALIZATION

Besides the three products described above, there are nine more products thought to be close to reach commercial market. These proteins are normally derived from animal organs and due to the possible risk of contamination with animal pathogens; there is a need for an alternative low cost supply of these products.

29.1. Aprotinin

Aprotinin is a serine protease inhibitor. This activity is known to modulate and lessen the systemic inflammatory response associated with cardio-pulmonary bypass surgery, which reduces the need for blood transfusions and decreases bleeding. Aprotinin is extracted from bovine lungs. With a growing concern regarding prions and other animal pathogens there is a need for alternative production source. Zhong et al. (1999) reported the expression of recombinant bovine aprotinin in corn seeds at levels of 0.068% TSP, but levels as high as 8.9% TSP were obtained by Delaney et al. (2002) using seed preferred promoter and maize codon optimized bovine aprotinin gene.

29.2. Collagen

Collagen is a structural protein currently derived from hooves and connective tissue of animals. Large quantity of this protein is consumed throughout the world in the form of gelatin. Collagen and gelatin are commonly used biomaterials in the medical, pharmaceutical and cosmetic industries. The first report of human collagen production in plants was by Ruggiero et al. (2000). The resultant protein is organized into a triple helix. Currently several companies have expressed their interest in bringing human collagen to market.

30. HUMAN GASTRIC LIPASE

Patients suffering from exocrine pancreatic insufficiency do not produce gastric lipase, which results in the inability in digestion of food lipids. The current supply of gastric lipase is from porcine pancreatic tissue. Meristem therapeutics, a France based company is advancing a corn derived mammalian lipase to clinical trials.

Gruber et al. (2001) reported the production of canine gastric lipase in transgenic tobacco.

31. HUMAN LACTOFERRIN

Human lactoferrin is a natural defense iron binding milk protein, which possess anti-bacterial, anti-viral, anti-fungal, anti-neoplastic and anti-inflammatory properties (Samyn-Petit et al., 2001). Recombinant human lactoferrin is produced in *Aspergillus* sp. for pharmaceutical purposes (Conneely et al., 2001). It was also expressed in tobacco cell suspension cultures by Mitra and Zhang (1994). Ventria Bioscience and Meristem therapeutic are attempting to commercialize this protein from rice and corn seeds respectively (Nandi et al., 2002, Legrand et al., 2003).

32. MONOCLONAL ANTIBODIES

These proteins are one of the important biopharmaceuticals in great demand and required in large quantities for therapeutic purposes. One of the first plant-derived monoclonal antibodies (mAb) expected to reach market is directed against dental caries designated as CaroRx. This secretory IgA inhibits the binding of oral pathogen *Streptococcus mutans* to teeth. The plant-derived mAb is extremely effective in reducing colonization of this bacteria using passive immunization and even prevented re-colonization for up to two years (Larrick et al., 2001). Phase I and Phase II clinical trials have been completed. Another mAb product expected to reach the market is one directed against genital herpes. Zeitlin et al. (1998) reported the production of anti-herpes humanized mAb in soybean and compared this with the mAb produced in mammalian cells in a mouse model. The two mAbs protected the mice against herpes simplex virus-2 equally. A USA based company Epicyte Pharmaceuticals produces this potential product in corn.

33. COST OF PRODUCTION OF RECOMBINANT PROTEINS IN PLANTS

Transgenic plants are considered as attractive expression systems as the cost of recombinant protein production and especially antibodies are much lower than the other available expression systems. The cost of antibody production in mammalian cell cultures range from \$106 to \$650 per gram, whereas the production cost in transgenic plants is in the range of \$0.1 to \$1 per gram depending on the expression levels of the protein and the tissue from which it is processed (Hood et al., 2002). The cost of producing an IgG from alfalfa grown in a 250 m^2 green house was estimated to be \$500 to 600 per gram compared to \$5000 per gram for the hybridoma produced antibody (Khoudi et al., 1999). Planet Biotechnology a USA based company compared the cost per gram of purified IgA by mammalian cell culture, transgenic goats and transgenic plants and found that the costs of production were below \$50 per gram when IgA was expressed in leaves compared to \$1000 using mammalian cell cultures or \$100 using transgenic animals (Daniell et al., 2001). Thus production of monoclonal antibodies from plants represents a nearly thousand-fold reduction in costs over manufacturing from mammalian cell culture systems (Molowa et al., 2002). Another example of cost effective production of recombinant protein is the production of avidin in transgenic corn. We need a ton of fresh eggs to produce 20 g avidin, which could roughly cost \$1000 whereas with the current expression levels of avidin only four bushel of corn can produce the same amount of avidin and costs about \$20 in total. It is one of the first proteins in Sigma catalogue that comes from transgenic plants.

34. CONCLUSIONS

Plant based molecular farming opens new opportunities to explore the rapid scalability, low cost manufacturing and the safer production of recombinant proteins on a large scale. The commercial success of plant derived recombinant proteins, which are available from sigma at a lower cost than the native proteins, is an indicative of the potential of this technology.

The synthesis and assembly of multimeric proteins including antibodies, vaccines and other therapeutic proteins produced in plants and their biological activity is a step towards initial success. Several other plant derived recombinant proteins are in the various stages of development for commercial applications. However, issues like low expression levels, improving glycosylation pattern, biosafety and acceptability need to be addressed. Recent success with the moss system holds promise to produce antibodies with glycosylation pattern similar to that of humans. Plastid expression technology may enable high levels of expressions to reach the industrial scale of production. Appropriate selection of hosts may also play an important role in developing this technology, which again depends on the type of product and its end use.

In vitro plant systems like plant cell suspension cultures and hairy roots should be further investigated to develop these systems for large-scale production of therapeutics under controlled conditions. These systems will provide containment and enable the establishment of cGMP and cGLP guidelines to produce products of similar consistency with precision. Microalgae systems have opened up new opportunities for molecular farming in closed conditions. Seeds offer advantages of long-term storage of proteins without the loss of biological activity. Fruits and vegetables enable production of edible vaccines for human and veterinary use.

Vaccination plays a very important role in the management of infectious diseases, as it is the safest and most economical way to control diseases. Vaccine and adjuvant production in edible tissues of the plant may offer economical production and formulation of oral vaccines to be used in mass immunization programs in the developing world. Considerable success has been achieved in the production of hepatitis B vaccine in plant systems. Further, immunogenicity studies have been carried out in mice and humans to show the efficacy of the plant derived vaccine.

Viral antigens require proper folding for their immunogenicity. ER retention seems to be one of the ways to achieve proper folding and assembly of multimeric proteins. Metabolic engineering of the plant to carry out special post translational modifications for certain therapeutic proteins such as sialyation of erythropoetin are the challenges to be addressed. Use of appropriate regulatory elements and promoters also contribute to the levels of expression. Gene silencing is one of the factors to be considered for stable and continuous production of recombinant protein in transgenic plants. Selection of lines with fewer copies of transgene, which can be obtained by *Agrobacterium* mediated transformation, use of matrix attachment regions and plant cell cultures, which are hemizygous and certain gene silencing triggers only in homozygous condition are some of the strategies to overcome transgene silencing. Post harvest inducible expression is an attractive tool to minimize the effects of transgene expression on the host plant's growth and development. Genetic background may also influence the levels of expression and may sometimes require breeding methods to optimize the recombinant protein yield. Another important factor is the costs involved in downstream processing. More emphasis is required to develop new technologies to minimize the cost of downstream processing. Established technologies include the oleosin partitioning and the use of affinity tags.

The costs of licensing the technology for genetic transformation; use of expression vectors and purification strategies from its patent assignees can contribute to the major costs, which warrants the need to develop own proprietary transformation and expression systems to make the products less expensive and more attractive. Regulatory guidelines for plant molecular farming need to be formulated. USDA has imposed stricter guidelines to avoid food products with traces of pharmaceutical proteins, as there were incidences of inadvertent contamination of food crops with plant derived pharmaceutical proteins (Watson et al., 2004). Non-food and non-feed crops like tobacco and other crops may be an alternative to overcome regulatory issues. Also the vegetatively propagated crops such as edible bananas may be suitable candidates to prevent the horizontal gene flow. Several gene containment methods are currently being explored, these include apomixis, incompatible genomes, control of seed dormancy or shattering, transgene mitigation, suicide genes, infertility barriers, male sterility and maternal inheritance. Plastid engineering is another approach to contain transgenes effectively with a few exceptions. An alternative strategy was to destroy the tapetum selectively during anther development by expressing RNAse gene and resulting in the development of male sterile plants (Daniell et al., 2001). Another concern expressed would be the use of antibiotic resistance genes for selection of transgenic plants especially in food crops. However, several methods are now available to excise the antibiotic resistance genes after selection or the use of positive selection methods.

With the initial success in the development of transgenic plants and the advances made in the area of molecular farming, several companies have started production of biopharmaceuticals in plant systems (Table 2). Venture capitalists are investing huge

Table 2. Host systems for biopharmaceutical production

Crop species	Recombinant protein/s expressed	Company
Alfalfa	Recombinant antibodies	Medicago, Canada
Corn	Industrial enzymes, vaccines, Recombinant antibodies	Largescale Biology Corporation, USA. Meristem Therapeutics, France.
Rice	Lactoferrin, Lysozyme	Ventria Biosystems Inc., USA.
Barley	Diagnostic antibody, Cellulase, human serum albumin, lactoferrin, lysozyme, thaumatin, human growth factor	Ventria Biosystems Inc., USA. ORF Genetics, Iceland. Maltagen, Germany.
Safflower/Rapeseed	Hirudin, Insulin	SemBioSys Genetics Inc., Canada.
Tobacco		
Moss	Antibodies, Vascular endothelial growth factor (VEGF)	Greenovation Biotech Gmbh, Germany.
Chlamydomonas	Monoclonal antibodies, Vaccines and Bioremediant delivery systems	Rincopharma, USA. Phycotransgenics, USA.

amounts of money in this sector. To cite an example, recently Paperboy Ventures, a Washington D.C. based private equity firm announced the $4.7 million investment in Rincon Pharmaceuticals, Inc. to develop recombinant protein therapeutics, such as monoclonal antibodies, by using eukaryotic algae. Plant molecular farming may emerge as a prospective technology for recombinant protein production especially, therapeutics/antibodies with the recent and on-going developments in this active field.

REFERENCES

Bakker H, Bardor M, Molthoff JW et al. (2001) Galactose-extended glycans of antibodies produced by transgenic plants, *Proc Natl Acad Sci USA*. 98: 2899–2904

Bardor M, Faveeuw C, Fitchette AC et al. (2003a) Immunoreactivity in mammals of two typical plant glyco-epitopes, core alpha 1,3-fucose and core xylose, *Glycobiology* 13: 427–434

Bardor M, Loutelier-Bourhis C, Paccalet T et al. (2003b) Monoclonal C5-1 antibody produced in transgenic alfalfa plants exhibits a N-glycosylation that is suitable for glyco-engineering into human-compatible structures. *Plant Biotechnol J* 1: 451–462

Barta A, Sommergruber K, Thomson D et al. (1986) The expression of a nopaline synthase- human growth hormone chimeric gene in transformed tobacco and sunflower callus tissue. *Plant Mol Biol* 6: 347–357

Baur A, Kaufmann F, Rolli H et al. (2005a) A fast and flexible PEG-mediated transient expression system in plants for high level expression of secreted recombinant proteins. *J Biotechnol* 119: 332–342

Baur A, Reski R, Gorr G, (2005b) Enhanced recovery of a secreted recombinant human growth factor using stabilizing additives and by co-expression of human serum albumin in the moss *Physcomitrella patens*. *Plant Biotechnol J* 3: 331–340

Bendich AJ (1987) Why do chloroplast and mitochondria contain so many copies of their genome?. *BioEssays* 6: 279–282

Boothe JG, Parmenter DL, Saponja JA (1997) Molecular farming in plants: oilseeds as vehicles for the production of pharmaceutical proteins. *Drug Develop Res* 42: 172–181

Borisjuk NV, Borisjuk LG, Logendra S et al. (1999) Production of recombinant proteins in plant root exudates. *Nature Biotech* 17: 466–469

Canizares MC, Nicholson L, Lomonossoff GP (2005) Use of viral vectors for vaccine production in plants. *Immunol Cell Biol* 83: 263–270

Chikwamba R, Cunnick J, Hathaway D et al. (2002) A functional antigen in a practical crop: LT-B producing maize protects mice against *Escherichia coli* heat labile enterotoxin (LT) and cholera toxin (CT). *Transgenic Res* 11: 479–493

Clendennen SK, Lopez-Gomez R, Gomez-Lim M et al. (1998) The abundant 31-kilodalton banana pulp protein is homologous to class III acidic chitinases. *Phytochemistry* 47: 613–619

Conneely OM, Headon DR, O'Malley BW et al. (2001) Production of recombinant lactoferrin and lactoferrin polypeptides using cDNA sequences in various organisms, US Patent 6, 228, 614, 2001

Cramer CL, Boothe JG, Oishi KK (1999) Transgenic plants for therapeutic proteins: Linking up stream and down stream strategies, In: Hammond J, McGarvey P, Yusibov V, (ed) Current topics Microbiology and Immunology, Plant biotechnology : New products and applications. Springer-Verlag, Berlin, p 95–118

Cramer CL, Weissenborn DL, (1997) HMG2 promoter expression system and post harvest production of gene products in plants and plant cell cultures. US patent No. 5,670, 349, 1997

Cramer CL, Weissenborn DL, Oishi KK et al. (1996) High level production of enzymatically active human lysosomal proteins in transgenic plants. In: Owen MRL, Pen J, (ed) Transgenic plants : A production system for industrial and pharmaceutical protein. Wiley, Chichester, p 299–310

Daniell H, Guda C, McPherson DT et al. (1996) Hyper expression of a synthetic protein-based polymer gene. *Meth Mol Biol* 63: 359–371

Daniell H, Khan MS, Allison L (2002) Milestones in chloroplast genetic engineering: an environmentally friendly era in biotechnology. *Trends Plant Sci* 7: 84–91

Daniell H, Streatfield SJ, Wycoff K, (2001) Medical molecular farming : production of antibodies, biopharmaceuticals and edible vaccines in plants. *TIPS* 6: 219–226

DeCosa B, Moar W, Lee SB et al. (2001) Hyper-expression of the Bt Cry-2 Aa2 operon in chloroplast leads to formation of insecticidal crystals. *Nature Biotechnol* 19: 71–74

Delaney DE (2002) Choice of crop species and development of transgenic product lines. In: Hood EE, Howard JA, (ed) Plants as factories for protein production. Kluwer, Dordrecht, p 139–158

Delaney D, Jilka J, Barker D et al. (2002) Production of aprotinin in transgenic maize seeds for the pharmaceutical and cell culture markets. In: Vasil IK, (ed) Plant biotechnology 2002 and beyond. Kluwer, Dordrecht, p 393–394

De Zoeten GA, Penwick JR, Horisberger MA et al. (1989) The expression, localization and effect of a human interferon in plants. *Virology* 172: 213–222

Dieryck W, Pagnier J, Poyart C et al. (1997) Human hemoglobin from transgenic tobacco. *Nature* 386: 29–30

Dogan B, Mason HS, Richter L et al. (2000) Process options in hepatitis B surface antigen extraction from transgenic potato. *Biotechnol Prog* 16: 435–441

Douce G, Giuliani MM, Giannelli V et al. (1998) Mucosal immunogenicity of genetically detoxified derivatives of heat labile toxin from *Escherichia coli*. *Vaccine* 16: 1065–1073

During K (1988) Wound-inducible expression and secretion of T4 lysozyme and monoclonal antibodies in Nicotiana tabacum. Ph.D Thesis. Mathematish-Naturwissenschaftlichen Fakultaet der Universitaet Zu Koeln

Edelbaum O, Stein D, Holland N et al. (1992) Expression of active human interferon in transgenic plants. *J Interferon Res* 12: 449–453

Elliott S, Lorenzini T, Asher S et al. (2003) Enhancement of therapeutic protein *in vivo* activities through glycoengineering. *Nature Biotechnol* 21: 414–421

Evangelista RL, Kusnadi AR, Howard JA et al. (1998) Process and economic evaluation of the extraction and purification of recombinant β-glucuronidase from transgenic corn. *Biotechnol Prog* 14: 607–614

Fiedler U, Conrad U (1995) High-level production and long term storage of engineered antibodies in transgenic tobacco seeds. *Biotechnology* 13: 1090–1093

Fiedler U, Phillips J, Artsaenko O et al. (1997) Optimization of ScFv production in transgenic plants. *Immunotechnology* 3: 205–216

Fischer R, Emans N (2000) Molecular farming of pharmaceutical proteins. *Transgenic Res* 9: 279–299

Fischoff DA (1992) Field performance of insect resistant crops. *J Cell Biochem Suppl* 16F: 199

Fraley RT, Rogers SG, Horsch RB et al. (1983) Expression of bacterial genes in plant cells. *Proc Natl Acad Sci USA* 80: 4803–4807

Franken E, Teuschel U, Hain R (1997) Recombinant proteins from transgenic plants. *Curr Opin Biotechnol* 8: 411–416

Franklin SE, Mayfield SP (2004) Prospects for molecular farming in the green alga *Chlamydomonas reinhardtii*. *Curr Opin Plant Biol* 7: 159–165

Freytag LC, Clements JD (2004) Mucosal adjuvants. *Vaccine* 23: 1804–1813

Ganz PR, Dudani AK, Tackaberry ES et al. (1996) Expression of human blood proteins in transgenic plants: The cytokine GM-CSF as a model protein.In: Owen MRL, Pen J (ed) Transgenic plants: A production system for industrial and pharmaceutical proteins. John Wiley and Sons, London, UK, p281–297

Gaume A, Komarnytsky S, Borisjuk N et al. (2003) Rhizosecretion of recombinant proteins from plant hairy roots. *Plant Cell Rep* 21: 1188–1193

Gleba Y, Klimyuk V, Marillonnet S (2005) Magnifection – a new platform for expressing recombinant vaccines in plants. *Vaccine* 23: 2042–2048

Gomord V, Chamberlain P, Jefferis R et al. (2005) Biopharmaceuticals production in plants : problems, solutions and opportunities. *TIBTECH* 23: 559–565

Gomord V, Denmat LA, Fitchette-Laine AC et al. (1997) The C-terminal HDEL sequence is sufficient for retention of secretory proteins in the endoplasmic reticulum(ER) but promotes vacuolar targeting of proteins that escape the ER. *Plant J* 11: 313–325

Gomord V, Sourrouille C, Fitchette AC et al. (2004) Production and glycosylation of Plant-made pharmaceuticals: the antibodies as a challenge. *Plant Biotechnol J* 2: 83–100

Gorr G, Jost W (2005) Glycosylation design in transgenic moss for better product efficacy. *BioProcessing J* 4:26–30

Grevich JJ, Daniell H (2005) Chloroplast genetic engineering: Recent advances and future perspectives. *Critical Rev Plant Sci* 24: 83–107

Gruber V, Berna PP, Arnaud T et al. (2001) Large-scale production of a therapeutic protein in transgenic tobacco plants: effect of subcellular targeting on quality of a recombinant dog gastric lipase. *Mol Breed* 7: 329–340

Hamamoto H, Sugiyama Y, Nakagawa N et al. (1993) A new tobacco mosaic virus vector and its use for the systemic production of angiotensin –1 converting enzyme inhibitor in transgenic tobacco and tomato. *Biotechnology* 11: 930–932

Haq TA, Mason HS, Clements JD et al. (1995) Oral immunization with a recombinant bacterial antigen produced in transgenic plants. *Science* 268: 714–716

Herbers K, Wilke I, Sonnewald U (1995) A thermostable xylanase from *Closteridium thermocellum* expressed at high levels in the apoplast of transgenic tobacco has no detrimental effects and is easily purified. *Biotechnology* 13: 63–66

Hiatt A, Cafferkey R, Bowdish K (1989) Production of antibodies in transgenic plants. *Nature* 342: 76–78

Higo K, Saito Y, Higo H, (1993) Expression of a chemically synthesized gene for human epidermal growth factor under the control of cauliflower mosaic virus 35S promoter in transgenic tobacco. *Biosci Biotech Biochem* 57: 1477–1481

Hood EE, Witcher DR, Maddock S et al. (1997) Commercial production of avidin from transgenic maize: characterization of transformant, production, processing, extraction and purification. *Mol Breed* 3: 291–306

Hood EE, Woodard SL, Horn ME (2002) Monoclonal antibody manufacturing in transgenic plants–myths and realities. *Curr Opin Biotechnol* 13: 630–635

Horn ME, Pappu KM, Bailey MR et al. (2004) Advantageous features of plant based systems for the development of HIV vaccines. *J Drug Targeting* 11: 539–545

Huang Z, Santi L, LePore K et al. (2006) Rapid, high-level production of hepatitis B core antigen in plant leaf and its immunogenicity in mice. *Vaccine* 24: 2506–2513

Huether CM, Lienhart O, Baur A et al. (2005) Glyco-engineering of moss lacking plant-specific sugar residues. *Plant Biol* 7: 292–299

Imani J, Berting A, Nitsche S et al. (2002) The integration of a major hepatitis B virus gene into cell-cycle synchronized carrot cell suspension cultures and its expression in regenerated carrot plants. *Plant Cell Tissue Org Cult* 71: 157–164

James EA, Wang C, Wang Z et al. (2000) Production and characterization of biologically active human GM-CSF secreted by genetically modified plant cells. *Prot Expr Purif* 19: 131–138

Jefferson RA, Kavanaugh TA, Bevan MW (1987) GUS fusion: β-glucuronidase as a sensitive and versatile gene fusion marker in higher plants. *EMBO J* 6: 3901–3907

Jilka J (1999) Have your vaccine and eat it too. In: Global Agbiotech progress Proceedings of the AgBiotech World Forum '99, Global AG BioTech Progress, Drug and Market Development Publications, p 89–95

Jost W, Link S, Horstmann V et al. (2005) Isolation and characterization of three moss-derived beta-tubulin promoters suitable for recombinant expression. *Curr Genet* 47: 111–120

Joung YH, Youm JW, Jeon JH (2004), Expression of the hepatitis B surface S and pre S2 antigens in tubers of *Solanum tuberosum. Plant Cell Rep* 22: 925–930

Kang TJ, Han SC, Kim MY et al. (2004) Expression of non-toxic mutant of *Escherichia coli* heat-labile enterotoxin in tobacco chloroplasts. *Protein Expr Purif* 38: 123–128

Kapila J, De Rycke R, van Montagu M et al. (1996) An *Agrobacterium* mediated transient expression system for intact leaves. *Plant Sci* 122: 101–108

Kapusta J, Modelska A, Figlerowicz M et al. (1999) A plant derived edible vaccine against hepatitis B virus. *FASEB J* 13: 1796–1799

Khoudi H, Laberge S, Ferullo JM et al. (1999) Production of a diagnostic monoclonal antibody in perennial alfalfa plants. *Biotechnol. Bioeng* 64: 135–143

Ko K, Tekoah Y, Rudd PM et al. (2003) Function and glycosylation of plant-derived antiviral monoclonal antibody. *Proc Natl Acad Sci USA* 10: 8013–8018

Komarnytsky S, Borisjuk NV, Borisjuk LG et al. (2000) Production of recombinant proteins in tobacco guttation fluid. *Plant Physiol* 124: 927–933

Kong Q, Richter L, Yang YF et al. (2001) Oral immunization with hepatitis B surface antigen expressed in transgenic plants. *Proc Natl Acad Sci USA* 98: 11539–11544

Koprivova A, Altmann F, Gorr G et al. (2003) N-glycosylation in the moss *Physcomitrella patens* is organized similarly to higher plants. *Plant Biol* 5: 582–591

Kumagai MH, Turpen TH, Weinzettl N et al. (1993) Rapid, high level expression of biologically active alpha-trichosanthin in transfected plant by an RNA viral vector. *Proc Natl Acad Sci USA* 90: 427–430

Kusnadi AR, Hood EE, Witcher DR et al. (1998) Production and purification of two recombinant proteins from transgenic corn. *Biotech Prog* 14: 149–155

Kusnadi AR, Niklov ZL, Howard JA (1997) Production of recombinant proteins in transgenic plants: Practical considerations. *Biotech Bioeng* 56: 473–484

Larrick JW, Yu L, Naftzger C et al. (2001) Production of secretory IgA antibodies in plants. *Biomol Engg* 18: 87–94

Lauterslager TGM, Florack DEA, van der Wal TJ et al. (2001) Oral immunization of naïve and primed animals with transgenic potato tubers expressing LT-B. *Vaccine* 19: 2749–2755

Legrand D, Salmon V, Spoik G et al. (2003) Recombinant lactoferrin, methods of production from plants and uses. US Patent 6 569 831

Magnuson NS, Linzmaier PM, Reeves R et al. (1998) Secretion of biologically active human interleukin-2 and interleukin-4 from genetically modified tobacco cells in suspension culture. *Prot Exp Purif* 13: 45–52

Masarik M, Kizek R, Kramer KJ et al. (2003) Application of avidin-biotin technology and adsorptive transfer tripping square-wave voltammetry for detection of DNA hybridization and avidin maize. *Anal Chem* 75: 2663–2669

Mason HS, Haq TA, Clements JD et al. (1998) Edible vaccine protects mice against *Escherichia coli* heat-labile enterotoxine (LT): potatoes expressing a synthetic LT-B gene. *Vaccine* 16: 1336–1343

Mason HS, Lam DMK, Arntzen CJ (1992) Expression of hepatitis B surface antigen in transgenic plants. *Proc Natl Acad Sci USA* 89: 11745–11749

Matsumoto S, Ikura K, Ueda M et al. (1995) Characterization of a human glycoprotein (erythropoietin) produced in cultured tobacco cells. *Pant Mol Biol* 27: 1163–1172

Mitchell VS, Philipose MM, Sanford JP (ed) (1993) The children's vaccine initiative: achieving the vision, *National Academy Press.* p23

Mitra A, Zhang Z (1994) Expression of a human lactoferrin cDNA in tobacco cells produces antibacterial protein(s). *Plant Physiol* 106: 997–981.

Molowa D, Schenouda M, Meyers A et al. (2002) The state of biologics manufacturing part II J.P. Morgan Market analysis of the state of manufacturing of biologics including mAbs. p 1–16

Muntz K (1998) Deposition of storage proteins. Deposition of storage proteins, *Plant Mol Biol* 38: 77–99

Nandi S, Suzuki YA, Huang J et al. (2002) Expression of human lactoferrin in transgenic rice grains for the application in infant formula. *Plant Sci* 163: 713–722

Newman T, Ohme-Takagi M, Taylor C et al. (1993) DST sequences, highly conserved among plant SAUR genes, target reporter transcripts for rapid decay in tobacco. *Plant Cell* 5: 701–714

Nikolov Z, Hammes D (2002) Production of recombinant proteins from transgenic crops, In: Hood EE, Howard J (ed) Plants as Factories for Protein Production. Kluwer Academic Publishers, Dordrecht, p 159–174

Palacpac NQ, Yoshida S, Sakai H et al. (1999) Stable expression of human 1,4-galactosyltransferase in plant cells modifies N-linked glycosylation patterns. *Proc Natl Acad Sci USA* 96: 4692–4697

Pen J, Sijmons PC, van Ooijen AJJ et al. (1993) Protein production in transgenic crops: analysis of plant molecular farming. In: Hiatt A (ed) Transgenic plants fundamentals and applications. Marcell Dekker, NY, p 239–254

Peng J, Song Q, Chengkui T (2002) Expression of hepatitis B surface antigen gene (HBsAg) in *Laminaria japonica* (Laminariales, Phaeophyta). *Chinese Sci Bullt* 47: 1438–1440

Peumans WJ, Proost P, Swennen RL et al. (2002) The abundant class III chitinase homolog in young developing banana fruits behaves as a transient vegetative storage protein and most probably serves as an important supply of aminoacids for the synthesis of ripening-associated proteins. *Plant Physiol* 130: 1063–1072

Rappuoli R, Pizza M, Douce G et al. (1999) Structure and mucosal adjuvanticity of cholera and *Escherichia coli* heat-labile enterotoxins. *Immunol Today* 20: 493–500

Richter L, Kipp PB (1999) Transgenic plants as edible vaccines. In: Hammond J, McGarvey P, Yusibov V (ed) Plant biotechnology-new products and applications. Springer-Verlag, New York, p159–176

Richter LJ, Thanavala Y, Arntzen CJ et al. (2000) Production of hepatitis B surface antigen in transgenic plants for oral immunization. *Nature Biotechnol* 18: 1167–1171

Ruggiero F, Exposito JY, Bournat P et al. (2000) Triple helix assembly and processing of human collagen produced in transgenic tobacco plants. *FEBS Lett* 469: 132–136

Rymerson RT, Menassa R, Brandle JE (2002) Tobacco, a platform for the production of recombinant proteins. In: Erickson L, Yu WJ, Brandle J, Rymerson R (ed) Molecular farming of plants and animals for human and veterinary medicine. Kluwer Academic Publishers, Netherlands, p1–31

Salmon V, Legrand D, Slomianny MC et al. (1998) Production of human lactoferrin in transgenic tobacco plants. *Prot Expr Purif* 13: 127–135

Samyn-Petit B, Gruber V, Flahaut C et al. (2001) N-Glycosylation potential of maize: the human lactoferrin used as a model. *Glycoconjugate J* 18: 519–527

Sandhu JS, Osadjan MD, Krasnyanski SF et al. (1999) Enhanced expression of the human respiratory syncytial virus: F gene in apple leaf protoplast. *Plant Cell Rep* 18: 394–397

Santi L, Giritch A, Roy CJ et al. (2006) Protection conferred by recombinant *Yersinia pestis* antigen produced by a rapid and highly scalable plant expression system. *Proc Natl Acad Sci USA* 103: 861–866

Schouten A, Roosien J, Van Engelen FA et al. (1996) The C-terminal KDEL sequence increases the expression level of a single-chain antibody designed to be targeted to both the cytosol and the secretory pathway in transgenic tobacco. *Plant Mol Biol* 30: 781–793

Shanks JV, Morgan J (1999) Plant 'hairy root' culture. *Curr Opin Biotechnol* 10: 151–155

Sijmons PC, Dekker BMM, Schrammeijer B et al. (1990) Production of correctly processed human serum albumin in transgenic plants. *Biotechnology* 8: 217–221

Smirnov SP, Teveroskaia EKH, Krasheninnikova LV et al. (1990) Design of an expression integrative vector and its application for introducing the human recombinant -interferon gene into plants. *Genetika* 26: 2111–2121

Smith ML, Keegan ME, Mason HS et al. (2002a) Factors Important in the Extraction, Stability and in vitro assembly of the hepatitis B surface antigen derived from recombinant plant systems. *Biotechnol Prog* 18: 538–550

Smith ML, Mason HS, Schuler ML (2002b) Hepatitis B surface antigen (HBsAg) expression in plant cell culture: kinetics of antigen accumulation in batch culture and its intracellular form. *Biotech Bioeng* 80: 812–822

Smith ML, Richter L, Arntzen CJ et al. (2003) Structural characterization of plant derived hepatitis B surface antigen employed in oral immunization studies. *Vaccine* 21: 4011–4021

Sojikul P, Buehner N, Mason HS (2003) A plant signal peptide- hepatitis B surface antigen fusion protein with enhanced stability and immunogenicity expressed in plant cells. *Proc Nat Acd Sci USA* 100: 2209–2214

Sriraman R, Bardor M, Sack M et al. (2004) Recombinant anti-hCG antibodies retained in the endoplasmic reticulum of transformed plants lack core-xylose and core- α(1,3)-fucose residues. *Plant Biotechnol J* 2: 279–288

Staub JM, Garcia B, Graves J et al. (2000) High yield production of a human therapeutic protein in tobacco chloroplasts. *Nat Biotechnol* 18: 333–338

Stevens LH, Stoopen GM, Elbers IJ et al. (2000) Effect of climate conditions and plant developmental stage on the stability of antibodies expressed in transgenic tobacco. *Plant Physiol* 124: 173–182

Stoger E, Ma JK, Fischer R et al. (2005) Sowing the seeds of success: pharamaceutical proteins from plants. *Curr Opin Biotechnol* 16: 1–7

Streatfeld SJ, Meyer JM, Barker DK et al. (2002) Development of an edible subunit vaccine in corn against enterotoxigenic strains of *Escherichia coli*. In *Vitro Cell Dev Biol Plant* 38: 11–17

Streatfield SJ (2005) Oral hepatitis B vaccine candidates produced and delivered in plant material. *Immunol Cell Biol* 83: 257–262

Sunil Kumar GB, Ganapathi TR, Bapat VA, (2004) Edible vaccines : Current Status and Future Prospects. *Physiol Mol Biol Plants* 10: 37–47

Sunil Kumar GB, Ganapathi TR, Revathi CJ et al. (2003) Expression of hepatitis B surface antigen in tobacco cell suspension cultures. *Prot. Expr. Purif.* 32: 10–17

Sunil Kumar GB, Ganapathi TR, Revathi CJ et al. (2005b) Expression of hepatitis B surface antigen in transgenic banana plants. *Planta* 222: 484–493

Sunil Kumar GB, Ganapathi TR, Srinivas L et al. (2005a) Secretion of hepatitis B surface antigen in transformed tobacco cell suspension cultures. *Biotechnol Lett* 27: 927–932

Sunil Kumar GB, Ganapathi TR, Srinivas L et al. (2006b) Expression of hepatitis B surface antigen in potato hairy roots. *Plant Sci* 170: 918–925

Sunil Kumar GB, Srinivas L, Ganapathi TR et al. (2006a) Hepatitis B surface antigen expression in transgenic tobacco (*Nicotiana tabacum*) plants using four different expression cassettes. *Plant Cell Tissue Org Cult* 84: 315–323

Sutton DW, Havstad PK, Kemp JD (1992) Synthetic cryIIIA gene from Bacillus thuringiensis improved for high expression in plants. *Transgenic Res* 1: 228–236

Tackaberry ES, Dudani AK, Prior F et al. (1999) Development of biopharmaceuticals in plant expression systems: Cloning, expression and immunological reactivity of cytomegalovirus glycoprotein B (UL55) in seeds of transgenic tobacco. *Vaccine* 17: 3020–3029

Takase, K, Hagiwara K (1998) Expression of human alpha-lactalbumin in transgenic tobacco. *J Biochem* 123: 440–444

Taylor C, Green P (1995) Identification and characterization of genes with unstable transcripts (GUTS) in tobacco. *Plant Mol Biol* 28: 27–38

Thanavala Y, Mahoney M, Pal S et al. (2005) Immunogenicity in humans of an edible vaccine for hepatitis B. *Proc Natl Acad Sci USA* 102: 3378–3382

Thanavala Y, Yang YF, Lyons P et al. (1995) Immunogenicity of transgenic plant-derived hepatitis B surface antigen. *Proc Natl Acad Sci USA* 92: 3358–3361

Twyman RM, Schillberg S, Fischer R (2005) Transgenic plants in the biopharmaceutical market. *Expert Opin Emerging Drugs* 10: 185–218

Twyman RM, Stoger E, Schillberg S et al. (2003) Molecular farming in plants : host systems and expression technology. *TIBTECH* 21: 570–578

Van den Akker F, Pizza M, Rappuoli R et al. (1997) Crystal structure of a non-toxic mutant of heat labile enterotoxin, which is a potent mucosal adjuvant. *Protein Sci* 6: 2650–2654

van Rooijen GJH, Moloney MM (1995) Structural requirements of oleosin domains for the subcellular targeting to the oil body. *Plant Physiol* 109: 1353–1361

Vandekerckhove J, Damme VJ, van Lijsebettens M et al. (1989) Enkephalins produced in transgenic plants using modified 2S seed storage proteins. *Biotechnology* 7: 929–932

Vaquero C, Sack M, Chandler J et al. (1999) Transient expression of a tumor specific single chain fragment and a chimeric antibody in tobacco leaves. *Proc Natl Acad Sci USA* 96: 11128–11133

von Schaewen A, Sturm A, O'Neill J et al. (1993) Isolation of a mutant *Arabidopsis* plant that lacks N-acetylglucosaminyltransferase 1 is unable to synthesize golgi modified complex N-linked glycans. *Plant Physiol* 102: 1109–1118

Wenderoth I, von Schaewen A (2000) Isolation and characterization of plant N-acetyl glucosaminyltransferase I (GntI) cDNA sequences. Functional analyses in the *Arabidopsis* cgl mutant and in antisense plants. *Plant Physiol* 123: 1097–1108

Watson J, Koya V, Leppla SH et al. (2004) Expression of *Bacillus anthracis* protective antigen in transgenic chloroplasts of tobacco, a non-food/feed crop. *Vaccine* 22: 4374–4384

WHO/Unicef/The World Bank. State of the world's vaccines and immunization 2002. Geneva: World Health Organization Department of Vaccines and Biologicals, 2002.

Witcher DR, Hood EE, Petersen D et al. (1998) Commercial production of β-glucuronidase (GUS): a model system for the production of proteins in plants. *Mol Breed* 4: 301–312

Woodard SL, Mayor JM, Bailey MR et al. (2003) Maize-derived bovine trypsin: characterization of the first large-scale commercial protein product from transgenic plants. *Biotechnol Appl Biochem* 38: 123–130

Woodleif WG, Chaplin JF, Campbell CR et al. (1981) Effect of variety and harvest treatments on protein yield of close grown tobacco. *Tob Sci* 25: 83–86

Wright A, Morrison SL (1998) Effect of C2-associated carbohydrate structure on Ig effector function: studies with chimeric mouse-human IgG1 antibodies in glycosylation mutants of Chinese hamster ovary cells. *J Immunol* 160: 3393–3402

Zeitlin L, Olmsted SS, Moench TR et al. (1998) A humanized monoclonal antibody produced in transgenic plants for immunoprotection of the vagina against genital herpes. *Nat Biotechnol* 16: 1361–1364

Zhong GY, Peterson D, Delaney DE et al. (1999) Commercial production of aprotinin in transgenic maize seeds Mol Breed 5: 345–356

CHAPTER 3

PLASTID PATHWAYS

Metabolic engineering via the chloroplast genome

TRACEY RUHLMAN AND HENRY DANIELL

Department of Molecular Biology & Microbiology, University of Central Florida, Biomolecular Science, Building #20, Room 336, Orlando, FL 32816-2364, USA

Abstract: Plant metabolic engineering has the potential to provide for the needs of an expanding population. Environmentally benign biosyntheisis of novel materials and pharmaceutical proteins along with the opportunity to improve the productivity and nutritive value of crop plants has focused considerable effort towards the genetic manipulation of crop species. The most important output traits that could be conferred through biotechnology often require the coordinated expression of several foreign genes. Conservative estimates predict some 3000 proteins are posttranslationally imported into plant plastids. Among them are the enzymes of various metabolic pathways such as those involved in the biosynthesis of the tocopherols (vitamin E) and carotenoids (vitamin A), branched chain and aromatic amino acids, and fatty acids. The ability of the chloroplast to integrate and express foreign sequences as operons makes this site an attractive alternative for genetic manipulations. Multigene engineering, high levels of recombinant protein accumulation and the security of transgene containment due to maternal inheritance of plastid genomes in most crop species are some of the features that contribute to the potential of the chloroplast system. Here we offer an overview of the fundamental characteristics of plastid protein expression and consider some possible candidate genes for the improvement of crop species through metabolic engineering of pathways compartmentalized within plastids

Keywords: chloroplast, oral delivery, genetic transformation, operon, UTR, protein expression, nutritional enhancement

To whom correspondence should be addressed. Henry Daniell, Pegasus Professor & Trustee Chair, University of Central Florida, 4000 Central Florida Blvd, Department Molecular Biology & Microbiology Biomolecular Science, Bldg # 20, Room 336, Orlando FL 32816-2364, Phone: 407-823-0952, Fax: 407-823-0956, E-mail: daniell@mail.ucf.edu

R. Verpoorte et al. (eds.), Applications of Plant Metabolic Engineering, 79–108.
© 2007 *Springer*.

1. INTRODUCTION

The current global population of 6.4 billion is expected to reach 10 billion by the year 2050. The rate of agricultural yield at present is not sufficient to meet this demand and already malnutrition and starvation are taking a toll world wide. In the past, productivity of primary producers, namely higher plants, has been accomplished through selective breeding programs but the successes in this area have reached a plateau. Bringing new acreage under cultivation is not a viable option as much of the unutilized lands in developing nations is of marginal quality and serious environmental consequences prohibit agricultural development on remaining fertile preserves.

The architects of the 'green revolution' envisioned that increased global carrying capacity would result from the development of new crop cultivars, the use of irrigation systems, and the application of chemical fertilizers and pesticides. Food production increased over 1000% from 1960 to 1990 but not without consequences in terms of production cost, dependence on chemical inputs, top soil erosion and salinization due to heavy fertilizer use and the development of pesticide-resistant species.

Nobel Peace Prize winner Norman Borlaug, considered the father of the green revolution, suggests that in biotechnology lies the potential to ameliorate environmental concerns, while meeting the rising demand for agricultural production (Borlaug, 2005). Indeed, through biotechnology many improvements have been made. Crop species have been genetically engineered to resist viral pathogens (Fitch *et al.*, 1992) and insect pests (Perlak *et al.*, 1990), tolerate drought (Shou *et al.*, 2004) and herbicide treatment (Padgette *et al.*, 1995; Chin *et al.*, 2003), and to enhance nutritional value (Goto *et al.*, 1999; Ye *et al.*, 2000; Datta *et al.*, 2003; Baisakh *et al.*, 2006) through the incorporation of novel DNA into the nuclear genome. Specific technical challenges related to the random integration of foreign DNA sequences such as transgene silencing (Fagard and Vaucheret, 2000) are further compounded by negative public sentiment partially fueled by the fear of transgene escape via pollen or seeds. Without the continued acceptance of regulatory agencies and the general public, recent advances made through biotechnology will be of limited use, and unable to facilitate much needed improvements in the field. The responsibility of finding new platforms for increased agricultural productivity without concomitant threat to the environment falls to science.

1.1. An Alternative Biotechnology Concept

Crop plants possess two genomes in addition to that of the nucleus, the organellar genomes of mitochondria and chloroplasts. Genetic engineering of higher plant chloroplasts may offer the potential to mitigate certain limitations of agricultural productivity. Technological advances, most notably the invention of the particle accelerator (Boynton *et al.*, 1988), and the ability to express foreign genes in plastids (Daniell *et al.*, 1987; 1990) , have provided the opportunity to explore the chloroplast genome as a new platform to address current and future demands for improved food production. The concept of chloroplast genetic engineering has been demonstrated to confer desirable plant traits including insect resistance (McBride *et al.*, 1995; De Cosa *et al.*, 2001), herbicide resistance (Daniell *et al.*, 1998; Iamtham and Day 2000), salt tolerance (Kumar *et al.*, 2004a), drought tolerance (Lee *et al.*, 2003), disease resistance (De Gray *et al.*, 2001), phytoremediation (Ruiz *et al.*, 2003) and reversible male sterility (Ruiz and Daniell, 2005).

1.1.1. Pharming is the future

Beyond the ability to increase the abundance and nutritional value of food sources lies the potential to express proteins of industrial interest in plant plastids. Accumulation of such diverse compounds as liquid crystal polymers (Viitanen *et al.*, 2004), vaccine antigens (Daniell *et al.*, 2005b; Koya *et al.*, 2005; Chebolu and Daniell, 2007) and other clinically relevant proteins (Staub *et al.*, 2000; Guda *et al.*, 2000; De Gray *et al.*, 2001; Fernandez-San Millan *et al.*, 2003; Leelavathi and Reddy 2003, Arlen *et al.*, 2007; Ruhlman *et al.*, 2007) has been achieved without concomitant loss of host plant viability. Concentrations at or beyond that which would be required for feasibility in subsequent refinement procedures have been obtained. Moreover the possibility to deliver pharmaceutical proteins orally has far reaching implications. Elimination of costs related to purification, cold shipping and storage and the need for sterile injection by medical professionals make the advancement of this technology very attractive. Local production of the source crop would be an additional advantage to developing nations adding to the allure of these systems as pharmaceutical platforms. The very exciting recent accomplishment of transmucosal delivery of a plant produced cholera toxin subunit B (CTB) green fluorescent protein (GFP) fusion via the ganglioside M1 (GM1) receptor on the cells of the intestinal epithelia (Limaye *et al.*, 2006) demonstrates the potential of a plant based oral delivery system. Inclusion of the recognition sequence for the ubiquitous protease furin between CTB and GFP allowed for intracellular cleavage of the fusion product and subsequent transport of GFP, but not CTB, via the mucosal vasculature to the liver and spleen of mice fed with pulverized transgenic leaf tissue (Limaye *et al.*, 2006).

Ongoing work in our lab employs the CTB/GM1 system to orally deliver plastid produced human proinsulin, as a fusion with CTB, to non-obese diabetic (NOD) mice. Our results indicate that presentation of the chloroplast derived antigen in the gut associated lymphoid tissue can prevent pancreatic insulitis in the NOD model (Devine 2005, Ruhlman *et al.*, 2007) supporting the application of this technology to the treatment of clinical conditions. Tobacco plants expressing the CTB-proinsulin fusion (CTB-Pins) accumulated foreign protein up to 16% of TSP in leaf tissue. Elevated expression of immunosuppressive cytokines such as interleukin 4 and 10 was observed following oral gavage of NOD mice with low doses of tobacco leaf material. Concomitantly significant preservation of the pancreatic islets of Langerhans was seen in NOD mice fed CTB-Pins expressing tobacco compared to controls. Furthermore we have demonstrated that the CTB-Pins fusion protein can be produced in lettuce chloroplasts *via* stable transformation of the lettuce plastome. CTB-Pins accumulated up to 2.5% of TSP in lettuce leaves when expression was driven by the tobacco plastid ribosomal operon promoter and the 5' translational control region of bacteriophage T7 gene 10 (T7g10; Ruhlman *et al.*, 2007). The accumulation of a pharmaceutically important protein such as CTB-Pins in non-toxic lettuce leaves brings us closer to the accomplishment of orally delivered therapies.

1.1.2. The protein producing potential of plastids

The genetic potential of the chloroplast lies in its negatively supercoiled, double stranded, circular DNA molecule referred to as the plastome. Within the angiosperms the plastome carries approximately 120 to 130 genes and ranges in size from 120 to 180 kilobases (kb) (Sugiura, 1992). Plastome molecules are clustered in nucleiods which are associated with plastid membranes and readily observed by fluorescent microscopy following DAPI staining (Mache and Lerbs-Mache, 2001). Of the estimated 3000 or so proteins found in the higher plant chloroplast (Colas des Francs-Small *et al.*, 2004; Richly and Leister, 2004), only a small fraction are encoded by the plastome. The bulk of the chloroplast proteome is nuclear encoded, translated on cytosolic ribosomes and subsequently translocated across the chloroplast envelopes (Zerges, 2000).

The plastome exists in a highly polyploid state with up to 100 identical copies present in each plastid of a mature leaf cell (Maier *et al.*, 2004). In a mature leaf, mesophyll cells carry up to 100 chloroplasts with the result that this genome alone can comprise up to 20% of the total cellular DNA content (Bendich, 1987). The plastome persists in all plastid differentiation types: the proplastids of meristematic tissues, green chloroplasts, red or yellow chromoplasts, the colorless plastids amyloplasts and leucoplasts (starch containing), and elaioplasts (oil containing).

The plastome is maternally inherited in most species of agricultural interest (Daniell, 2002; Hagemann, 2004). In maternal inheritance systems, paternal transmission of plastids is impeded during either the first pollen mitosis *via* unequal plastid distribution, or during generative or sperm cell development *via* plastid degeneration (Birky, 2001). Therefore, the generative and sperm cells in mature pollen tend to be free of plastids. Confinement of transgenic plastids in maternal tissues abrogates the ability of recombinant sequences to disperse to weedy relatives, or nearby agricultural stands. An additional level of control is offered by the recently developed system for reversible male sterility via light regulated expression of the *phaA* gene of *Acinetobacter* sp. Encoding β-ketothiolase in the chloroplast genome (Ruiz and Daniell, 2005).

2. THE CHLOROPLAST GENOME

2.1. The Molecule and Its Genetic Transformation

Plastid genome organization and structural features are conserved among eukaryotic photosynthetic organisms (Sugiura, 1992; Raubeson and Jansen, 2005). The circular molecule (figure 1) can be divided into three distinct domains: large single copy (LSC), small single copy (SSC) and the inverted repeat (IR) which is present in

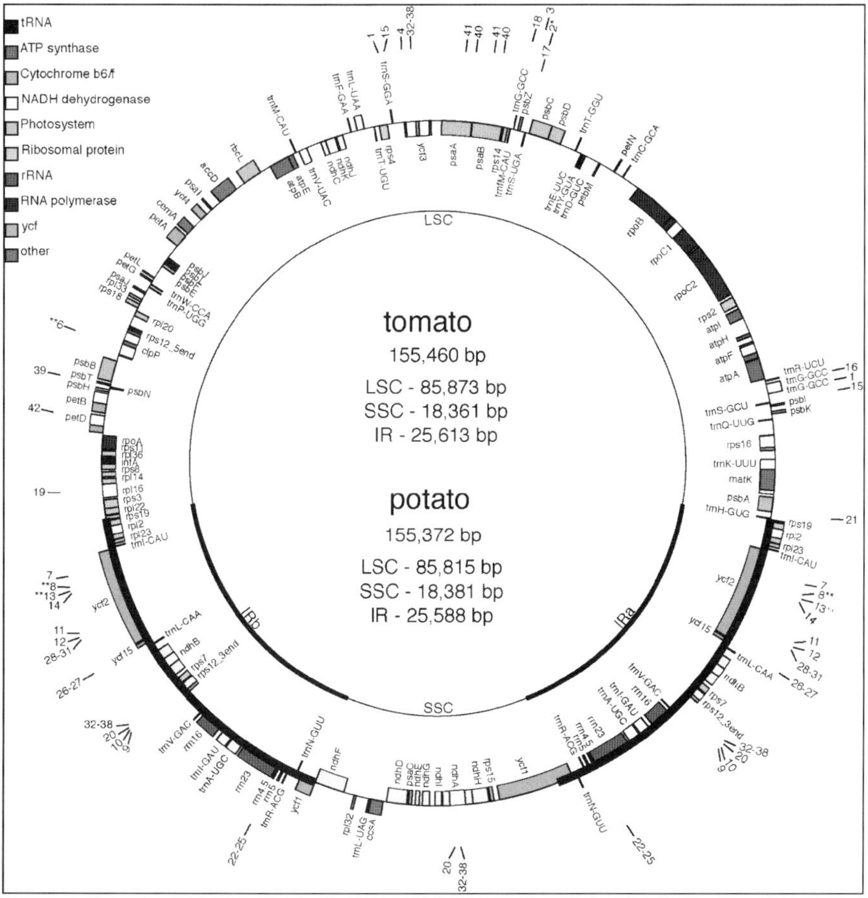

Figure 1. The Chloroplast Genome: Schematic representation of the plastid genomes of two Solanaceae species, *Solanum tuberosum* (potatoe) and *Lycopersicion esculentum* (tomatoe)

exact duplicate separated by the two single copy regions. Restriction fragment length polymorphism (RFLP) analysis indicates that the molecule exists in two orientations present in equimolar proportions within a single plant (Palmer, 1983).

The circular molecule undergoes interconversion to a dumbbell-shaped conformation that is believed to be facilitated by the presence of the IR. Concerted evolution within the IR (Kolodner and Tewari, 1975; Kolodner *et al.*, 1976) suggests intramolecular recombination between the repeats is a possible mechanism. The plastid RecA homolog has demonstrated DNA strand transfer activity (Cerutti and Jagendorf, 1993) and is thought to be responsible for the site specific integration of foreign DNA sequences in the plastid genome by homologous recombination. Through successive rounds of regeneration on selective media the iteration of the transplastome is favored as plastids carrying the resistance marker, and in turn the cells that harbor these plastids, are preferentially maintained as plastome molecules are divided up between daughter chloroplasts and subsequently as plastids are partitioned between daughter cells at mitosis (Moller and Moller, 2005). Over time all plastid genome copies, regardless of plastid type, carry the transplastome to the exclusion of detectable wild type copies, a condition referred to as homoplasmy.

2.2. Operons in Plastids

Many chloroplast genes of crop plants are co-transcribed from operons producing polycistronic primary transcripts. Several monocistrons are also transcribed including *psbA* and *rbcL*, encoding the D1 core polypeptide of photosystem II, and Rubisco large subunit respectively, as well as most of the 30 tRNA genes (Bonen *et al.*, 2004).

2.2.1. *Translational regulation*

Transcript availability is not the rate limiting factor for protein accumulation in mature leaf chloroplasts. Transcript stability, maturation and subsequent translation into protein products are influenced by several features of the mRNA. Chloroplast mRNAs include 5' and 3' untranslated regions (UTRs), both of which confer on the molecule distinctive elements necessary for the eventual production of chloroplast proteins from mono- or polycistronic transcription units (Monde *et al.*, 2000). Accumulation of polycistronic mRNAs as well as their efficient translation leading to high levels of foreign protein has been established (Jeong *et al.*, 2004; Quesada-Vargas *et al.*, 2005).

Native and heterologous elements have been successfully employed for the regulation of foreign protein expression in chloroplast transformation experiments. The ability to drive protein accumulation in the plastid system has lead to the implementation of the *psbA* 5' UTR in many transformation experiments where high levels of foreign product is desired (De Cosa *et al.*, 2001; Fernandez-San Milan *et al.*, 2003; Leelavathi and Reddy 2003; Dhingra *et al.*, 2004; Molina *et al.*, 2004; Watson *et al.*, 2004; Devine 2005; Chebolu and Daniell 2007; Viitanen *et al.*, 2004). Due to its dependence on light for activation of translation this element has an obvious limitation in non green tissues. An alternative to this highly utilized sequence that has been employed successfully for expression in green (Guda *et al.*, 2000; Staub *et al.*, 2000) and nongreen (Kumar *et al.*, 2004a) tissues is the 5' translation control region of bacteriophage T7 gene 10 (T7g10). In addition to its ability to associate with plastid ribosomes in a light-independent manner this sequence element should be free of developmental regulation in plant plastids. Utilization of heterologous translational elements may be advantageous in several regards. Chloroplast transformation for multi-gene constructs will require the implementation non-repetitive translation sequences for transgenes to avoid looping out events leading to the elimination of inserted sequences. Furthermore foreign signals should not detract from or have to compete with the expression of endogenous proteins. This could be a benefit for foreign protein accumulation and possibly deter possible interference with the expression of native proteins.

3. PLASTID OPERONS AND MULTIGENE ENGINEERING

That the plastid expression system allows for the transcription of operons from a single promoter to produce translatable polycistronic mRNAs (De Cosa *et al.*, 2001; Ruiz *et al.*, 2003; Quesada-Vargas *et al.*, 2005) offers great potential for metabolic engineering. The ability to transform the plastome for multiple genes in a single recombination event makes possible the expression of multi-enzyme pathways in the first transformed generation eliminating the need to cross lines recombinant for individual genes. Simultaneous integration of selectable markers along with genes of interest assures that regenerants expressing said markers will harbor the entire transformation cassette. Multi-gene engineering permitted the 'double gene single selection' system (Kumar *et al.*, 2004b) that facilitated the generation of homoplasmic cotton transformants through the ability to apply selective pressure in green as well as non-green stages of development.

3.1. Chaperones

While the plastid is less protease rich than the cytoplasm, it is not free from proteolytic activity (Adam and Clarke, 2002). Inclusion of the native chaperonin of the *Bacillus thuringiensis* (Bt) *cry2Aa2* operon led to an abundance (about 46% of TSP) of Cry protein in transplastomic tobacco plants (De Cosa *et al.*, 2001), the highest level of transgene expression reported for plants. An upstream sequence of the operon, *orf2*, encodes a protein involved in Cry folding into cuboidal crystals. In its crystalline form Cry is highly resistant to proteolytic activity (Crickmore and Ellar, 1992; Staples *et al.*, 2001). As co-integration of such folding factors is feasible through plastid transformation, a worthwhile approach would be to identify native or general factors for proteins of interest which are susceptible to degradation.

Operonal expression conferred high levels of tolerance to the organomercuial compound phenylmertcuric acetate (PMA) on chloroplast transformants for the *merA* and *merB* genes. Originating from bacteria, mercuric ion reductase and organomercurial lyase are encoded by *merA* and *merB* respectively. These enzymes convert toxic methyl-mercury (CH_2Hg) to elemental mercury (Hg(0)). The latter is much less toxic, readily volatilized and should theoretically be released by plants to the atmosphere through transpiration (Rugh *et al.*, 1996; Bizily *et al.*, 1999; Bizily *et al.*, 2000). Previous studies employing nuclear transformation and subsequent breeding to establish both genes in transgenic *Arabidopsis* demonstrated that expression of both genes was required to confer significant resistance to toxic mercury (Bizily *et al.*, 1999; Bizily *et al.*, 2000). Measurements of Hg(0) evolution correlated well with the ability of these plants to survive on the highest concentration reported in this work: 5 μM of PMA. Chloroplast transformants expressing the native bacterial sequence as an operon with SD-type signal driving translation were not susceptible to 100 μM of PMA and were able to survive at concentrations up to 400 μM of PMA (Ruiz *et al.*, 2003).

3.2. Vitamins

3.2.1. Vitamin A

Improvement of the nutritive value of seeds and vegetative tissues has been a goal of genetic engineers for some time. In Southeast Asia, it is estimated that up to half a million children go blind each year because of vitamin A deficiency (VAD). An estimated 140–250 million children suffer subclinical VAD which exacerbates afflictions such as diarrhea, respiratory ailments, and other childhood diseases such as measles. Supplementation with vitamin A reduces mortality from measles by 50% (WHO). As rice is the predominant staple in Southeast Asia it has been a primary target for engineering of vitamin A.

The development of a nuclear transgenic rice lines expressing three separate genes involved in provitamin A synthesis: phytoene synthase, phytoene desaturase, ξ-carotene desaturase (the two latter activities, separate in plants, were mediated by the single bacterial enzyme CrtI) and lycopene β-cyclase, has come as the result of many years of effort and required transformation with three different vectors (Ye *et al.*, 2000, Datta *et al.*, 2003; Baisakh *et al.*, 2006). The precursor for β-carotene, geranylgeranyl-diphosphate, is synthesized in wild type rice endosperm plastids, the site to which the enzymes were targeted. The ability to introduce several genes in a single transformation event through plastid genetic engineering could be used to express metabolic pathways, such as provitamin A synthesis, directly in plastids.

3.2.2. Vitamin E

The world vitamin E market was valued at more than one billion dollars in 1995 with synthetically produced comprising approximately 88% of the supply (Herbers, 2003). This essential nutrient is thought to provide protection from some cancers due to its antioxidant properties (Maeda *et al.*, 1992; Pastori *et al.*, 1998; Santillo and Lowe, 2006; Shiau *et al.*, 2006). The tocopherols and tocotrienols that comprise the group of lipid species collectively known as vitamin E are formed and accumulate in the photosynthetic plastids and cyannobacteria. Several experiments employing nuclear transformation to overexpress various enzymes involved in vitamin E synthesis have been carried out. These enzymes are nuclear encoded and imported to plastids, with the exception of 4-hydroxyphenylpyruvate dioxygenase (HPD) which is active in the cytoplasm catalyzing a single step in the pathway leading to tocopherol.

Efforts to increase substrate availability through expression of plastid localized, feedback insensitive enzymes (Falk *et al.*, 2005) and overexpression of homogentisate phytyltransferase (HPT) (Collakova and DellaPenna, 2003a, 2003b), whose activity represents the first committed step in tocopherol synthesis, have demonstrated the possibility to improve vitamin E availability in plants. In a recent study the HPD enzyme from barley was expressed in tobacco chloroplasts. These transplastomic plants accumulated more than twice as much α-tocopherol in leaves than wild type plants, while no significant difference was detected in seeds (Falk *et al.*, 2005). The authors speculate that HPD product HGA must interact with HPT through a membrane interaction and that HGA produced in the plastid is restricted in this association. This is a reasonable assumption as in the wild type HGA migrates back across the plastid envelope for continuation of the pathway. Unfortunately the tobacco plants in these experiments are not knocked out for endogenous HPD activity and therefore a definitive conclusion cannot be drawn from these results. Investigation with nuclear transformants overexpressing HPT has demonstrated that this activity is limiting in the pathway leading to vitamin E (Collakova and DellaPenna, 2003a). This suggests that simply flooding HPT with substrate may not necessarily lead to an enhancement of tocopherol synthesis.

In a recent review on metabolic engineering for nutritional enhancement of plants, it was proposed that at least five genes in the vitamin E pathway may have to be upregulated in some way to significantly enhance its accumulation in an oilseed crop (Kinney, 2006). Transformation of the plant chloroplast with any or all of these genes via multi-gene constructs would allow their expression to be controlled at several levels. All of these genes could be transcribed as an operon from a single promoter, with tailoring of translation made possible through the use of UTRs specific for developmental stages or light environment. Especially promising may be the use of transplastomic soybean in future studies as plastid expression is prolonged in the mature seed (Dufourmantel *et al.*, 2005) compared to other non-photosynthetic oil seeds.

3.3. Amino Acids

In addition to vitamins, several amino acids are at least partially synthesized in the plant chloroplast by enzymes which are nuclear encoded (see Table 1), providing an insight into how we may better approach nutritional enhancement of plants. The ability to transfer groups of genes, for example enzymatic pathways, in single transformation construct makes the chloroplast an attractive site for the engineered expression of essential amino acids.

Methionine, lysine, threonine and isoleucine constitute the aspartate family of amino acids. With the exception of methionine these amino acids are synthesized entirely in the plastid from a parent molecule, aspartic acid, imported from the cytosol. Methionine synthesis is completed by enzymes localized outside the plastid. The biosynthesis pathways of methionine and threonine diverge from O-phosphohomoserine (OPHS), an intermediate metabolite in the aspartate family of amino acids (Figure 2). OPHS represents the common substrate for both Threonine Synthase (TS) and Cystathionine-γ-Synthase (CGS). OPHS is either directly converted to threonine by TS, or, in a three-step mechanism, to methionine through condensation of cysteine and OPHS to cystathionine, which is subsequently

Table 1. Abbreviated list of nuclear-encoded metabolic enzymes post-translationally imported into plastids. Adapted from http://www.chloroplast.net

Amino Acid Metabolism	
Aspartate Kinase (EC 2.7.2.4)	Lys and Thr; involved in generation of homoserine for Met, Gly, Ser metabolism
Homoserine Dehydrogenase (EC 1.1.1.3)	Threonine, Isoleucine and Methionine
Aspartate Semialdehyde Dehydrogenase (EC 1.2.1.11)	produces branchpoint intermediate to Lys and Thr or Met synthesis
Diaminopimelate Decarboxylase (EC 4.1.1.20)	final enzyme in Lysine synthesis
Diaminopimelate Epimerase (EC 5.1.1.7)	Lysine synthesis
Dihydrodipicolinate Reductase (EC 1.3.1.26)	Lysine; formation of tetrahydrodipiolinate
Dihydrodipicolinate Synthase (EC 4.2.1.52)	first reaction unique to Lys synthesis; senitive to Lys feedback inhibition
Homoserine Kinase (EC 2.7.1.39)	forms branchpoint intermediate in pathways for Met and Thr
Succinyldiaminopimelate Transaminase (EC 2.6.1.17)	Lysine (putative)
Threonine Synthase (EC 4.2.3.1)	first committed step in Thr synthesis
2-isopropylmalate Synthase (EC 2.3.3.13)	introductory enzyme in Leu synthesis
3-isopropylmalate Dehydratase, large and small subunit (EC 4.2.1.33)	branched chain amino acid synthesis
3-isopropylmalate Dehydrogenase (EC 1.1.1.85)	Leucine synthesis
Acetolactate Reductoisomerase (ketol-acid reductoisomerase) (EC 1.1.1.86)	reduces acetohydroxyacids to dihydroxyacids (branched chain pathway)
Acetolactate Synthase (Acetohydroxy Acid Synthase) catalytic and regulatory subunit (EC 2.2.1.6)	entry point to branched chain pathway; herbicide target (ALS inhibitors)
Branched-chain amino acid Aminotransferase (EC 2.6.1.42)	Valine, Leucine and Isoleucine (last in synthesis, first in degredation)
Dihydroxy Acid Dehydratase (EC 4.2.1.9)	branched chain pathway, third enzyme
Threonine Deaminase (Threonine Dehydratase) (EC 4.3.1.19)	Ile synthesis; conversion of L-threonine to α-ketobutyrate
3-phosphoglycerate Dehydrogenase (EC 1.1.1.95)	Glycine and Serine; entry point to Ser, Gly pathway
3-phosphoserine Phosphatase (EC 3.1.3.3)	3-phosphoserine to serine: final step
Phosphoserine Aminotransferase (EC 2.6.1.52)	Serine
3-dehydroquinate Dehydratase (EC 4.2.1.10)/ Shikimate 5-dehydrogenase (EC 1.1.1.25)	Tryptophan, Tyrosine and Phenylalanine (bifunctional enzyme)
3-dehydroquinate Synthase (EC 4.2.3.4)	pre-chorismate pathway; aromatic amino acids
3-deoxy-7-phosphoheptulonate Synthase (EC 2.5.1.54)	first enzyme of the shikimate pathway
Shikimate Kinase (EC 2.7.1.71)	phosporylation of shikimate; aromatic amino acid synthesis
5-enolpyruvylshikimate-3-phosphate (EPSP) synthase (EC 2.5.1.19)	target of glyphosate herbicides
Chorismate Synthase (EC 4.2.3.5)	product chorismate is last common precursor to numerous aromatic compounds
Anthranilate Phosphoribosyltransferase (EC 2.4.2.18)	Tryptophan biosynthesis
Anthranilate Synthase α- and β-subunits (EC 4.1.3.27)	chorismate to anthranilate
Pretyrosine (arogenate) Dehydrogenase (EC 1.3.1.43)	Tyrosine biosynthesis

Chorismate Mutase (EC 5.4.99.5)	shikimate pathway, first enzyme leading to Tyr and Phe
Indole-3-glycerol-phosphate Synthase (EC 4.1.1.48)	Tryptophan
Phosphoribosylanthranilate Isomerase (EC 5.3.1.24)	Tryptophan
Prephenate Dehydratase (EC 4.2.1.51)	Phenylalanine biosynthesis; prephenate to phenylpyruvate
Tryptophan synthase α- and β-subunits (EC 4.2.1.20)	final step in Tryptophan synthesis
Glutamate-cysteine Ligase (EC 6.3.2.2)	first step in glutathione synthesis
Glutathione Synthetase (EC 6.3.2.3)	Cysteine/sulfur metabolism; Glutathione synthesis
ATP Sulfurylase (sulfate adenylyltransferase) (EC 2.7.7.4)	Cysteine and Methionine; sulfur assimilation
Cystathionine beta lyase (EC 4.4.1.8)	Cysteine and Methionine; sulfur metabolism (cystathionine to homocysteine)
Cystathionine gamma Synthase (EC 2.5.1.48)	Cysteine and Methionine; sulfur metabolism (cystathionine from homocysteine)
O-acetylserine (thiol)-lyase (OAS TL) (EC 2.5.1.47)	Cysteine and Methionine; sulfur metabolism (Cysteine Synthase complex)
Serine O-acetyltransferase (EC 2.3.1.30)	Cysteine and Methionine; sulfur metabolism (Cysteine Synthase complex)
Acetylornithine Deacetylase (EC 3.5.1.16)	Arginine metabolism; last step from glutamate to ornithine (catabolism)
Acetyl Ornithine Transaminase (EC 2.6.1.11)	Arginine; glutamate to ornithine (catabolism)
Delta-1-pyrroline-5-carboxylate Reductase (EC 1.5.1.2)	final step in Proline synthesis
N-acetylglutamate Kinase (EC 2.7.2.8)	key regulatory step in Arginine synthesis
N-acetylglutamate Synthase (EC 2.3.1.1)/N-acetylornithine Glutamate Acetyltransferase (EC 2.3.1.35)	Ornithine synthesis
Ornithine Carbamoyltransferase (EC 2.1.3.3)	Arginine synthesis
Carbamoylphosphate Synthetase A and B subunits (EC 6.3.5.5)	Pyrimidine and Arginine metabolism
1-(5-phosphoribosyl)-5-[(5-phosphoribosylamino)methylideneamino] imidazole-4-carboxamide isomerase (EC 5.3.1.16)	Histidine synthesis
ATP-phosphoribosyl Transferase (EC 2.4.2.17)	first step in Histidine synthesis
Glutamine Amidotransferase (EC 2.4.2.-) / Cyclase (EC 4.1.3.-)	Histidine synthesis
Imidazoleglycerol-phosphate Dehydratase (EC 4.2.1.19)	Histidine synthesis; target for the triazole phosphonate herbicides
Phosphoribosyl-ATP pyrophosphohydrolase (EC 3.6.1.31) / phosphoribosyl-AMP cyclohydrolase (EC 3.5.4.19)	Histidine synthesis
Fatty Acid Metabolism	
Glutathione Peroxidase Phospholipid Pydroperoxide (EC 1.11.1.12) (EC 1.11.1.9)	sulfur metabolism; development and stress response
Acetyl-CoA Carboxylase (EC 6.4.1.2)	Acetyl-CoA, malonyl-CoA synthesis

(*continued*)

Table 1. (continued)

Fatty Acid Metabolism	
heteromeric form	
Biotin carboxyl carrier protein	
Carboxyltransferase alpha subunit	
Biotin Carboxylase (EC 6.3.4.14) subunit	
Acetyl-CoA Synthetase (EC 6.2.1.1)	
[acyl carrier protein] S-malonyltransferase (EC 2.3.1.39)	Elongase
Pyruvate Dehydrogenase α and β subunits (EC 1.2.4.1)	
Dihydrolipoamide S-acetyltransferase (EC 2.3.1.12)	Pyruvate Dehydrogenase subunit
1-acylglycerol-3-phosphate O-acyltransferase /lysophosphatidic acid acyltransferase (EC 2.3.1.51)	synthesis of phosphatidic acid
Chlorophyll Synthesis	**Chlorophyll Synthesis**
Glutamyl-tRNA Synthetase (EC 6.1.1.17)	L-glutamate to glutamate 1-semialdehyde (Mg-tetrapyrrol synthesis); amino acid synthesis
Aminolevulinate Dehydratase (EC 4.2.1.24)	chlorophyll synthesis
Chlorophyll a Oxygenase	conversion of a methyl group to a formyl side group (chlorophyll b synthesis)
Chlorophyll Synthetase (EC 2.5.1.62)	chlorophyll a synthesis; attachment of phytyl side chain
Coproporphyrinogen III Oxidase (EC 1.3.3.3)	formation of protoporphyrin IX
Glutamate-1-semialdehyde Aminotransferase (EC 5.4.3.8)	formation of delta-aminolevulinate from glutamate 1-semialdehyde
Magnesium Chelatase, Chl D, H, I subunits (EC 4.99.1.-)	branchpoint in chlorophyll/heme synthesis
Magnesium Protoporphyrin IX Methyltranserase (EC 2.1.1.11)	chlorophyll metabolism; Mg-dependent pathway
Protochlorophyllide Reductase (EC 1.3.1.33) subunits A, B and C	(POR) conversion to chlorophyllide
Uroporphyrinogen III Synthase (EC 4.2.1.75)	chlorophyll metabolism
Carotenoid and Xanthopyll Synthesis	
α−carotene epsilon-hydroxylase	lutein synthesis
β−carotene Hydroxylase (EC 1.14.13.-)	zeaxanthan synthesis
Lycopene epsilon-cyclase	α-and β-carotene synthesis
9-cis-epoxycarotenoid Dioxygenase (neoxanthin cleavage enzyme)	carotenoids to ABA through xanthoxin
Carotenoid Isomerase	lycopene synthesis
Lycopene β-cyclase (EC 1.14.-.-)	α-and β-carotene synthesis
Phytoene Desaturase (EC 1.3.99.-)/Phytofluene Desaturase	pytoene to γ-carotene
Phytoene Synthase (EC 2.5.1.32)	first committed step in carotenoid synthesis
Violaxanthin De-epoxidase	violaxanthin to zeaxanthin
Zeaxanthin Epoxidase	zeaxanthin to violaxanthin, entrance to ABA biosynthetic pathway
ζ−carotene Desaturase (EC 1.14.99.30)	lycopene synthesis
Carotenoid Associated Protein	

Pigments, Carotenoid and Plastoquinone	
1-deoxy-D-xylulose 5-phosphate Reductoisomerase (EC 1.1.1.267)	MVA- independent pathway; synthesis of C5 isoprenoid intermediates isopentyl diphosphate, dimethylallyl diphosphate
2C-methyl-D-erythritol 2,4-cyclodiphosphate Synthase (EC 4.6.1.12)	Geranylgeranyl diphosphate synthesis
4-diphosphocytidyl-2C-methyl-D-erythritol Kinase (EC 2.7.1.148)	Geranylgeranyl diphosphate synthesis
4-diphosphocytidyl-2C-methyl-D-erythritol Synthase (EC 2.7.7.60)	Geranylgeranyl diphosphate synthesis
4-hydroxy-3-methylbut-2-en-1-yl diphosphate Synthase (EC 1.17.4.3)	Geranylgeranyl diphosphate synthesis
4-hydroxy-3-methylbut-2-enyl diphosphate Reductase (EC 1.17.1.2)	Geranylgeranyl diphosphate synthesis
Geranyl Diphosphate Synthase (EC 2.5.1.1)	short chain prenyltransferase
Geranylgeranyl Diphosphate Synthase (EC 2.5.1.29)	branchpoint to gibberillins, carotenes, vitamin K2 and E, chloropyll, tyrosine, ABA
Isopentenyl Diphosphate Isomerase (EC 5.3.3.2)	Diverts isopentenyl diphosphate to cytokinin pthwy
Monoterpene Sesquiterpene Synthase like proteins including taxa-4(5),11(12)-diene synthase, a diterpene cyclase	Isoprenoid synthesis; terpenes including TAXOL/ PACLITAXIL Plastidal entry point to gibberellin pathway
3-oxo-5-alpha-steroid 4-dehydrogenase (EC 1.3.99.-)	carotenoids to brassinolide (via squalene)
Cinnamate 4-hydroxylase (EC 1.14.13.11)	anthocyanin pathway; flavanoid synthesis
Phenylalanine Ammonia Lyase (EC 4.3.1.5)	anthocyanin pathway; flavanoid synthesis; first enzyme
Secondary Metabolism	
Adenine Phosphoribosyltransferase (EC 2.4.2.7)	caffeine synthesis; purine nucleotide synthesis
Adenosine Kinase (EC 2.7.1.20)	caffeine synthesis; purine nucleotide synthesis
Groporphyrin III Methylase (EC 2.1.1.107)	Siroheme synthesis
Vitamins	
4-methyl-5(b-hydroxyethyl)-thiazole Monophosphate Biosynthesis Protein	Thiamine synthesis (vitamin B1)
Hydroxyethylthiazole Kinase (EC:2.7.1.50)	Vitamin B1
Phosphomethylpyrimidine Kinase (EC 2.7.4.7)/Thiamine-phosphate Pyrophosphorylase (EC 2.5.1.3)	Vitamin B1
Thiamin Biosynthesis Protein	Vitamin B1
Geranylgeranyl Reductase	branchpoint to vitamins K2 and E, and chlorophyll
Homogentisate Phytyltransferase	Vitamin E - α-Tocopherol
Tocopherol Cyclase	Vitamin E - α-Tocopherol
Gamma Tocopherol Methyltransferase (EC 2.1.1.95)	Vitamin E - α-Tocopherol
1.4-dihydroxy-2-naphthoate (DHNA) Phytyltransferase (EC 2.5.1.-)	Vitamin K2 – Phylloquinone
2-Oxoglutarate Decarboxylase (EC 4.1.1.71) / SHCHC synthase (EC 4.1.3.-)	Vitamin K2 – Phylloquinone
C-methyltransferase (EC 2.1.1.-)	vitamin K2 – Phylloquinone; putative ubiE homologue
Isochorismate Synthase (EC 5.4.99.6)	shikimate pathway
O-succinylbenzoate Co-A Ligase (EC 6.2.1.26)	Vitamin K2 – Phylloquinone

converted to homocysteine and then methionine. Thus, the enzymes CGS in the methionine pathway and TS in the threonine pathway compete for the common substrate, OPHS (Avraham and Amir, 2005). The fact that plant TS and CGS are branch point enzymes competing for the same substrate demands the effective regulation of the respective enzymatic activities.

3.3.1. Enhancing Sulfur amino acid accumulation

Over-expression of CGS in potato increased methionine levels in the leaves, roots and tubers 6 fold (Di *et al.*, 2003). Antisense of TS in potato transgenic lines reduced threonine to 45% compared to wild-type controls, whereas methionine levels increased up to 239-fold depending on the transgenic line and environmental conditions (Zeh *et al.*, 2001). Increased levels of homoserine and homocysteine indicate increased carbon allocation into the aspartate pathway. Tubers of TS antisense potato plants contained a methionine level increased by a factor of 30 but showed reduction in threonine. These plants offer a major biotechnological advance toward the development of crop plants with improved nutritional quality.

Methionine is considered the most limiting amino acid and efforts to improve the sulfur content of seed and forage crops has fostered the body of work exploring ways to achieve this through the accumulation of methionine and cysteine in vegetative tissues and storage organs. Sulfur-rich seed storage proteins such as the 2S albumins from sunflower seed (SSA), Brazil nut, and pea vicilin have been employed in

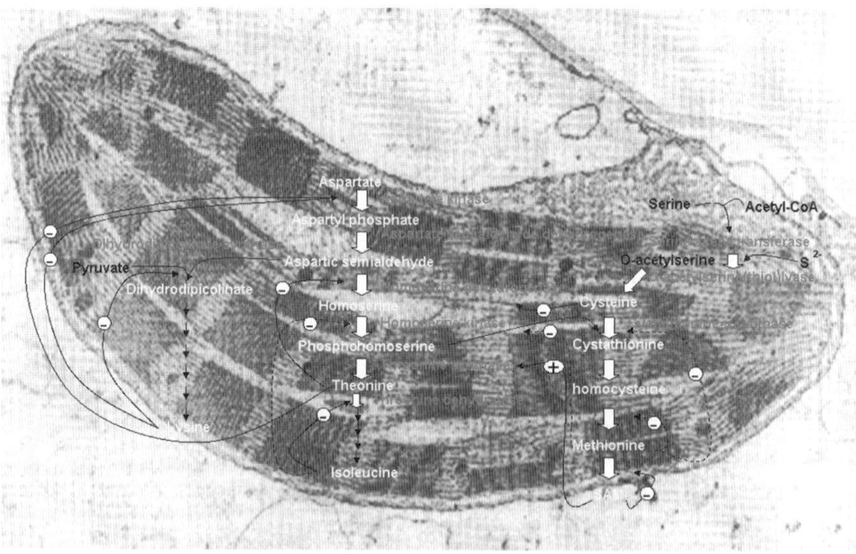

Figure 2. Amino Acid Biosynthesis in Chloroplasts: Many metabolic pathways, including several for the synthesis of essential amino acids are, localized to plastids. Black lines indicate feedback regulation (+ or −). Dashed lines indicate this regulation has not been observed in all species investigated, modified from http://www.compulink.co.uk/~argus/Dreambio/photosynthesis/chloroplast 4.gif

nuclear transformation experiments. Initial experiments with the Brazil nut albumin resulted in detectable expression in all tissues and organs examined due to the inclusion of a constitutive promoter. The gene product was localized predominantly to mesophyll vacuoles where it could not accumulate to high levels (Muntz *et al.*, 1998). The vegetative vacuole is a protease rich compartment and this may have led to rapid degradation of the foreign protein. In subsequent work with SSA and pea vicilin, the fusion of a C-terminal endoplasmic reticulum (ER) targeting signal (KDEL) has facilitated varying degrees of accumulation in nuclear transgenic plants. In tobacco plants expressing the pea vicilin gene addition of the KDEL sequence resulted in a 100-fold enhancement of this recombinant protein and extended its half life by a factor of 10 (Wandelt et al., 1992). For both tobacco and alfalfa, ER targeting of SSA produced "easily detectable" expression of foreign protein in western analysis as compared to transformants for the SSA gene alone in which no SSA was detected (Tabe *et al.*, 1995), again confirming the contrasting results observed in leaves & seeds. Similar results were found for SSA expression in *Trifolium subterraneum* L. (Khan *et al.*, 1996).

It has been suggested that the availability of soluble methionine limits the potential accumulation of SSA in nuclear transgenic lines. Experiments have been carried out with the methionine-rich zein genes of maize to address the same limitation for its accumulation in alfalfa (Bagga *et al.*, 2004; Amira *et al.*, 2005) and tobacco (Amira *et al.*, 2005). Transgenic alfalfa and tobacco lines expressing ß-zein were crossed with those expressing *Arabidopsis* cystathionine-γ-synthase (atCGS). Soluble methionine levels were reduced when compared to plants expressing atCGS alone in both species and ß-zein expression was significantly enhanced in alfalfa. In tobacco leaf samples from plants co-expressing atCGS and ß-zein, significant accumulation of ß-zein was not observed compared to those expressing ß-zein alone, suggesting that cytosolic translation of this mRNA is not limited by methionine availability.

Proteins with high sulfur content are likely to be stabilized by disulfide bridges. Improved accumulation of sulfur amino acids in ER targeting experiments suggests this may be the case. This compartment contains functional PDI for the establishment of such stabilizing bonds (Denecke *et al.*, 1992; Levitan *et al.*, 2005). Furthermore those products targeted to the ER would not be subject to the protease-rich environment of the cytoplasm. Chloroplast may prove to be an ideal site for the expression of sulfur rich seed storage proteins as PDI is active in this compartment (Trebitsh *et al.*, 2001; Kim and Mayfield, 2002; Levitan *et al.*, 2005; Alergand *et al.*, 2006). In addition export of proteins translated on plastid ribosomes across the organelle's double membrane has not been reported suggesting they are retained within the chloroplast.

3.3.2. *Improvement of lysine and threonine content*

Cereal grains represent a major constituent in the diet of humans and livestock world wide. Lysine and threonine are limiting amino acids in cereal grains and improving the lysine content of fodder for poultry and swine has received considerable attention. Lysine metabolism in plants is regulated primarily by feed back inhibition of plastid localized dihydrodipicolinate synthase (DHPS). To address this limitation on lysine accumulation, the feed back insensitive bacterial DHPS has been transferred to the nuclear genome of tobacco (Shaul and Galili, 1993; Karchi *et al.*, 1994), soybean and canola (Falco *et al.*, 1995; Mazur *et al.*, 1999). Constitutive expression facilitated significant accumulation of free lysine in tobacco leaves but not without serious deleterious effects (Perl *et al.*, 1992). Expression from a seed specific promoter provided amelioration of this effect at the expense of lysine production (Karchi *et al.*, 1994). Moderate accumulation of free lysine was observed in developing seeds of transgenic tobacco via the expression of two bacterial enzymes, DHPS and feedback insensitive Aspartate Kinase but content was not significantly higher than the wild type when seeds reached maturity. It was found that the presence elevated levels of free lysine stimulated the catabolic activity of the cytosolic enzyme Lysine-Ketoglutarate Reductase (Karchi *et al.*, 1994; Tang *et al.*, 1997; Zhu *et al.*, 2001). Subsequent seed specific expression of the bacterial DHPS in soybean and canola demonstrated a significant enrichment of lysine over wild type seeds. Unfortunately this resulted in problems with seed germination and the presence of various catabolites of lysine was noted (Falco *et al.*, 1995). A more successful approach for lysine accumulation, particularly in cereal grains, has been transformation with genes encoding proteins with a high proportion of this amino acid in its sequence. Several genes have been employed in maize to relative success. Constructs incorporating the barley high lysine (BHL) or hordothionine led to an elevation of lysine content in transgenic seed (in Galili *et al.*, 2005). Soybean vegetative storage protein β-subunit (S-VSPβ), also rich in lysine, was co-expressed constitutively with the bacterial DHPS in transgenic tobacco. S-VSPβ accumulation was higher in these plants than in those expressing the DHPS alone with total lysine level increased by 30% over wild type (Guenoune *et al.*, 2003).

Designing experiments to include both the enzymatic source of desired amino acids as well as a proteolytically stable "sink" polypeptide or polymer may be the route to their stable accumulation in plant tissues. Chloroplast transformation may support this end in several ways. Multi-gene engineering, various benefits facilitated by compartmentalization (i.e. relief of co suppression, protection from potential degredative influences in the cytoplasm) and the presence of plastid PDIs are among the features that make plastid biotechnology a worthwhile approach for explorations such as those described herein.

4. CHLOROPLASTS AS BIOREACTORS: CAN WE REPLACE PLANTS WITH PLANTS?

Technology has brought our society a long way in terms of our ability to produce desirable products on a large scale in industrial plant operations. The mandate for future productivity includes the development of technologies that meet the demands of a growing population without concomitant threat to the environment. In the plastids of green plants we are finding the potential to express desired protein products, and, through the engineering of biosynthetic pathways, accumulate important non-protein molecules in a cost-effective manner.

4.1. Pharmaceutical Products

A number of pharmaceutical proteins have been successfully expressed in transgenic plastids including various therapeutic agents and vaccine antigens (Daniell *et al.*, 2001a,b; Fernandez San-Millan *et al.*, 2003; Tregoning *et al.*, 2004; Watson *et al.*, 2004; Koya *et al.*, 2005). Expression of foreign proteins as fusion partners has facilitated significant accumulation of human somatotropin (hST) and interferon gamma (IFN-g) in transplastomic tobacco. While plastid transformants expressing IFN-g alone were found to accumulate 100 times as much of the foreign protein than nuclear transformants for the same gene, up to 0.1% of TSP (Leelavathi and Reddy, 2003), for feasible purification of therapeutic proteins from plant tissues accumulation will have to reach approximately 1% of TSP (Daniell 2006). Transformation of tobacco plastids with a new construct which carried the *ifnG* gene fused to the *uidA* gene for β-glucaronadase (GUS), a much more stable protein, allowed the IFN-g fusion to accumulate to an estimated level of 6% of TSP (Leelavathi and Reddy, 2003). Previously, the fusion approach had facilitated the accumulation of human somatotropin (hST). Initial expression of hST reached up to 0.2% of TSP in plastid transformants but when expressed as a fusion with Ubiquitin, hST accumulation in tobacco leaf tissues reached up to 7% of TSP, more than 300 times the level seen in nuclear transformants for the same gene (Staub *et al.*, 2000).

Although plastids are not thought to encode any disulfide bonded proteins, many nuclear encoded examples are ultimately expressed within the organelle. Plastid-localized PDI catalyzes disulfide bond formation for imported proteins and its presence may have contributed to the increased stability of hST in plastids. Human Serum Albumin (HSA) is the major component in blood and is used to replace blood volume in many clinical and trauma situations. In its mature form the multimeric globular protein is stabilized by 17 disulfide bonds. Potato nuclear transformants in which HSA expression was targeted to the tuber apoplast have been reported to accumulate up to ten times more of the foreign protein than transgenic plants expressing HSA in the cytoplasm (Farran *et al.*, 2002). Expression in the protease-poor apoplastic space can provide protection for foreign proteins susceptible to proteolytic degradation. Where expression of HSA was driven by *psbA* 5'UTR transplastomic plants accumulated more than 50 times as much protein as nuclear tranasforments localizing HSA to the apoplast (Fernandez-San Milan *et al.*, 2003). The ability to carry out post-translational modifications, such as disulfide bond formation, is likely a major contributing factor in limiting HSA degredation by proteases in the plastid. Abundant local expression of HSA in tobacco plastids lead to the formation of inclusion bodies which allowed for a facile purification strategy yielding a recovery of 0.25 mg HSA/g fresh weight, well within the range of industrial-scale feasibility.

In addition to added stability, correct disulfide bond formation in some therapeutic proteins is an absolute requirement for functional, biologically active molecules. For example, the CTB/GM1 system described above relies on the ability of CTB monomers to assemble in the pentameric form for GM1 binding and subsequent internalization. Each monomer contains an intramolecular disulfide bridge which is essential to subunit association (Ludwig *et al.*, 1985; Dertzbaugh and Cox, 1998). *In vitro* GM1 binding and *in vivo* epithelial absorption following oral delivery of plastid-derived CTB fusion proteins, such as CTB-GFP and CTB-Pins, demonstrates that the integrity of the pentamer is preserved in minimally processed plant tissues (Limaye *et al.*, 2006; Ruhlman *et al.*, 2007).

One of the limitations to the production of affordable therapeutic proteins by bacterial fermentation is the requirement for post-harvest modification to achieve correct structural conformation. Treatment of hepatitis C with interferon-α2b (IFN-α2b), currently produced in *E. coli*, is estimated to cost up to $26,000 per year. Biologically active INF-α2b contains two disulfide bonds and complete loss of antiviral activity results from their reduction (Morehead *et al.*, 1984; Bodo and Maurer-Fogy, 1986). When expressed in tobacco plastids INF-α2b accumulated up to 20% of tsp, or 3 mg g^{-1} of fresh leaf weight. *In vitro* and *in vivo* assays using crude transplastomic leaf extracts and purified plastid-derived INF-α2b demonstrated biological activity comparable to the commercial product produced in *E. coli* (Arlen *et al.*, 2007). The possibility to introduce several genes as a single construct allows for the expression of multisubunit complexes with transcription of said genes under the control of a single promoter. Having a readily available mRNA pool may help to maximize the potential for correct association into functional units and fine tuning for stoichiometry could be accomplished through manipulation of 5' translation elements.

Operon engineering in plastids has been employed to express the heavy (H) and light (L) chain of the Guy's 13 antibody. This human monoclonal antibody recognizes streptococcal antigen I/II (SA I/II), a major cell surface glycoprotein of *Streptococcus mutans*. In clinical trials this antibody has been shown to prevent colonization in the human oral cavity. The H and L chain were expressed in plastids where they were shown to assemble to the 160 kDa antibody by non-reducing western analysis with no concomitant deleterious effects observed in the transformants. The H and L chains associated through disulfide bridges formed by the plastid apparatus (Daniell *et al.*, 2005b). In nuclear transformants, after several reproductive crosses to achieve co-expression, inclusion of the J chain and the secretory component along with H and L led to the accumulation of total assembled IgA/G of 5% of TSP (Ma *et al.*, 1995; Ma *et al.*, 1998). This result suggests chloroplast operonal expression for secretory antibodies such as Guy's 13 merits further investigation.

4.2. Biopolymer Production

4.2.1. Polyhydroxybutyrate

The pleiotropic effects that have constrained cytoplasmic expression of certain foreign proteins have not been limited to enzymes involved in amino acid and vitamin synthesis. Nocent phenotypes have been observed in transformation scenarios aimed at the accumulation of engineered biopolymers (Nawrath *et al.*, 1994; Wrobel *et al.*, 2004). Nuclear transformation experiments have utilized the plastid as a storage vessel in such instances. Polyhydroxybutyrate (PHB) is synthesized from acetyl-coenzyme A by the consecutive activity of three enzymes of bacterial origin: β-ketothiolase, acetoacetyl-CoA reductase, and PHB synthase. In the bacteria PHB serves as a carbon storage molecule but it has attracted considerable interest from industry due to its potential application in biodegradable plastics and elastic polymers. Crosses made of parental lines transformed for two of the three enzymes, *phaB* and *phaC*, in initial experiments with Arabidopsis (in Nawrath *et al.*, 1994) constituted the first report of PHB production in plants. Despite low yield of PHB these plants presented severely stunted phenotypes. Subsequent experiments fused the constitutive 35S CMV promoter and the transit peptide sequence of the small subunit of pea ribulose-bisphosphate carboxylase to the genes for each of the three enzymes. A similar strategy of sexual crosses led to Arabidopsis plants expressing substantial amounts of PHB in chloroplasts (Nawrath *et al.*, 1994). An alternative to this approach would be to introduce this pathway, in a single transformation event, directly into the plastid genome for localized expression. While initial attempts resulted in tandem integration of the operon, only very low abundance of PHB in the leaves of transplastomic tobacco plants (Nakashita *et al.*, 2001). An improved vector carrying a multigene construct delivered to the Arabidopsis nuclear genome facilitated abundant accumulation of PHB in chloroplasts, unfortunately with concomitant deleterious effects on plant fitness (Bohmert *et al.*, 2000). Later experiments integrating the operon into the inverted repeat region of the tobacco plastome demonstrated enhanced PHB accumulation but phenotypic abnormalities including growth inhibition and sterility accompanied the improvement (Lossl *et al.*, 2003). In order to address nocuous side effects of PHB and other foreign proteins in the chloroplast recent investigations have sought to establish an inducible system for control of transgene expression (Lossl *et al.*, 2005; Muhlbauer and Koop, 2005).

Among the many attractive aspects of chloroplast transformation technology is the level of transgene containment achieved through the maternal inheritance of plastids in most agronomically important species. In an effort to diagnose the impediment to effective expression of the *phb* operon in transplastomic tobacco Ruiz and Daniell (2005) have discovered a novel and reversible instrument for induction of cytoplasmic male sterility (CMS). CMS is commonly associated with incompatability between nuclear and mitochondrial genomes and has been addressed in the literature as a deterrent to transgene escape via pollen and the production of commercial F1 hybrid seed (Saeglitz *et al.*, 2000; Havey, 2004; Chase, 2006). During the course of their investigation into the light regulated expression of β-ketothiolase in plastid transformants none of the severe pleiotropic effects, such as chlorosis and retarded growth, which had been found in plants expressing the entire operon were observed. The persistent characteristic in these plants was the abberant morphology of the male reproductive structures including shortened filaments and pollen grains that were collapsed or absent altogether. This phenotype persisted in the T1 plants generated by fertilization with wild type pollen. It was hypothesized that the male sterile effect resulted from a depletion of the acetyl-CoA pool in plastids expressing β-ketothiolase. Restoration of male fertility was accomplished by continuous illumination of transplastomic plants. The reducing environment facilitiated by continuous exposure to light favored the activity of ACCase enabling it to compete with β-ketothiolase for the available substrate, acetyl-CoA, restoring fatty acid synthesis and male feritilty (Ruiz and Daniell, 2005).

4.2.2. *p-hydroxybenzoic acid*

Bacterial and synthetic genes have been explored for polymer production applications in plastid transformants. The shikimate pathway, responsible for the synthesis of aromatic amino acids (reviewed in Herrmann and Weaver, 1999), has gained considerable attention from biotechnology through the years. Inhibition at chloroplast localized enolpyruvylshikimate-3-phosphate synthase (EPSPS) is the activity that makes Roundup® constituent glyphosate (N-phosphonomethylglycine) an effective herbicide. Resistance to glyphosate is engineered into a plant by adding a gene from a soil bacterium (ie *Agrobacterium* sp), or even another plant species, that encodes a version of the EPSPS that is resistant to the herbicide (reviewed in Dill, 2005). The last of seven steps in the shikimate pathway (which occurs exclusively in plastids and bacteria) results in chorismate. This compound is of particular interest as it provides abundant substrate for a bacterial enzyme which has been expressed in transplastomic plants to produce p-hydroxybenzoic acid (pHBA). The *E. coli ubiC* gene encodes chrosimate pyruvate-lyase (CPL) which catalyzes the conversion of chorismate to pHBA, the major monomer in liquid crystal polymers. Introducing this gene to the plastid genome resulted in accumulation of pHBA polymer to extraordinary levels, up to 26.5% of the dry weight (Viitanen *et al.*, 2004). Although this represents a 50-fold enhancement of the best values reported for nuclear transformants transplastomic plants were indistinguishable from wild types.

4.2.3. *Protein based polymers*

Naturally occurring proteins, such as the silks of insects and mammalian elastin, exhibit properties which make their controlled production desirable. Genetically engineered expression of protein based polymers (PBPs) offers the potential to precisely program attributes like molecular mass and amino acid composition that dictate the functional properties of the end product (Meyer and Chilkoti, 2002). In order for the production of PBPs for industrial applications to be feasible in transgenic plants, high levels of accumulation will have to be obtained. The synthetic variations of PBPs designed and tested to date have demonstrated an extraordinary degree of biocompatibility making them attractive for medical applications (Betre *et al.*, 2002).

The protein sequence of mammalian elastin, one of the strongest known natural fibers, consists of repeated blocks of five amino acids. It is from this sequence that synthetic analogs are designed. One such analog is the protein based elastomer $GVGVP_{121}$ (amino acids Gly-Val-Gly-Val-Pro repeated 121 times). This concatomer sequence has been expressed in several systems including *E.coli* (Daniell *et al.*, 1997) *Aspergillus nidulans* (Herzog *et al.*, 1997) tobacco NT1 cells (Zhang *et al.*, 1995), and whole tobacco plants (Zhang *et al.*, 1996; Guda *et al.*, 2000). One attractive feature of these types of polymers is that they display inverse temperature transition (Urry, 1988). While $GVGVP_{121}$ is soluble in water at room temperature it forms a more ordered aggregate as temperature increases. This reversible process could lend itself to the eventual purification of the polymer from plant tissues. Although chloroplast transformants expressing $GVGVP_{121}$ accumulated up to 100-fold more transcript for the polymer than the nuclear transformant, the polymer itself was not abundant. Western analyses revealed polypeptides of a smaller molecular weight than $GVGVP_{121}$. Because this sequence contains no known protease cleavage sites the authors reasonably suggest that these products were the result of incomplete polymerization. Certainly transcription was not limiting. The unusually heavy demand for constituent amino acids may have been the factor hindering continued polymer synthesis through all 121 repeats. Glycine, which comprises 40% of the final sequence regardless of length, is synthesized in mitochondria by two mechanisms. Serine may be converted to glycine by the action of Serine Hydroxymethyl Transferase in what amounts to a reversal of serine synthesis. Thought to predominate is the synthesis of glycine from CO_2 and NH_4 with N_5N_{10}-mthylene tetrahydrofolate (THF) acting as a donor of one carbon units via glycine synthase. Could these activities be established in plastids coupled to polymer expression? The ability to engineer such constructs for incorporation in the plastome in a single transformation event makes it a feasible investigation.

5. PROSPECTS FOR ENHANCING PLANT PRODUCTIVITY

Chloroplast biotechnology has allowed the examination of a variety of plastid functions that prior to its inception had seemed elusive. Ribulose-1,5-bisphosphate carboxylase/oxygenase (Rubisco) has attracted considerable attention and presents a target for metabolic engineers interested in improving agricultural productivity (Spreitzer and Salvucci, 2002). One way to increase the radiation use efficiency (RUE) in plants could be to enhance Rubisco's affinity for CO_2 over its competitor O_2 in reaction with ribulose-1,5-bisphosphate (RuBP). A 20% increase in light saturated net carbon exchange (A_{max}) could result from doubling the enzymes affinity for CO_2 in the reaction and abating photorespiration (Reynolds *et al.*, 2000). Neither substrate binds directly to the enzyme making manipulation of Rubisco for specificity especially challenging (reviewed in Parry *et al.*, 2003). Further complicating genetic modification of the hexadecameric holoenzyme has been the location of the genes for the large (*rbcL*) and small (*rbcS* family) subunits in the plastid and nucleus respectively (Rodermel, 1999). The large subunit (LSU), where the active site is located, is well conserved in crop plants (~90% identity) whereas the small subunits (SSU), to which a function has not yet been ascribed, are less so (~70% identity). The divergence seen in the small subunit may account for variation in efficiency among Rubiscos from different species (Spreitzer, 1999). Several approaches have been undertaken to introduce subunit genes from cyanobacteria, algae, sunflower and pea to tobacco or Arabidopsis through nuclear and plastid transformation (reviewed in Parry *et al.*, 2003). A recent effort to recover Rubisco activity in tobacco plants with antisense-ablated SSU expression (Rodermel *et al.*, 1988) has demonstrated that, with the use of appropriate regulatory signals, it is possible to effectively relocate *rbcS* to the plastid genome in a land plant. Employing *rbcS* constructs fused to T7g10 RBS or the *psbA* 5' UTR facilitated SSU protein accumulation to 60% and 106% of the wild type control respectively. Moreover these transplantomic plants not only accumulated abundant SSU, they were able to fully complement the antisense phenotype assembling functional Rusbisco (Dhingra *et al.*, 2004).

5.1. Future Perspectives

5.1.1. *Sequencing crop plastomes*

The future of chloroplast biotechnology for a range of applications, including metabolic engineering gives us reason to be optimistic. There are however several areas of concern that will require attention if the potential of this technology is to be realized. Where it was once thought that plastome sequences held little variation from one species to the next, recently sequenced genomes are revealing a richness of diversity among plastid genomes that was not expected (Kanamoto *et al.*, 2005; Kim *et al.*, 2005; Daniell *et al.*, 2006). While overall gene content and order is highly conserved in land plants this same conservation is not observed in non-coding sequence such as introns and intergenic spacers (IGS) which, along with the UTRs of genes comprise up to 50% of the plastome (Saski *et al.*, 2005; Daniell *et al.*, 2006; Jansen *et al.*, 2006; Lee *et al.*, 2006). Foreign genes are targeted to IGS regions as not to disrupt endogenous gene function. Integration of foreign sequences is dependent on homologous recombination between the transformation vector and the plastid genome. It is possible to achieve integration without complete homology but recombination and hence transformation efficiency is impaired (DeGray *et al.*, 2001; Zubko *et al.*, 2004). Additionally, evaluations of UTRs from different species indicates the need to employ species specific regulatory elements such as promoters and translation sequences to elevate the level of foreign protein expression (Kramzar *et al.*, 2006). The last year has seen a remarkable effort in this regard. Grevich and Daniell (2005) reported just six crop genomes had been published. Since then a number of agronomically important genomes such as cotton (Lee *et al.*, 2006; Ibrahim *et al.*, 2006), coffee (Samson *et al.*, 2007) grape (Jansen *et al.*, 2006), carrot (Ruhlman *et al.*, 2006), cucumber (Kim *et al.*, 2006), sweet orange (Bausher *et al.*, 2006), potato (Chung *et al.*, 2006) and tomato (Daniell *et al.*, 2006), barley, sorghum and turfgrass (Saski *et al.*, 2007) have been added (see a complete list in http://megasun.bch.umontreal.ca/ogmp/projects/other/cp.list.html) and methods have been published to enable expansion of this list (Dhingra and Folta, 2005; Jansen *et al.*, 2005).

5.1.2. *Alternative methods of DNA delivery*

If this interest in plastid genomics stems from a desire to bring new crop species to chloroplast biotechnology it will need to be accompanied by an understanding of the somatic embryogenesis system to enable selective pressure sufficient to establish homoplasmy in systems where adventitious shoot formation is not an option for regeneration (Daniell *et al.*, 2005a). The elucidation of protoplast to plant culture techniques may facilitate plastid transformation in recalcitrant species. The use of polyethylene glycol to introduce foreign DNA into plant chloroplasts has proven feasible in isolated protoplasts of lettuce and tobacco (Eibl *et al.*, 1999; Lelivelt *et al.*, 2005).

5.1.3. Eliminating antibiotic resistance markers

Ultimately acceptance of regulatory agencies and the marketplace will demand the elimination of antibiotic resistance markers in transgenic plants. Several approaches are being explored in this regard. Implementation of the CRE recombinase system in plastid transformants takes advantage of functional lox sites within the plastome (Corneille *et al.*, 2001). While this system was highly efficient in marker elimination it relies on the use of nuclear transgenic plants harboring the recombinase gene as opposed to the wild type background. Another system which relies on recombination for marker removal employs engineered direct repeats flanking the sequence to be discarded. Marker excision by this method was automatic and highly efficient without the need to engineer two genomes (Day *et al.*, 2005). Ideally an operative system that utilizes native plant sequences for markers, such as betaine aldehyde dehydrogenase (BADH) from spinach (Daniell *et al.*, 2001c), rather than antibiotic resistance would be employed. Work should be encouraged to identify possible positive selective strategies incorporating environmentally benign markers.

Transformation of crop plant plastids holds promise as a means to address some of the many and varied needs of the global population. What could be accomplished through the insightful application of this technology may only be limited by our own preconceptions of what is possible. Among the many benefits associated with plastid transformation, the ability to cointegrate several genes as an operon is especially attractive to metabolic engineering. What is the maximum number of genes, or length of sequence that can be transferred to and subsequently expressed in transgenic plastids? Presently we do not know what the upper limit. When considering the possibilities keep in mind the words of William Blake: "What is now proved was once only imagined…"

REFERENCES

Adam Z, Clarke AK (2002) Cutting edge of chloroplast proteolysis. *Trends Plant Sci* 7: 451.

Alergand T, Peled-Zehavi H, Katz Y, Danon A (2006) The chloroplast protein disulfide isomerase RB60 reacts with a regulatory disulfide of the RNA-binding protein RB47. *Plant Cell Physiol* 47 (4): 540–548.

Amira G, Ifat M, Tal A, Hana B, Shmuel G, Rachel A (2005) Soluble methionine enhances accumulation of a 15 kDa zein, a methionine-rich storage protein, in transgenic alfalfa but not in transgenic tobacco plants. *J Exp Bot* 56: 2443–2452.

Arlen PA, Falconer R, Cherukumilli S *et al.* (2007) Field production and functional evaluation of chloroplast-derived interferon alpha 2b. *Plant Biotechnol* J doi:10.1111/j.1467–7652.2007.00258.X

Avraham T, Amir R (2005) The expression level of threonine synthase and cystathionine-gamma-synthase is influenced by the level of both threonine and methionine in *Arabidopsis* plants. *Transgenic Res* 14: 299–311.

Bagga S, Armendaris A, Klypina N, Raya I, Ghoshroy S, Endress M, Sutton D, Kempb J, Sengupta-Gopalan C (2004) Genetic engineering ruminal stable high methionine protein in the foliage of alfalfa. *Plant Sci* 166: 273–283.

Bendich AJ (1987) Why do chloroplasts and mitochondria contain so many copies of their genome? *Bioessays* 6: 279.

Bausher MG, Singh ND, Lee SB *et al.* (2006) The complete chloroplast genome sequence of *Citrus sinensis* (L.) Osbeck var 'Ridge Pineapple': organization and phylogenetic relationships to other angiosperms. *BMC Plant Bio* 30:6–21

Betre H, Setton LA, Meyer DE, Chilkoti A (2002) Characterization of a genetically engineered elastin-like polypeptide for cartilaginous tissue repair. *Biomacromolecules* 3: 910–916.

Birch-Machin I, Newell CA, Hibberd JM *et al.* (2004) Accumulation of rotavirus VP6 protein in chloroplasts of transplastomic tobacco is limited by protein stability. *Plant Biotech* J 2:261–270

Birky CW, Jr. (2001) The inheritance of genes in mitochondria and chloroplasts: laws, mechanisms, and models. *Annu. Rev. Genet.* 35: 125.

Bizily SP, Rugh CL, Meagher RB (2000) Phytodetoxification of hazardous organomercurials by genetically engineered plants. *Nat Biotechnol* 18: 213–217.

Bizily SP, Rugh CL, Summers AO, Meagher RB (1999) Phytoremediation of methylmercury pollution: *merB* expression in *Arabidopsis thaliana* confers resistance to organomercurials. *Proc Natl Acad Sci USA* 96: 6808–6813.

Bohmert K, Balbo I, Kopka J, Mittendorf V, Nawrath C, Poirier Y, Tischendorf G, Trethewey RN, Willmitzer L (2000) Transgenic *Arabidopsis* plants can accumulate polyhydroxybutyrate to up to 4% of their fresh weight. *Planta* 211: 841–845.

Bonen L (2004) Translational machinery in plant organelles. *In* Molecular Biology and Biotechnology of Plant Organelles: Chloroplasts and Mitochondria. Edited by Daniell H, Chase C. Springer, Dordrecht, The Netherlands, p 323.

Borlaug N (2005) Biotechnology and the Green Revolution: Interview with Norman Borlaug. www.actionbioscience.org/biotech/borlaug.html.

Boynton JE, Gillham NW, Harris EH, Hosler JP, Johnson AM, Jones AR, Randolph-Anderson BL, Robertson D, Klein TM, Shark KB (1988) Chloroplast transformation in *Chlamydomonas* with high velocity microprojectiles. *Science* 240: 1534.

Cerutti H, Jagendorf AT (1993) DNA strand-transfer activity in pea (*Pisum sativum* L.) chloroplasts. *Plant Physiol* 102: 145.

Chakraborty S, Chakraborty N, Datta A (2000) Increased nutritive value of transgenic potato by expressing a nonallergenic seed albumin gene from *Amaranthus hypochondriacus*. *Proc Natl Acad Sci USA* 97: 3724–3729.

Chase CD (2006) Genetically engineered cytoplasmic male sterility. *Trends Plant Sci* 11: 7.

Chebolu S and Daniell H (2007) Stable expression of GAL/GALNAc lectin of *Entamoeba histolytica* in transgenic chloroplast and immunogenicity in mice towards vaccine development for amebiasis. *Plant Biotech J* 2:230–239

Chin HG, Kim GD, Marin I, Mersha F, Evans TC, Jr., Chen L, Xu MQ, Pradhan S (2003) Protein trans-splicing in transgenic plant chloroplast: Reconstruction of herbicide resistance from split genes. *Proc Natl Acad of Sci USA*100: 4510.

Chung HJ, Jung JD, Park HW *et al.* (2006) The complete chloroplast genome sequences of *Solanum tuberosum* and comparative analysis with Solanaceae species identified the presence of a 241 bp deletion in cultivated potato chloroplast DNA sequence. *Plant Cell Rep* 25:1360–1379

Colas des Francs-Small C, Szurek B, Small I (2004) Proteomics, bioinformatics and genomics applied to plant organelles. *In* Molecular Biology and Biotechnology of Plant Organelles: Chloroplast and Mitochondria. Edited by Daniell H, Chase C Springer, Dordrecht, The Netherlands, p 179.

Collakova E, DellaPenna D (2003a) Homogentisate Phytyltransferase activity is limiting for tocopherol biosynthesis in *Arabidopsis*. *Plant Physiol.* 131: 632–642.

Collakova E, DellaPenna D (2003b) The role of Homogentisate Phytyltransferase and other tocopherol pathway enzymes in the regulation of tocopherol synthesis during abiotic stress. *Plant Physiol.* 133: 930–940.

Corneille S, Lutz K, Svab Z, Maliga P (2001) Efficient elimination of selectable marker genes from the plastid genome by the CRE-lox site-specific recombination system. *Plant J.* 27: 171.

Crickmore N, Ellar DJ (1992) Involvement of a possible chaperonin in the efficient expression of a cloned *cryIIA* delta-endotoxin gene in *Bacillus thuringiensis*. *Mol Microbiol* 6: 1533–1537.

Daniell H (2002) Molecular strategies for gene containment in transgenic crops. *Nat Biotechnol* 20: 581–586.

Daniell H (2006) Production of biophamaceuticals and vaccines in plants *via* the chloroplast genome. *Biotechnol J* 1:1071–1079

Daniell H, Chebolu S, Kumar S, Singleton M, Falconer R (2005) Chloroplast-derived vaccine antigens and other therapeutic proteins. *Vaccine* 23: 1779–1783.

Daniell H, Datta R, Varma S *et al.* (1998) Containment of herbicide resistance through genetic engineering of the chloroplast genome. *Nat Biotechnol* 16:345–348

Daniell H, Guda C, McPherson DT, Zhang X, Xu J, Urry DW (1997) Hyperexpression of a synthetic protein-based polymer gene. *Methods Mol Biol* 63: 359–371.

Daniell H, Kumar S, Dufourmantel N (2005) Breakthrough in chloroplast genetic engineering of agronomically important crops. *Trends Biotechnol* 23: 238.

Daniell H, Lee S-B, Panchal T, Wiebe PO (2001a) Expression of the native cholera toxin B subunit gene and assembly as functional oligomers in transgenic tobacco chloroplasts. *J Mol Biol 311*: 1001.

Daniell H, Lee SB, Grevich J, Saski C, Quesada-Vargas T, Guda C, Tomkins J, Jansen RK (2006) Complete chloroplast genome sequences of *Solanum bulbocastanum*, *Solanum lycopersicum* and comparative analyses with other Solanaceae genomes. *Theor Appl Genet* 112: 1503–1518.

Daniell H, McFadden BA (1987) Uptake and expression of bacterial and cyanobacterial genes by isolated cucumber etioplasts. *Proc Natl Acad Sci USA* 84:6349–53

Daniell H, Muthukumar B, Lee SB (2001c) Marker free transgenic plants: engineering the chloroplast genome without the use of antibiotic selection. *Curr Genet* 39: 109–116.

Daniell H, Ruiz ON, Dhingra A (2005) Chloroplast genetic engineering to improve agronomic traits. *Methods Mol. Biol* 286: 111.

Daniell H, Streatfield SJ, Wycoff K (2001b) Medical molecular farming: production of antibodies, biopharmaceuticals and edible vaccines in plants. *Trends Plant Sci* 6: 219–226.

Daniell H, Vivekananda J, Nielsen BL *et al.* (1990) Transient foreign gene expression in chloroplasts of cultured tobacco cells after biolistic delivery of chloroplast vectors. *Proc Natl Acad Sci USA* 87:88–92

Day A, Kode V, Madesis P, Iamtham S (2005) Simple and efficient removal of marker genes from plastids by homologous recombination. *Methods Mol. Biol* 286: 255.

Dertzbaugh MT, Cox LM (1998) The affinity of cholera toxin for Ni2+ ion. *Protein Eng* 11:557–581

De Cosa B, Moar W, Lee SB, Miller M, Daniell H (2001) Overexpression of the *Bt cry2Aa2* operon in chloroplasts leads to formation of insecticidal crystals. *Nat Biotechnol* 19: 71–74.

DeGray G, Rajasekaran K, Smith F, Sanford J, Daniell H (2001) Expression of an antimicrobial peptide via the chloroplast genome to control phytopathogenic bacteria and fungi. *Plant Physiol* 127: 852–862.

Denecke J, De Rycke R, Botterman J (1992) Plant and mammalian sorting signals for protein retention in the endoplasmic reticulum contain a conserved epitope. *EMBO J* 11: 2345–2355.

Devine A (2005) Oral administration of chloroplast-derived cholera toxin B-proinsulin fusion protein protects against development of insulitis in non-obese diabetic mice. Thesis, University of Central Florida

Dhingra A, Folta KM (2005) ASAP: amplification, sequencing & annotation of plastomes. *BMC Genomics* 6: 176.

Dhingra A, Portis AR, Jr., Daniell H (2004) Enhanced translation of a chloroplast-expressed RbcS gene restores small subunit levels and photosynthesis in nuclear RbcS antisense plants. *Proc Natl Acad Sci USA* 101: 6315.

Di R, Kim J, Martin MN, Leustek T, Jhoo J, Ho CT, Tumer NE (2003) Enhancement of the primary flavor compound methional in potato by increasing the level of soluble methionine. *J Agric Food Chem* 51: 5695–5702.

Dill G (2005) Glyphosate-resistant crops: history, status and future. *Pest Manag Sci* 61: 219–224.

Dufourmantel N, Tissot G, Goutorbe F, GarÃ§on F, Muhr C, Jansens S, Pelissier B, Peltier G, Dubald M (2005) Generation and analysis of soybean plastid transformants expressing *Bacillus thuringiensis* Cry1Ab Protoxin. *Plant Mol Biol* 58: 659.

Eibl C, Zou Z, Beck a, Kim M, Mullet J, Koop HU (1999) *In vivo* analysis of plastid *psbA*, *rbcL* and *rpl32* UTR elements by chloroplast transformation: tobacco plastid gene expression is controlled by modulation of transcript levels and translation efficiency. *Plant J* 19: 333.

Fagard M, Vaucheret H (2000) (Trans)gene silencing in plants: How many mechanisms? *Annu Rev Plant Physiol Plant Mol. Biol* 51: 167.

Falco SC, Guida T, Locke M, Mauvais J, Sanders C, Ward RT, Webber P (1995) Transgenic canola and soybean seeds with increased lysine. *Biotechnology* 13: 577–582.

Falk J, Brosch M, Schafer A, Braun S, Krupinska K (2005) Characterization of transplastomic tobacco plants with a plastid localized barley 4-hydroxyphenyl-pyruvate dioxygenase. *J of Plant Physiol* 162: 738.

Farran I, Sánchez-Serrano JJ, Medina JF *et al.* (2002) Targeted Expression of Human Serum Albumin to Potato Tubers. *Transgenic Res* 11:337–346

Fernandez-San Millan A, Mingo-Castel A, Miller M, Daniell H (2003) A chloroplast transgenic approach to hyper-express and purify human serum albumin, a protein highly susceptible to proteolytic degradation. *Plant Biotechnol J* 1: 71.

Fitch MMM, Manshardt RM, Gonsalves D *et al.*, (1992) Virus resistant papaya plants derived form tissues bombarded with the coat protein gene of papaya ringspot virus. *Nat Biotechnol* 10: 1466–1472

Galili G, Amir R, Hoefgen R, Hesse H (2005) Improving the levels of essential amino acids and sulfur metabolites in plants. *Biol Chem* 386: 817–831.

Goto F, Yoshihara T, Shigemoto N, Toki S, Takaiwa F (1999) Iron fortification of rice seed by the soybean ferritin gene. *Nat Biotechnol* 17: 282.

Guda C, Lee SB, Daniell H (2000) Stable expression of a biodegradable protein-based polymer in tobacco chloroplasts. *Plant Cell Rep* 19: 257.

Guenoune D, Amir R, Badani H, Wolf S, Galili S (2003) Coexpression of the soybean vegetative storage protein beta subunit (S-VSP β) either with the bacterial feedback-insensitive dihydrodipicolinate synthase or with S-VSP a stabilizes the S-VSP β transgene protein and enhances lysine production in transgenic tobacco plants. *Transgenic Res* 12: 123–126.

Hagemann R (2004) The sexual inheritance of plant organelles. *In* Molecular Biology and Biotechnology of Plant Organelles: Chloroplast and Mitochondria. Edited by Daniell H, Chase C. Springer, Dordrecht, The Netherlands, p 93.

Havey M (2004) The use of cytoplasmic male sterility for hybrid seed produciton. *In* Molecular Biology and Biotechnology of Plant Organelles. Edited by Daniell H, Chase C. Springer, Dordrecht, The Netherlands, pp 617–628.

Herbers K (2003) Vitamin production in transgenic plants. *J Plant Physiol* 160: 821.

Herrmann KM, Weaver LM (1999) The Shikimate Pathway. *Ann Rev Plant Physiol Plant Mol Biol* 50: 473–503.

Herzog RW, Singh NK, Urry DW, Daniell H (1997) Expression of a synthetic protein-based polymer (elastomer) gene in *Aspergillus nidulans*. *Appl Microbiol Biotechnol* 47: 368–372.

Iamtham S, Day A (2000) Removal of antibiotic resistance genes from transgenic tobacco plastids. *Nat Biotechnol* 18:172–176

http://megasun.bch.umontreal.ca/ogmp/projects/other/cp.list.html.

Ibrahim RI, Azuma J, Sakamoto M (2006) Complete nucleotide sequence of the cotton (*Gossypium barbadense* L.) chloroplast genome with a comparative analysis of sequences among 9 dicot plants. *Genes Genet Syst* 81:311–321

Jansen RK, Kaittanis C, Lee SB, Saski C, Tomkins J, Alverson AJ, Daniell H (2006) Phylogenetic analyses of *Vitis* (Vitaceae) based on complete chloroplast genome sequences: effects of taxon sampling and phylogenetic methods on resolving relationships among rosids. *BMC Evol Biol* 6: 32.

Jansen RK, Raubeson LA, Boore JL, dePamphilis CW, Chumley TW, Haberle RC, Wyman SK, Alverson AJ, Peery R, Herman SJ, Fourcade HM, Kuehl JV, McNeal JR, Leebens-Mack J, Cui L (2005) Methods for obtaining and analyzing whole chloroplast genome sequences. *Methods Enzymol* 395: 348–384.

Jeong SW, Jeong WJ, Woo JW, Choi DW, Park YI, Liu JR (2004) Dicistronic expression of the green fluorescent protein and antibiotic resistance genes in the plastid for selection and tracking of plastid-transformed cells in tobacco. *Plant Cell Rep* 22: 747.

Kanamoto H, Yamashita A, Okumura S, Hattori M, Tomizawa K (2005) The complete genome sequence of the *Lactuca sativa* (lettuce) chloroplast. Unpublished.

Karchi H, Shaul O, Galili G (1994) Lysine synthesis and catabolism are coordinately regulated during tobacco seed development. *Proc Natl Acad Sci USA* 91: 2577–2581.

Khan MR, Ceriotti A, Tabe L, Aryan A, McNabb W, Moore A, Craig S, Spencer D, Higgins TJ (1996) Accumulation of a sulfur-rich seed albumin from sunflower in the leaves of transgenic subterranean clover (*Trifolium subterraneum* L.). *Transgenic Res* 5: 179–185.

Kim J, Mayfield SP (2002) The active site of the thioredoxin-like domain of chloroplast protein disulfide isomerase, RB60, catalyzes the redox-regulated binding of chloroplast poly(A)-binding protein, RB47, to the 5' untranslated region of *psbA* mRNA. *Plant Cell Physiol* 43: 1238.

Kim JS, Jung JD, Lee JA, Park HW, Oh KH, Jeong WJ, Choi DW, Liu JR, Cho KY (2006) Complete sequence and organization of the cucumber (*Cucumis sativus* L. cv. Baekmibaekdadagi) chloroplast genome. *Plant Cell Rep* 25:334–340.

Kinney AJ (2006) Metabolic engineering in plants for human health and nutrition. *Curr Opin Biotechnol* 17(2): 130–138.

Kolodner R, Tewari KK (1975) The molecular size and conformation of the chloroplast DNA from higher plants. *Biochim Biophys Acta* 402: 372.

Kolodner R, Tewari KK, Warner RC (1976) Physical studies on the size and structure of the covalently closed circular chloroplast DNA from higher plants. *Biochim Biophys Acta* 447: 144.

Koya V, Moayeri M, Leppla SH, Daniell H (2005) Plant-based vaccine: mice immunized with chloroplast-derived anthrax protective antigen survive anthrax lethal toxin challenge. *Infect Immun* 73: 8266–8274.

Kramzar LM, Mueller T, Erickson B, Higgs DC (2006) Regulatory sequences of orthologous *petD* chloroplast mRNAs are highly specific among *Chlamydomonas* species. *Plant Mol Biol* 60: 405–422.

Kumar S, Dhingra A, Daniell H (2004a) Plastid-expressed betaine aldehyde dehydrogenase gene in carrot cultured cells, roots, and leaves confers enhanced salt tolerance. *Plant Physiol* 136: 2843.

Kumar S, Dhingra A, Daniell H (2004b) Stable transformation of the cotton plastid genome and maternal inheritance of transgenes. *Plant Mol Biol* 56: 203–216.

Lee SB, Kaittanis C, Jansen RK, Hostetler JB, Tallon LJ, Town CD, Daniell H (2006) The complete chloroplast genome sequence of *Gossypium hirsutum*: organization and phylogenetic relationships to other angiosperms. *BMC Genomics* 7: 61.

Lee SB, Kwon HB, Kwon SJ *et al.* (2003) Accumulation of trehalose within transgenic chloroplasts confers drought tolerance. *Mol Breeding* 11: 1–13.

Leelavathi S, Reddy VS (2003) Chloroplast expression of His-tagged GUS-fusions: a general strategy to overproduce and purify foreign proteins using transplastomic plants as bioreactors. *Mol Breed* 11: 49.

Lelivelt C, McCabe M, Newell C, deSnoo C, Dun K, Birch-Machin I, Gray J, Mills K, Nugent J (2005) Stable plastid transformation in lettuce (*Lactuca sativa* L.). *Plant Mol Biol* 58: 763.

Levitan A, Trebitsh T, Kiss V, Pereg Y, Dangoor I, Danon A (2005) Dual targeting of the protein disulfide isomerase RB60 to the chloroplast and the endoplasmic reticulum. *Proc Natl Acad Sci USA* 102: 6225–6230.

Limaye A, Koya V, Samsam M, Daniell H (2006) Receptor-mediated oral delivery of a bioencapsulated green fluorescent protein expressed in transgenic chloroplasts into the mouse circulatory system. *FASEB J* 20: 959–961.

Lossl A, Bohmert K, Harloff H, Eibl C, Muhlbauer S, Koop HU (2005) Inducible trans-activation of plastid transgenes: expression of the *R. eutropha phb* operon in transplastomic tobacco. *Plant Cell Physiol* 46: 1462–1471.

Lossl A, Eibl C, Harloff HJ, Jung C, Koop HU (2003) Polyester synthesis in transplastomic tobacco (*Nicotiana tabacum* L.): significant contents of polyhydroxybutyrate are associated with growth reduction. *Plant Cell Rep* 21: 891–899.

Ludwig DS, Holmes RK, Schoolnik GK (1985) Chemical and Immunochemical studies on the receptor binding domain of cholera toxin B subunit. *J Biol Chem* 260:12528–12534

Ma JK, Hiatt A, Hein M, Vine ND, Wang F, Stabila P, van Dolleweerd C, Mostov K, Lehner T (1995) Generation and assembly of secretory antibodies in plants. *Science* 268: 716–719.

Ma JK, Hikmat BY, Wycoff K, *et al.* (1998) Chacterization of a recombinant plant monoclonal secretory antibody and preventative immunotherapy in humans. *Nat Med* 4:601–606

Mache R, Lerbs-Mache S (2001) Chlroplast genetic system of higher plants: Chromosome replication, chloroplast division and elements of the transcriptional apparatus. *Curr Sci* 80: 217.

Maeda H, Katsuki T, Akaike T, Yasutake R (1992) High correlation between lipid peroxide radical and tumor-promoter effect: suppression of tumor promotion in the Epstein-Barr virus/B-lymphocyte system and scavenging of alkyl peroxide radicals by various vegetable extracts. *Jpn J Cancer Res* 83: 923–928.

Maier RM, Schmitz-Linneweber C (2004) Plastid Genomes. *In* Molecular Biology and Biotechnology of Plant Organelles: Chloroplasts and Mitochondria. Edited by Daniell H, Chase C. Springer, Dordrecht, The Netherlands, pp 115–150.

Mazur B, Krebbers E, Tingey S (1999) Gene discovery and product development for grain quality traits. *Science* 285: 372–375.

McBride KE, Svab Z, Schaff DJ *et al.* (1995) Amplification of a chimeric Bacillus gene in chloroplasts leads to an extraordinary level of insecticidal protein in tobacco. *Biotechnology* 13:362–365

Meyer DE, Chilkoti A (2002) Genetically encoded synthesis of protein-based polymers with precisely specified molecular weight and sequence by recursive directional ligation: Examples from the elastin-like polypeptide system. *Biomacromolecules* 3: 357–367.

Molina A, Herva-Stubbs S, Daniell H, et al. (2004) High yield expression of a viral peptide animal vaccine in transgenic tobacco chloroplasts. *Plant Biotech J* 2:141–153

Moller SG, Moller SG (2005) Plastid division in higher plants. *In* Plastids. Edited by Roberts J. Blackwell, Oxford, UK, p 126.

Monde RA, Schuster G, Stern DB (2000) Processing and degradation of chloroplast mRNA. *Biochimie* 82: 573.

Muhlbauer SK, Koop H-U (2005) External control of transgene expression in tobacco plastids using the bacterial *lac* repressor. *The Plant J* 43: 941–946.

Muntz K, Christov V, Saalbach G, Saalbach I, Waddell D, Pickardt T, Schieder O, Wustenhagen T (1998) Genetic engineering for high methionine grain legumes. *Nahrung* 42: 125–127.

Nakashita H, Arai Y, Shikanai T, Doi Y, Yamaguchi I (2001) Introduction of bacterial metabolism into higher plants by polycistronic transgene expression. *Biosci Biotechnol Biochem* 65: 1688–1691.

Nawrath C, Poirier Y, Somerville C (1994) Targeting of the polyhydroxybutyrate biosynthetic pathway to the plastids of *Arabidopsis thaliana* results in high levels of polymer accumulation. *Proc Natl Acad Sci USA* 91: 12760–12764.

Padgette SR, Kolacz KH, Delannay X *et al.* (1995) Development, identificaiton and characterization of a glyphosate-tolerant soybean line. *Crop Sci* 35:1451–1461

Palmer JD (1983) Chloroplast DNA exists in two orientations. *Nature* 301: 92.

Parry MA, Andralojc PJ, Mitchell RA, Madgwick PJ, Keys AJ (2003) Manipulation of Rubisco: the amount, activity, function and regulation. *J Exp Bot* 54: 1321–1333.

Pastori M, Pfander H, Boscoboinik D, Azzi A (1998) Lycopene in association with alpha-tocopherol inhibits at physiological concentrations proliferation of prostate carcinoma cells. *Biochem Biophys Res Commun* 250: 582–585.

Perl A, Shaul O, Galili G (1992) Regulation of lysine synthesis in transgenic potato plants expressing a bacterial dihydrodipicolinate synthase in their chloroplasts. *Plant Mol Biol* 19: 815–823.

Perlak FJ, Deaton RW, Armstrong TA *et al.* (1990) Insect resistant cotton. *Biotechnology* 8:939–943

Quesada-Vargas T, Ruiz ON, Daniell H (2005) Characterization of heterologous multigene operons in transgenic chloroplasts: transcription, processing, and translation. *Plant Physiol* 138: 1746–1762.

Raubeson LA, Jansen RK (2005) Chloroplast genomes of land plants. In: Henry RJ (ed) Plant diversity and evolution: genotypic variation in higher plants. CAB International, Wallingford UK

Reynolds MP, van Ginkel M, Ribaut JM (2000) Avenues for genetic modification of radiation use efficiency in wheat. *J Exp Bot* 51 Spec No: 459–473.

Richly E, Leister D (2004) NUMTs in sequenced eukaryotic genomes. *Mol.Biol Evol.* 21: 1081.

Rodermel S (1999) Subunit control of Rubisco biosynthesis: a relic of an endosymbiotic past? *Photosynth Res* 59: 10.

Rodermel SR, Abbott MS, Bogorad L (1988) Nuclear-organelle interactions: nuclear antisense gene inhibits ribulose bisphosphate carboxylase enzyme levels in transformed tobacco plants. *Cell* 55: 673–681.

Rugh CL, Wilde HD, Stack NM, Thompson DM, Summers AO, Meagher RB (1996) Mercuric ion reduction and resistance in transgenic *Arabidopsis thaliana* plants expressing a modified bacterial *merA* gene. *Proc Natl Acad Sci USA* 93: 3182–3187.

Ruhlman T, Ahangari R, Devine A *et al.* (2007) Expression of cholera toxin B-proinsulin fusion protein in lettuce and tobacco chloroplast – oral administration protects against development of insulitis in non-obese diabetic mice. *Plant Biotech J* doi:10.1111/j. 1467–7652.2007.00259.X

Ruiz ON, Daniell H (2005) Engineering cytoplasmic male sterility via the chloroplast genome by expression of β-ketothiolase. *Plant Physiol* 138: 1232–1246.

Ruiz ON, Hussein HS, Terry N, Daniell H (2003) Phytoremediation of organomercurial compounds via chloroplast genetic engineering. *Plant Physiol* 132: 1344.

Saeglitz C, Pohl M, Bartsch D (2000) Monitoring gene flow from transgenic sugar beet using cytoplasmic male-sterile bait plants. *Mol Ecol* 9: 2035–2040.

Samson N, Bausher MG, Lee SB et al. (2007) The complete nucleotide sequence of the coffee (*Coffea arabica* L.) chloroplast genome: organization and implications for biotechnology and phylogenetic relationships amongst angiosperms. *Plant Biotech J* 5:339–353

Santillo V, Lowe F (2006) Role of vitamins, minerals and supplements in the prevention and management of prostate cancer. *Internat Braz J Urol* 32: 3–14.

Saski C, Lee SB, Daniell H, Wood TC, Tomkins J, Kim HG, Jansen RK (2005) Complete chloroplast genome sequence of *Gycine max* and comparative analyses with other legume genomes. *Plant Mol Biol* 59: 309–322.

Saski C, Lee SB, Fjelheim S *et al.* (2007) Complete chloroplast genome sequence of *Hordeum vulgare, Sorghum bicolor* and *Agrostis stolonifera*. *Theor Appl Genet*, in press

Shaul O, Galili G (1993) Concerted regulation of lysine and threonine synthesis in tobacco plants expressing bacterial feedback-insensitive aspartate kinase and dihydrodipicolinate synthase. *Plant Mol Biol* 23: 759–768.

Shiau C-W, Huang J-W, Wang D-S, Weng J-R, Yang C-C, Lin C-H, Li C, Chen C-S (2006) a-Tocopheryl succinate induces apoptosis in prostate cancer cells in part through inhibition of BCL-XL/BCL-2 function. *J Biol Chem*: 281(17):11819–25.

Shou H, Bordallo P, Wang K (2004) Expression of the *Nicotiana* protein kinase (NPK1) enhanced drought tolerance in transgenic maize. *J Exp Bot* 55: 1013.

Spreitzer RJ (1999) Questions about the complexity of chloroplast ribulose-1,5-bisphosphate carboxylase/oxygenase. *Photosynthesis Res* 60: 29–42.

Spreitzer RJ, Salvucci ME (2002) Rubisco: structure, regulatory interactions, and possibilities for a better enzyme. *Annu Rev Plant Biol* 53: 449–475.

Staples N, Ellar D, Crickmore N (2001) Cellular localization and characterization of the *Bacillus thuringiensis orf2* crystallization factor. *Curr Microbiol* 42: 388–392.

Staub JM, Garcia B, Graves J, Hajdukiewicz PT, Hunter P, Nehra N, Paradkar V, Schlittler M, Carroll JA, Spatola L, Ward D, Ye G, Russell DA (2000) High-yield production of a human therapeutic protein in tobacco chloroplasts. *Nat. Biotechnol* 18: 333.

Sugiura M (1992) The chloroplast genome. *Plant Mol. Biol* 19: 149.

Tabe LM, Wardley-Richardson T, Ceriotti A, Aryan A, McNabb W, Moore A, Higgins TJ (1995) A biotechnological approach to improving the nutritive value of alfalfa. *J Anim Sc* i 73: 2752–2759.

Tang G, Miron D, Zhu-Shimoni JX, Galili G (1997) Regulation of lysine catabolism through lysine-ketoglutarate reductase and saccharopine dehydrogenase in *Arabidopsis*. *Plant Cell* 9: 1305–1316.

Trebitsh T, Meiri E, Ostersetzer O, Adam Z, Danon A (2001) The protein disulfide isomerase-like RB60 is partitioned between stroma and thylakoids in *Chlamydomonas reinhardtii* chloroplasts. *J Biol Chem* 276: 4564.

Tregoning J, Maliga P, Dougan G, Nixon PJ (2004) New advances in the production of edible plant vaccines: chloroplast expression of a tetanus vaccine antigen, TetC. *Phytochemistry* 65: 989.

Urry DW (1988) Entropic elastic processes in protein mechanisms. I. Elastic structure due to an inverse temperature transition and elasticity due to internal chain dynamics. *J Protein Chem* 7: 1–34.

Viitanen PV, Devine AL, Khan MS, Deuel DL, Van Dyk DE, Daniell H (2004) Metabolic engineering of the chloroplast genome using the *Echerichia coli ubiC* gene reveals that chorismate is a readily abundant plant precursor for p-hydroxybenzoic acid biosynthesis. *Plant Physiol* 136(4): 4048–4060.

Wandelt CI KM, Craig S, Schroeder HE, Spencer D, Higgins TJ (1992) Vicilin with carboxy-terminal KDEL is retained in the endoplasmic reticulum and accumulates to high levels in the leaves of transgenic plants.. *The Plant J* 2(2): 181.

Watson J, Koya V, Leppla SH, Daniell H (2004) Expression of *Bacillus anthracis* protective antigen in transgenic chloroplasts of tobacco, a non-food/feed crop. *Vaccine* 22: 4374.

Wrobel M, Zebrowski J, Szopa J (2004) Polyhydroxybutyrate synthesis in transgenic flax. *J Biotechnol* 107: 41–54.

Ye X, Al Babili S, Kloti A, Zhang J, Lucca P, Beyer P, Potrykus I (2000) Engineering the provitamin A (beta-carotene) biosynthetic pathway into (carotenoid-free) rice endosperm. *Science* 287: 303.

Zeh M, Casazza AP, Kreft O, Roessner U, Bieberich K, Willmitzer L, Hoefgen R, Hesse H (2001) Antisense inhibition of threonine synthase leads to high methionine content in transgenic potato plants. *Plant Physiol* 127: 792–802.

Zerges W (2000) Translation in chloroplasts. *Biochimie* 82: 583.

Zhang X, Guda C, Datta R, Dute R, Urry D, Daniell H (1995) Nuclear expression of an environmentally friendly synthetic protein-based polymer in tobacco cells. *Biotechnol. Lett* 17: 1279–1284.

Zhang X, Urry DW, Daniell H (1996) Expression of an environmentally friendly synthetic protein-based polymer gene in transgenic tobacco plants. *Plant Cell Rep* 16: 174.

Zhu X, Tang G, Granier F, Bouchez D, Galili G (2001) A T-DNA insertion knockout of the bifunctional lysine-ketoglutarate reductase/saccharopine dehydrogenase gene elevates lysine levels in *Arabidopsis* seeds. *Plant Physiol* 126: 1539–1545.

Zubko MK, Zubko EI, Zuilen Kv, Meyer P, Day A (2004) Stable transformation of petunia plastids. *Transgenic Res* 13: 523.

CHAPTER 4

METABOLIC ENGINEERING OF THE ALKALOID BIOSYNTHESIS IN PLANTS: FUNCTIONAL GENOMICS APPROACHES

KIRSI-MARJA OKSMAN-CALDENTEY*, SUVI T. HÄKKINEN AND HEIKO RISCHER

VTT Technical Research Centre of Finland
P.O. Box 1000
FI – 02044 VTT (Espoo)
Finland

Abstract: Numerous pharmaceuticals currently on the market are based on plant-derived compounds. Many of these compounds are still isolated from whole plants, this being the only feasible production method. The exploitation of cell culture systems and biotechnological production of these complex molecules has been limited by the limited knowledge on their biosynthesis. Understanding the complexity of the regulation of plant metabolism has deepened in recent years, due to major advances in plant genomics and metabolomics. A general problem encountered when characterizing the plant metabolome is the extreme diversity of the compounds which sets a challenge to analytical methods. Modern systems biology tools, together with the development of large plant genomics and metabolomics databases will dramatically facilitate the advance in our knowledge of gene-to-metabolite networks in plants.

Here we describe recent progress in studies on nicotine, terpenoid indole and tropane alkaloid pathways, and introduce the technology platform which has been developed for the exploration of poorly understood biosynthetic pathways in medicinal plants. This approach, based on functional genomics, has been applied to identify genes involved in alkaloid pathways. Furthermore, it is shown how combinatorial biochemistry can be used for creating entirely novel plant-derived compounds. The great advantage of this technology is that it's applicable to any plant species, this being particularly important when it comes to exotic medicinal plants. Better understanding of metabolite synthesis and its regulation will be of crucial importance for improving the efficiency and sustainability of plant secondary metabolite production

Keywords: nicotine alkaloids, terpenoid indole alkaloids, tropane alkaloids, functional genomics, pathway engineering

* Correnspondence email: kirsi-marja.oksman@vtt.fi

R. Verpoorte et al. (eds.), Applications of Plant Metabolic Engineering, 109–127.
© 2007 *Springer*.

1. INTRODUCTION

Plants are excellent organic chemists in nature and constantly respond to environmental changes by adjusting their capacity to produce natural products. Low molecular weight products, so-called secondary metabolites, exhibit many biological functions vital for survival such as responding to stress (Hirai et al., 2004) or mediating pollination (Osbourn et al., 2003). However, for most of the secondary metabolites the exact function in plants still remains unknown. Since the discovery of the opium alkaloid morphine almost two centuries ago, alkaloids are still one of the most studied groups of plant secondary metabolites. Currently more than 15 000 different alkaloids are known (Verpoorte, 2000) and they are classified into several subclasses according to their chemical structures. Contrary to e.g. plant phenolics which are abundant throughout the whole plant kingdom, alkaloids are often restricted to certain plant families or even certain plant species. The reason why alkaloids have been of so wide interest can be explained by their strong physiological properties leading to their use as pharmaceuticals. Furthermore the isolation of alkaloids from plant matrices is relatively simple compared to many other plant compounds. This has facilitated the analysis and isolation of very small amounts of various alkaloids using different chromatographical methods (e.g. LC, GC) and their structure elucidations by spectroscopical systems (MS, NMR).

Although many alkaloids have been used for decades by humans surprisingly little is known how these valuable plant secondary metabolites are formed biosynthetically in plant tissues. However, the past years have provided a lot of new molecular information on the biosynthesis of some of the most studied alkaloids. Their biosynthesis often starts from amino acids and involves typically dozens of intermediates and enzymes (van der Heijden et al., 2004). Many of these enzymes have been localized in subcellular compartments other than the cytosol (Burlat et al., 2004). Not only enzymatic genes but also master regulators e.g. transcription factors have an important role in controlling the overall machinery of the biosynthetic pathways (van der Fits and Memelink, 2000). Often various biosynthetic steps take place in different plant organs, cell types or organelles and extensive intra- or intercellular translocation is not unusual in alkaloid biosynthesis (St-Pierre et al., 1999). The first alkaloid biosynthetic pathway which was fully characterized was the route to the benzophenanthridine alkaloid berberine. Meanwhile the elucidation of the morphine pathway is approaching completion, too (Ounaroon et al., 2003), (Weid et al., 2004). Characterization of many other alkaloid biosynthetic pathways is currently going on by several research groups worldwide using either classical biochemical or functional genomics approaches (Weid et al., 2004), (Goossens et al., 2003a), (Hirai et al., 2004), (Rischer et al., 2006). We believe, anyhow, that successful metabolic engineering of secondary metabolite pathways is dependent on a thorough knowledge of the pathway in question and a detailed understanding of the regulatory mechnisms controlling the onset and flux of the desired metabolites (Oksman-Caldentey and Inzé, 2004).

Development of functional genomics tools together with the availability of the entire genome sequence of *Arabidopsis thaliana* (Arabidopsis Genome Initiative, 2000), followed by rice (Yu et al., 2002) and other model or crop plants soon to become available, allows elucidating the primary and secondary metabolite synthesis by a systems biology approach (Oksman-Caldentey and Inzé, 2004), (Oksman-Caldentey and Saito, 2005), (Hirai et al., 2004). The mining and exploitation of the enormous amount of data obtained from genomics and metabolomics will certainly bring us into a new era of understanding plants and other biological systems (Oksman-Caldentey and Inzé, 2004), (Hirai et al., 2004), (Rischer et al., 2006).

In this chapter recent advances on studies of nicotine, terpenoid indole and tropane alkaloid pathways are described. We present a novel technology platform based on functional genomics which has been successfully developed and applied to discover gene-to-metabolite networks in medicinal plants in order to engineer plant cells for the production of known and novel alkaloids. The role of transporter genes for the accumulation and secretion of alkaloids is also highlighted.

2. THE DEVELOPMENT OF PLANT GENOMICS AND METABOLOMICS

Rational engineering of complicated metabolic pathways involved in the production of plant secondary metabolites has been greatly hampered by our poor knowledge of the biosynthetic pathways and their regulatory mechanisms leading to the formation of the desired compounds e.g. alkaloids. Furthermore flux charts of these metabolic pathways are almost not existing. Labelled precursor feedings have on the one hand been successfully performed to draw chemical reaction schemes of the pathways but still there are unknown steps in many alkaloid pathways (Robins et al., 1994a), (Oksman-Caldentey and Inzé, 2004). On the other hand low abundance of certain alkaloids and their respective biosynthetic enzymes in plant tissues has made it difficult to elucidate alkaloid pathways by classical approaches. Peptide sequences have been determined from purified enzymes in reverse genetics approaches for example which then allows detecting the right genes encoding these enzymes. Alternatively homology based screening of candidate clones in combination with expression pattern analysis can be used. These approaches have allowed the discovery of dozens of enzymatic genes catalyzing different steps of alkaloid biosynthesis (Facchini et al., 2004), (Facchini and St-Pierre, 2005).

In contrast to traditional biochemical analysis with a limited number of genes and metabolites, genome-wide methods are now available for the holistic analysis of some model plant species. Unfortunately for many interesting and valuable medicinal plants genomics tools, based on either complete genome data or expressed sequence tag (EST) databases, are not easily applicable. However, a transcriptional profiling method called cDNA-AFLP (amplified fragment length polymorphism) transcriptional profiling, offers a viable alternative (Breyne and Zabeau, 2001),

(Goossens et al., 2003a), (Rischer et al., 2006). One of the greatest advantages of cDNA-AFLP is that prior sequence data is not needed and thus is very suitable for exotic, rare medicinal plants. Moreover it gives quantitative expression profiles and allows identification of novel genes as well as distinguishes between the isoforms of encoded proteins.

Metabolomics which means the ability to measure large numbers of biomolecular components in parallel has only recently been recognized as an important element of post-genome science (Oksman-Caldentey and Saito, 2005). Metabolic fingerprinting as a phenotyping tool dates already back to the 1970s (Jellum, 1977) but metabolomics made the first breakthrough in microbes (Tweeddale et al., 1998) and only afterwards it showed its potential in studying plant metabolites, too (Trethewey et al., 1999). However until recently almost all plant metabolomics studies have basically comprised primary metabolites only.

General problems encountered when characterizing the plant metabolome are the highly complex nature and the enormous chemical diversity of the compounds (Oksman-Caldentey et al., 2004). This is especially obvious for secondary metabolite analysis which is far more complex than metabolite profiling of primary metabolites. Plants produce about 200,000 or more metabolites (Fiehn, 2002), and within a particular species the number of metabolites is estimated to be 5000–25000 which is comparable in order of magnitude to the number of genes (Trethewey, 2004). It has been estimated that 25-30% of the genes of *Arabidopsis* encode enzymes of metabolism (Arabidopsis Genome Initiative, 2000). One of the key challenges is to find an optimal balance between the coverage of the metabolites and the accuracy of the metabolite measurements (Oksman-Caldentey et al., 2004). The instrumentation has developed rapidly during the past decade and allows now to increase the coverage of metabolites within a single analysis (Fiehn, 2002). However, if one applies only one extraction system, it is very likely that many metabolites remain in the plant matrix and cannot be profiled. Therefore we propose the use of at least two different metabolomics platforms (e.g. LC-MS and GC-MS). Compounds sharing common chemical properties, extraction conditions, chromatographic separation and subsequent instrumental analysis are analysed together. A recently published study (Tikunov et al., 2005) nicely illustrates that the non-targeted exploration of specific fractions – in this case tomato fruit volatiles in the headspace – is very rewarding.

Hirai and coworkers (Hirai et al., 2004) showed significant progress exploring cellular processes by combining genome-wide transcriptomics and metabolomics under deficiency of sulfur and nitrogen in the model plant *Arabidopsis thaliana*. This study was one of the first pioneering papers successfully bringing our knowledge closer to understanding the important link between genomic data and the function of metabolites in plants. Microarray analysis was combined with targeted quantitative metabolite analysis in their study, and the data handling required the development of novel bioinformatics tools. Even though this study dealt mainly with primary metabolites, interesting observations were also made for secondary metabolites such as glucosinolates and alliins (Hirai et al., 2004).

3. STUDIES TO REVEAL MISSING LINKS IN NICOTINE ALKALOID PATHWAY

Nicotine and related alkaloids have been extensively studied during the past decades, *e.g.* by precursor feeding approaches. In 1983, Leete (1983) estimated that more than 300 feeding experiments involving labelled putative precursors of nicotine alkaloids (pyrrolidine alkaloids) had been described. More than 2500 compounds have been identified in tobacco, and extensive reviews on the field have been published (Robins et al., 1994a), (Nugroho and Verpoorte, 2002). The nicotine alkaloid biosynthesis starts from the amino acids ornithine and arginine, and several genes encoding for enzymes involved in the early part of the pathway have been cloned, including ADC (arginine decarboxylase) and ODC (ornithine decarboxylase) (Wang et al., 2000), PMT (putrescine *N*-methyltransferase) (Hibi et al., 1994), (Hashimoto et al., 1998), (Riechers and Timko, 1999), (Winz and Baldwin, 2001), MPO (methylputrescine oxidase) (Heim et al., 2007), (Katoh et al., 2007), and QPRT (quinolinate phosphoribosyltransferase) (Sinclair et al., 2000). PMT is an important enzyme in the overall regulation of the pathway (Robins et al., 1994a) driving the flow of nitrogen away from polyamine biosynthesis to alkaloid biosynthesis (Fig. 1). The product of this reaction, *N*-methylputrescine, is further converted to *N*-methylamino butanal catalysed by diamine oxidase DAO, or *N*-methylputresine oxidase MPO (EC 1.4.3.6). It was first discovered from tobacco roots (Mizusaki et al., 1972) and has been purified and characterised later (McLauchlan et al., 1993). However, the corresponding gene encoding for MPO was only very recently discovered in tobacco (Heim et al., 2007), (Katoh et al., 2007) has not yet been identified. Interestingly, MPO has shown to catalyse the oxidation of cadaverine (Mizusaki et al., 1972), (Hashimoto et al., 1990), thus accepting also other substrates from the pathway. Spontaneous cyclisation of *N*-methylaminobutanal leads to the formation of *N*-methylpyrrolinium ion, which donates the pyrrolidine ring structure to nicotine (Leete et al., 1980). Nicotine is formed in a condensation reaction of *N*-methylpyrrolinium and 3,6-dihydronicotinic acid, the latter providing the pyridine ring (Fig. 1).

The expression patterns of two tobacco isoflavone reductase-like (IRL) genes was studied by Shoji and co-workers (Shoji et al., 2002) and it was suggested that the condensation reaction between nicotinic acid and *N*-methylpyrrolinium is catalysed by a certain NAPDH-dependent reductase called A622. It was shown that *A622* is jasmonate-inducible (Hibi et al., 1994), (Shoji et al., 2000) and the expression pattern of *A622* is highly similar to that of *PMT* (Shoji et al., 2002). However, the conclusive finding concerning the final step in nicotine formation has not been obtained, yet.

Nicotine is further demethylated to yield nornicotine (Fig. 1). It was suggested that this reaction involves cytochrome P-450 (Chelvarajan et al., 1993), (Imaishi et al., 1995), (Hao and Yeoman, 1998) which was confirmed recently by the characterization of CYP82E4 (Siminszky et al., 2005). Nornicotine is converted to myosmine via a presumably irreversible reaction (Leete and Chedekel, 1974). Even though the enzymes participating in the early biosynthetic steps of alkaloid metabolism are relatively well known, the regulation of the formation of various nicotinic acid derived alkaloids, such as anabasine, anatabine and anatalline has still to be elucidated. Anabasine is synthesized from nicotinic acid and Δ^1-piperideine deriving from lysine. Concerning the biosynthetic origin of anatabine, it was suggested that anatabine is formed by a dimerization reaction of a metabolite of nicotinic acid (Leete and Slattery, 1976). Anatalline was first isolated by Kisaki and co-workers (Kisaki et al., 1968) from roots of *N. tabacum* and it was found in high concentrations in methyl jasmonate elicited BY-2 cell cultures of tobacco in our recent studies (Goossens et al., 2003a). Anatalline was shown to accumulate in relatively high amounts in two isomeric forms, however it has never been found in the leaves of tobacco, where other nicotinic acid derived alkaloids accumulate (Häkkinen et al., 2004). Also in the case of anatalline the biosynthetic origin and metabolism still need to be elucidated.

The genetic regulation of nicotine alkaloid biosynthesis in *N. tabacum* is determined by the *A* and *B* loci (Legg et al., 1969), (Legg and Collins, 1971). It has been postulated that the *A* and *B* alleles are regulatory genes coordinating alkaloid biosynthetic genes (Saunders and Bush, 1979), thus *A* and *B* loci have sometimes

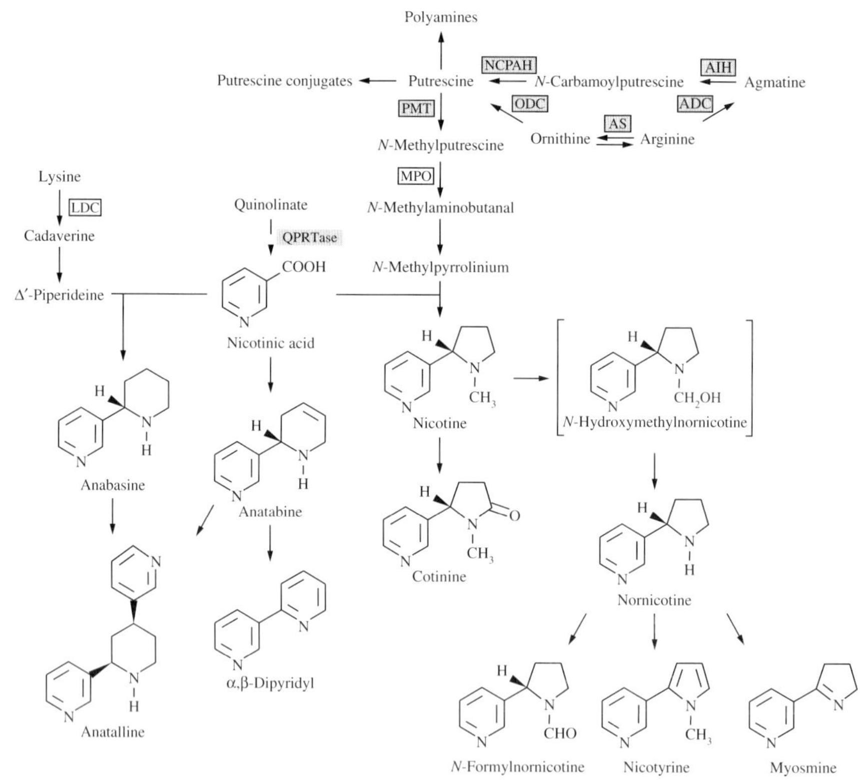

Figure 1. Biosynthetic pathway of nicotine alkaloids

been referred as *NIC1* and *NIC2*, respectively (Hibi et al., 1994). Recently, it was shown, however, that *A* and *B* loci are not alleles coordinating specifically only the alkaloid biosynthesis, but regulate a rather diverse set of gene families which only represent a small part of the nicotine biosynthetic genes (Kidd et al., 2006).

Recent advances in functional genomics approaches have led to interesting discoveries in understanding alkaloid biosynthesis. Among the first described in 1994 was the differential screening by substraction hybridisation of cDNA libraries by Hibi and co-workers (Hibi et al., 1994), who isolated two genes, *A411* encoding for PMT, and *A622* (see above). More recently, T-DNA activation tagging was assessed to investigate alkaloid metabolism in tobacco, resulting in five-fold nicotine accumulation in mutated plants compared to wild-type plants (Littleton et al., 2005).

We studied the nicotine alkaloid regulation in tobacco BY-2 cell cultures, by applying a novel approach which combines targeted metabolite profiling with cDNA-AFLP–based transcript profiling (Goossens et al., 2003a). From 20 000 analyzed gene tags, in total 591 tags were modulated by methyl jasmonate (MJM). Homology searches performed with the unique gene tags revealed that 58% of the tags displayed similarity with genes with known functions, however no homology was found for 26% of the tags. All genes, except one, known to be involved in nicotine alkaloid biosynthesis were found in a single experiment and they were shown to cluster together with various novel genes or genes encoding proteins with unknown functions. These novel genes, either structural or regulatory are thought to aid to fill up the gaps in nicotine alkaloid pathway. At the metabolite level, besides nicotine alkaloids, also highly induced accumulation of *N*-methylputrescine was observed after treating the cells with methyl jasmonate. Since nicotine was only a minor alkaloid accumulating in elicited BY-2, we are convinced that the limiting step in the pathway could be in the conversion of *N*-methylputrescine into nicotine, which includes the activity of *N*-methylputrescine oxidase (MPO). Indeed, we could find several putative oxidases clustering together with known nicotine alkaloid biosynthetic genes (Goossens et al., 2003a). Investigation of these novel genes is being continued in our laboratory by expressing them in both undifferentiated and differentiated cell culture systems to reveal their function in nicotine alkaloid biosynthesis. Recently, a further examination of these gene tags was performed to find potential regulators of the genes functioning in the nicotine alkaloid pathway (De Sutter et al., 2005). Based on these studies, two novel tobacco AP2-domain transcription factors named NtORC1 and NtJAP1 were shown to positively regulate the putrescine methyltransferase promoter.

4. NOVEL GENE-TO-METABOLITE NETWORKS FOR TERPENOID INDOLE ALKALOID BIOSYNTHESIS

Catharanthus roseus L. G. Don, the Madagascar periwinkle, is a medicinal plant which contains over 120 terpenoid indole alkaloids (TIAs) some of which are in clinical use as important anticancer drugs. Vinblastine and vincristine, the two most important alkaloids, are produced only in very low amounts in *C. roseus* and plants still remain the current source for isolating these compounds for industrial use (van der Heijden et al., 2004).

Biosynthetically TIAs are condensation products of building blocks derived from two pathways (Fig. 2). The shikimate pathway (Herrmann and Weaver, 1999)

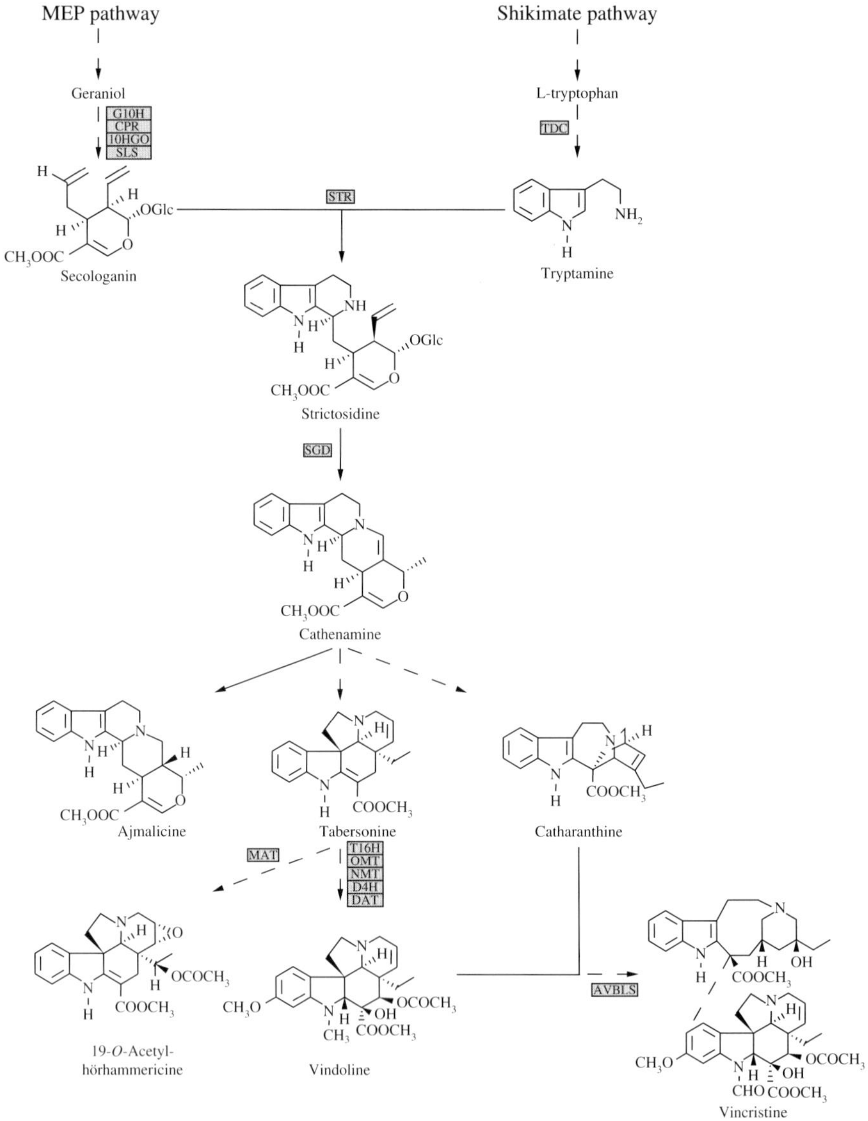

Figure 2. Biosynthetic pathway of terpenoid indole alkaloids

provides the amino acid tryptophan which is decarboxylated by tryptophan decarboxylase (TDC) to form the indole moiety tryptamine. The terpenoid moiety secologanin is synthesized along the terpenoid pathway from geraniol which in turn is synthesized in the plastidic non-mevalonate (MEP) pathway starting from glyceraldehyde-3-phosphate and pyruvate (Dubey et al., 2003). The well-characterized enzyme strictosidine synthase (STR) catalyzes the condensation of tryptamine and secologanin to form strictosidine (Kutchan, 1993) a central alkaloid precursor not only for TIAs present in *Catharanthus* but also for many other molecules such as the prominent anticancer compound camptothecin for example (Lorence and Nessler, 2004). Hydrolysis of strictosidine is catalysed by strictosidine glucosidase (SGD). The resulting, very reactive, aglucone cathenamine can be interconverted so that the basic skeletons of diverse alkaloids derive from it (Lounasmaa and Hanhinen, 1998), and are then further modified by the addition of substituents.

Pharmaceutically interesting bisindole alkaloids such as vincristine are formed by the condensation of vindoline, which is derived from tabersonine, and catharanthine in a series of reactions involving tabersonine 16-hydroxylase (T16H), *O*-methyltransferase (OMT), *N*-methyltransferase (NMT), desacetoxyvindoline 4-hydroxylase (D4H) and deacetylvindoline 4-*O*-acetyltransferase (DAT). It has to be noted that both enzymes and intermediates require extensive intra- and intercellular translocation and some reactions take place only in specific compartments (Burlat et al., 2004). In addition to the spatial requirements, TIA biosynthesis is under strict developmental and environmental control e.g. vindoline accumulation is light dependent and influenced by jasmonate (Vasquez-Flota and De Luca, 1998). The coordinated expression of the biosynthetic genes in response to external and internal signals is controlled by transcription factors. Recently, members of the plant-specific AP2-domain family e.g. octadecanoid-responsive *Catharanthus* AP2-domain ORCA2 and ORCA3 have been isolated by T-DNA activation tagging (van der Fits and Memelink, 2000), (Memelink et al., 2001).

Our current knowledge on the TIA biosynthesis as summarized above is based on approximately 50 years of research. Originally an outline of the chemical reactions and metabolites involved has been gained by labelling experiments often employing cell cultures, as already shown for the nicotine alkaloid pathway. Later, biosynthetic enzymes have been isolated and characterized by the classical one-by-one approach. With the advent of systems biology powerful functional genomics tools became now available allowing comprehensive investigations at an accelerated pace. Employing cDNA-AFLP-based transcript profiling (Breyne et al., 2003) we were recently able to monitor the expression of all but two of the so far known genes involved in TIA biosynthesis in a single experiment in response to elicitation (Rischer et al., 2006). This was only possible because cDNA-AFLP is not restricted to sequenced model plants in contrast to standard transcript profiling methods such as e.g. microarray analysis. Altogether 417 differentially-expressed transcript tags with unique sequences were obtained. As shown by comparison of these sequences with the 236 currently available *C. roseus* entries in the public EMBL database only less than 10% matched perfectly confirming that the majority of tags contain new sequence information.

Overall 37% of the tags were not similar to any known plant genes. Additionally the transcript sequences were compared to a recently performed *C. roseus* proteome data set (Jacobs et al., 2005). A corresponding tag could only be found for nine out of 58 functionally annotated protein spots in total. Three of them are known to be involved in TIA biosynthesis namely two STR isoforms and the PS protein.

In an attempt to link gene expression with metabolite accumulation we performed non-targeted metabolic profiling of the same samples used for transcript profiling. The final set of 178 metabolites included nine compounds known from the TIA pathway (Rischer et al., 2006). A principal component analysis of the combined transcript and metabolite profiles revealed a clear separation of the treatments and the sampling time. Correlation network analysis (Oresic et al., 2004) was then used to visualize the complex networks of secondary metabolite biosynthesis in *C. roseus* cells (Rischer et al., 2006). The constructed networks allow for example the fast identification of those genes most likely to be involved in TIA metabolism which is usually a critical step in the creation of priority lists for further functional characterization. On the other hand the networks assist the elucidation of unknown pathways by linking the accumulation patterns of unknown and known metabolites. Since the first kinetic metabolic models for *Catharanthus* are currently being developed (Morgan and Shanks, 2002), (Leduc et al., 2006) it is anticipated that the integration of metabolic profiling data into flux analysis will ultimately create an enormous opportunity to gain detailed insight into pathway organisation and will finally allow predictive metabolic engineering.

5. TRANSPORTER GENES AND THEIR FUNCTIONS IN PLANT SECONDARY METABOLISM

Besides enhancement and accumulation of secondary metabolites in the cultured cells secretion of the metabolites from the cells to the culture medium is an important element from the process point of view. The engineering of transporter systems should be thus an effective way to increase production of secondary metabolites as well as to facilitate pumping them out of the cells. Both would benefit considerably the large-scale production and down-stream processing of secondary metabolites with potential pharmaceutical value.

Transport of various chemical compounds across the biological membranes is an essential physiological process in all living organisms. Membrane transport is involved in the uptake of nutrients as well as in the efflux of toxic compounds, cell surface macromolecules, end products and intermediates of metabolism. In addition, transport systems function in ion homeostasis, communication between the cell and its environment, and also provide constituents of energy-generating systems. Plant cells are capable of producing a wide variety of secondary compounds, which may exert toxicity to plant cells, thus it is crucial that these compounds can be efficiently removed from cytosol. They can be transported into the apoplastic space where further chemical modifications may occur, or they can be transported into the vacuole. There are two possibilities how the active transport into the vacuole may proceed: by proton pumps or via ABC (ATP binding cassette) transporters (Theodoulou, 2000), (Yazaki, 2005).

Among all multicellular organisms which have been sequenced so far, plants have the largest representation of ABC transporter proteins encoded in their genome. Altogether 129 ORFs have been found from the *Arabidopsis* genome encoding ABC proteins (Sánchez-Fernández et al., 2001). The reason for such a high number of transpoters in plants is still unclear, but they have been suggested to play a role in adaptive functions against environmental stress. The ABC superfamily contains transport systems of both uptake and efflux. Members of both transporter groups generally cluster phylogenetically together (Saier and Paulsen, 2001). ATP hydrolysis without protein phosphorylation is used for an active energy source of the transport. Structurally ABC transporters are characterized by one or two cytosolically oriented nucleotide-binding folds (NBFs) or ATP-binding cassettes (ABCs) linked to multiple hydrophobic transmembrane-spanning domains (TMD). The substrate crosses the membrane via the TMD domains and in some cases the substrate specificity is determined by TMD domains (Theodoulou, 2000). NBFs provide the site for ATP hydrolysis. ABC transporters are classified in four main subfamilies: MDR (multidrug resistance), MRP (MDR-associated protein), AOH (ABC one homolog) and PDR (pleiotropic drug resistance).

Plant alkaloids are often transported by ABC transporters in microbes, but only few of them have been characterized from plants (Yazaki, 2005), (Yazaki, 2006). Nicotine alkaloids, which are synthesized in root tissue, are further transported to leaves, where they have the defensive role for the survival of the plant. The transporter protein responsible for this transport process has not yet been identified, but based on studies performed with tobacco hornworm it was suggested that a P-glycoprotein mediated (MDR-like) transport is involved (Gaertner, 1998). In different species, the same compounds can be transported by different transporter systems. Berberine uptake in *Coptis japonica* was shown to involve MDR-type ABC-transporter (Sakai et al., 2002), (Shitan et al., 2003), however, in *Arabidopsis thaliana*, which does not produce endogenous berberine, the transport proceeds by another type of transporter called MATE (multi antimicrobial extrusion/multidrug and toxin extrusion) (Li et al., 2002).

Plant plasma membrane transporters might function in the secretion of endogenous metabolites that play a role in defence reactions, as suggested by the findings of Jasinski and co-workers (Jasinski et al., 2001). They cloned a *NpABC1* (PDR5 homolog) from *Nicotiana plumbaginifolia* and localized it in the plasma membrane. *NpABC1* was shown to be strongly expressed after addition of sclareolide or the actual antifungal diterpene sclareol. The plant PDR family remains still poorly investigated. PDR5 is the yeast *Saccharomyces cerevisiae* ABC transporter conferring resistance to several unrelated drugs, and substrates of the transport process include also plant derived compounds, such as taxol, indole alkaloids, and flavonoids (Kolaczkowski et al., 1996), (Kolaczkowski et al., 1998). It was also shown that PDR5 is capable of using tropane alkaloids as substrates (Goossens et al., 2003b). Further, the effect was demonstrated with tobacco cells, where PDR5-carrying cells showed decreased cell death after administration of hyoscyamine. The PDR family remains an interesting target when it comes to secretion of plant secondary metabolites and applications in biotechnological production systems using plant cells, which are often encountered by poor secretion of the produced secondary metabolites (Goossens et al., 2003b).

6. COMBINATORIAL BIOCHEMISTRY APPROACH TO ENGINEER TROPANE ALKALOID PATHWAY

A limited number of key genes encode enzymes which are responsible for the synthesis of the pivotal backbone structures that constitute the characteristics of the different classes of secondary metabolites. Great progress has been made in the identification of theses genes (Kutchan, 1995). The subsequent "decoration" of the basic skeleton structures generates the huge chemical and pharmacological diversity of plant secondary metabolites. The large majority of these processes are again mediated by a limited number of enzyme classes, such as methyl- glycosyl-, acyltransferases, all encoded by multigene families. Enzymes are often substrate-, regio- and/or stereospecific in nature (Chau et al., 2004).

The Combinatorial Biochemistry concept is based on the idea that since different plant species can synthesize structurally closely related compounds (e.g. species specific alkaloids), it is possible that the enzymes acting in the biosynthesis of these compounds in one plant could use related substrates when introduced in another plant. This kind of approach is also likely to generate completely new substances, as was shown in 1996 by Laurila and co-workers (Laurila et al., 1996). They generated somatic hybrids from two *Solanum* species, *S. tuberosum* and *S. brevidens*. The resulting hybrid plants were shown to produce not only tomatidine and solanidine, which are commonly synthesized in *S. brevidens* and *S. tuberosum*, respectively, but also a novel glycoalkaloid demissidine, which has not been reported in either of the parental plants. An another example was given by the group of Toni Kutchan, who studied the substrate specificity of plant recombinant *O*-methyltransferases and observed that these enzymes could even accept substrates from different secondary metabolite classes (Frick and Kutchan, 1999), (Frick et al., 2001). Four different *Thalictrum tuberosum O*-methyltransferases were tested for their respective substrate specificity, and it was observed that while all four were able to *O*-methylate caffeic acid, two of the four enzymes accepted also the alkaloidal intermediate norcoclaurine, without clear indication of the preferred substrate.

An interesting option to study the Combinatorial Biochemistry idea involves two Solanaceae species, *Nicotiana* and *Hyoscyamus*, which partly share the common biosynthetic pathway. Tropane alkaloids, in particular hyoscyamine and scopolamine, are widely used in medicine as anticholinergic agents, acting on the parasympathetic nervous system. Enhancement of scopolamine production in various plant cell culture systems has gained a lot of interest, scopolamine being more valuable and preferred for its higher physiological activity and fewer side effects on the central nervous system. Both nicotine and tropane alkaloids share the common biosynthesis up to *N*-methylpyrrolium (Fig. 1 and 3). Only a few functional enzymes in the lower part of the tropane alkaloid pathway have been characterized (Fig. 3), namely tropinone reductase I (TRI) (Nakajima et al., 1993), tropinone reductase II (TRII) (Nakajima et al., 1993), (Keiner et al., 2002), and hyoscyamine 6β-hydroxylase (H6H) (Matsuda et al., 1991). The role of tropinone in tropane alkaloid biosynthesis was unclear for long time, since in 1990 Landgrebe and Leete showed its incorporation in the tropane ester moiety (Landgrebe and Leete, 1990). Tropinone is further converted into tropine and pseudotropine by the reactions catalysed by two distinct enzymes, TRI and TRII. Littorine, a positional isomer of hyoscyamine, is formed from tropine and phenyllactic acid, the latter deriving from phenylalanine. Hyoscyamine was shown to be directly synthesized from littorine by Robins and co-workers (Robins et al., 1994b). Hyoscyamine-6β-hydroxylase

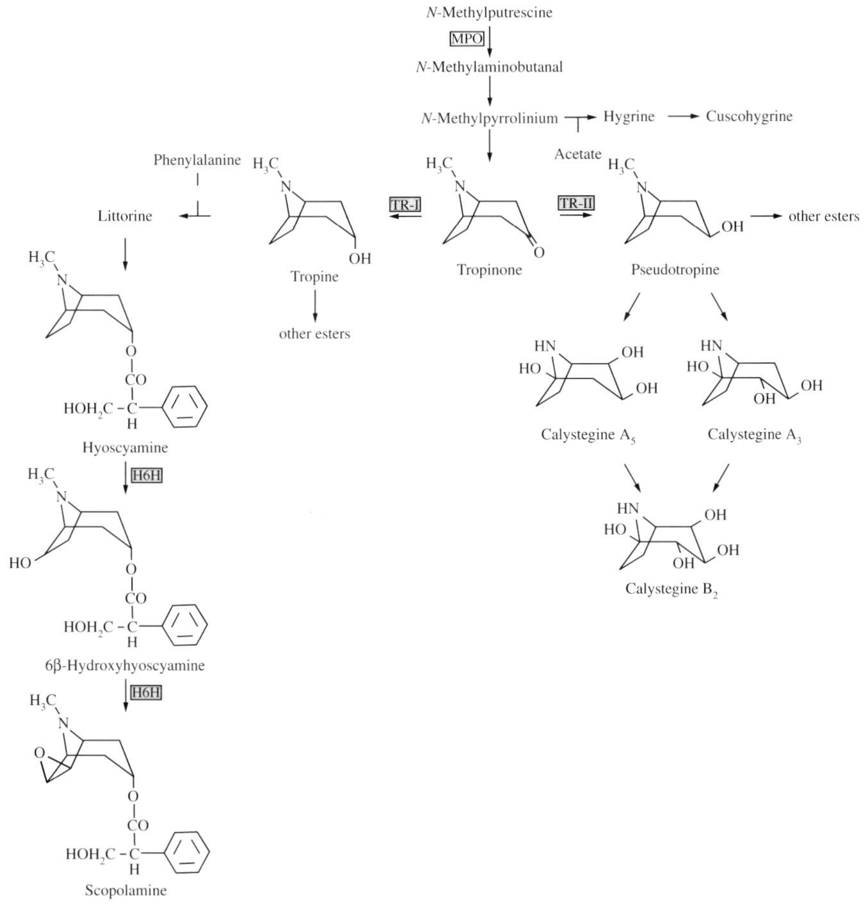

Figure 3. Biosynthetic pathway of tropane alkaloids

(EC 1.14.11.11) catalyzes both the hydroxylation of hyoscyamine leading to 6β-hydroxyhyoscyamine and the epoxidation of the latter leading to scopolamine (Hashimoto and Yamada, 1986).

Scopolamine production can be increased in transgenic plants of hyoscyamine-producing plants by overexpressing *h6h*, as has been reported first by Hashimoto and co-workers (Hashimoto et al., 1993) and Yun and co-workers (Yun et al., 1992). In these studies hyoscyamine was converted entirely to scopolamine. Later on it was shown that also scopolamine content can be increased in *Hyoscyamus muticus* hairy roots cultures over 100-fold compared to the controls but the contents of hyoscyamine were not reduced (Jouhikainen et al., 1999). Recently, simultaneous overexpression of *pmt* and *h6h* in *H. niger* resulted in a scopolamine production of 411 mg l^{-1}, which is the highest level hitherto reported from cultivated hairy roots (Zhang et al., 2004).

Calystegines, which are synthesized in the other branch of the tropane pathway (Fig. 3), are polyhydroxy nortropane alkaloids, possessing strong glycosidase inhibitory activity (Asano et al., 2000). The compounds were originally found in *Calystegia sepium* (Goldmann et al., 1990), however they have been detected in many Solanaceae, also in species which were not thought to possess the tropane alkaloid pathway, e.g. *Solanum tuberosum* (Dräger et al., 1995). Calystegines are formed from tropinone in a reaction catalysed by TRII (Dräger et al., 1994). Recently, it was shown that the ratios of the tropine and pseudotropine–derived alkaloids were altered by overexpressing the *trI* or *trII* in *Atropa belladonna* (Richter et al., 2005). Rocha and co-workers (Rocha et al., 2002) introduced two genes involved in tropane alkaloid biosynthesis, *trI* and *h6h*, from *H. niger* into *N. tabacum*. Surprisingly, besides accumulating the expected reaction products of these enzymes, also elevated levels of nicotine and related alkaloids were found, suggesting an induction of the activity of overall nicotine alkaloid pathway caused by the genetic engineering of the pathway. Also in our studies with *N. tabacum* hairy roots carrying *h6h*, elevated nicotine alkaloid production was observed (Häkkinen et al., 2005). However, rather than being caused by the introduction of the foreign gene, we suggest that resulting alterations in the endogenous alkaloid production were caused by the elicitor effect of exogenously applied hyoscyamine as a foreign substrate. These roots were effectively taking up hyoscyamine and 85% of the converted scopolamine was released into the medium.

The value of metabolic engineering using a Combinatorial Biochemistry approach (Oksman-Caldentey and Inzé, 2004) is shown in our current studies. Selected genes from *N. tabacum*, which were assumed to have a role in alkaloid metabolism (Goossens et al., 2003a) were overexpressed in the related species *Hyoscyamus*, as well as in the unrelated species *Catharanthus*. Subsequently, the metabolite profiles of the generated transgenic lines were studied. We were able to find tobacco genes, which caused an elevated production of untypical polyamines and pseudotropine-derived tropanes (*e.g.* 3β-acetoxytropane) in *Hyoscyamus muticus* hairy roots. One of these genes caused also a different alkaloid spectrum in *Catharanthus* roots, and the further characterization of these terpenoid indole alkaloid derivatives not present in control hairy roots is currently in process (unpublished data).

7. FUTURE OUTLOOK

Plant cells offer an attractive biotechnological production system for high-value plant-derived complex molecules but unfortunately to date this has had only very little commercial success due to our limited knowledge on the biosynthetic routes of these compounds. Therefore a better understanding of metabolite biosynthesis and the regulation thereof will be increasingly important for improving the sustainability and efficiency of useful metabolite production. Encouraging results have been obtained already for some non-model plant species. Especially the development of functional genomics tools has opened new possibilities to link genomics, transcriptomics and metabolomics data, and thus to use the biochemical capacity of plants more efficiently to produce known and novel metabolites through metabolic engineering.

REFERENCES

Arabidopsis Genome Initiative (2000) *Nature* 408: 796–826.

Asano N, Nash RJ, Molyneux RJ, Fleet GW (2000) Sugar-mimic glycosidase inhibitors: natural occurrence, biological activity and prospects for therapeutic action. *Tetrahedron Asymmetry* 11: 1645–1680.

Breyne P, Zabeau M (2001) Genome-wide expression analysis of plant cell cycle modulated genes. *Curr Opin Plant Biol* 4: 136–142.

Breyne P, Dreesen R, Cannoot B, Rombauts D, Vandepoele K, Rombauts S, Vanderhaeghen R, Inzé D, Zabeau M (2003) Quantitative cDNA-AFLP analysis for genome–wide expression studies Mol Gen Genom 269: 173–179.

Burlat V, Oudin A, Courtois M, Rideau M, St-Pierre B (2004) Co-expression of three MEP pathway genes and *geraniol 10-hydroxylase* in internal phloem parenchyma of *Catharanthus roseus* implicates multicellular translocation of intermediates during the biosynthesis of monoterpene indole alkaloids and isoprenoid-derived primary metabolites. *Plant J* 38: 131–141.

Chau MD, Walker K, Long R, Croteau R (2004) Regioselectivity of taxoid-*O*-acyltransferases: heterologous expression and characterization of a new taxadien-5α-ol-*O*-acyltransferase. *Arch Biochem Biophys* 430: 237–246.

Chelvarajan RL, Fannin FF, Bush LP (1993) Study of nicotine demethylation in *Nicotiana otophora*. *J Agric Food Chem* 41: 858–862.

De Sutter V, Vanderhaegen R, Tilleman S, Lammertyn F, Vanhoutte I, Karimi M, Inzé D, Goossens A, Hilson P (2005) Exploration of jasmonate signalling via automated and standardized transient assays in tobacco cells. *Plant J* 44: 1065–1076.

Dräger B, Funck C, Hoehler A, Mrachatz G, Nahrstedt A, Portsteffen A, Schaal A, Schmidt R (1994) Calystegines as a new group of tropane alkaloids in Solanaceae. *Plant Cell Tiss Organ Cult* 38: 235–240.

Dräger B, van Almsick A, Mrachatz G (1995) Distribution of calystegines in several Solanaceae. *Planta Med* 61: 577–579.

Dubey, VS, Bhalla R, Luthra R (2003) An overview of the non-mevalonate pathway for terpenoid biosynthesis in plants. *J Biosci* 28: 637–646.

Facchini, PJ, Bird, DA, St-Pierre, B (2004) Can *Arabidopsis* make complex alkaloids? Trends Plant Sci 9:116–122.

Facchini PJ, St-Pierre B (2005) Synthesis and trafficking of alkaloid biosynthetic enzymes. *Curr Opin Plant Biol* 8: 657–666.

Fiehn O (2002) Metabolomics – the link between genotypes and phenotypes Plant Mol Biol 48: 155–171.

Frick S, Kutchan T (1999) Molecular cloning and functional expression of *O*-methyltransferases common to isoquinoline alkaloid and phenylpropanoid biosynthesis. *Plant J* 17: 329–339.

Frick S, Ounaroon A, Kutchan TM (2001) Combinatorial biochemistry in plants: the case of *O*-methyltransferases. *Phytochemistry* 56: 1–4.

Gaertner LS, Murray CL, Morris CE (1998) Transepithelial transport of nicotine and vinblastine in isolated Malpighian tubules of the tobacco hornworm (*Manduca sexta*) suggestes a P-glycoprotein-like mechanism. *J Exp Biol* 201: 2637–2645.

Goldmann A, Milat M-L, Ducrot P-H et al. (1990) Tropane derivatives from *Calystegia sepium*. *Phytochemistry* 29: 2125–2127.

Goossens A, Häkkinen ST, Laakso I, Seppänen-Laakso T, Biondi S, deSutter V, Lammertyn F, Nuutila AM, Söderlund H, Zabeau M, Inzé D, Oksman-Caldentey K-M (2003a) A functional genomics approach toward the understanding of secondary metabolism in plant cells. *Proc Natl Adac Sci USA* 100: 8595–8600.

Goossens A, Häkkinen ST, Laakso I, Oksman-Caldentey K-M, Inzé D (2003b) Secretion of secondary metabolites by ATP-binding cassette transporters in plant cell suspension cultures. *Plant Physiol* 131: 1161–1164.

Hao D-Y, Yeoman MM (1998) Evidence in favour of an oxidative *N*-demethylation of nicotine to nornicotine in tobacco cell cultures. *J Plant Physiol* 152: 420–426.

Hashimoto T, Yamada Y (1986) Hyoscyamine 6β-hydroxylase, a 2-oxoglutarate-dependent dioxygenase, in alkaloid-producing root cultures. *Plant Physiol* 81: 619–625.

Hashimoto T, Mitani A, Yamada Y (1990) Diamine oxidase from cultured roots of *Hyoscyamus niger*. *Plant Physiol* 93: 216–221.

Hashimoto T, Yun D-J, Yamada Y (1993) Production of tropane alkaloids in genetically engineered root cultures. *Phytochemistry* 32: 713–718.

Hashimoto T, Shoji T, Mihara T, Oguri H, Tamaki K, Suzuki K, Yamada Y (1998) Intraspecific variability of the tandem repeats in *Nicotiana* putrescine *N*-methyltransferases. *Plant Mol Biol* 37: 25–37.

Herrmann KM, Weaver LM (1999) The shikimate pathway. *Annu Rev Plant Physiol Plant Mol Biol* 50:473–503.

Hibi N, Higashiguchi S, Hashimoto T, Yamada Y (1994) Gene expression in tobacco low-nicotine mutants. *Plant Cell* 6: 723–735.

Hirai MY, Yano M, Goodenowe DB, Kanaya S, Kimura T, Awazuhara M, Arita M, Fujiwara T, Saito K (2004) Integration of transcriptomics and metabolomics for understanding of global responses to nutritional stresses in *Arabidopsis thaliana* Proc Natl Acad Sci USA 101: 10205–10210.

Häkkinen ST, Rischer H, Laakso I, Maaheimo H, Seppänen-Laakso T, Oksman-Caldentey K-M (2004) Anatalline and other methyl jasmonate inducible nicotine alkaloids from *Nicotiana tabacum* cv. BY-2 cell cultures. *Planta Med* 70: 936–941.

Häkkinen ST, Moyano E, Cusidó RM, Palazón J, Piñol MT, Oksman-Caldentey K-M (2005) Enhanced secretion of tropane alkaloids in *Nicotiana tabacum* hairy roots expressing heterologous hyoscyamine-6β-hydroxylase. *J Exp Bot* 56: 2611–2618.

Heim WG, Sykes KA, Hildreth SB, Sun J, Lu RH, Jelesko JG (2007) Cloning and characterization of a *Nicotiana tabacum* methylputrescine oxidase transcript. *Phytochemistry* 68: 454–463.

Imaishi H, Yamada T, Ohkawa H (1995) Purification and immunochemical characteristics of NADPH-cytochrome P-450 oxidoreductase from tobacco cultured cells. *Biochim Biophys Acta* 1246: 53–60.

Jacobs DI, Gaspari M, van der Greef J, van der Heijden R, Verpoorte R (2005) Proteome analysis of the medicinal plant *Catharanthus roseus*. *Planta* 221: 690–704.

Jasinski M, Stukkens Y, Degand H, Purnelle B, Marchand-Brynaert J, Boutry M (2001) A plant plasma membrane ATP binding cassette-type transporter is involved in antifungal terpenoid secretion. *Plant Cell* 13: 1095–1107.

Jellum E (1977) Profiling human body fluids in healthy and disease states using gas chromatography and mass spectrometry with special reference to organic acids. *J Chromatogr* 143: 427–462.

Jouhikainen K, Lindgren L, Jokelainen T, Hiltunen R, Teeri TH, Oksman-Caldentey K-M (1999) Enhancement of scopolamine production in *Hyoscyamus muticus* L. hairy root cultures by genetic engineering. *Planta* 208: 545–551.

Katoh A, Shoji T, Hashimoto T (2007) Molecular cloning of *N*-methylputrescine oxidase from tobacco. *Plant Cell Physiol* 48: 550–554.

Keiner R, Kaiser H, Nakajima K, Hashimoto T, Dräger B (2002) Molecular cloning, expression and characterization of tropinone reductase II, an enzyme of the SDR family in *Solanum tuberosum* (L.). *Plant Mol Biol* 48: 299–308.

Kidd SH, Melillo AA, Lu R-H, Reed DG, Kuno N, Uchida K, Furuya M, Jelesko JG (2006) The *A* and *B* loci in tobacco regulate a network of stress response genes, few of which are associated with nicotine biosynthesis. *Plant Mol Biol* 60: 699–716.

Kisaki T, Mizusaki S, Tamaki E (1968) Phytochemical studies on tobacco alkaloids - XI. A new alkaloid in *Nicotiana tabacum* roots. *Phytochemistry* 7: 323–327.

Kolaczkowski M, Kolaczkowska A, Luczynski J, Witek S, Goffeau A (1998) *In vivo* characterization of the drug resistance profile of the major ABC transporters and other components of the yeast pleiotropic drug resistance network. *Microb Drug Resist* 4: 143–158.

Kolaczkowski M, van der Rest M, Cybularz-Kolaczkowska A, Soumillion JP, Konings WN, Goffeau A (1996) Anticancer drugs, ionophoric peptides, and steroids as substrates of the yeast multidrug transporter PDR5p. *J Biol Chem* 271: 31543–31548.

Kutchan, TM (1995) Alkaloid biosynthesis – the basis for metabolic engineering of medicinal plants. *Plant Cell* 7: 1059–1079.

Kutchan TM (1993) Strictosidine - from Alkaloid to Enzyme to Gene. *Phytochemistry* 32: 493–505.

Landgrebe ME, Leete E (1990) The metabolism of tropinone in intact *Datura innoxia* plants. *Phytochemistry* 29: 2521–2524.

Laurila J, Laakso I, Valkonen JPT, Hiltunen R, Pehu E (1996) Formation of parental-type and novel glycoalkaloids in somatic hybrids between *Solanum brevidens* and *S. tuberosum*. *Plant Sci* 118: 145–155.

Leduc M, Tikhomiroff C, Cloutier M, Perrier M, Jolicoeur M (2006) Development of a kinetic metabolic model: application to *Catharanthus roseus* hairy root. *Bioproc Biosyst Eng* 28: 295–313.

Leete E (1980) Alkaloids derived from ornithine, lysine and nicotinic acid. In: Bell EA, Charlwood BV (eds) Encyclopedia of plant physiology, new series, secondary plant products, vol 8. Springer-Verlag, Berlin, pp 65–91.

Leete E (1983) Biosynthesis and metabolism of the tobacco alkaloids. In: Pelletier SW (ed) Alkaloids: chemical and biological perspectives, John Wiley & Sons, New York, pp 85–152.

Leete E, Chedekel MR (1974) Metabolism of nicotine in *Nicotiana glauca*. *Phytochemistry* 13: 1853–1859.

Leete E, Slattery SA (1976) Incorporation of [2-^{14}C]- and [6-^{14}C]nicotinic acid into the tobacco alkaloids. Biosynthesis of anatabine and α-β-dipyridyl. *J Am Chem Soc* 98: 6326–6330.

Legg, PD, Chaplin JF, Collins GB (1969) Inheritance of percent total alkaloids in *Nicotiana tabacum* L.; populations derived from crosses of low alkaloid lines with burley and flue-cured varieties. *J Hered* 60: 213–217.

Legg PD, Collins GB (1971) Inheritance of percent total alkaloids in *Nicotiana tabacum* L. II. genetic effects of two loci in Burley 21 X LA Burley 21 populations. *Can J Cyt* 13: 287–291.

Li L, He Z, Girdhar KP, Tsuchiya T, Luan S (2002) Functional cloning and characterisation of a plant efflux carrier for multidrug and heavy metal detoxification. *J Biol Chem* 277: 5360–5368.

Littleton J, Rogers T, Falcone D (2005) Novel approaches to plant drug discovery based on high throughput pharmacological screening and genetic manipulation. *Life Sci* 78: 467–475.

Lorence A, Nessler CL (2004) Camptothecin, over four decades of surprising findings. *Phytochemistry* 65: 2735–2749.

Lounasmaa M, Hanhinen P (1998) Biomimetic formation and interconversion in the heteroyohimbine series. *Heterocycles* 48: 1483–1492.

Matsuda J, Okabe S, Hashimoto T, Yamada Y (1991) Molecular cloning of hyoscyamine 6β-hydroxylase, a 2-oxoglutarate-dependent dioxygenase, from cultured roots of *Hyoscyamus niger*. *J Biol Chem* 266: 9460–9464.

McLauchlan WR, McKee RA, Evans DM (1993) The purification and immunocharacterisation of *N*-methylputrescine oxidase from transformed root cultures of *Nicotiana tabacum* l. *cv SC58*. *Planta* 191: 440–445.

Memelink J, Verpoorte R, Kijne JW (2001) ORCAnization of jasmonate - responsive gene expression in alkaloid metabolism Trends Plant Sci. 6: 212–219.

Mizusaki S, Tanabe Y, Noguchi M, Tamaki E (1972) *N*-methylputrescine oxidase from tobacco roots. *Phytochemistry* 11: 2757–2762.

Morgan JA, Shanks JV (2002) Quantification of metabolic flux in plant secondary metabolism by a biogenetic organizational approach. *Metabolic Eng* 4: 257–262.

Nakajima K, Hashimoto T, Yamada Y (1993) Two tropinone reductases with different stereospecifities are short-chain dehydrogenases evolved from a common ancestor. *Proc Natl Acad Sci USA* 90: 9591–9595.

Nugroho LH, Verpoorte R (2002) Secondary metabolism in tobacco. *Plant Cell Tissue Organ Cult* 68: 105–125.

Oksman-Caldentey K-M, Inzé D (2004) Plant cell factories in the post-genomic era: new ways to produce designer secondary metabolites. *Trends Plant Sci* 9: 433–440.

Oksman-Caldentey K-M, Inzé D, Oresic M (2004) Connecting genes to metabolites by a systems biology approach. *Proc Natl Acad Sci USA* 101: 9949–9950.

Oksman-Caldentey K-M, Saito K (2005) Integrating genomics and metabolomics for engineering plant metabolite pathways. *Curr Opin Biotech* 16: 174–179.

Oresic M, Clish CB, Davidov EJ, Verheij E, Vogels JTWE, Havekes LM, Neumann E, Adourian A, Naylor S, van der Greef J, Plasterer T (2004) Phenotype characterization using integrated gene transcript, protein and metabolite profiling. *Appl Bioinformatics* 3: 205–217.

Osbourn AE, Qi XQ, Townsend B, Qin B (2003) Dissecting plant secondary metabolism - constitutive chemical defences in cereals. *New Phytol* 159: 101–108.

Ounaroon A, Decker, G, Schmidt J, Lottspeich F, Kutchan TM (2003) (*R,S*)-Reticuline 7-*O*-methyltransferase and (*R,S*)-norcoclaurine 6-*O*-methyltransferase of *Papaver somniferum* – cDNA cloning and charcterization of methyl transfer enzymes of alkaloid biosynthesis in opium poppy. *Plant Cell* 36: 808–819.

Richter U, Rother G, Fabian A-K, Rahfeld B, Dräger B (2005) Overexpression of tropinone reductases alters alkaloid composition in *Atropa belldonna* root cultures. *J Exp Bot* 56: 645–652.

Rischer H, Oresic M, Seppänen-Laakso T, Katajamaa M, Lammertyn F, Ardiles-Diaz W, Van Montagu MCE, Inzé D, Oksman-Caldentey K-M, Goossens A (2006) Gene-to-metabolite networks for terpenoid indole alkaloid biosynthesis in *Catharanthus roseus* cells. *Proc Natl Acad Sci USA* 103: 5614–5619.

Riechers DE, Timko MP (1999) Structure and expression of the gene family encoding putrescine *N*-methyltransferase in *Nicotiana tabacum*: new clues to the evolutionary origin of cultivated tobacco. *Plant Mol Biol* 41: 387–401.

Robins RJ, Bachmann P, Woolley JG (1994a) Biosynthesis of hyoscyamine involves an intramolecular rearrangement of littorine. *J Chem Soc Perkin Trans* 1: 615–619.

Robins RJ, Walton NJ, Parr AJ, Aird ELH, Rhodes MJC, Hamill JD (1994b) Progress in the genetic engineering of the pyridine and tropane alkaloid biosynthetic pathways of Solanaceous plants. In: Ellis BE, Kuroki GW, Stafford HA (eds) Genetic engineering of plant secondary metabolism. Plenum Press, New York, pp 1–33.

Rocha P, Stenzel O, Parr A, Walton N, Christou P, Dräger B, Leech MJ (2002) Functional expression of tropinone reductase I (*trI*) and hyoscyamine-6β-hydroxylase (*h6h*) from *Hyoscyamus niger* in *Nicotiana tabacum*. *Plant Sci* 162: 905–913.

Saier MHJr, Paulsen I (2001) Phylogeny of multidrug transporters. *Sem Cell Dev Biol* 12: 205–213.

Sakai K, Shitan N, Sato F, Ueda K, Tazaki K (2002) Characterization of berberine transport into *Coptis japonica* cells and the involvement of ABC protein. *J Exp Bot* 53: 1879–1886.

Sánchez-Fernández R, Davies TGE, Coleman JOD, Rea PA (2001) The *Arabidopsis thaliana* ABC protein superfamily, a complete inventory. *J Biol Chem* 276: 30231–31244.

Saunders JQ, Bush LP (1979) Nicotine biosynthetic enzyme activities in *Nicotiana tabacum* L. genotypes with different alkaloid levels. *Plant Physiol* 64: 236–240.

Shitan N, Bazin I, Dan K, Obata K, Kigawa K, Ueda K, Sato F, Forestier C, Yazaki K (2003) Involvement of CjMDR1, a plant multidrug-resistance-type ATP-binding cassette protein, in alkaloid transport in *Coptis japonica*. *Proc Natl Acad Sci USA* 100: 751–756.

Shoji T, Nakajima K, Hashimoto T (2000) Ethylene suppresses jasmonate-inducible gene expression in nicotine biosynthesis. *Plant Cell Physiol* 41: 1072–1076.

Shoji T, Winz R, Iwase T, Nakajima K, Yamada Y, Hashimoto T (2002) Expression patterns of two tobacco isoflavone reductase-like genes and their possible roles in secondary metabolism in tobacco. *Plant Mol Biol* 50: 427–440.

Siminszky B, Gavilano L, Bowen SW, Dewey RE (2005) Conversion of nicotine to nornicotine in *Nicotiana tabacum* is mediated by CYP82E4, a cytochrome P450 mono-oxygenase. *Proc Natl Acad Sci USA* 102: 14919–14924.

Sinclair SJ, Murphy KJ, Birch CD, Hamill JD (2000) Molecular characterization of quinolinate phosphoribosyltransferase (QPRTase) in *Nicotiana*. *Plant Mol Biol* 44: 603–617.

St-Pierre B, Vazquez-Flota FA, De Luca V (1999) Multicellular compartmentation of *Catharanthus roseus* alkaloid biosynthesis predicts intercellular translocation of a pathway intermediate. *Plant Cell* 11: 887–900.

Theodoulou FL (2000) Plant ABC transporters. *Biochim Biophys Acta* 1465: 79–103.

Tikunov Y, Lommen A, de Vos CHR, Verhoeven HA, Bino RJ, Hall RD, Bovy AG (2005) A novel approach for non-targeted data analysis for metabolomics. Large-scale profiling of tomato fruit volatiles. *Plant Physiol* 139: 1125–1137.

Trethewey R (2004) Metabolite profiling as an aid to metabolic engineering. *Curr Opin Plant Biol* 7: 196–201.

Trethewey RN, Krotzky AJ, Willmitzer L (1999) Metabolic profiling: a Rosetta stone for genomics? Curr Opin Plant Biol 2: 83–85.

Tweeddale H, Notley-McRobb L, Ferenci T (1998) Effect of slow growth on metabolism of *Escherichia coli*, as revealed bu global metabolite pool ("Metabolome") analysis. *J Bacteriol* 180: 5109–5116.

van der Fits L, Memelink J (2000) ORCA3, a jasmonate-responsive transcriptional regulator of plant primary and secondary metabolism. *Science* 289: 295–297.

van der Heijden R, Jacobs DI, Snoeijer W, Hallard D, Verpoorte R (2004) The *Catharanthus* alkaloids: pharmacognosy and biotechnology Curr Med Chem 11: 607–628.

Vazquez-Flota FA, De Luca V (1998) Jasmonate modulates development- and light-regulated alkaloid biosynthesis in *Catharanthus roseus*. *Phytochemistry* 49:395–402.

Verpoorte R (2000) Secondary metabolism. In: Verpoorte R, Alfermann AW (eds) Metabolic engineering of plant secondary metabolism. Kluwer Academic Publisher, Dordrech, pp 1–29.

Wang JM, Sheehan M, Brookman H, Timko MP (2000) Characterization of cDNAs differentially expressed in roots of tobacco (*Nicotiana tabacum* cv Burley 21) during the early stages of alkaloid biosynthesis. *Plant Sci* 158: 19–32.

Weid M, Ziefler J, Kutchan TM (2004) The roles of latex and the vascular bundle in morphine biosynthesis in the opium poppy, *Papver somniferum*. *Proc Natl Acad Aci USA* 101: 13957–13962.

Winz RA, Baldwin IT (2001) Molecular interactions between the specialist herbovore *Manduca sexta* (Lepidoptera, Sphingidae) and its natural host *Nicotiana attenuata*. IV. Insect-induced ethylene reduces jasmonate-induced nicotine accumulation by regulating putrescine *N*-methyltransferase transcripts. *Plant Physiol* 125: 2189–2202.

Zhang L, Ding R, Chai Y, Bonfill M, Moyano E, Oksman-Caldentey K-M, Xu T, Pi Y, Wang Z, Zhang H, Kai G, Liao Z, Sun X, Tang K (2004) Engineering tropane biosynthetic pathway in *Hyoscyamus niger* hairy root cultures. *Proc Natl Acad Sci USA* 101: 6786–6791.

Yazaki K (2006) ABC transporters involved in the transport of plant secondary metabolites. *FEBS Lett* 580: 1183–1191.

Yazaki K (2005) Transporters of secondary metabolites. *Curr Opin Plant Biol* 8: 301–307.

Yu J, Hu S, Wang J, Wong GK-S, Li S, Liu B, Deng Y, Dai L, Zhou Y, Zhang X et al. (2002) A draft sequence of the rice genome (*Oryza sativa* l. ssp. *indica*). *Science* 296: 79–92.

Yun DJ, Hashimoto T, Yamada Y (1992) Metabolic engineering of medicinal plants: transgenic *Atropa belladonna* with an improved alkaloid composition. *Proc Natl Acad Sci USA* 89: 11799–11803.

CHAPTER 5

POLYAMINE BIOSYNTHETIC PATHWAY: A POTENTIAL TARGET FOR ENHANCING ALKALOID PRODUCTION

Polyamines in alkaloid production

ESHA BHATTACHARYA, AND M. V. RAJAM[1]

[1] *Department of Genetics, University of Delhi – South Campus, Benito Juarez Road, New Delhi*

[2]*NFCL (Nagarjuna Fertilizers Corporation Ltd) Panjagulta, Hyderabad 500 082, India*

Abstract: Plant metabolic engineering is quite a young science in plants, however, in the past decade it has generated a lot of interest. The reason being that plant secondary metabolites are economically very important as they find applications in pharmaceuticals, insecticides, flavours, fragrances and dyes but their production in plants are in very low quantities. With the development of basic molecular biology and genetic engineering techniques a lot of strategies and tools have become available for targeted improvement of quality and quantity of secondary metabolites. Plant alkaloids constitute the second largest group of secondary metabolites and provide many pharmacologically active compounds. The knowledge of secondary metabolite pathways is however very limited and is the major constraint for successful application of metabolic engineering. Polyamines are low molecular weight polycationic molecules, ubiquitous in nature and are known to play an important role in the regulation of plant growth and development. They also act as precursors of many of the economically important secondary metabolites. Polyamine biosynthetic pathway provides an attractive model for such studies as the various steps in the pathway are very well worked out and most of the genes involved in the pathway have been cloned. Further, polyamine biosynthesis has also been engineered for enhanced alkaloid content by over-expression of ornithine decarboxylase and arginine decarboxylase genes. The present review highlights the prospects of engineering of polyamine biosynthesis for enhanced alkaloid content and also provides a forward looking perspective of this exciting field over the coming years

Keywords: polyamines; metabolic engineering; arginine decarboxylase; ornithine decarboxylase; putrescine-*N*-methyltransferase; alkaloid

Abbreviations: ADC, arginine decarboxylase; DAO, diamine oxidase; HSS, homospermidine synthase; LDC, lysine decarboxylase; ODC, ornithine decarboxylase; PA, polyamine; PMT, putrescine-*N*-methyltransferase; Put, putrescine; SAM, S-adenosylmethionine; SAMDC, S-adenosylmethionine decarboxylase; Spd, spermidine; Spm, spermine

R. Verpoorte et al. (eds.), Applications of Plant Metabolic Engineering, 129–143.
© 2007 *Springer.*

1. INTRODUCTION

Plants synthesize an extensive array of low molecular weight organic compounds known as secondary metabolites, whose major role is in plant defense system and they also help the plant to survive in its environment. Presently, about 1,00,000 such compounds have been isolated from the plants (Verpoorte, 2000). The production of these compounds is often low (less than 1% dry weight) and depends greatly on the physiological and developmental stage of the plants (Oksman-Caldentey and Inze, 2004). These secondary metabolites are economically very important as they find applications in pharmaceuticals, insecticides, flavours, fragrances and dyes. Thus they form an interesting target for metabolic engineering.

Currently almost a quarter of all prescribed pharmaceuticals are directly or indirectly plant-derived. Many plant species containing such economically important compounds are rare or have become endangered because of over-harvesting. Thus biotechnological means for enhancing the production of such economically important secondary metabolites in plant cells or organ cultures provides an attractive alternative strategy. However, this strategy has had limited success as they often do not lead to production of sufficient amounts of the required metabolite. Molecular engineering of secondary metabolism has the potential to increase production and improve the product composition. In the recent years efforts have been done to elucidate the basic biochemistry and molecular biology of some of the secondary metabolite pathways, and these studies elucidate that the diversification of secondary metabolism originated from the elaboration of a few central intermediates that could be modified through oxidation, followed by the protection of hydroxyl groups by acylation, glucosylation, methylation, or among others, prenylation (de Luca and St. Pierre, 2000).

Polyamine (PA) biosynthetic pathway has been very well worked out and offers an interesting case for manipulation of genes involved in this pathway for the modulation of secondary metabolite production as the diamine putrescine (Put), the triamine spermidine (Spd) and the tetra-amine spermine (Spm) are important precursors for some of the most important secondary metabolites. Also the availability of the cloned genes for PA metabolism offers a wide range of opportunities to be exploited. In the present review different strategies have been discussed, with special emphasis on PA biosynthetic pathway, which have in the past few years helped considerably in increasing our understanding about the complexity of alkaloid metabolism.

2. POLYAMINES

PAs are low molecular weight polycationic molecules, ubiquitous in nature and are known to play an important role in the regulation of plant growth and development (Rajam et al., 1985; Galston and Kaur-Sawhney, 1990; Rajam, 1997; Urano et al., 2005; Kumar et al., 2006). The common aliphatic PAs are Put (1,4-diaminobutane), Spd and Spm, however, besides these common PAs, there are also some uncommon ones, such as norspermidine, norspermine and homospermidine, which have been associated with the capacity of some biological systems to grow and function under extreme environments (Slocum and Flores, 1991). The diamine cadaverine (Cad-1,5-diaminopentane) is the penta homologue of Put and it is present in many biological systems. However, the existence of this diamine in plants is sporadic and less common than its homologue Put. PAs exist in three forms in the cell, as free cations, covalently bound to low molecular weight phenolic compounds like hydroxycinnamic acids (conjugated form of PAs) and bound to macromolecules or membranes (bound form of PAs). In animals, protozoa and most of the fungi, the diamine Put, the precursor of the higher PAs, Spd and Spm is synthesized directly from ornithine by the enzyme ornithine decarboxylase (ODC). Plants and bacteria have an alternative route to the production of Put by decarboxylation of arginine that is catalyzed by arginine decarboxylase (ADC) (Fig. 1). Additional reactions convert Put into Spd and Spm. These steps are catalyzed by Spd and Spm synthases respectively, which add propyl amino groups to the decarboxylated S-adenosylmethionine(SAM) which is generated from SAM by SAM decarboxylase (SAMDC). Cad is mainly found in the Leguminosae family and it originates by the decarboxylation of lysine through the action of lysine decarboxylase (LDC), a pyridoxal-phosphate dependent enzyme that has not been as extensively studied as ODC (Slocum et al., 1984; Rajam, 1997).

The PA biosynthetic pathway genes *adc*, *odc*, *samdc*, *spd synthase* and *spm synthase* have been cloned from various systems, including bacteria, fungi, plants and animals (Rajam *et al.*, 1998). Though there are many reports on cloning of PA metabolic genes, only a few reports from plant sources are mentioned here.

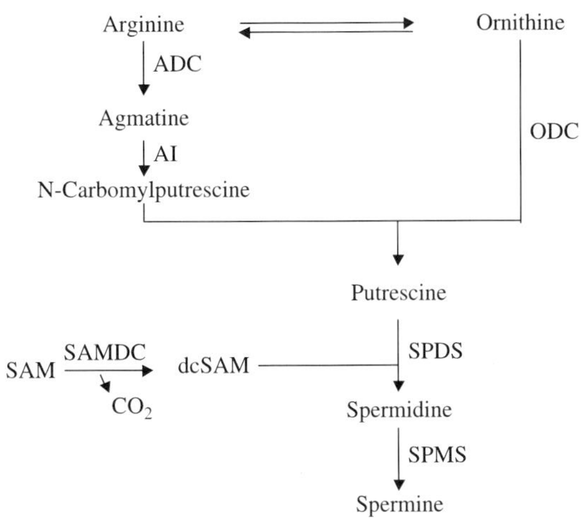

Figure 1. Polyamine biosynthetic pathway in plants. Enzymes and intermediates involved in the pathway are indicated: ADC, Arginine decarboxylase; AI, Agmatine iminohydrolase; ODC, Ornithine decarboxylase; SAM, S-adenosyl methionine, SAMDC, SAM decarboxylase; dcSAM, decarboxylated SAM; SPDS, Spermidine synthase; SPMS, Spermine synthase

The *adc* gene has been cloned from different plant species like tomato (Rastogi et al., 1993), pea (Perez-Amador et al., 1995), *Arabidopsis* (Watson and Malmberg, 1996) and rice (Chattopadhyay et al., 1997). The *odc* gene has been cloned from plant sources like *Datura* (Michael et al., 1996), tobacco (Mallik et al., 1996) and tomato (Alabadi and Carbonell, 1998). The *samdc* gene has been cloned from *Arabidopsis*, *Datura*, potato (Taylor et al., 1992), *Catharanthus* (Schroeder and Schroeder, 1995), tomato, tobacco (Kumar et al., 1997; Park et al., 1998) and rice (Li and Chen, 2000). Plant *spd synthase* has been cloned from a few species like tobacco, *Arabidopsis* and *Hyoscyamus niger* (Hashimoto et al., 1998). The *Spm synthase* gene has been cloned from *E. coli* and humans (Korhonen et al., 1995) but not from plant species.

2.1. Polyamines as Precursor for Some Important Secondary Metabolites

Plant secondary metabolites have been classified into major classes like polyketides, isoprenoids, alkaloids, phenylpropanoids and flavonoids, which can be further subdivided into several subclasses (Fig. 2). Plant alkaloids constitute the second largest group of secondary metabolites after terpenoids, providing many pharmacologically active compounds. Over 12000 different alkaloids have been described so far and they are found in about 20% of the plant species (de Luca and St. Pierre, 2000). These are low molecular weight nitrogenous compounds derived from a restricted number of amino-acid precursors like ornithine, lysine, tyrosine, tryptophan and phenylalanine, that are converted to versatile central intermediates from which the diversity of alkaloid structures found in plants are derived. In the

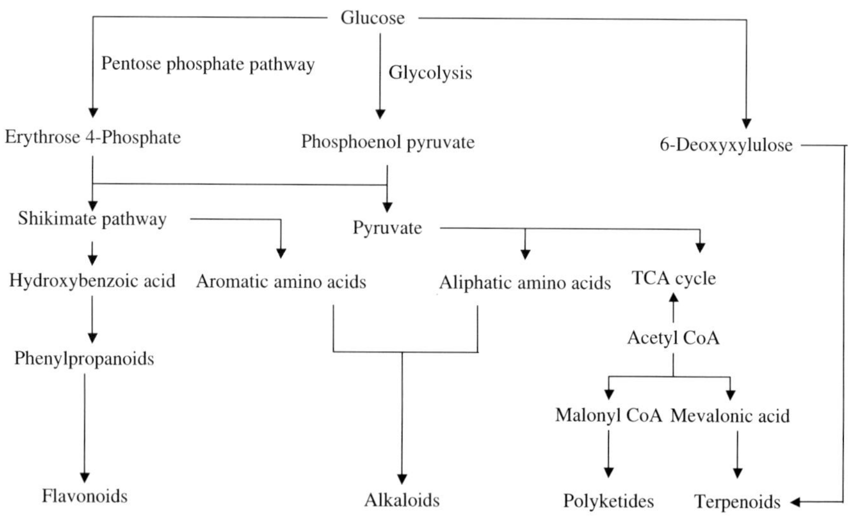

Figure 2. Pathways of secondary metabolism in plants

recent years studies have shown that alkaloid synthesis is not a random process, but is highly ordered with respect to plant development, controlling the expression of pathways within organs, within specific cells or within organelles inside those cells (de Luca and St. Pierre, 2000).

The diamine Put acts as precursor for a wide variety of alkaloids like the pyrrolidine rings of nicotine, nornicotine and tropane alkaloids (hyoscyamine, scopolamine) while Cad and Spd are precursors of quinolizidine and lunarine alkaloids respectively (Ghosh, 2000; Kumar et al., 2006) (Fig. 3). The PAs Put and Spd are also essential biosynthetic precursor of pyrrolizidine alkaloids (Graser and Hartmann, 2000). Besides this Put has also been implicated as the precursor of the calystegines, which are polyhydroxylated nortropane alkaloids that occur predominantly in Convolvulaceae and Solanaceae (Drager, 2004). A few genera of the plant family Solanaceae including *Hyoscyamus*, *Duboisia*, *Atropa* and *Scopolia* are able to produce biologically active nicotine and tropane alkaloids simultaneously (Hashimoto and Yamada, 1994). Nicotine breeding efforts have focused on increasing the levels of this alkaloid in tobacco for smoking purposes as its rapid absorption, transport and binding to acetylcholine receptors in the brain ensures the pleasure obtained from it. Besides this, it is also highly toxic and makes an effective insecticide. The tropane alkaloids are used to control motion sickness, as powerful bronchodilators to treat chronic bronchitis, as antimuscarinic drugs for the control of Parkinson's disease, or as midriatics, which are used to dilate the pupil of the eye to facilitate surgery (de Luca and St. Pierre, 2000).

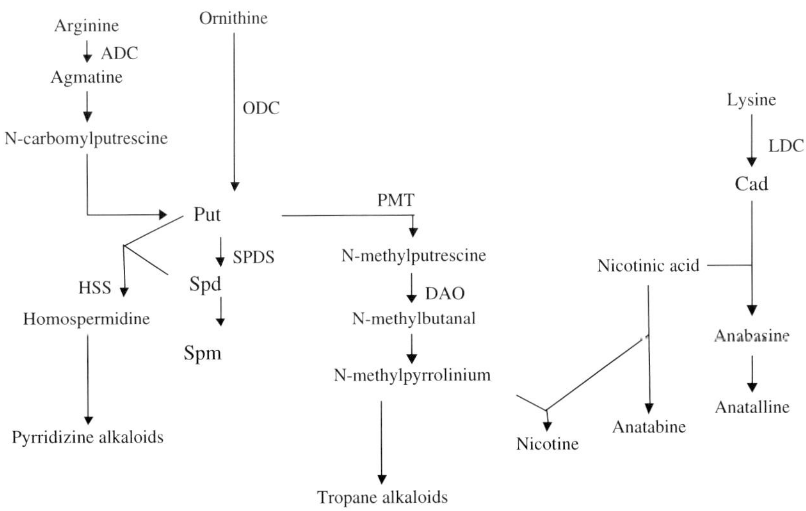

Figure 3. Correlation between the polyamines and the alkaloid biosynthesis pathway. The enzymes involved are indicated: ADC, Arginine decarboxylase; DAO, Diamineoxidase; LDC, Lysine decarboxylase; PMT, Putrescine *N*-methyltransferase; ODC, Ornithine decarboxylase; SPDS, Spermidine synthase

The biosynthesis of both nicotine and tropane alkaloids involves the formation of a pyrrolidine ring that is derived from Put via the sequential action of a SAM dependent putrescine-*N*-methyltransferase (PMT), a diamine oxidase (DAO) and a spontaneous chemical rearrangement. Put acts as the precursor for both the PAs, such as Spd and Spm and pyridine/tropane alkaloids. Nicotine and members of the tropane class have the *N*-methyl- Δ-pyrrolinium cation as the starting point involved in tropane biosynthesis. *N*-Methyl-Δ-pyrrolinium cation formation begins with the decarboxylation of ornithine and/or arginine by ODC and ADC respectively. These enzymes are involved in the formation of Put either directly by ODC, or via agmatine and *N*-carbamoyl putrescine in the case of ADC. The first committed step in both the nicotine and tropane biosynthesis is catalyzed by PMT. It is the enzyme involved in the removal of Put from the PA pool and it catalyzes the *N*-methylation of Put to form *N*-methylputrescine. The presence of PMT is unique to alkaloid producing plants and it appears to have evolved from *spd synthase*, which catalyzes the transfer of the amino propyl moiety of decarboxylated SAM to form Spd (Hibi et al., 1992). It has been proposed that PMT evolved from *spd synthase* after restricted alterations of critical decarboxylated SAM binding amino acid residues (Hashimoto et al., 1998). Tobacco PMT differs from *Spd synthase* by the addition of tandem repeats of eleven amino acid residues at the N-terminus, the repeat element is not required for the enzymatic activity.

PMT catalyzes the transfer of the methyl moiety of SAM to Put. Subsequently, *N*-methylputrescine is oxidatively deaminated by DAO to form the reactive *N*-methyl-Δ-pyrrolinium cation. The *N*-methyl-Δ-pyrrolinium cation is thought to condense with acetoacetic acid to yield hygrine as a precursor of the tropane ring, or with nicotinic acid to form nicotine (Facchini, 2001) although the enzymology of these steps is not known. The biosynthetic pathway of tropane alkaloids is not yet clear. Hyoscyamine is produced by the condensation of tropine and the phenylalanine derived intermediate tropic acid. Hyoscyamine can be converted to its epoxide scopolamine by 6β-hydroxylation of the tropane ring followed by intramolecular epoxide formation catalyzed by the enzyme oxoglutarate dependent dioxygenase, 2-hyoscyamine 6β-hydroxylase (H6H).

Besides nicotine, anatabine and anabasine are two tobacco alkaloids, which are derived in a manner that parallels the origin of nicotine. Anabasine is a minor alkaloid accompanying nicotine in *Nicotiana tabacum* L. However, it is the major alkaloid in *N. glauca* where it is accompanied by moderate amounts of nicotine. The unity of alkaloid biosynthesis in the *Nicotiana* genus is indicated by the utilization of nicotinic acid as precursor of the pyridine rings of nicotine and of anabasine. Variations in the reduced ring were accounted for by using ornithine and its homolog lysine as precursors. The production of both nicotine and anabasine involves the condensation of the primary metabolite nicotinic acid with another nitrogen containing metabolite. In the case of nicotine this is *N*-methyl pyrrolinium (Leete, 1979; Feth et al., 1986; Wagner et al., 1986), and in the case of anabasine, nicotinic acid is condensed with *Δ*-piperidinium (Leete, 1979; Walton and Belshaw, 1988). Each of these metabolites is derived from a separate area of primary metabolism. Besides these alkaloids accumulation of calystegines has been observed in Solanaceae members, especially in *Solanum tuberosum*, particularly in the tuber and also in lower levels in the whole plant (Keiner and Drager 2000) and in the family Convolvulaceae. Calystegines are polyhydroxylated nortropane alkaloids and since they were identified, their biosynthetic pathway has been a matter of debate. Considering their nortropane structure, it was postulated that calystegine had descended from Put like the tropane alkaloids (Goldmann et al., 1990). There is now evidence that the biosynthesis of calystegines proceeds via the first steps of the tropane alkaloid pathway (Stenzel et al., 2006).

PAs also act as precursor for the pyrrolizidine group of alkaloids. These alkaloids are highly toxic to vertebrates causing liver damage and are also strong feeding deterrents for most herbivores. They are found within a scattered and restricted set of species within the families Asteraceae, Boraginaceae, Fabaceae and Orchidaceae (de Luca and Laflamme, 2001). Spd is an essential biosynthetic precursor of pyrrolizidine alkaloids (Graser and Hartmann, 2000). It provides its aminobutyl group, which is transferred, to Put yielding homospermidine, the specific building block of the necine base moiety of pyrrolizidine alkaloids. *Homospermidine* synthase (HSS) catalyzes the formation of homospermidine from Put and Spd.

Cad acts as an important intermediate in different alkaloid production, like the quinolizidine and piperidine alkaloids, which are derived from lysine via a Cad intermediate. Despite some similarities in overall structure to pyrrolizidine alkaloids, a quite distinctive pathway results in the production of quinolizidine alkaloids. The conjugated PAs often conjugate with phenolic acids, fatty acids or amino acids. The naturally occurring PA conjugates are often linear molecules but also poly-azamacrocyclic derivatives have been isolated from different plant families and show a broad structural complexity. These conjugated PAs contain one or two carboxylic acid moieties, often cinnamic acid, *p*-coumaric acid and/or its methoxylated derivatives besides the PA chain. Rare classes of macrobicyclic Spm alkaloids have been identified in the plant families Acanthaceae, Scrophulariaceae and especially in *Aphelandra* plants (Nezbedova et al., 2001). The main alkaloids in *Aphelandra* plants are aphelandrine and orantine (Bosshardt et al., 1978).

3. METABOLIC ENGINEERING OF THE ALKALOID BIOSYNTHESIS GENES

The basic aim of genetic engineering of a secondary metabolic pathway is to either increase or decrease the quantity of a certain compound or group of compounds. Usually two general approaches are followed where the production of a compound / compounds has to be increased. Either the expression of one or a few genes is changed to overcome the specific rate-limiting steps in the pathway, to shut down the competitive pathways, or to decrease the catabolism of the product of interest. Alternatively attempts have been made to change the expression of the regulatory genes that control multiple biosynthesis genes. In those cases when the production of certain unwanted compounds have to be decreased several genetic engineering approaches are possible. An enzymatic step in the pathway can be knocked out, by reducing the level of the corresponding mRNA via antisense, co-suppression or RNA interference technologies, or by over-expressing an antibody against the enzyme.

Considerable work on engineering single step has already been performed in plants, with different degrees of success, as described in the following examples. In most cases the natural yields of the tropane alkaloids hyoscyamine and scopolamine are too low for commercialization. Though genetic engineering approach has been tried to increase the production of nicotine, the conversion of hyoscyamine to the much more valuable scopolamine has been the major goal of these studies. Over-expression of yeast *odc* gene in *Nicotiana rustica* was shown to moderately increase leaf nicotine levels in this plant (Hamill et al., 1990). CaMV35S promoter drove the gene with an upstream duplicated enhancer sequence. The presence of the transgene enhanced ODC activity and there was enhanced accumulation of both Put and the alkaloid nicotine. The magnitude of change even in the presence of the strong promoter was found to be only 2-fold, which suggested that regulatory factors must be acting to limit the potential increase in metabolic flux caused by this manipulation. However, it also demonstrated for the first time that PA biosynthetic pathway genes could be manipulated to elevate the levels of the secondary metabolites. Though both ODC and ADC ultimately result in the production of Put it has been suggested that the route via arginine production supplies most of the Put for alkaloid biosynthesis (Hashimoto and Yamada, 1992). Tiburcio et al. (1985) had studied the correlation between PAs and pyrrolidine alkaloids in developing tobacco callus. They found that specific inhibition of ADC activity had a corresponding strong inhibition of pyrrolidine alkaloid levels, whereas specific inhibition of ODC was less effective in decreasing the levels of the alkaloid. In an earlier experiment oat ADC cDNA had been introduced in tobacco plants to see whether its over-expression had any effect on the levels of the PAs and subsequently in the levels of the tobacco alkaloid nicotine (Burtin and Micheal, 1997). They however found that though agmatine production was increased no corresponding increase was there in the PA levels and in nicotine levels.

Recently Narula et al. (2004) introduced two genes of the PA biosynthetic pathway, *odc* and *adc*, separately into anther-derived calli of *Datura innoxia* through *Agrobacterium tumefaciens*. *Datura*, a member of the Solanaceae family, is important, as it is a primary source of scopolamine and hyoscyamine. The transgenic lines showed higher PA levels, mainly in the Put titre and such lines also yielded a high level of the alkaloid, hyoscyamine, which is usually found in traces in the normal plants.

Nicotine and tropane alkaloids are mainly produced in roots and accumulate mostly within the vacuoles. PMT enzyme activity, the first committed step in the production of both nicotine and tropane alkaloids has primarily been detected within roots of tropane alkaloid producing *Atropa belladonna*, *Hyoscyamus niger* and *Datura*. Histochemical analysis of transgenic *A. belladonna* expressing β-glucuronidase (GUS) behind the PMT promotor showed that GUS is expressed specifically in root pericycle cells (Suzuki et al., 1999). The PMT gene under the control of CaMV35S promoter was used to raise transgenic *Nicotiana sylvestris* plants. The transgenic tobacco lines accumulated N-methyl Put and nicotine content increased by 40% as compared to wild type plants (Sato et al., 2001). Also it was observed that in a co-suppression line in which PMT expression level was decreased as compared to that of the wild type, showed a corresponding drop of nicotine levels. The low nicotine line accumulated high amounts of Put and Spd, indicating that the efficient inhibition of PMT activity shifted the nitrogen flow from nicotine synthesis to PA formation and the plants showed abnormalities which were a result of elevated levels of PAs in the plants (Katoh et al., 2005). This shows that manipulating even a single step in a biosynthetic pathway can have major effects on not only the secondary metabolite accumulation but also on plant growth and development. In recent years many studies have been done to see the effect of over-expression of *pmt* and *h6h* genes on the levels of tropane alkaloids, however, it is beyond the scope of this review to mention all such examples. Strategies involving the *pmt* and *h6h* genes have been reviewed quite extensively earlier (Facchini, 2001; Zhang et al., 2005).

Anabasine is a minor alkaloid accompanying nicotine in *N. tabacum*. Nicotine production, strictly correlated rate wise with growth, and occurs in the root tips, while anabasine production seems to occur primarily in matured root tissues. The enzyme LDC is an important rate-limiting step in anabasine synthesis in *Nicotiana*. Feeding of Cad to *N. rustica* root cultures led to the production of increased anabasine, at the expense of nicotine (Walton et al., 1988). As Cad originates by the decarboxylation of lysine through the action of LDC, over-expression of a bacterial gene encoding LDC was used to transform several hairy root cultures of *N. tabacum* (Fecker et al., 1993). The construct used for transformation had two direct repeats of a bacterial *ldc* gene controlled by CaMV35S promoter. The transgenic lines showed high LDC activity and such activity was sufficient to increase Cad levels of the best line. Some of the overproduced Cad of this line was used for the formation of anabasine, as shown by a 3-fold increase of this alkaloid. In transgenic lines with lower LDC activity the changes of Cad and anabasine levels were correspondingly lower and sometimes hardly distinguishable from controls. Feeding of lysine to root cultures, even to those with low LDC activity, greatly enhanced Cad and anabasine levels, while the amino acid had no or very little effect on controls and LDC negative lines. Together, these results highlight the close link between nicotine and anabasine synthesis, which apparently competes for nicotinic acid. In another experiment, enhanced accumulation of Cad and anabasine was observed as LDC was fused to the RBCS transit peptide (Herminghaus et al., 1996).

Engineering of some genes for the production of unusual PAs which are intermediates in the pyrrolizidine alkaloid biosynthetic pathway have also been done, for example the *homospermidine synthase* (HSS) gene. It is a branch-point enzyme that links the secondary pathway (pyrrolizidine alkaloids) to primary metabolism i.e. the polyamines (Kaiser et al., 2002). The unique precursor in the formation of the necine base moiety of pyrrolizidine alkaloids is the PA homospermidine which is formed from Spd and Put in an NAD^+ dependent reaction catalyzed by HSS (Graser and Hartmann, 1997). Since the diamine Put is a precursor of homospermidine and nicotine a bacterial *hss* gene was expressed in *N. tabacum* and the effect on free and conjugated PA levels was determined. The *hss* gene from *Rhodopseudomonas viridis* under the control of the CaMV35S promoter was cloned in *Agrobacterium tumefaciens* Ti-plasmid in sense and antisense orientation and both these *hss* gene constructs were then used to transform tobacco plants. Two of the transgenic sense lines were generated which showed weak expression of *hss* gene. These transgenic sense plants showed a significantly decreased content of free Spd while the pool of conjugated Spd was not affected. The two sense transgenic plants exhibited a range of abnormal phenotypes such as dwarfness and stunted growth, however the effect of engineering the pyrrolizidine alkaloid biosynthetic pathway gene on the alkaloid production was not tested.

Kinetic studies with recombinant HSS from *Senecio vernalis* (Ober and Hartmann 1999) revealed that Put could not substitute for Spd in its function as amino butyl donor as assumed previously with partially purified enzyme preparations (Bottcher et al., 1994). The formation of homospermidine is the only known example in which a functional moiety of Spd is channeled into a secondary biosynthetic pathway. There are, however, examples showing the incorporation of the intact Spd molecule into secondary compounds such as lunarine in seeds of *Lunaria annua* (Sagner et al., 1998) and the inandenines of *Oncinotis* spp. or palustrines of *Equisetum* spp. (Guggisberg and Hesse 1983).

Alternative strategies like antisense RNA technology have also been tried to increase the accumulation of the secondary metabolites. Chintapakorn and Hamill (2003) used antisense approach to down-regulate activity of the PMT enzyme in transformed roots of *N. tabacum* to determine the effect on alkaloid metabolism. Transformed root lines contained markedly reduced PMT activity with a concomitant reduction in nicotine content compared to the control plants. Several of the antisense-PMT transformed root lines showed a substantial increase in anatabine content relative to the controls however, no negative effects upon growth were observed. According to the authors the elevated anatabine levels in the antisense-PMT lines were a direct consequence of a relative oversupply of nicotinic acid that in the absence of adequate levels of *N*-methyl-Δ-pyrrolinium cation (the ultimate product of PMT activity) was used to synthesize anatabine directly. Their experiment showed that such alternative strategies could also be tried to not only increase the production of economically important secondary metabolite which are produced in very less amount in the plants under natural conditions, but also they help us in understanding how manipulating a single step in the pathway can influence the other intermediates in the pathway.

In the past few years strategies for metabolic engineering are striving to fortify the biosynthetic pathway by over-expressing multiple biosynthetic genes, by manipulating regulatory genes that control the expression of multiple pathway enzyme genes, or both (Sato et al., 2001). These results strengthen the idea that manipulating multiple steps in biosynthetic pathway can increase the yield of secondary metabolites. Another example is in case of henbane *(Hyoscyamus niger)* where the gene for PMT and H6H was simultaneously introduced and over-expressed in hairy root cultures (Zhang et al., 2004). Transgenic hairy root lines expressing both *pmt* and *h6h* genes produced significantly higher levels of scopolamine as compared to the wild types and the transgenic lines harboring either single *pmt* or *h6h* gene. The best line produced over nine-fold greater scopolamine than the wild type and twice the amount in the highest scopolamine-producing *h6h* single gene transgenic line.

Attempts to increase the yield of secondary metabolites have also focused on changing the expression of regulatory genes that control multiple biosynthesis genes. Transcription factors are attractive alternatives for metabolic engineering as the central transcription factors control a wide array of genes responsible for the metabolic differentiation of plant cells. The biosynthetic pathway genes are under the control of *trans*-acting regulatory factors and thus these factors strictly regulate the production of secondary metabolites. Transcription factors are sequence-specific DNA-binding proteins that interact with the promoter regions of target genes and modulate the rate of initiation of mRNA synthesis by RNA polymearse II (Gantet and Memelink, 2002). How the expressions of these transcription factors are regulated is not clearly understood. Regulatory genes have been shown to be more generally applicable as useful tools to identify genes from secondary metabolite pathways and for up- or down-regulation of a pathway or part of a pathway (Verpoorte and Memelink, 2002). There are quite a few reports about the promoter analysis to identify the regulators of alkaloid biosynthesis as they provide a feasible alternative approach for the isolation of transcription factors. However, there have been no reports yet regarding the effect of expressing such factors on the yield of alkaloids.

There has also been a recent trend for a functional genomics approach towards understanding secondary metabolism in plant cells. Goossens et al. (2003) tried combined targeted metabolite analysis with cDNA-amplified fragment length polymorphism based transcript profiling of methyl jasmonate (MeJA)-elicited tobacco Bright yellow-2 (BY-2) cells. Their study helped in in-depth comparisons with previously reported transcriptome analyses and revealed the presence of all, except one, of the genes known to be involved in the biosynthesis of nicotine alkaloids in *Nicotiana* species. They found that the BY-2 cells showed a marked increase in the content of various nicotinic acid derived alkaloids after MeJA treatment. Besides nicotine, two anatalline isomers and anatabine also accumulated more after MeJA treatment. Most of these genes were induced early by MeJA in tobacco BY-2 cells and clustered together with novel gene or genes encoding proteins with still unknown functions, which could be the probable candidates to encode for the missing links in nicotine biosynthesis. ADC and ODC (structural genes for the biosynthesis of *N*-methylpyrrolinium moiety), aspartate oxidase and quinolinate phosphoribosyltransferase (structural genes for the biosynthesis of nicotinic acid moiety) and SAM synthetase (structural gene for the catalysis of the methyl-group donor for nicotine biosynthesis) responded to MeJA treatment. Also the expression of the *pmt* gene was up-regulated, as early as 1 h after jasmonate treatment thereby demonstrating the co-regulation of the *pmt* gene with the other nicotine metabolic genes. Their findings correlate well with earlier reports on jasmonate inducibility of the genes encoding ODC and PMT (Imanishi et al., 1998).

4. CONCLUSIONS AND FUTURE PROSPECTS

Genetic engineering provides the means, either alone or in combination with traditional breeding, to develop novel plant sources with quantitatively and qualitatively improved pharmacological properties. It is known that plants, either closely or distantly related, synthesize structurally similar but nevertheless diverse molecules. Thus by introducing genes involved in the biosynthesis of a given compound isolated from one plant and introduced into another plant synthesizing related molecules, may produce new chemical structures not previously found in nature as the enzymes may share certain substrate specificity. Though rapid progress has been made in our understanding of the biochemistry and molecular biology of secondary metabolites in plants the picture is far from complete. The knowledge of secondary metabolite pathways is very limited and is the major constraint for successful application of metabolic engineering. Despite little knowledge presently available some very interesting results have already been obtained. The pace of gene discovery in plant metabolism has increased dramatically in the past few years and this holds great promise for the future of plant metabolic engineering however, there are major challenges to be faced in the coming years. It is imperative to obtain more information about the various steps involved in secondary metabolism, right from finding out the enzymes involved, isolating and cloning of the related genes, and also to understand the mechanism of transport, compartmentalization, and accumulation of the secondary metabolites. This understanding will open the way for successful strategies for altering the accumulation of the economically important compounds.

Presently a major constraint in engineering plant secondary metabolite production is that only a few genes of these pathways are known. Though transgenic plants over-expressing certain genes of some important biosynthetic pathways have shown some promise, they have often failed to significantly increase the levels of desired secondary metabolite. The conclusion drawn from these studies emphasizes that majority of the enzymes in a pathway have to be augmented in a coordinated manner to achieve a considerable boost in secondary metabolite production as the steps in the pathway are under strict control. Thus manipulation of the regulatory genes, especially the *trans*-acting factors specifically involved in secondary metabolite production is an attractive idea but much remains to be learned before these can be used to control native gene expression patterns. Subsequently understanding the mechanism of intra- and inter-cellular transport of metabolites and enzymes might also reveal some important information that might facilitate engineering of specific transporters which may deliver increased amounts of secondary metabolites to their proper accumulation sites. As has been seen from the examples, PA biosynthetic pathway provides an attractive model for such studies as the various steps in the pathway are known and PAs also act as precursors of many economically important secondary metabolites. With continuing studies in the field of metabolic engineering and with such an array of strategies and tools at our disposal we can assume to see exciting new discoveries and also increased yield in commercial production of secondary metabolites in the foreseeable future.

REFERENCES

Alabadi D, Carbonell J (1998) Expression of ornithine decarboxylase is transiently increased by pollination, 2,4-Dichlorophenoxyacetic acid and gibberellic acid in tomato ovaries. *Plant Physiol* 118:323–328.

Bosshardt H, Guggisberg A, Johne S et al. (1978) Über alkaloide der genera *Aphelandra* und *Encephalosphaera* (Acanthaceae). *Pharm Acta Helv* 53:355–357.

Bottcher F, Ober D, Hartmann T (1994) Biosynthesis of pyrrolizidine alkaloids: putrescine and spermidine are essential substrates of enzymatic homospermidine formation. *Can J Chem* 72:80–85.

Burtin D, Michael AJ (1997) Over-expression of arginine decarboxylase in transgenic plants. *Biochem J* 325:331–337.

Chattopadhyay MK, Gupta S, Sengupta DN et al. (1997) Expression of arginine decarboxylase in seedlings of *indica* rice (*Oryza sativa* L.) cultivars as affected by salinity stress. Plant Mol Biol 34:477–483.

Chintapakorn Y, Hamill JD (2003) Antisense-mediated down regulation of putrescine *N*-methyltransferase activity intransgenic *Nicotiana tabacum* L. can lead to elevated levels of anatabine at the expense of nicotine. *Plant Mol Biol* 53:87–105.

De Luca V, Laflamme P (2001) The expanding universe of alkaloid biosynthesis. *Curr Opin Plant Biol* 4:225–233.

De Luca V, St-Pierre B (2000) The cell and developmental biology of alkaloid biosynthesis. *Trends Plant Sci* 4:168–73.

Drager B (2004) Chemistry and biology of calystegines. Nat Prod Rep 21:211–223.

Facchini PJ (2001) Alkaloid biosynthesis in plants: Biochemistry, cell biology, molecular regulation, and metabolic engineering applications. *Annu Rev Plant Physiol Plant Mol Biol* 52:29–66.

Fecker LF, Rugenhagen C, Berlin J (1993) Increased production of cadaverine and anabasine in hairy root cultures of *Nicotiana tabacum* expressing a bacterial lysine decarboxylasegene. *Plant Mol Biol* 23: 11–21.

Feth F, Wagner R, Wagner KG (1986) Regulation in tobacco callus of enzyme activities of the nicotine pathway. I. The route ornithine to methyl pyrroline. *Planta* 168:402–407.

Galston AW, Kaur-Sawhney R (1990) Polyamines in plant physiology. Plant Physiol 94: 406–410.

Gantet P, Memelink J (2002) Transcription factors: Tools to engineer the production of pharmacologically active plant metabolites. *Trends Pharmacol Sci* 23: 563–569.

Ghosh B (2000) Polyamines and plant alkaloids. *Indian J Expt Bot* 38:1086–1091.

Goldmann A, Milat ML, Ducrot PH et al. (1990) Tropane derivatives from *Calystegia sepium*. *Phytochemistry* 29:2125–2128.

Goossens A, Hakkinen ST, Laakso I (2003) A functional genomics approach toward the understanding of secondary metabolism in plant cells. *Proc Natl Acad Sci USA* 100: 8595–8600.

Graser G, Hartmann T (1997) Biosynthetic incorporation of the aminobutyl group of spermidine into pyrrolizidine alkaloids. *Phytochemistry* 45:1591–1595.

Graser G, Hartmann T (2000) Biosynthesis of spermidine, a direct precursor of pyrrolizidine alkaloids in root cultures of *Senecio vulgaris* L. *Planta* 211:239–245.

Guggisberg A, Hesse M (1983) Putrescine, spermidine, spermine, and related polyamine alkaloids. *Alkaloids* 22:85–188.

Hamill JD, Robins RJ, Parr AJ et al. (1990) Over-expression of a yeast ornithine decarboxylase gene in transgenic roots of *Nicotiana rustica* can lead to enhanced nicotine accumulation. *Plant Mol Biol* 15: 27–38.

Hashimoto T, Yamada Y (eds) (1992) Tropane alkaloid biosynthesis:regulation and application. In: Proc Annu Penn State Symp Plant Physiol, 7th Rockville: Am. Soc. Plant Physiol Press, pp 122–134.

Hashimoto T, Yamada Y (1994) Alkaloid biogenesis: Molecular aspects. *Annu Rev Plant Physiol Plant Mol Biol* 45:257–285.

Hashimoto T, Tamaki K, Suzuki K et al. (1998) Molecular cloning of plant spermidine synthases. *Plant Cell Physiol* 39:73–79.

Herminghaus S, Tholl D, Rugenhagen C et al. (1996) Improved metabolic action of a bacterial lysine decarboxylase gene in tobacco hairy root cultures by its fusion to a *rbcS* transitpeptide coding sequence. *Transgenic Res* 5:193–201.

Hibi N, Fujita T, Hatano M (1992) Putrescine *N*-methyltransferase in cultured roots of *Hyoscyamusalbus*. *Plant Physiol* 100: 826–835.

Imanishi S, Hashizume K, Nakakita M (1998) Differential induction by methyl jasmonate of genes encoding ornithine decarboxylase and other enzymes involved in nicotine biosynthesis in tobacco cell cultures.Plant Mol Biol 38:1101–1111.

Kaiser A, Sell S, Hehl R (2002) Heterologous expression of a bacterial homospermidine synthase gene in transgenic tobacco: effects on the polyamine pathway. *Arch Pharm (Weinheim)* 335:143–51.

Katoh A, Ohki H, Inai K et al. (2005) Molecular regulation of nicotine biosynthesis Plant Biotechnol 23: 389–392.

Keiner R, Drager B (2000) Calystegine distribution in potato (*Solanum tuberosum*) tubers and plants. *Plant Sci* 150:171–179.

Korhonen VP, Halmekyto M, Kauppinen L (1995) Molecular cloning of a cDNA encoding human spermine synthase. *DNA Cell Biol* 14:841–847.

Kumar A, Altabella T, Taylor MA et al. (1997) Recent advances in polyamine research. *Trends Plant Sci* 2:124–130.

Kumar SV, Sharma ML Rajam MV (2006) Polyamine biosynthetic pathway as a novel target for potential applications in plant biotechnology. Physiol Mol Biol Plants 12:53–58.

Leete E (1979) The alkaloids: alkaloids derived from ornithine,lysine and nicotinic acid, In: EA Bell and BV Charlwood (ed) Encyclopedia of Plant Physiology, New Series, Secondary Plant Products, Vol. 8, Springer-Verlag, Berlin, pp 65–91.

Li ZY, Chen SY (2000) Differential accumulation of the S-adenosylmethionine decarboxylase transcript in rice seedlings in response to salt and drought stress. *Theor Appl Genet* 100: 782–788.

Mallik V, Watson MB, Malmberg RL (1996) A tobacco ornithine decarboxylase partial cDNA clone. *J Plant Biochem Biotech* 5:109–112.

Micheal AJ, Furze JM, Rhodes MJC et al. (1996) Molecular cloning and functional identification of a plant ornithine decarboxylase cDNA. Biochem J 314: 24–248.

Narula A, Kumar SV, Pande D et al. (2004) *Agrobacterium* mediated transfer of arginie decarboxylase and ornithine decarboxylase genes to *Datura innoxia* enhances shoot regeneration and hyoscyamine biosynthesis. *J Plant Biochem Biotech* 13:127–130.

Nezbedová L, Hesse M, Drandarov K et al. (2001) Phenol oxidative coupling in the biogenesis of the macrocyclic spermine alkaloids aphelandrine and orantine in *Aphelandra* sp. *Planta* 213: 411–417.

Ober D, Hartmann T (1999) Homospermidine synthase, the first pathway-specific enzyme of pyrrolizidine alkaloid biosynthesis,evolved from deoxyhypusine synthase. *Proc Natl Acad Sci USA* 96:14777–14782.

Oksman-Caldentey KM, Inze D (2004) Plant cell factories in the post-genomic era: new ways to produce designer secondary metabolites. *Trends Plant Sci* 9: 433–440.

Park WK, Lee SH, Park KY (1998) Cloning and characterization of genome clone (Accession no. U64927) encoding S-adenosyl-L-methionine decarboxylase whose gene expression was regulated by light in morning glory (*Ipomea nil* L.). *Plant Physiol* 116: 867–872.

Perez-Amadour MA, Carbonell J (1995) Arginine decarboxylase and putrescine oxidase in ovaries of *Pissum sativum* L. changes during ovary senescence and early fruit development. *Plant Physiol* 107: 865–872.

Rajam MV (1997) Polyamines. In: MNV Prasad (ed) Plant Ecophysiology, John Wiley and Sons, New York, pp 343–374.

Rajam MV, Weinstein LH, Galston AW (1985) Prevention of a plant disease by specific inhibition of fungal polyamine biosynthesis. *Proc Natl Acad Sci USA* 82:6874–6878.

Rajam MV, Shoeb F, Yadav JS (1998) Polyamines as modulators of plant regeneration in tissue cultures. In: P.S. Srivastava (ed) Plant Tissue Culture and Molecular Biology: Applications and Prospects,Narosa Publishing House, New Delhi, India, pp 620–641.

Rastogi R, Dulson J, Rothstein SJ (1993) Cloning of tomato (*Lycopersicum esculentum* Mill.) arginine decarboxylase gene and its expression during fruit ripening. *Plant Physiol* 103:829–834.

Sagner S, Shen ZW, Deus-Neumann B et al. (1998) The biosynthesis of lunarine in seeds of *Lunaria annua*. *Phytochemistry* 47:375–387.

Sato F, Hashimoto T, Hachiya A et al. (2001) Metabolic engineering of plant alkaloid biosynthesis. *Proc Natl Acad Sci USA* 98:367–372.

Schroeder G, Schroeder J (1995) cDNA for S-adenosyl-L-methionine decarboxylase from *Catharanthus roseus*, heterologous expression, identification of the proenzyme processing site, evidence for the presence of both subunits in the active enzyme and a conserved region in the 5' messenger RNA leader. *Eur J Biochem* 228: 74–78.

Slocum RD, Kaur-Sawhney R, Galston AW (1984) The physiology and biochemistry of polyamines in plants. *Arch Biochem Biophys* 235:283–303.

Slocum RD, Flores HE (1991) Uncommon polyamines in plants and other organisms. In: RD Slocum, Flores HE (eds) Biochemistry and Physiology of Polyamines in Plants. CRC Press, Boca Raton, FL,pp121–136.

Stenzel O, Teuber M, Drager B (2006) Putrescine N-methyltransferase in *Solanum tuberosum* L., a calystegine-forming plant. *Planta* 223: 200–212.

Suzuki K, Yamada Y, Hashimoto T (1999) Expression of *Atropa belladonna* putrescine *N*-methyltransferase gene in root pericycle. *Plant Cell Physiol* 40: 289–297.

Taylor MA, Mad Arif SA, Kumar A et al. (1992) Expression and sequence analysis of cDNAs induced during the early stages of tuberization in different organs of potato plant (*Solanum tuberosum* L.). *Plant Mol Biol* 20: 641–651.

Tiburcio AF, Kaur-Sawhney R, Ingersoll RB et al. (1985) Correlation between polyamines and pyrrolonide alkaloids in developing tobacco callus. *Plant Physiol* 78: 323–326.

Urano K, Hobo T, Shinozaki K (2005) *Arabidopsis* ADC genes involved in polyamine biosynthesis are essential for seed development. *FEBS Letters* 579:1557–1564.

Verpoorte R (2000) Secondary metabolism. In: Verpoorte R, Alfermann AW (eds) Metabolic Engineering of Plant Secondary Metabolism, Kluwer Academic Publishers, pp1–29.

Verpoorte R, Memelink J (2002) Engineering secondary metabolite production in plants. *Curr Opin Biotechnol* 13:181–187.

Wagner R, Feth F, Wagner KG (1986) Regulation in tobacco callus of enzyme activities of the nicotine pathway. II. The pyridine nucleotide cycle. *Planta* 168: 408–413.

Walton NJ, Robins RJ, Rhodes MJC (1988) Perturbation of alkaloid production by cadaverine in hairy root cultures of *Nicotiana rustica*. *Plant Sci* 54: 125–131.

Walton NJ, Belshaw NJ (1988) The effect of cadaverine on the formation of anabasine from lysine in hairy root cultures of *Nicotiana hesperis*. *Plant Cell Rep* 7:115–118.

Watson MB, Malmberg RL (1996) Regulation of *Arabidopsis thaliana* (L.) arginine decarboxylase by potassium deficiency stress. *Plant Physiol* 111: 1077–1083.

Zhang L, Ding RX, Chai YR et al. (2004) Engineering tropane biosynthetic pathway in *Hyoscyamus niger* hairy root cultures. *Proc Natl Acad Sci USA* 101:6786–6791.

Zhang L, Kai GY, Lu BB (2005) Metabolic engineering of tropane alkaloid biosynthesis in plants. *J Int Plant Biol* 47:136–143.

CHAPTER 6

METABOLIC ENGINEERING IN ALKALOID BIOSYNTHESIS: CASE STUDIES IN TYROSINE- AND PUTRESCINE-DERIVED ALKALOIDS

Molecular engineering in alkaloid biosynthesis

FUMIHIKO SATO[1], KOJI INAI[2] AND TAKASHI HASHIMOTO[2]

[1] *Department of Plant Gene and Totipotency, Graduate School of Biostudies, Kyoto University, Kyoto, 606-8502, Japan*

[2] *Graduate School of Biological Sciences, Nara Institute of Science and Technology, 8916-5 Takayama, Ikoma, Nara 630-0192, Japan*

Abstract: Higher plants produce divergent classes of metabolites. Metabolic engineering offers tremendous potential to improve the productivity and quality of these metabolites. We present here case studies on two types of alkaloids. Nicotine and tropane alkaloids, such as hyoscyamine and scopolamine, which are synthesized from putrescine in several solanaceous plants, and have a common evolutional origin. These alkaloids are synthesized in limited cell types in the root and then transported to aerial parts where they accumulate. On the other hand, isoquinoline alkaloids, such as morphine, sanguinarine, and berberine are synthesized from tyrosine in Magnoliaceae, Ranunculaceae, Berberidaceae, Papaveraceae, and many other species. While this biosynthesis may have a monophyletic origin, it is regulated in a more complicated manner. In this review, we summarize the enzymes and genes in biosynthesis, and the potentials and pitfalls in metabolic engineering for alkaloid production

Keywords: biosynthetic pathway, cell-specific gene expression, expressed sequence tag (EST), isoquinoline alkaloid, metabolic engineering, nicotine, quality control, quantity improvement, RNA interference (RNAi), tropane alkaloid, secondary metabolite, transport

Abbreviations: acetylcoenzyme A:salutaridinol-7-*O*-acetyltransferase, SAT; Arginine decarboxylase; ADC, berberine bridge enzyme, BBE; β-glucuronidase, GUS; canadine synthase, CDS or CYP719; Cauliflower mosaic virus 35S, CaMV35S; codeinone reductase, COR; decarboxylated S-adenosylmethionine, dSAM; diamine oxidase (DAO); 3′ hydroxy

* To whom correspondence should be addressed. Fumihiko Sato for tyrosine-derived alkaloids and general topics, *Department of Plant Gene and Totipotency, Graduate School of Biostudies, Kyoto University, Kyoto, 606-8502, Japan*, E-mail:fsato@lif.kyoto-u.ac.jp, Takashi Hashimoto for putrescine-derived alkaloids, *Graduate School of Biological Sciences, Nara Institute of Science and Technology, 8916-5 Takayama, Ikoma, Nara 630-0192, Japan*, E-mail:hasimoto@bs.naist.ac.jp

R. Verpoorte et al. (eds.), Applications of Plant Metabolic Engineering, 145–173.
© 2007 *Springer*.

N-methylcoclaurine 4′-*O*-methyltransferase, 4′OMT; hyoscyamine 6β-hydroxylase, H6H; methyl jasmonate, MeJA; norcoclaurine 6-*O*-methyltransferase, 6OMT; norcoclaurine synthase, NCS; ornithine decarboxylase, ODC; polymerase chain reaction, PCR; putrescine *N*-methyltransferase, PMT; *S*-adenosyl methionine, SAM; scoulerine 9-*O*-methyltransferase, SMT; spermidine synthase, SPDS; tropinone reductase I/II, TR-I/II; tyrosine/dopa decarboxylase, TYDC

1. INTRODUCTION

Higher plants produce a wide range of chemicals; more than 25,000 terpenoids, and about 12,000 alkaloids and 8,000 phenolic substances have been identified thus far (Croteau et al., 2000). These chemicals have a variety of functions in plants. Many secondary metabolites in plants promote human health not only as pharmaceuticals, but also as dietary supplements and functional foods (Briskin, 2000, Raskin et al., 2002). Alkaloids, which are a diverse family of alkaline nitrogen-containing compounds, include many biologically active chemicals such as berberine, caffeine, colchicine, emetine, hyoscyamine, morphine, nicotine, scopolamine, tubocurarine, and others. We focus here on the alkaloids, due to their relatively high biological activities, and especially on tyrosine-derived isoquinoline alkaloids and putrescine-derived tropane and nicotine alkaloids, which are currently the subjects of intensive study (Verpoorte and Memelink, 2002, Hashimoto and Yamada, 2003, Facchini and St-Pierre, 2005, Kutchan, 2005a,b, Zhao et al., 2005).

2. GENERAL STRATEGY FOR METABOLIC ENGINEERING

The establishment of genetic engineering techniques has made possible the modification of metabolic pathways and the genetic regulation of secondary metabolism. In fact, the industrial production of secondary metabolites is better feasible if we can improve productivity, either by reducing the requirements of the growth period or by increasing the level of metabolites (Misawa, 1994). Furthermore, it may be possible to improve the quality of metabolites by reducing undesired pathways or introducing new pathways to produce novel compounds, or by completely blocking a pathway to accumulate intermediates. Such metabolic engineering would modify plant cells to become "green chemical factories" to obtain the desired compounds. In the sections below, we discuss how metabolic engineering offers the high potential for the production of metabolites in intact plants or cell cultures; general strategies are shown in Fig. 1 (also see refs. Croteau et al., 2000, Verpoorte and Memelink, 2002).

2.1. Isolation of Biosynthetic and Regulatory Genes in Secondary Metabolism

The availability of biosynthetic and regulatory genes is important for modifying a given pathway, however our current knowledge on these subjects is still limited. Thus, we begin by explaining how to isolate desired genes. Since the biosynthetic

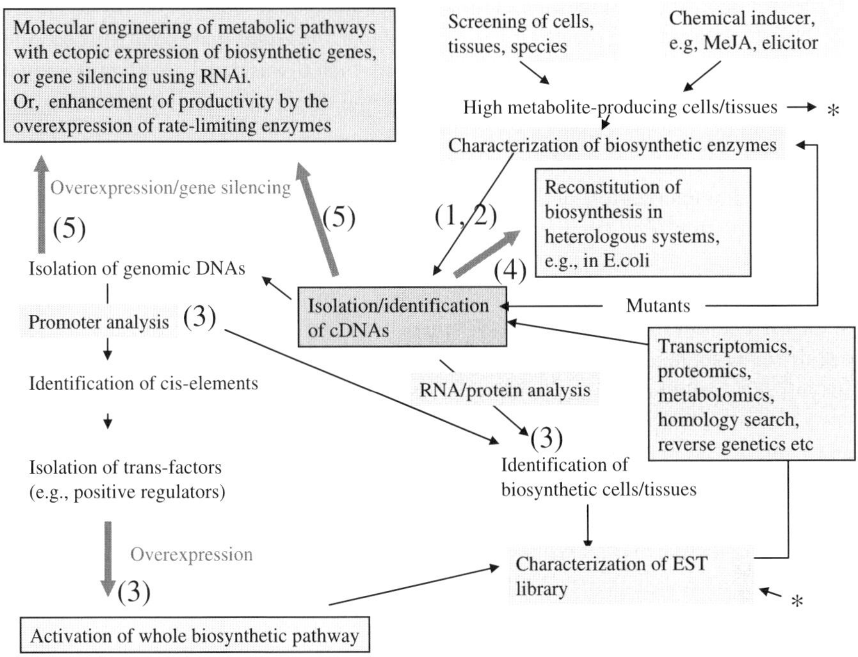

Figure 1. Strategies for overcoming the limitations in the production of metabolites in plant cells
1) characterization of biosynthetic enzymes and isolation of biosynthetic genes,
2) hetrologous expression of biosynthetic enzymes, charecterization of their enzymatic properties and identification of the rate-limiting step,
3) characterization of the regulation of the gene expression of biosynthetic enzymes and isolation of transcriptional regulators, regulation of the whole pathway with transcriptional regulation (see Kato et al., 2007 for recent progress)
4) reconstruction of biosynthesis in heterologous systems
5) metabolic engineering with the creation of a new branch pathway to produce novel compounds, reducing an undesired pathway to enhance the metabolic flow to a desired path

genes in secondary metabolism in plants are usually not clustered, the establishment and/or isolation of high-metabolite-producing cells is required to characterize the pathway and isolate the enzymes and genes (Hashimoto and Yamada, 2003). For example, the establishment of high berberine-producing *Coptis japonica* cells and/or the isolation of oil-gland cells in mint plants has been shown to be very useful for isolating biosynthetic genes (Lange et al., 2000, Morishige et al., 2002). On the other hand, the characterization of mutant or plant species with differences in metabolite productivity, such as low-nicotine mutants or *Papaver* species with differences in morphine productivity, has been used to identify the gene(s) specific to alkaloid production (Hibi et al., 1994, Ziegler et al., 2005). Another strategy for isolating a desired gene without mutants is to use chemicals (e.g., elicitors or methyljasmonate) to activate secondary metabolism (Verpoorte and Memelink, 2002). Alternatively, if we could isolate general transcriptional factor(s) in the target secondary metabolism, we may be able to use transgenic cells with enhanced gene expression in secondary metabolism as a starting material; e.g., *Arabidopsis thaliana* that overexpressed the PAP1 gene encoding an MYB transcriptional factor was used to identify novel genes involved in flavonoid biosynthesis using the integration of metabolomics and transcriptomics (Tohge et al., 2005).

In either case, the biochemical characterization of biosynthetic enzymes or the molecular characterization of ESTs by microarray or RNA blot-analysis, as well as the cDNA-amplified fragment length polymorphism (AFLP) method (Goossens et al., 2003b), should be effective for evaluating the functional linkage of genes in combination with targeted metabolite analysis. A proteomics approach can also be useful for examining proteins by two-dimensional gel electrophoresis and internal peptide microsequencing. Using this approach, a representative enzyme in morphine biosynthesis could be detected within the serum fraction of latex from opium poppy (Decker et al., 2000). As shown above, a recent trend is to avoid biochemical purification and to directly isolate candidate clones by a combination of expression pattern analysis and reverse-genetics approaches as well as homology-based screening (Hashimoto and Yamada, 2003). Thus, there has been a rapid increase in the availability of molecular tools (Weckwerth and Fiehn, 2002, Hashimoto and Yamada, 2003, Facchini and St-Pierre, 2005, Kutchan, 2005a,b), whereas we face the limits of biochemical characterization.

2.2. Whole Modification of Biosynthetic Activity Through Transcriptional Regulation and/or Regulation of Rate-Limiting Process(es) in the Pathway

Due to the difficulty of studying biosynthetic pathways in secondary metabolism, researchers have in general isolated few biosynthetic genes and characterized their promoter sequences, except for anthocyanin biosynthesis (Winkel-Shirley, 2001). The molecular characterization of anthocyanin biosynthesis is a limited successful example of the application of transcriptional regulation to metabolite production; i.e., the introduction of transcriptional regulator-R and C1 in a heterologous system (Lloyd et al., 1992) and recent successes with activation tagging methods (Borevitz et al., 2000) show the high potential and bright future of these approaches. Especially, the successful isolation of ORCA3, a transcriptional factor with a jasmonate (JA)-responsive AP2 domain, in indole alkaloid biosynthesis in *Catharanthus* using activation tagging has demonstrated the effectiveness of this approach in alkaloid biosynthesis (Van Der Fits and Memelink, 2000).

However, no terpenoid indole alkaloids were induced by the ectopic expression of ORCA3 in cultured *C. roseus* cells, although this increases the expression of the terpenoid indole alkaloid biosynthetic genes and the accumulation of significantly more tryptophan and tryptamine. This result indicates that several transcription factors are involved in alkaloid biosynthesis, especially for indole and morphinan alkaloids.

Over-expression of the rate-limiting enzyme is an alternative method for increasing the accumulation of the desired compounds, if we can only identify the rate-limiting step and isolate the target gene. The pioneering work by Yun et al. (1992) is an example of such a successful application in scopolamine biosynthesis in transgenic *Atropa* with *Hyoscyamus* hyoscyamine 6-β-hydroxylase. The crucial point is identification of the rate-limiting step. Whereas an early step is speculated to be rate-limiting for the whole pathway, simple over-expression of the early-step enzyme in secondary metabolism is usually not sufficient for inducing the biosynthesis of the desired end-product due to a lack of sufficient induction of late-step enzymes. For example, over-expression of tryptophan decarboxylase in *Catharanthus roseus* cell cultures resulted in higher levels of the immediate product tryptamine, but not increased levels of alkaloids, whereas higher levels of alkaloids were noted in the over-expression of strictosidine synthase (Canel et al., 1998). In general, there are multiple rate-limiting processes. However, over-expression of the gene encoding deoxyxylulose phosphate reductoisomerase in mint resulted in an almost 50% increase in essential oil (monoterpenoid) production (Mahmoud and Croteau, 2001, 2002). Furthermore, the recent successful production of dhurrin in *Arabidopsis* by the coordinated expression of two cytochrome P450 genes with a glucosyltransferase suggests that such multiple rate-limiting steps can be overcome by the pyramiding of metabolic engineering (Kristensen et al., 2005).

2.3. Qualitative Control of Metabolites

For the successful industrial application of secondary metabolite production in plant cells, both the quantity and quality of metabolites must be improved, i.e., control of the metabolite profile is important. A simple composition of metabolites is desirable for industrial production. The simplest approach to produce desired compounds is to re-construct entire biosynthetic processes in vitro (Rathbone and Bruce, 2002, Ro et al., 2006). So far, a considerable number of genes involved in alkaloid biosynthesis have been cloned and expressed in *E. coli* or insect cells (De Luca and Laflamme, 2001, Facchini, 2001, Rathbone and Bruce, 2002, Verpoorte and Memelink, 2002, Hashimoto and Yamada, 2003). Microbial enzymes have been shown to be useful for the biotransformation of chemicals. Plant enzymes are much more likely to be biocatalysts, since they have rather high substrate specificity, e.g., a P450, CYP719 (Ikezawa et al., 2003). The combination of microbial and plant enzymes as well as the use of artificially synthesized chemicals also increase the production of more divergent chemicals. Thus, while microbial cells have less capacity to store metabolites, heterologous systems could be useful for the production of plant-derived metabolites.

Whereas the re-construction of biosynthetic pathways in vitro would be preferable for biotransformation, the limited availability of genes as well as the limited supply of substrate in biotransformation support the notion that the modification of plant metabolism is a more promising approach. The creation of a new branch to produce novel compounds by the introduction of a novel gene has been successfully demonstrated (Sato et al., 2001). A crucial point in creating a new branch in a pre-existing pathway is the substrate-affinity and/or reaction specificity of the introduced enzyme(s).

A more promising approach is the down-regulation of gene expression for an undesired pathway. Down-regulation using antisense RNA and co-suppression have been shown to be effective, since mint plants transformed with the antisense version of menthofuran synthase cDNA produced less than half of the undesired monoterpene oil component than did wild type (Mahmoud and Croteau, 2001). Suppression of a P450 hydroxylase gene in plant trichome glands has also been associated with the accumulation of cembratriene-ol and enhanced resistance against aphids (Wang et al., 2001). The recent development of the RNA-interference (RNAi) method using double-stranded (ds)RNA-induced post-transcriptional and/or transcriptional silencing is a more efficient method for modifying a pathway to shut down (Wesley et al., 2001, Wang and Waterhouse, 2002, Waterhouse and Helliwell, 2003). However, we should be careful in selecting the target sequence, since only a 22 nucleotide-long perfect match was sufficient to silence homologous genes with an identical sequence (Ishihara et al., 2005).

Successful applications of these strategies are described in more detail in the later section on isoquinoline alkaloids. Note that these pathway modifications may also induce the activation of new/pre-existing pathways since more divergent chemicals have been found to accumulate after pathway modification.

2.4. Transport, Accumulation and Storage

Since secondary metabolites are produced in specialized plant cells and accumulate in organs apart from the site of biosynthesis, transport and accumulation/storage or secretion have physiological importance. However, little is known about these processes. Only some attempts to modify transport activity by the ectopic expression of yeast PDR5 (pleiotrophic drug resistance-type ATP-binding cassette transporter) genes in transgenic tobacco (*Nicotiana tabacum* L.), and stimulation of the secretion of secondary metabolites in plant cell cultures have been reported (Goossens et al., 2003a). Recent progress is discussed in the section on nicotine, tropane and isoquinoline alkaloids. The importance of transport is discussed in the section on the over-expression in transgenic *Atropa* of hyoscyamine 6 β-hydroxylase, which converts hyoscyamine to scopolamine during the transport process.

3. CASE STUDIES

3.1. Nicotine and Tropane Alkaloids

3.1.1. Biosynthetic pathway and structural genes

Putrescine, a symmetrical diamine, is formed from basic amino acids, ornithine and/or arginine, and is metabolized to higher polyamines in all organisms and to particular alkaloids in a few plant species (Hashimoto and Yamada, 1994). Putrescine is metabolized to nicotine in tobacco and other *Nicotiana* and related species, and to pharmacologically active tropane alkaloids in some medicinal solanaceous plants (Fig. 2). Interestingly, *Duboisia* species synthesize high levels of both nicotine and tropane alkaloids. Apparently, nicotine and tropane alkaloids share the same evolutionary origin during diversification of the Solanaceae. Other classes of alkaloids derived from putrescine include nortropane alkaloids, calystegines, and pyrrolizidine alkaloids, such as retronecine.

It is not yet clear which pathway, from ornithine or arginine, is used to synthesize these alkaloids via putrescine. L-Ornithine is converted to putrescine in one step

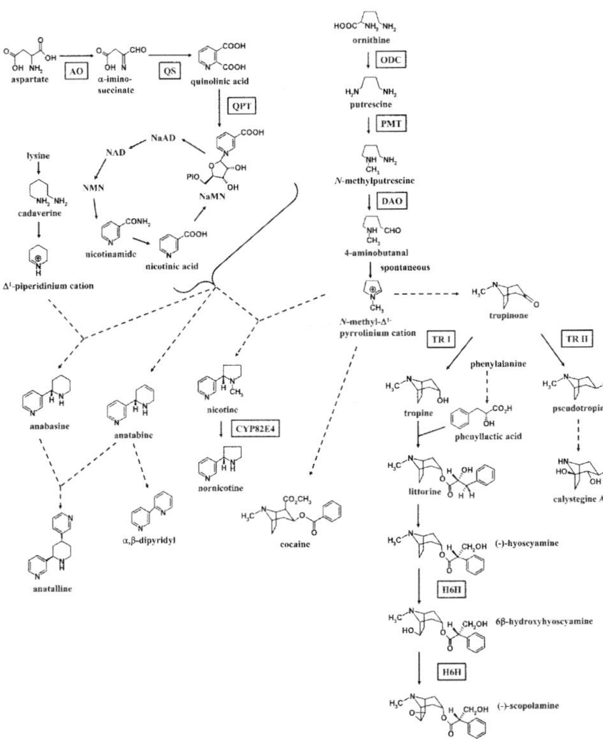

Figure 2. Biosynthetic pathways of nicotine and tropane alkaloids
Nicotine is synthesized by condensation of an intermediate in the NAD salvage pathway and the methylpyrrolinium cation derived from ornithine via putrescine. This cation is also used for the biosynthesis of tropane alkaloids, such as hyoscyamine and scopolamine. Metabolites derived from the pathways are also shown. Enzymes involved in the synthesis of these alkaloids are indicated: ODC; ornithine decarboxylase, PMT; putrescine *N*-methyltransferase, DAO; diamine oxidase, AO; aspartate oxidase, QS; quinolinate synthase, QPT; quinolinate phosphoribosyltransferase, CYP82E4; nicotine-demethylating cytochrome P450 monooxygenase, TR-I; tropine-forming tropinone reductase, TR-II; pseudotropine-forming tropinone reductase, and H6H; hyoscyamine 6β-hydroxylase

by ornithine decarboxylase (ODC; EC 4.1.1.17), whereas L-arginine is first decarboxylated to agmatine by arginine decarboxylase (ADC; EC 4.1.1.19), then to putrescine via *N*-carbamoylputrescine. Since arginine is converted to ornithine by arginase, feeding studies using labeled amino acids do not give definite conclusions. The addition of a specific suicide inhibitor of ADC (α-difluoromethylagrinine) or of ODC (α-difluoromethylornithine) in culture medium of tobacco callus showed that the ADC inhibitor more effectively decreased nicotine contents (Tiburicio and Galston, 1986). On the other hand, a tobacco ODC gene is coordinately regulated with other enzyme genes in nicotine biosynthesis, while a tobacco ADC gene is not (Imanishi et al., 1998, Hashimoto et al., unpublished results).

Biosynthetic pathways specific to nicotine and tropane alkaloids start with the *N*-methylation of putrescine, which is catalyzed by putrescine *N*-methyltransferase (PMT; EC 2.1.1.53). The amino acid sequence of PMT is highly homologous to that of spermidine synthase (SPDS), which transfers the amino-propyl moiety of decarboxylated *S*-adenosylmethionine (dSAM) to putrescine (Hibi et al., 1994). PMT catalyzes a transfer of the methyl moiety of *S*-adenosylmethionine (SAM) to putrescine. It has been proposed that PMT evolved from SPDS after restricted alterations of critical dSAM binding amino acid residues (Hashimoto et al., 1998a). Although tobacco PMT differs from SPDS by the presence of tandem repeats of 11 amino acid residues at the *N*-terminus, the repeat element is not required for enzymatic activity. Five PMT genes of tobacco possess variable repeat numbers, whereas the tandem repeats are absent in PMTs from *Solanaceae* plants that produce tropane alkaloids and calystegines (Hashimoto et al., 1998b, Stenzel et al., 2006). The tobacco tandem repeats have been used for molecular confirmation of the presumed origin of cultivated tobacco (Hashimoto et al., 1998b, Riecher and Timko, 1999).

Diamine oxidase (DAO; EC 1.4.3.6) catalyzes the deamination of *N*-methylputrescine to give 4-methylaminobutanal, which is spontaneously cyclized to *N*-methylpyrrolinum cation. The DAOs involved in nicotine and tropane alkaloid biosynthesis have higher affinity for *N*-methylputrescine than for putrescine and other symmetrical diamines (Hashimoto et al., 1990, Walton and McLauchalan, 1990, Haslam and Young, 1992), and thus are often referred to as *N*-methylputrescine oxidase. In contrast, pea and pig DAOs bind *N*-methylputrescine with low affinity. DAOs in alkaloid-producing plants may have evolved from DAOs through the optimization of substrate specificity.

In tobacco, *N*-methylpyrrolinum cation condenses with nicotinic acid, or its derivative, to give nicotine. Although labeled nicotinic acid was incorporated into the pyridine ring of nicotine when administered to tobacco (Leete and Liu, 1973, Leete, 1979, 1983), it is not known whether nicotinic acid itself or a metabolite derived from it is the direct precursor of nicotine. Nicotinic acid is a metabolite in the salvage pathway of NAD. In *Arabidopsis thaliana*, and probably in other dicotyledonous species as well, NAD is synthesized from L-aspartic acid (Kato et al., 2006). The early NAD biosynthetic pathway consists of aspartate oxidase (EC 1.4.3.16), quinolinate synthase, and quinolinic acid phosphoribosyl transferase

(EC 2.4.2.19), all of which are localized in the plastid and are coordinately regulated with nicotine biosynthesis (Sinclair et al., 2000, Kato et al., 2006).

A tobacco cytochrome P450 monooxygenase converted nicotine to nornicotine when expressed heterologously in yeast, and suppressed nornicotine formation when its gene expression was suppressed by RNAi-induced silencing in "converter" tobacco plants (Siminszky et al., 2005). Thus, this CYP82E4 monooxygenase qualifies as a nicotine demethylase.

N-Methylpyrrolinum cation is also metabolized to tropinone in tropane alkaloid-producing plants, but the enzymes and genes involved have not been identified. Tropinone is then reduced either to tropine or to pseudo-tropine, by two closely related tropinone reductases with distinct stereospecificity (Dräger, 2005). The reductases are NADPH-dependent short-chain dehydrogenases with a conserved NADPH-binding domain and a somewhat divergent tropinone-binding domain (Nakajima et al., 1998). Several amino acid residues in the substrate-binding domain have been identified to position tropinone for stereospecific hydride transfer from NADPH (Nakajima et al., 1999, Yamashita et al., 2003).

Tropine is coupled to phenyllactate to give littorine (Robins et al., 1994), which is then rearranged to hyoscyamine by a littorine mutase (Ollagnier et al., 1998). The enzymes responsible for these reactions have not been purified and the corresponding genes are yet to be cloned. Two-step epoxidation reactions from hyoscyamine to scopolamine via 6β-hydroxyhyoscyamine are catalyzed by a bifunctional enzyme, hyoscyamine 6β-hydroxylase (H6H; EC 1.14.11.11) (Hashimoto et al., 1993). H6H belongs to the 2-oxoglutarate-dependent dioxygenase family, and is endowed with strong hydroxylase activity toward hyoscyamine and comparatively weak epoxidase activity toward 6β-hydroxyhyoscyamine. It has been demonstrated that H6H alone is sufficient for the efficient conversion of hyoscyamine to its final product scopolamine in a transformed *E. coli* strain, transgenic tobacco plants, and transgenic *Atropa belladonna* plants that ectopically expressed H6H (Yun et al., 1992, Hashimoto et al., 1993, Yun et al., 1993). Cloned structural genes that are specifically involved in the biosynthesis of nicotine and tropane alkaloids have been summarized previously (Hashimoto and Yamada, 2003).

3.1.2. *Site of alkaloid formation and transport*

Both nicotine and tropane alkaloids are synthesized in the root, but the tissue types involved in their synthesis are different, as has been well illustrated based on the expression patterns of PMT, an enzyme used for both types of alkaloids. Immunohistochemistry and promoter::GUS fusion reporters showed that tobacco PMT genes, as well as an A622 oxidoreductase gene which has been implicated in nicotine biosynthesis, are expressed strongly in epidermis and cortex cells of the tobacco root tip, and moderately in the outermost layer of the cortex, and in parenchyma cells surrounding the xylem of the differentiated region of the root (Shoji et al., 2000, Shoji et al., 2002). On the other hand, the PMT genes of *Atropa belladonna*, a tropane alkaloid-producing plant, are specifically expressed in the pericycle at the differentiating region of the root (Suzuki et al., 1999a). H6H, which catalyzes the late steps of scopolamine biosynthesis, is also localized in the root pericycle, mainly in the pericycle cells facing the xylem, of scopolamine-producing plants (Hashimoto et al., 1991, Kanegae et al., 1994, Suzuki et al., 1999b, Nakajima and Hashimoto 1999), whereas tropinone reductase-I, which catalyzes an intervening step between PMT and H6H, is localized in the endodermis and cortex, but not in the pericycle, of the roots of *Hyoscyamus niger* (Nakajima and Hashimoto, 1999). The difference in the localization of some enzymes in different cell types indicates that pathway intermediates for scopolamine biosynthesis are transported between different root cell layers.

The localization of H6H in pericycle cells next to the xylem is a reasonable site for the efficient transport of scopolamine into the xylem. The difference in the localization patterns of PMT and A622 at the tip and the differentiated region of the tobacco roots can also be interpreted in terms of the transport of nicotine into the xylem. The cortex cells of the root tip have not differentiated in the Casparian band and the apoplastic flow of metabolites from the cortex to the stele is presumably not restricted. In the root differentiation zone, the suberin layer in the Casparian band prevents free apoplastic flow. In this region, nicotine synthesis in the parenchyma cells surrounding the xylem may facilitate nicotine loading into the xylem. Once in the xylem, nicotine and scopolamine will be transported to the aerial parts as constituents of the xylem sap.

Nicotine and tropane alkaloids which are unloaded to the leaf and other aerial tissues accumulate in the vacuole (Wink and Roberts, 1998). It is not known whether specific transporters are required to load or unload these alkaloids into or from the xylem and take them up to the vacuole from the cytoplasm.

3.1.3. Metabolic engineering

With the available cloned structural genes in alkaloid biosynthesis at hand, the simplest strategies for metabolic engineering are to over-express them with strong, constitutive promoters, such as the CaMV35 promoter, and to suppress them by RNAi or antisense techniques. The results from over-expression experiments are variable; when the target enzyme is not rate-limiting, there is no or only a modest increase in the final pathway products, whereas enhancement of the critical step leads to the efficient conversion of pathway intermediates to the end products.

The most successful example of the metabolic engineering of plant alkaloids is the generation of scopolamine-type *Atropa* plants (Yun et al., 1992). *Atropa belladonna* is a hyoscyamine-rich medicinal plant. Although this species has its own H6H genes in the genome, endogenous H6H genes are rather poorly expressed, resulting in low scopolamine accumulation. Constitutive expression of the *Hyoscyamus niger* H6H gene under the CaMV35S promoter significantly enhanced the conversion efficiency in transgenic *Atropa* plants, which contained scopolamine almost exclusively in the leaf. However, the root tissues of the transgenic plants (Yun et al., 1992) and *Atropa* root cultures over-expressing H6H (Hashimoto et al., 1993) showed insufficient conversion of hyoscyamine to scopolamine, and accumulated the intermediate,

6β-hydroxyhyoscyamine. The fact that the conversion efficiency dramatically improved as alkaloids were transported from the root to the leaf suggests that ectopically expressed H6H in the aerial parts is capable of conversion during the transport process.

The structural genes introduced into transgenic plants, root cultures, or cultured cells, encode ODC, lysine decarboxylase, PMT, tropinone reductase, H6H, and cytochrome P450 monooxygenases from plant, mammalian, or bacterial sources (Tables 1 and 2). An expressed enzyme usually catalyzes the expected reaction in the plant host cells, but the impact of reinforcement of a particular step in the alkaloid pathway depends on the subsequent metabolic flow to the final product and on the availability of the substrate. The simultaneous expression of multiple enzymes in the pathway may be necessary, as in metabolic engineering in microbes (Ro et al., 2006).

The inhibition of gene expression of a pathway enzyme often causes an effective reduction of metabolites downstream of the pathway (Table 1). For example, a co-suppression line, in which the PMT expression level was decreased to 16% of the wild-type level, accumulated nicotine at a level only 2% of that in the wild type (Sato et al., 2001). The low-nicotine line accumulated high amounts of putrescine and spermidine, indicating that the efficient inhibition of PMT activity shifted the nitrogen flow from nicotine synthesis to polyamine formation. The co-suppression line also showed several distinct morphological phenotypes: neighboring leaves were fused at their bases, forming a continuous spiral sheet along the stem, inflorescent stems were often branched, and self-pollinated flower produced only a small seed set (less than 10% of wild type). These abnormalities may be caused by the increased accumulation of polyamines, which possess hormonal functions for plant development (Galston and Kaur-Sawhney, 1995).

Lowering gene expression to moderate levels may not significantly affect the accumulation of final products of the pathway. The expression level of tobacco A622 oxidoreductase, for example, needs to be reduced to less than 10% of the wild-type level to see consistent reductions in the subsequent metabolites (Hashimoto et al., unpublished results). Moreover, when the expression of A622 gene was highly reduced, growth of tobacco roots was considerable inhibited, possibly because of the accumulation of pathway intermediates. In such cases, conditional gene-suppression technologies in which RNAi constructs are expressed upon inducer treatment (Moore et al., 2006) work well to shut down the target pathway.

3.2. Case Studies for Isoquinoline Alkaloid Biosynthesis

3.2.1. Enzymes/genes and the site of biosynthesis

Isoquinoline alkaloids are a large and diverse group of alkaloids with ~2500 defined structures. They include the analgesic morphine from *Papaver somniferum*; the antigout colchicine from *Colchicum autumnale*;the emetic and antiamoebic

Table 1. Metabolic engineering of nicotine alkaloids

plant species	transgenic material	transgene*1 (source)	metabolites		reference
			increased	decreased	
1) overexpression					
N. rustica	hairy root	*35S::ODC(Saccharomyces cereviseae)*	methylputrescine nicotine		Hamill et al., 1990
N. sylvestris	plant	*35S::PMT(N. tabacum)*	nicotine		Sato et al., 2001
N. tabacum	plant, callus	*35S::ODC, 35S::3′-truncated ODC*(mouse)	putrescine		DeScenzo and Minocha, 1993
	hairy root	*2x35S::bacterial LDC (Hafnia alvei)*	cadaverine anabsine	putrescine	Fecker et al., 1993
	plant	*35S::TR I, 35S::H6H (H.niger)*	nicotine nornicotine anabascine anatabine		Rocha et al., 2002
	plant	*tetracycline inducible promoter::ADC (Avena sativa)*	putrescine		Panicot et al., 2002
	plant	*35S::CYP2A6 (human cytochrome P450 2A6)*	cotinine		Dueckershoff et al., 2005
	plant	*35S::CYP82E4v1 (N.tabacum)*	nornicotine	nicotine	Siminszky et al., 2005

2) downregulation					
N. sylvesitris	plant	*35S::PMT(N. tabacum)cosuppression*		nicotine putrescine	Sato et al., 2001
	plant	*35S::antisense PMT (N. attenuata)*		nicotine	Voelckel et al., 2001
N. attenuata	plant	*TRV-sense PMT*		nicotine	Saedler and Baldwin, 2004
		TRV-antisense PMT		nicotine	
		TRV-PMT/RNAi		nicotine	
	plant	*35S::PMT/RNAi(N.attenuata)*	anatabine	nicotine	Steppuhn et al., 2004
N. tabacum	hairy root	*35S::antisense PMT*	anatabine	nicotine	Chintapakorn and Hamill, 2003
	plant	*35S::CYP82E2 family/RNAi (N.tabacum)*	nicotine	nornicotine	Siminszky et al., 2005
		35S::CYP82E2 cosuppression (N.tabacum)	nicotine	nornicotine	

*1 Overexpression was achieved by using cauliflower mosaic virus 35S promoter, 35S dual enhancers, or tetracycline-inducible promoter. Antisense expression or RNAi methodology was used to downregulate target genes.

Table 2. Metabolic engineering of tropane alkaloids

plant spieces	transgenic material	transgene*1 (source)	metabolites		reference
			increased	decreased	
Atropa belladonna	plant	*35S::H6H*(*H. niger*)	scopolamine	hyoscyamine	Yun et al., 1992
	hairy root	*35S::H6H* (*H. niger*)	scopolamine	hyoscyamine	Hashimoto et al., 1993
			6β-hydroxyhyoscyamine		
	plant	*35S::PMT* (*N.tabacum*)	methylputrescine		Sato et al., 2003
	hairy root		methylputrescine		
	plant	*35S::PMT* (*N. tabacum*)	methylputrescine		Rothe et al., 2003
	root culture		methylputrescine		
	hairy root	*35S::TR I* (*D. stramonium*)	tropine hyoscyamine scopolamine	 calystegine	Richer et al., 2005
		35S::TR II (*D. stramonium*)	pseudotropine calystegine		
Datura metel	hairy root	*35S::PMT* (*N. tabacum*)	hyoscyamine scopolamine		Moyano et al., 2003
Duboisia hybrid*2	hairy root	*35S::PMT*(*N. tabacum*)	methylputrescine		Moyano et al., 2002
	hairy root	*35S::H6H* (*H. niger*)	scopolamine	hyoscyamine	Palazon et al., 2003

Hyoscyamus muticus	hairy root	*35S::H6H(H. niger)*	scopolamine		Jouhikainen et al., 1999
	hairy root	*35S::PMT* (*N. tabacum*)	hyoscyamine		Moyano et al., 2003
	hairy root	*35S::H6H* (*H. niger*)	hyoscyamine		
			6β-hydroxyhyoscyamine		Hakkinen et al., 2005
			scopolamine		
H. niger	hairy root	*35S::H6H* (*H. niger*)	scopolamine		Zhang et al., 2004
		35S::PMT (*N. tabacum*), *35S::H6H* (*H. niger*)	scopolamine	hyoscyamine	
Scopolia parviflora	hairy root	*35S::H6H* (*H. niger*)	hyoscyamine scopolamine		Kang et al., 2005
	hairy root	*35S::PMT* (*N. tabacum*)	hyoscyamine scopolamine		Lee et al., 2005

*1 All transgenes were expressed under the cauliflower mosaic virus 35S promoter.

*2 *Duboisia* hybrid is a hybrid between *Duboisia myoporoides* and *D. leichhardtii*.

emetine from *Cephaelis ipecacuanha*; the skeletal muscle relaxant tubocurarine from *Chondodendron* spp.; and the antimicrobial compounds berberine and sanguinarine from divergent plant species including *Berberis* spp., *Sanguinaria* spp., and *Coptis* spp..

Isoquinoline alkaloid biosynthesis begins with the conversion of tyrosine to both dopamine and 4-hydroxyphenylacetaldehyde by decarboxylation, *ortho*-hydroxylation, and deamination (Fig. 3) (Facchini, 2001, Sato and Yamada, 2007). Among these early steps, only tyrosine/dopa decarboxylase (an aromatic L-amino acid decarboxylase; TYDC), which converts tyrosine and dopa to their corresponding amines, has been purified and characterized. This small family of genes

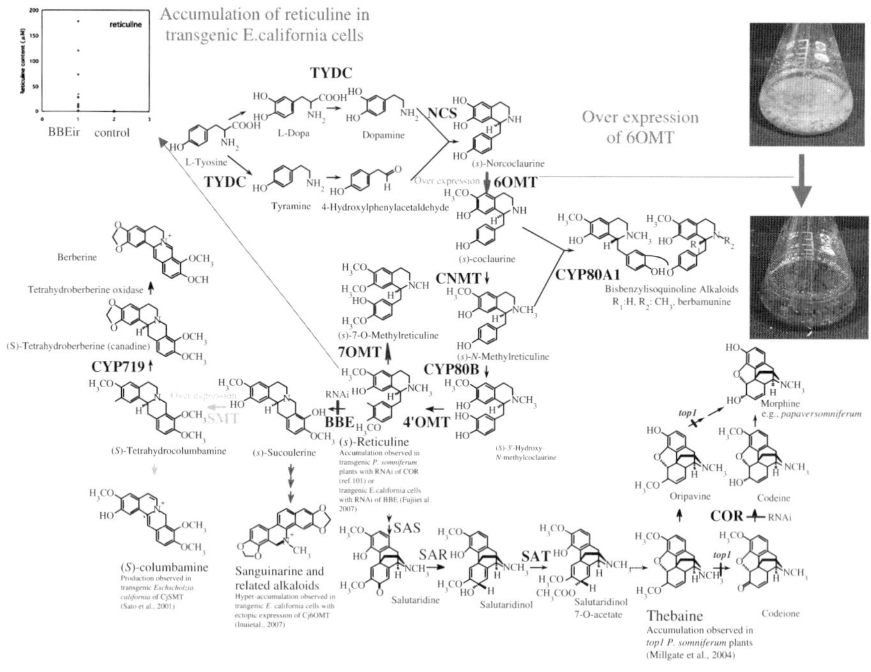

Figure 3. Biosynthetic pathways to various isoquinoline alkaloids with some examples of metabolic engineering
Unbroken arrows indicate single enzymatic conversions and broken arrows indicate multiple enzymatic steps. Enzymes for which the corresponding genes have been cloned are indicated in bold. tyrosine/dopa decarboxlyase, TYDC; norcoclaurine synthase, NCS; norcoclaurine 6-*O*-methyltransferase, 6OMT; coclaurine *N*-methyltransferase, CNMT; bebamunine synthase, CYP80A1; *N*-methylcoclaurine 3′-hydroxylase, CYP80B1; 3′-hydroxy *N*- methylcoclaurine 4′-*O*-methyltransferase 4′OMT; berberine bridge enzyme, BBE; reticuline 7-O-methyltransferase, 7OMT; canadine synthase (methylene dioxybridge-forming enzyme), CYP719; scoulerine 9-*O*-methyltransferase, SMT; salutaridine synthase, SAS; salutaridine reductase, SAR; acetylcoenzyme A: salutaridinol-7-*O*-acetyltransferase, SAT; codeinone reductase, COR; berbamunine synthase, CYP80A1;

(~15 genes) was isolated from opium poppy (*Papaver somniferum*) and each subfamily has been shown to have distinct developmental and inducible expression patterns (Facchini and De Luca, 1994, 1995, Park et al., 1999).

Dopamine and 4-hydroxyphenylacetaldehyde are condensed by norcoclaurine synthase (NCS) to yield (*S*)-norcoclaurine, which is the central precursor to all isoquinoline alkaloids. Recently, NCS has been purified and characterized (Samanami and Facchini, 2002) from cultured *Thalictrum flavum* spp., and a TfNCS cDNA belonging to the PR10 family was isolated from *Thalictrum flavum* (Samanani et al., 2004), whereas a novel dioxygenase-like protein (CjNCS) from cultured *Coptis japonica* cells was also shown to catalyze this NCS reaction (Minami et al., 2007). The presence of TYDC- (Facchini, 2001), TfNCS (Samanani et al., 2004) and CjNCS-homologues (Minami et al., 2007) in *Arabidopsis* or rice suggests that these genes either have other basic biological roles or that the isoquinoline biosynthesis pathway is relatively universal in the plant kingdom, although the sequence homology of TfNCS and CjNCS with Arabidopsis or rice homologues was relatively low (less than 20% on an amino acid basis) (Liscombe et al., 2005, Minami et al., 2007) and no isoquinoline alkaloid has been found in *Arabidopsis* or rice.

(*S*)-Norcoclaurine is sequentially converted to coclaurine by *S*-adenosyl methionine (SAM)-dependent norcoclaurine 6-*O*-methyltransferase (6OMT) (Morishige et al., 2000), to *N*-methylcoclaurine by coclaurine *N*-methyltransferase (Choi et al., 2002), to 3′-hydroxy-*N*-methyl coclaurine by P450 hydroxylase (Pauli and Kutchan, 1998), and then to (*S*)-reticuline by 3′-hydroxy *N*-methylcoclaurine 4′-*O*-methyltransferase (4′OMT, see Fig. 3) (Morishige et al., 2000). All of the cDNAs for these reactions have been isolated and functional recombinant proteins have been produced. Detailed biochemical studies using recombinant enzymes have shown their strict reaction specificities, and these enzymes regulate biosynthesis sequentially and in a coordinated manner. For example, CNMT prefers coclaurine to 6-*O*-methylnorlaudanosoline and 4‘OMT prefers an *N*-methylated substrate, which suggests that the pathway in Fig. 3 has a sequence of *N*-methylation, hydroxylation and 4’-*O*-methylation. On the other hand, *Thalictrum* cells may show some variation, since *Thalictrum O*-methyltransferases can form heterodimers and exhibit broad substrate specificity (Frick and Kutchan, 1999). Current data also indicate that all of these enzymes, except the membrane-bound P450 CYP80B1, are located in the cytosol.

Whereas dimeric bisbenzylisoquinoline alkaloids, such as berbamunine and tubocurarine, are produced from the intermediates of the (*S*)-reticuline pathway, by the action of a phenol coupling P450-dependent oxidase (berbamunine synthase, CYP80A1) (Krause and Kutchan, 1995), reticuline is the central intermediate in branch pathways that lead to benzophenanthridine alkaloids (e.g. sanguinarine and marcarpine), protoberberine alkaloids (e.g. berberine and palmatine), and morphinan alkaloids (e.g. morphine and codeine) (Fig. 3). Many of the enzymes involved in these branch pathways have been purified and the corresponding cDNAs have been cloned.

The first committed step in the biosynthesis of benzophenanthridine, a protoberberine alkaloid, involves conversion of the *N*-methyl group of (*S*)-reticuline into the methylene bridge moiety of (*S*)-scoulerine by the berberine bridge enzyme (BBE) (Dittrich and Kutchan, 1991). This unique enzyme is soluble but localized in vesicles (Bock et al., 2002). Immunocytological staining of *Papaver somniferum* tissue with antibodies against BBE led to a characteristic labeling of electron-dense aggregates in idioblasts that are not connected to the laticifer system, which demonstrates that benzophenanthridine and morphine biosyntheses show strict cytological separation within this plant (Bock et al., 2002). In benzophenanthridine alkaloid biosynthesis, (*S*)-scoulerine can be converted to (*S*)-stylopine by two P450-dependent oxidases, (*S*)-chelanthifoline synthase and (*S*)-stylopine synthase, which result in the formation of two methylenedioxy groups (see Facchini, 2001, Sato and Yamada, 2007, Ikezawa et al., 2007).

On the other hand, in protoberberine biosynthesis, (*S*)-scoulerine is converted to (*S*)-tetrahydrocolumbamine by the SAM-dependent scoulerine 9-*O*-methyltransferase (SMT) (Takeshita et al., 1995), and then to tetrahydroberberine (canadine) by a P450-dependent canadine synthase (CDS or CYP719) (Ikezawa et al., 2003). The isolation and characterization of these enzymes and the corresponding cDNAs has confirmed that berberine biosynthesis proceeds via canadine and not via columbamine. Again, the enzyme substrate specificity shows a clear preference for this pathway. Note that the CYP719 family, like CYP80, is not found in *Arabidopsis* or rice and is very unique for benzylisoquinoline alkaloid biosynthesis (Nelson et al., 2004).

In morphinan alkaloid biosynthesis, (*S*)-reticuline is converted to its (*R*)-enantiomer via the stereospecific reduction of 1,2-dehydroreticuline with NADPH-dependent cytosolic 1,2-dehydroreticuline reductase. Subsequent intramolecular carbon-carbon phenol coupling of (*R*)-reticuline by a P450-dependent salutaridine synthase results in the formation of salutaridine. Salutaridine:NADPH 7-oxidoreductase then reduces salutaridine to (7*S*)-salutaridinol. Transformation of salutaridinol into thebaine involves the closure of an oxide bridge between C-4 and C-5 by acetylcoenzyme A:salutaridinol-7-O-acetyltransferase (SAT) (Grothe et al., 2001). Furthermore, thebaine can be converted to codeinone, and then reduced to codeine by cytosolic NADPH-dependent codeinone reductase (COR) (Unterlinner et al., 1999). Finally, codeine is demethylated to give morphine. Interestingly, COR genes have been found in some *Papaver* species that do not produce morphine (Unterlinner et al., 1999), whereas SAT transcript was detected in *Papaver* species that accumulate alkaloids with a morphinan nucleus, consistent with the expected distribution (Unterlinner et al., 1999, Grothe et al., 2001). Note that a recent isolation of *top1* poppy mutant illustrated that thebaine can be demethylated in two steps through either codeinone or oripavine to morphine (Millgate et al., 2004).

Northern blot analysis using the eight available genes in morphinan alkaloid biosynthesis showed that while all of the transcripts are detected in every organ, the highest levels are seen in stems and flower buds and the lowest levels are seen in leaves (Unterlinner et al., 1999, Gothe et al., 2001). The accumulation of each transcript, with the exception of *COR*, was markedly induced in response to treatment with an elicitor or wounding of cultured cells. All known enzymes in the morphine pathway have been detected in cultured cells (Unterlinner et al., 1999, Huang and Kutchan, 2000, Grothe et al., 2001, Facchini and Park, 2003). On the other hand, differences in the cell-type-specific localization of biosynthetic enzymes have been reported in capsules and stems of intact opium poppy plants. In situ localization of alkaloid biosynthetic gene transcripts indicated that seven biosynthetic enzymes (6OMT, CNMT, CYP80B, 4′OMT and BBE in reticuline biosynthesis, and SAT and COR in a morphine pathway) are localized in sieve elements in opium poppy and corresponding gene transcripts were localized in the supporting companion cells (Bird et al., 2003, Facchini and St-Pierre, 2005). However, another immunocytochemical analysis clearly showed that 4′OMT and SAT were localized in phloem parenchyma cells, and COR, the penultimate step in morphine biosynthesis, is localized in laticifers, which is the site of morphinan alkaloid accumulation (Weid et al., 2004). Although this difference in the cell-type-specific localization of enzymes remains to be clarified, it is noteworthy that both studies showed the different localization of biosynthetic gene transcripts and corresponding enzymes.

Cell-type-specific expression was also recently reported in protoberberine alkaloid biosynthesis in *Thalictrum flavum* spp (Samanani et al., 2005). Whereas gene transcripts for biosynthetic enzyme were most abundant in rhizomes, they were also detected at lower levels in roots and other organs. Further in situ RNA hybridization analysis revealed that all transcripts were mainly localized in the immature endodermis and pericycle in roots, while they were localized in the protoderm of leaf primordia in rhizomes. These data and an analysis of alkaloid accumulation clearly indicated that distinct and different cell types are involved in the biosynthesis and accumulation of benzylisoquinoline alkaloids in *T. flavum* and *P. somniferum*.

As mentioned above, plant alkaloids are often translocated from the source organ to a sink organ (Hashimoto and Yamada, 1994). Berberine in *C. japonica* is also translocated from root tissues to rhizome, which involves transport across several membranes. Recent identification of the function of CjMDR (a multidrug resistance gene) for uptake in rhizome could be the first step in understanding the mechanism of accumulation and detoxification of these biologically active metabolites (Shitan et al., 2005). The further characterization of the vacuolar transport of berberine by H^+/berberine antiporter clearly indicates that multiple transporters are also involved in subcellular and intercellular transport (Ohtani et al., 2005). Differences in the inhibition of berberine transport by its analogs suggest that transporter activity might be critical for the biosynthesis. Whereas the transporter activity of *C. japonica* vacuoles suggest the high affinity for reticuline, transgenic California poppy cells with RNAi of BBE showed the accumulation of large amounts of reticuline in the culture medium, which suggest a low accumulation activity for this compound (Fujii et al., 2007). When we consider the intercellular network in biosynthesis, especially of morphinan alkaloids, these transporter activities would be important determinants of such a network.

3.2.2. Metabolic engineering

Since almost all biosynthetic genes from norcoclaurine to berberine have been isolated and their activities have been demonstrated in microbial systems (see Fig. 3, Dittrich and Kutchan, 1991, Takeshita et al., 1995, Pauli and Kutchan, 1998, Morishige et al., 2000, Facchini, 2001, Choi et al., 2002, Ikezawa et al., 2003, Samanani et al., 2004, Minami et al., 2007), the complete reconstruction of the biosynthetic pathway of berberine or of biosynthetic intermediates should be feasible in the near future. The bioconversion of chemicals would also be a promising goal. While plant enzymes have high substrate specificity and relatively low activity, such high specificity, especially stereo-specificity, should be useful for the production of more stereo-specific and site-specific conversion. A specific reaction in biosynthesis, such as phenol-coupling with CYP80A1 to produce bisbenzylisoquinoline alkaloids from norcoclaurine (Kraus and Kutchan, 1995) and coclaurine, or methylene-dioxy bridge formation with CYP719, should be a unique tool for bioconversion (Ikezawa et al., 2003, 2007). On the other hand, a plant enzyme also may have rather broad specificity, as seen for CNMT of *C. japonica*, which can use even simple isoquinoline derivatives as a substrate for *N*-methylation (Morishige et al., 2004).

cDNA resources provide more opportunities to modify the metabolic pathways in plant cells, although there are few published results. The first molecular engineering in an isoquinoline alkaloid was performed in California poppy (*Eschscholzia californica)* with the overexpression of *C. japonica* scoulerine *O*-methyltransferase (SMT) cDNA. A complete modification of the alkaloid profile with the introduction of *C j*SMT cDNA into California poppy cells was observed; i.e., the alkaloid profile changed from sanguinarine (benzophenanthridine-type) to columbamine (berberine-type). Whereas the addition of a new branch seems simple and easy to achieve, we believe that this successful conversion of alkaloid profiles depended on the enzyme properties; i.e., overexpressed *C.j* SMT should be superior to *E. californica* chelanthifoline synthase activity (Sato et al., 2001) (Fig. 3). A pitfall to this approach, while it might be promising for future metabolic engineering, is that a newly introduced pathway can provide the substrate for further enzymatic conversion and produce novel compounds which are not detected in wild-type cells. Whereas the identification of new compounds is still underway (Takemura et al. unpublished data), more than 10 new peaks, which were not found in non-transformed cells, have been detected (Sato et al., 2001).

The over-expression of pre-existing enzyme encoding genes has also been examined in California poppy cells with two *O*-methyltransferase cDNAs isolated from *Coptis japonica* cells. Our data suggested that ectopic expression of Cj6OMT markedly increased the production of benzophenanthridine alkaloid biosynthesis in California poppy cells, whereas that of 4‘OMT showed only a marginal increase in alkaloid production (Inui et al., 2007). Our results provide the experimental support for an earlier speculation regarding which early step in isoquinoline alkaloid biosynthesis is rate-limiting; high berberine-producing *C. japonica* cells showed higher enzyme activities in early steps than low-producing cells (Sato et al., 1994). Similarly, an increase in berberine biosynthesis with the addition of methyljasmonate was accompanied by an increase in early-step OMTs and CNMT (Frick and Kutchan, 1999).

The modification of alkaloid profiles is also desired in isoquinoline alkaloid biosynthesis. In the production of morphinan alkaloids, the *top1* mutant of poppy overproducing thebaine (Millgate et al., 2004) has great economic interests. Thus, the last step for morphine biosynthesis catalyzed by COR was knocked down with RNAi techniques (Allen et al., 2004). Unexpectedly, RNAi with full-length cDNA of COR not only down-regulated COR but also stopped other biosynthetic activities and induced the accumulation of reticuline. While some authors suggested that this phenomenon might be induced by the disruption of enzyme complex for morphine alkaloid biosynthesis through the loss of COR, the inhibition of biosynthesis with COR RNAi could occur by the off-target silencing effects through a 22bp-match with a sequence in chimeric hairpin RNA (see Ishihara et al., 2005).

Reticuline is an important intermediate in isoquinoline alkaloid biosynthesis, since many types of isoquinoline alkaloids, including morphine, codeine, papaverine, berberine and sanguinarine originate from this compound. While poppy plants with COR RNAi accumulated reticuline accidentally (Allen et al., 2004), direct gene silencing of BBE in a pathway with RNAi also enabled the accumulation of reticuline in transgenic California poppy cells (Fujii et al., 2007). In the case of BBE RNAi, no further interference with biosynthesis was observed. However, transgenic *E. californica* cells with antisense BBE RNA lost alkaloid productivity with no accumulation of the intermediate reticuline (Park et al., 2002). This result clearly shows that the RNAi technique is useful for metabolic engineering and that plant cells are capable of accumulating intermediates even if the metabolism has been altered. Note that even RNAi is not sufficient to induce the accumulation of intermediate in some cases; e.g., low-caffeine coffee that had been transformed with caffeine synthase RNAi vector did not accumulated any intermediate, such as theobromine (Ogita et al., 2003). In California poppy cells, produced reticuline was mainly found in the medium (Fujii et al., 2007). This secretion/leakage of metabolite would enable it to escape from further catabolism. The accumulation of 7-*O*-methylreticuline in California poppy cells with BBE RNAi as well as opium poppy with COR RNAi suggests that the accumulation of chemical(s) would induce the activation of a pre-existing pathway or the creation of a new pathway from an intermediate.

As the above examples suggest, metabolic engineering in isoquinoline alkaloid biosynthesis is at a stage where it can be practically useful for increasing the quality and quantity of metabolites.

4. FUTURE PERSPECTIVES

The most common problems for the commercial application of metabolite production in plant systems are low productivity and the high cost of purification due to the production of structurally related compounds in biosynthesis. Field harvests may also contribute to the risk of fluctuation in productivity due to climate change, pest infection and contamination with undesired wild plants. Molecular engineering is a powerful tool, and may be able to overcome these problems, especially when cell culture systems are used efficiently. However, cultured cells with high metabolite productivity have not been accepted as being equivalent to natural harvests, since it has often been reported that in vitro-cultured cells/ tissues have different metabolite profiles than intact tissues (Fu, 1998, Yamazaki et al., 2003). Similarly, metabolically engineered transgenic plants and cultured cells also require additional quality certificates compared to *natural harvests*. Over-expression of the rate-limiting step and/or the activation of general transcriptional regulation system increase the pool of metabolites for further metabolic modification. The creation of a new branch or trimming of a pathway may create a new pathway from the currently produced end chemicals, while the trimming of undesired branches would considerably reduce the costs of tedious downstream processing to purify the product. Functional evaluation of the metabolic profile in plant cells would be crucial for the future engineering of medicinal compounds.

5. SUMMARY

Many secondary metabolites that have been isolated from higher plants are used as important natural resources for pharmaceuticals. Whereas our knowledge of the basic mechanism of biosynthesis has increased over the past 30 years, practical applications are still limited due to relatively low productivity. Recent advances in molecular biology in plant sciences, with the development of comprehensive analysis of expressed genes in biosynthesis-specialized cells and integrated analyses of expression profiles, as well as the development of metabolic profiling analysis, have considerably stimulated the development of metabolic engineering in plants, even in the realm of secondary metabolism. The identification of many biosynthetic genes and characterization of the spatial and developmental regulation of their expression have clarified their physiological importance in the biosynthesis of secondary metabolites, and revealed bottlenecks in their production. The molecular engineering of secondary metabolism may lead to a new era for the production of medicinal compounds in transgenic plants as well as cell/tissue cultures. Both the reconstruction of biosynthetic processes in heterologous systems and the modification of metabolic processes in pre-existing metabolism should increase the productivity and purity (or diversity) of the chemicals produced, and the metabolic profiles in newly developed transgenic plants should be characterized. Metabolic engineering techniques surely offer high and broad potential for the industrial production of useful secondary metabolites in plants.

ACKNOWLEDGEMENTS

We thank Dr. Yasuyuki Yamada, Professor Emeritus of Nara Institute of Science and Technology, for his continuous support and encouragement. This research was supported in part by a Research for the Future Program Grant from the Japan Society for the Promotion of Science (JSPS-RFTF00L01605 to T.H. and JSPS-RFTF00L01606 to F.S.) and by a Grand-in-Aid for Scientific Research from the Ministry of Education, Sports, and Culture of Japan (to F.S.).

REFERENCES

Allen RS, Millgate AG, Chitty JA et al. (2004) RNAi-mediated replacement of morphine with the nonnarcotic alkaloid reticuline in opium poppy. *Nat Biotechnol* 22: 1559–1566.

Bird DA, Franceschi VR, Facchini PJ (2003) A tale of three cell-types: alkaloid biosynthesis is localized to sieve elements in opium poppy. *Plant Cell* 15: 2626–2635.

Bock A, Wanner G, Zenk MH (2002) Immunocytological localization of two enzymes involved in berberine biosynthesis. *Planta* 216: 57–63.

Borevitz JO, Xia YJ, Blount J et al. (2000) Activation tagging identifies a conserved MYB regulator of phenylpropanoid biosynthesis. *Plant Cell* 12: 2383–2393.

Briskin DP (2000) Medicinal plants and phytomedicines. Linking plant biochemistry and physiology to human health. *Plant Physiol 124:* 507–514.

Canel C, Lopes-Cardoso MI, Whitmer S et al. (1998) Effects of over-expression of strictosidine synthase and tryptophan decarboxylase on alkaloid production by cell cultures of *Catharanthus roseus*. *Planta* 205: 414–419.

Chintapakorn Y, Hamill JD (2003) Antisense-mediated downregulation of putrescine *N*-methyltransferase activity in transgenic *Nicotiana tabacum* L. can lead to elevated levels of anatabine at the expense of nicotine. *Plant Mol Biol* 53: 87–105.

Choi KB, Morishige T, Shitan N et al. (2002) Molecular cloning and characterization of coclaurine *N*-methyltransferase from cultured cells of *Coptis japonica*. *J Biol Chem* 277: 830–835.

Croteau R, Kutchan TM, Lewis NG (2000) Natural products (Secondary metabolites). In: Buchanan BB, Gruissem W, Jones RL (eds)Biochemistry & Molecular Biology of Plants, Am Soc Plant Physiol,Maryland.

Decker G, Wanner G, Zenk MH et al. (2000) Characterization of proteins in latex of the opium poppy (*Papaver somniferum*)using two-dimensional gel electrophoresis and microsequencing. *Electrophoresis* 21: 3500–3516.

De Luca V, Laflamme P (2001) The expanding universe of alkaloid biosynthesis. Curr. Opin. *Plant Biol* 4: 225–233.

DeScenzo RA, Minocha SC (1993) Modulation of cellular polyamines in tobacco by transfer and expression of mouse ornithine decarboxylase cDNA. *Plant Mol Biol* 22: 113–127.

Dittrich H, Kutchan TM (1991) Molecular cloning, expression, and induction of berberine bridge enzyme, an enzyme essential to the formation of benzophenanthridine alkaloids in the response of plants to pathogen attack. *Proc Natl Acad Sci USA* 88: 9969–9973.

Dräger B (2005) Tropinone reductases, enzymes at the branch point of tropane alkaloid metabolism. *Phytochemistry* 67: 327–337.

Dueckershoff K, Ungerb M, Frank A et al. (2005) Modified nicotine metabolism in transgenic tobacco plants expressing the human cytochrome P450 2A6 cDNA. *FEBS Lett* 579: 2480–2484.

Facchini P (2001) Alkaloid biosynthesis in plants: biochemistry,cell biology, molecular regulation, and metabolic engineering applications. *Annu Rev Plant Physiol Plant Mol Biol* 52:29–66.

Facchini PJ, De Luca V (1994) Differential and tissue-specific expression of a gene family for tyrosine/dopa decarboxylase in opium poppy. *J Biol Chem* 269: 26684–26690.

Facchini PJ, De Luca V (1995) Phloem-specific expression of tyrosine/dopa decarboxylase genes and the biosynthesis of isoquinoline alkaloids in opium poppy. *Plant Cell* 7: 1811–1821.

Facchini PJ, Park SU (2003) Developmental and inducible accumulation of gene transcripts involved in alkaloid biosynthesis in opium poppy. *Phytochemistry* 64: 177–186.

Facchini PJ, St-Pierre B (2005) Synthesis and trafficking of alkaloid biosynthetic enzymes. Curr. Opin. *Plant Biol* 8: 657–666.

Fecker LF, Rügenhagen C, Berlin J (1993) Increased production of cadaverine and anabasine in hairy root cultures of *Nicotiana tabacum* expressing a bacterial lysine decarboxylase gene. *Plant Mol Biol* 23: 11–21.

Frick S, Kutchan TM (1999) Molecular cloning and functional expression of *O*-methyltransferases common to isoquinoline alkaloid and phenylpropanoid biosynthesis. *Plant J* 17: 329–339.

Fu T-J (1998) Safety considerations for food ingredients produced by plant cell and tissue culture. *CHEMTECH* 28: 40-46 http://pubs.acs.org/hotartcl/chemtech/98/.

Fujii N, Inui T, Iwasa K et al. (2007) Knockdown of berberine bridge enzyme by RNAi accumulates (*S*)-reticuline and activates a silent pathway in cultured California poppy cells. *Transgenic Res* 16: 363–375.

Galston AW, Kaur-Sawhney R (1995) Polyamines as endogenous growth regulators. In: Davies PJ (ed) Plant Hormones. Kluwer Academic Pub, Dordrecht, Netherlands.

Goossens A, Haekkinen ST, Laakso I et al. (2003a) Secretion of secondary metabolites by ATP-binding cassette transporters in plant cell suspension cultures. *Plant Physiol* 131: 1161–1164.

Goossens A, Haekkinen ST, Laakso I et al. (2003b) A functional genomics approach toward the understanding of secondary metabolism in plant cells. *Proc Natl Acad Sci USA* 100: 8595–8600.

Grothe T, Lenz R, Kutchan TM (2001) Molecular characterization of the salutaridinol-7-O-acetyltransferase involved in morphine biosynthesis in opium poppy *Papaver somniferum*. *J Biol Chem* 276: 30717–30723.

Häkkinen ST, Moyano E, Cusidó RM et al. (2005) Enhanced secretion of tropane alkaloids in *Nicotiana tabacum* hairy roots expressing heterologous hyoscyamine-6β-hydroxylase. *J Exp Bot* 56: 2611–2618.

Hamill JD, Robins RJ, Parr AJ (1990) Over-expressing a yeast ornithine decarboxylase gene in transgenic roots of *Nicotiana rustica* can lead to enhanced nicotine accumulation. *Plant Mol Biol* 15: 27–38.

Hashimoto T, Mitani A, Yamada Y (1990) Diamine oxidase from cultured roots of *Hyoscyamus niger*. *Plant Physiol* 93:216–221.

Hashimoto T, Hayashi A, Amano Y et al. (1991) Hyoscyamine 6β-hydroxylase, an enzyme involved in tropane alkaloid biosynthesis, is localized at the pericycle of the root. *J Biol Chem* 266: 4648–4653.

Hashimoto T, Matsuda J, Yamada Y (1993) Two-step epoxidation of hyoscyamine to scopolamine is catalyzed by bifunctional hyoscyamine 6β-hydroxylase. *FEBS Lett* 329: 35–39.

Hashimoto T, Shoji T, Mihara H et al. (1998a) Intraspecific variability of the tandem repeats in Nicotiana putrescine N-methyltransferases. *Plant Mol Biol* 37: 25–37.

Hashimoto T, Tamaki K, Suzuki K et al. (1998b) Molecular cloning of plant spermidine synthase. *Plant Cell Physiol* 39:73–79.

Hashimoto T, Yamada Y (1994) Alkaloid biogenesis: Molecular aspects. *Annu Rev Plant Physiol Plant Mol Biol* 45:257–285.

Hashimoto T, Yamada Y (2003) New genes in alkaloid metabolism and transport. Curr. Opin. *Biotechnol* 14: 163–168.

Haslam SC, Young TW (1992) Purification of *N*-methylputrescine oxidase from *Nicotiana rustica*. *Phytochemistry* 31: 4075–4079.

Hibi N, Higashiguchi S, Hashimoto T et al. (1994) Gene expression in tobacco low-nicotine mutants. *Plant Cell* 6: 723–735.

Huang F-C, Kutchan TM (2000) Distribution of morphinan and benzo[*c*]phenanthridine alkaloid gene transcript accumulation in *Papaver somniferum*. *Phytochemistry* 53: 555–564.

Ikezawa N, Tanaka M, Nagayoshi M et al. (2003) Molecular cloning and characterization of CYP719, a methylenedioxy bridge-forming enzyme that belongs to a novel P450 family, from cultured *Coptis japonica* cells. *J Biol Chem* 278: 38557–38565.

Ikezawa N, Iwasa K, Sato F (2007) Molecular cloning and characterization of methylenedioxy bridge-forming enzymes involved in stylopine biosynthesis in *Eschscholzia californica*.FEBS J 274: 1019–1035.

Imanishi S, Hashizume K, Nakakita M (1998) Differential induction of methyl jasmonate of genes encoding ornithine decarboxylase and other enzymes involved in nicotine biosynthesis in tobacco cell cultures. *Plant Mol Biol* 38: 1101–1111.

Inui T, Tamura K, Fujii N, et al. (2007) Overexpression of *Coptis japonica* norcoclaurine 6-*O*-methyltransferase overcomes the rate-limiting step in benzylisoquinoline alkaloid biosynthesis in cultured *Eschscholzia californica*. *Plant Cell Physiol* 48: 252–262.

Ishihara S, Yamamoto Y, Ifuku K et al. (2005) Functional analysis of four members of the PsbP family in photosystem II in *Nicotiana tabacum* using differential RNA interference. *Plant Cell Physiol* 46:1885–1893.

Jouhikainen K, Lindgren L, Jokelainen T et al. (1999) Enhancement of scopolamine production in *Hyoscyamus muticus* L. hairy root cultures by genetic engineering. *Planta* 208: 545–551.

Kanegae T, Kajiya H, Amano, Y et al. (1994) Species-dependent expression of the hyoscyamine 6β-hydroxylase gene in the pericycle. *Plant Physiol* 105: 483–490.

Kang YM, Lee OS, Jung HY (2005) Overexpression of hyoscyamine 6β-hydroxylase(h6h) gene and enhanced production of tropane alkaloids in *Scopolia parviflora* hairy root lines. *J Micro Biotech* 15 91–98.

Kato A, Uenohara K, Akita M et al. (2006) Early steps in the biosynthesis of NAD in Arabidopsis start with aspartate and occur in the plastid. *Plant Physiol* 141: 851–857.

Kato K, Dubouzet E, Kokabu Y, et al. (2007) Identification of a WRKY protein as a transcriptional regulator of benzylisoquinoline alkaloid biosynthesis in *Coptis japonica*. *Plant Cell Physio* l48:8–18.

Kraus PFX, Kutchan TM (1995) Molecular-cloning and heterologous expression of a cDNA-encoding berbamunine synthase, a C-O-phenol-coupling cytochrome-P450 from the higher-plant *Berberis stolonifera*. *Proc Natl Acad Sci USA* 92 (6): 2071–2075.

Kristensen C, Morant M, Olsen CE et al. (2005) Metabolic engineering of dhurrin in transgenic Arabidopsis plants with marginal inadvertent effects on the metabolome and transcriptome. *Proc. Natl. Acad. Sci. USA* 102(5): 1779–1784.

Kutchan T.M (2005a) A role for intra-and intercellular translocation in natural product biosynthesis. Curr. Opin. *Plant Biol* 8: 292–300.

Kutchan T.M (2005b) Predictive metabolic engineering in plants:still full of surprises. *Trends Biotechnol* 23: 381–383.

Lange BM, Wildung MR, Stauber EJ et al. (2000) Probing essential oil biosynthesis and secretion by functional evaluation of expressed sequence tags from mint glandular trichomes. *Proc Natl Acad Sci USA* 97: 2934–2939.

Lee OS, Kang YM, Jung HY et al. (2005) Enhanced production of tropane alkaloids in *Scopolia parviflora* by introducing the *PMT* (putrescine *N*-methyltransferase) gene. In Vitro Cell Dev. *Biol-Plant* 41: 167–172.

Leete E (1979) The alkaloids: alkaloids derived from ornithine,lysine and nicotinic acid. In: Bell EA, Charlwood BV (eds)Encyclopedia of Plant Physiology, New Series. Vol 8,Springer-Verlag, Berlin.

Leete E (1983) Biosynthesis and metabolism of the tobacco alkaloids. In: Pelletier SW (ed) Alkaloids: Chemical and Biological Perspectives. Vol 1, John Wiley, New York.

Leete E, Liu Y-Y (1973) Metabolism of [2-^{3}H]- and [6-^{3}H]- nicotinic acid in intact *Nicotiana tabacum* plants. *Phytochemistry* 12: 593–596.

Liscombe DK, Macleod BP, Loukanina N et al. (2005) Evidence for the monophyletic evolution of benzylisoquinoline alkaloid biosynthesis in angiosperms. *Phytochemistry* 66: 1374–1393.

Lloyd AM, Walbot V, Davis RW (1992) *Arabidopsis* and *Nicotiana* anthocyanin production activated by maize regulator-r and regulator-C1. *Science* 258: 1773–1775.

Mahmoud SS, Croteau RB (2001) Metabolic engineering of essential oil yield and composition in mint by altering expression of deoxyxylulose phosphate reductoisomerase and menthofuran synthase.Proc Natl Acad Sci USA 98: 8915–8920.

Mahmoud SS, Croteau RB (2002) Strategies for transgenic manipulation of monoterpene biosynthesis in plants. *Trends Plant Sci* 7: 366–373.

Millgate AG, Pogson BJ, Wilson IW et al. (2004) Morphine-pathway block in *top1* poppies. *Nature* 431: 413–414.

Minami M, Dubouzet E, Iwasa K et al. (2007) Functional analysis of norcoclaurine synthase in *Coptis japonica*, *J Biol Chem* 282:6274–6282.

Misawa M (1994) Plant tissue culture: an alternative for production of useful metabolite. In:FAO AGRICULTURAL SERVICES BULLETIN No. 108, M-06, ISBN 92-5-103391-9,http://www.fao.org/docrep/t0831e/t0831e00.htm#con

Moore I, Samalova M, Kurup S (2006) Transactivated and chemically inducible gene expression in plants. *Plant J* 45: 651–683.

Morishige T, Tsujita T, Yamada Y et al. (2000) Molecular characterization of the *S*-adenosyl-L-methionine: 3′hydroxyl-*N*-methylcoclaurine 4 ′-*O*-methyltransferase involved in isoquinoline alkaloid biosynthesis in *Coptis japonica*. *J Biol Chem* 275:23398–23405.

Morishige T, Dubouzet E, Choi K.B et al. (2002) Molecular cloning of columbamine *O*-methyltransferase from cultured *Coptis japonica* cells. *Eur. J. Biochem* 269: 5659–5667.

Morishige T, Choi KB, Sato F (2004) In vivo bioconversion of tetrahydroisoquinoline by recombinant coclaurine N-methyltransferase. *Biosci Biotechnol Biochem* 68(4): 939–941.

Moyano E, Fornalé S, Palazón J et al. (2002) Alkaloid production in *Duboisia* hybrid hairy root cultures overexpressing the *pmt* gene. *Phytochemistry* 59: 697–702.

Moyano E, Jouhikainen K, Tammela P et al. (2003) Effect of *pmt* gene overexpression on tropane alkaloid production in transformed root cultures of *Datura metel* and *Hyoscyamus muticus*. *J Exp Bot* 54: 203–221.

Nakajima K, Yamashita A, Akama H et al. (1998) Crystal structures of two tropinone reductases: different reaction stereospecificities in the same protein fold. *Proc Natl Acad Sci USA* 95: 4876–4881.

Nakajima K, Kato H, Oda J et al. (1999) Site-directed mutagenesis of putative substrate-binding residues reveals a mechanism controlling the different stereospecificities of two tropinone reductases. *J Biol Chem* 274: 16563–16568.

Nakajima K, Hashimoto T (1999) Two tropinone reductases, that catalyze opposite stereospecific reductions in tropane alkaloid biosynthesis, are localized in plant root with different cell-specific patterns. *Plant Cell Physiol* 40: 1099–1107.

Nelson DR, Schuler MA, Paquette SM et al. (2004) Comparative genomics of rice and arabidopsis. Analysis of 727 cytochrome P450 genes and pseudogenes from a monocot and a dicot. *Plant Physiol* 135: 756–772.

Ogita S, Uefuji H, Yamaguchi Y et al. (2003) Production of decaffeinated coffee plants. *Nature* 423: 823.

Ohtani M, Shitan N, Sakai K et al. (2005) Characterization of vacular transport of the endogenous alkaloid berberine in *Coptis japonica*. *Plant Physiol* 138: 1939–1946.

Ollagnier S, Kervio E, Rétey J (1998) The role and source of 5'-deoxyadenosyl radical in a carbon skeleton rearrangement catalyzed by a plant enzyme. *FEBS Lett* 437: 309–312.

Palazón J, Moyano E, Rcusidó RM et al. (2003) Alkaloid production in *Duboisia* hybrid hairy roots and plants overexpressing the *h6h* gene. *Plant Sci* 165: 1289–1295.

Panicot M, Masgrau C, Borrell A et al. (2002) Effects of putrescine accumulation in tobacco transgenic plants with different expression levels of oat arginine decarboxylase. *Physiol Plant* 114: 281–287.

Park SU, Johnson AG, Penzes-Yost C et al. (1999) Analysis of promoters from tyrosine/dihydroxyphenylalanine decarboxylase and berberine bridge enzyme genes involved in benzylisoquinoline alkaloid biosynthesis in opium. *Plant Mol Biol* 40: 121–131.

Park SU, Yu M, Facchini PJ (2002) Antisense RNA-mediated suppression of benzophenanthridine alkaloid biosynthesis in transgenic cell cultures of California poppy. *Plant Physiol* 128: 696–706.

Pauli HH, Kutchan TM (1998) Molecular cloning and functional heterologous expression of two alleles encoding (*S*)-*N*-methylcoclaurine 3′-hydroxylase (CYP80B1), a new methyl jasmonate-inducible cytochrome P-450-dependent mono-oxygenase of benzylisoquinoline alkaloid biosynthesis. *Plant J* 13: 793–801.

Raskin I, Ribnicky DM, Komarnytsky S et al. (2002) Plants and human health in the twenty-first century. *Trends Biotechnol* 20: 522–531.

Rathbone DA, Bruce NC (2002) Microbial transformation of alkaloids. *Curr Opin Microbiol* 5: 274–281.

Richer U, Rothe G, Fabian AK et al. (2005) Overexpression of tropinone reductases alters alkaloid composition in *Atropa belladonna* root cultures. *J Exp Bot* 56: 645–652.

Riecher DE, Timko MP (1999) Structure and expression of the gene family encoding putrescine *N*-methyltransferase in *Nicotiana tabacum*: new clues to the evolutionary origin of cultivated tobacco. *Plant Mol Biol* 41: 387–401.

Ro DK, Paradise EM, Ouellet M et al. (2006) Production of the antimalarial drug precursor artemisinic acid in engineered yeast. *Nature* 440: 940–943.

Robins RJ, Bachmann P, Woolley JG (1994) Biosynthesis of hyoscyamine involves an intramolecular rearrangement of littorine.J Chem Soc Perkin Trans 1994: 615–619.

Rocha P, Stenzel O, Parr A et al. (2002) Functional expression of tropinone reductase I (*trI*) and hyoscyamine-6β-hydroxylase (*h6h*) from Hyoscyamus niger in *Nicotiana tabacum*.Plant Sci 162: 905–913.

Rothe G, Hachiya A, Yamada Y et al. (2003) Alkaloids in plants and root cultures of *Atropa belladonna* overexpressing putrescine *N*-methyltransferase. *J Exp Bot* 54: 2065–2070.

Saedler R, Baldwin IT (2004) Virus-induced gene silencing of jasmonate-induced direct defences, nicotine and trypsin proteinase-inhibitors in *Nicotiana attenuata*. *J Exp Bot* 55:151–157.

Samanani N, Facchini PJ (2002) Purification and characterization of norcoclaurine synthase. *J Biol Chem* 277: 33878–33883.

Samanani N, Liscombe DK, Facchini PJ (2004) Molecular cloning and characterization of norcoclaurine synthase, an enzyme catalyzing the first committed step in benzylisoquinoline alkaloid biosynthesis. *Plant J* 40: 302–313.

Samanani N, Park S-U, Facchini PJ, (2005) Cell type-specific localization of transcripts encoding nine consecutive enzymes involved in protoberberine alkaloid biosynthesis. *Plant Cell* 17:915–926.

Sato F, Takeshita N, Fujiwara H et al. (1994) Characterization of *Coptis japonica* cells with different alkaloid productivities. *Plant Cell Tissue Organ Cult* 38: 249–256.

Sato F, Hashimoto T, Hachiya A et al. (2001) Metabolic engineering of plant alkaloid biosynthesis. *Proc Natl Acad Sci USA* 98: 367–372.

Sato F, Yamada Y (in press) Engineering medicinal compounds in cell cultures. In: Bohnert HJ, Nguyen HT (eds) Bioengineering and Molecular Biology of Plant Pathways Volume 1. Elsevier, Amsterdam.

Shitan N, Bazin I, Dan K et al. (2003) Involvement of CjMDR1, a plant multidrug-resistance-type ATP-binding cassette protein, in alkaloid transport in *Coptis japonica*. *Proc Natl Acad Sci USA* 100: 751–756.

Shoji T, Yamada Y, Hashimoto T (2000) Jasmonate induction of putrescine *N*-methyltransferase genes in the root of *Nicotiana sylvestris*. *Plant Cell Physiol* 41: 831–839.

Shoji T, Winz R, Iwasa T et al. (2002) Expression patterns of two tobacco isoflavone reductase-like genes and their possible roles in secondary metabolism in tobacco. *Plant Mol Biol* 50: 427–440.

Siminszky B, Gavilano L, Bowen SW et al. (2005) Conversion of nicotine to nornicotine in *Nicotiana tabacum* is mediated by CYP82E4, a cytochrome P450 monooxygenase. *Proc Natl Acad Sci USA* 102: 14919–14924.

Sinclair SJ, Murphy KJ, Birch CD et al. (2000) Molecular characterization of quinolinate phosphoribosyltransferase (QPRTase) in *Nicotiana*. *Plant Mol Biol* 44: 603–617.

Stenzel O, Teuber M, Dräger B (2006) Putrescine *N*-methyltransferase in *Solanum tuberosum* L., a calystegine-forming plant. *Planta* 223: 200–212.

Steppuhn A, Gase K, Krock B et al. (2004) Nicotine's defensive function in nature. *PLOS Biol* 2: 1074–1079.

Suzuki K, Yamada Y, Hashimoto T (1999a) Expression of *Atropa belladonna* putrescine *N*-methyltransferase gene in root pericycle.Plant Cell Physiol 40: 289–297.

Suzuki K, Yun DY, Chen XY et al. (1999b) An *Atropa belladonna* hyoscyamine 6β-hydroxylase gene is differentially expressed in the root pericycle and anthers. *Plant Mol Biol* 40: 141–152.

Takeshita N Fujiwara H, Mimura H et al. (1995) Molecular cloning and characterization of *S*-adenosyl-L-methionine:scoulerine-9-*O*-methyltransferase from cultured cells of *Coptis japonica*. *Plant Cell Physiol* 36: 29–36.

Tiburicio AF, Galston AW (1986) Arginine decarboxylase as the source of putrescine for tobacco alkaloids. *Phytochemistry* 25:107–110.

Tohge T, Nishiyama Y, Hirai MY et al. (2005) Functional genomics by integrated analysis of metabolome and transcriptome of *Arabidopsis* plants over-expressing an MYB transcription factor.Plant J 42: 218–235.

Unterlinner B, Lenz R, Kutchan TM (1999) Molecular cloning and functional expression of codeinone reductase: the penultimate enzyme in morphine biosynthesis in the opium poppy *Papaver somniferum*. *Plant J* 18: 465–475.

Van Der Fits L, Memelink J (2000) ORCA3, a jasmonate-responsive transcriptional regulator of plant primary and secondary metabolism. *Science* 289: 295–297.

Verpoorte R, Memelink J (2002) Engineering secondary metabolite production in plants, Curr. *Opin. Biotechnol* 13: 181–187.

Voelckel C, Kruegel T, Gase K et al. (2001) Anti-sense expression of putrescine *N*-methyltransferase confirms defensive role of nicotine in *Nicotiana sylvestris* against *Manduca sexta*. Chemoecology 11: 121–126.

Walton NJ, McLauchlan WR (1990) Diamine oxidase and alkaloid production in transformed root cultures of *Nicotiana tabacum*. *Phytochemistry* 29: 1455–1457.

Wang E, Wang R, Deparasis J et al. (2001) Suppression of a P450 hydroxylase gene in plant trichome glands enhances natural-product-based aphid resistance. *Nature Biotech* 19: 371–374.

Wang MB, Waterhouse PM (2002) Application of gene silencing in plants. *Curr Opin Plant Biol* 5(2):146–150.

Waterhouse PM, Helliwell CA (2003) Exploring plant genomes by RNA-induced gene silencing. *Nat Rev Genet* 4: 29–38.

Weckwerth W, Fiehn O (2002) Can we discover novel pathways using metabolomic analysis? Curr. *Opin. Biotechnol* 13: 156–160.

Weid M, Ziegler J, Kutchan TM (2004) The roles of latex and the vascular bundle in morphine biosynthesis in the opium poppy, *Papaver somniferum*. *Proc Natl Acad Sci USA* 101: 13957–13962.

Wesley SV, Helliwell CA, Smith NA et al. (2001) Construct design for efficient, effective and high-throughput gene silencing in plants. *Plant J* 27: 581–590.

Wink M, and Roberts MF (1998) Compartmentation of alkaloid synthesis, transport, and storage. In: Roberts MF, Wink M (eds) Alkaloids; Biochemistry, Ecology, and Medicinal Applications. Plenum Press, New York.

Winkel-Shirley B (2001) Flavonoid biosynthesis. A colorful model for genetics, biochemistry, cell biology, and biotechnology. *Plant Physiol* 126 (2): 485–493.

Yamashita A, Endo M, Higashi T et al. (2003) Capturing enzyme structure prior to reaction initiation: tropinone reductase-II-substrate complexes. *Biochemistry* 42: 5566–5573.

Yamazaki Y, Urano A, Sudo H et al. (2003) Metabolite profiling of alkaloids and strictosidine synthase activity in camptothecin-producing plants *Phytochemistry* 62: 461–470.

Yun DJ, Hashimoto T, Yamada Y (1992) Metabolic engineering of medicinal-plants – transgenic *Atropa belladonna* with an improved alkaloid composition. *Proc Natl Acad Sci USA* 89:11799–11803.

Yun DJ, Hashimoto T, Yamada Y (1993) Transgenic tobacco plants with two consecutive oxidation reactions catalyzed by hyoscyamine 6β-hydroxylase, Biosci Biotech Biochem 57: 502–503.

Zhang L, Ding R, Chai Y et al. (2004) Engineering tropane biosynthetic pathway in *Hyoscyamus niger* hairy root cultures. *Proc Natl Acad Sci USA* 101: 6786–67910.

Zhao J, Davis LC, Verpoorte R (2005) Elicitor signal transduction leading to production of plant secondary metabolites. *Biotech. Adv 23:* 283–333.

Ziegler J, Diaz-Chavez ML, Karamell R et al. (2005) Comparative macroarray analysis of morphine-containing *Papaver somniferum* and eight morphine-free *Papaver* species identifies an *O*-methyltransferase involved in benzylisoquinoline biosynthesis. *Planta* 222: 458–471.

Two recent papers suggested the existence of additional rate-limiting steps in isoquinoline alkaloid biosynthesis; i.e., (*S*)-*N*-methylcoclaurine 3′-hydroxylase (CYP80B3)(Frick S, Kramell R, Kutchan TM (2007) Metabolic engineering with a morphine biosynthetic P450 in opium poppy surpasses breeding. Metab Eng 9(2):169–76.) and codeinone reductase (Larkin PJ, Miller JA, Allen RS, et al. (2007) Increasing morphinan alkaloid production by over-expressing codeinone reductase in transgenic *Papaver somniferum*. Plant Biotechnol J 5(1):26–37.)

CHAPTER 7

APPLICATION OF METABOLIC ENGINEERING TO VANILLIN BIOSYNTHETIC PATHWAYS IN *VANILLA PLANIFOLIA*

DAPHNA HAVKIN-FRENKEL AND FAITH C. BELANGER
The Biotechnology Center for Agriculture and the Environment, School of Environmental and Biological Science, Rutgers, The State University of New Jersey, New Brunswick, NJ, 08903, USA
Phone 732-932-8165x304
Fax 732-932-6535

Keywords: Vanillin, flavour, P-hydroxybenzaldehyde, beta-glucosidase, glycosyl transferase

1. INTRODUCTION

Vanilla is the most popular flavor in the world, the most widely used on both a dollar and tonnage basis. Natural vanilla flavor is obtained from beans of the climbing vanilla orchid, *Vanilla planifolia*. The annual world consumption of vanilla beans, on average, is estimated to be over 2,000 tons (Ranadive 2003, 2006). Vanilla, next to saffron is the most expensive spice. The major flavor compound is vanillin (4-hydroxy-3-methoxybenzaldehyde), although over 250 different other compounds have been isolated from vanilla beans, including 4-hydroxybenzaldehyde, 4-hydroxybenzoic acid and 4-hydroxy-3-methoxybenzoic acid (Guarino and Brown, 1985; Adedji, et al., 1993, Hartman, 1992, 2003; Ranadive, 2003, 2006). Out of about 110 known vanilla species only two are in cultivation, *Vanilla planifolia* and *Vanilla tahitensis*, and allowed to be used in food (US FDA regulation, 21CFR 169.175.).

The tropical climbing orchid *Vanilla planifolia* is native to the southeastern region of Mexico. Vanilla is grown commercially in Madagascar, Comoros, India, Indonesia, Uganda, Mexico, Papua New Guinea and Tahiti. The latter two cultivate *Vanilla tahitensis*. Many other countries produce small amounts of vanilla beans (Fig. 1). Vanilla beans are harvested up to 8–9 months post-pollination. At this stage, the

R. Verpoorte et al. (eds.), Applications of Plant Metabolic Engineering, 175–196.
© 2007 *Springer*.

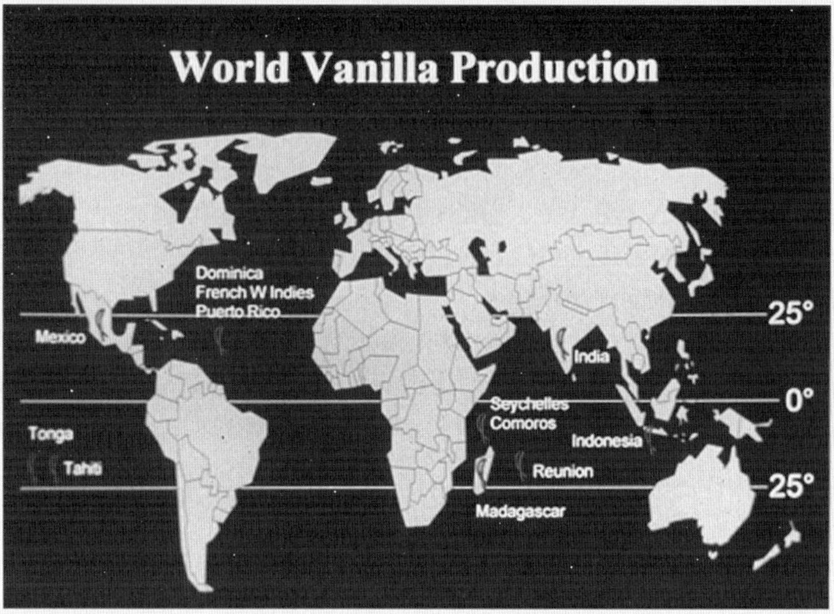

Figure 1. World Vanilla Production

green beans are flavorless but contain large quantities of glucosides of the various flavor compounds. The characteristic flavor develops during "curing" of the beans, a process that can last for as long as 6 months. During the curing process various glycosides are hydrolyzed, and various other constituents undergo oxidation or polymerization (Havkin-Frenkel et al., 2005). Vanillin, the major flavor component is, in addition, a nutraceutical because of its properties as an antioxidant and anti-microbial agent (Fitzgerald et al., 2004; Fitzgerald et al., 2005, Gasson et al., 1998; Lopez-Malo et al., 1995; Kometani et al., 1993). Although vanillin is one of the most studied chemicals, its biosynthetic pathway(s) in plants are still a controversial issue, because no one as of yet has identified and characterized all the steps in the proposed pathways. For more detailed information on vanilla cultivation, curing, the market of vanilla, and flavor and fragrance applications, please see Vanilla International Congress, 1st, Princeton NJ, USA, Nov 11–12, 2003, Allured Publication Corp.

Metabolic engineering can provide better and more appealing features in plant growth and development, an increase in desirable metabolites and thus, better crops. However, in vanilla, the immediate need is for plants that are resistant to the soil pathogen *Fusarium*. Since all the vanilla in cultivation is propagated with the use of cuttings or by micropropagation, cultivated vanilla plants are genotypcially identical, and all cultivated plants are sensitive to *Fusarium*. There is therefore, a danger of losing all the vanilla in cultivation if the *Fusarium* becomes more aggressive. Breeding programs combined with metabolic engineering can help solve this problem and prevent the disappearance of the crop in the event that the disease becomes aggressive. Until today there is no reported information in the literature about metabolic engineering of vanilla or a reliable transformation system. However the potential of such a system is enormous.

Vanilla flavor has ignited the imagination and interest of many scientists for many decades. The growing awareness for natural products has increased the demand for natural vanillin and has emphasized the need to understand the biosynthetic pathway to vanillin. This chapter will deal with the vanillin pathway in *Vanilla planifolia* and the potential application of metabolic engineering to the vanilla industry. The vanillin biosynthetic pathway can be divided into 2 parts: 1. Biosynthesis of glucovanillin, the parent compound of vanillin in the green vanilla beans, and 2. Hydrolysis of glucovanillin to vanillin in the curing process. In this chapter we will discuss the biosynthesis of glucovanillin. The hydrolysis of glucovanillin as been reviewed previously (Havkin-Frenkel et al., 2005).

2. VANILLIN BIOSYNTHETIC PATHWAY IN *VANILLA PLANIFOLIA*

There is an agreement in the literature that vanillin is a product of the shikimic acid pathway. In this pathway phenyalanine or tyrosine undergo deamination to a C_6-C_3 phenylpropanoid, which then serves as a precursor for vanillin. A general view on the metabolic origin of vanillin is outlined in (Fig 2A, B).

Although there is an agreement that vanillin originates from phenylpropanoid C_6-C_3 compounds, there are two major views as to how a phenylpropanoid precursor is converted to vanillin. One school of thought, proposed by Zenk (1965) suggested that the aromatic ring on C_6-C_3 compounds (*trans*-cinnamic acid, *p*-coumaric acid) undergoes hydroxylation and methylation giving rise to ferulic acid. The later may then undergo chain shortening to vanillin. This scheme is termed the 'ferulate pathway'. Another view argues that chain shortening of a phenylpropanoid is the first metabolic event, followed by hydroxylation and methylation of the aromatic

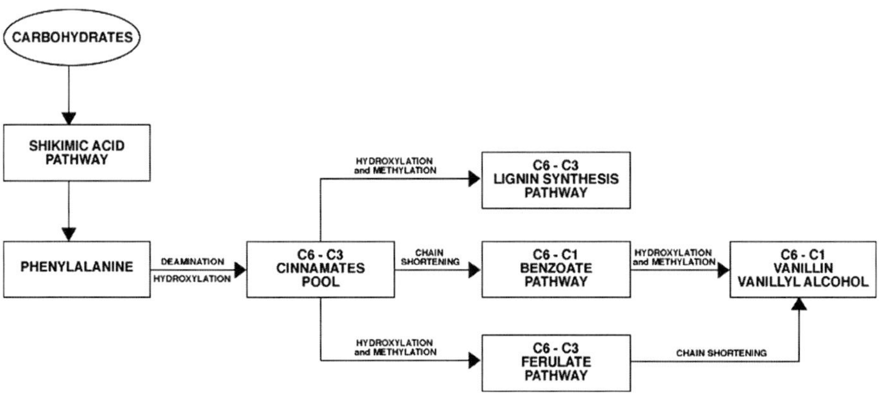

Figure 2A. Possible Route to Vanillin Biosynthesis in Vanilla, showing the Ferulate and Benzoate Pathways

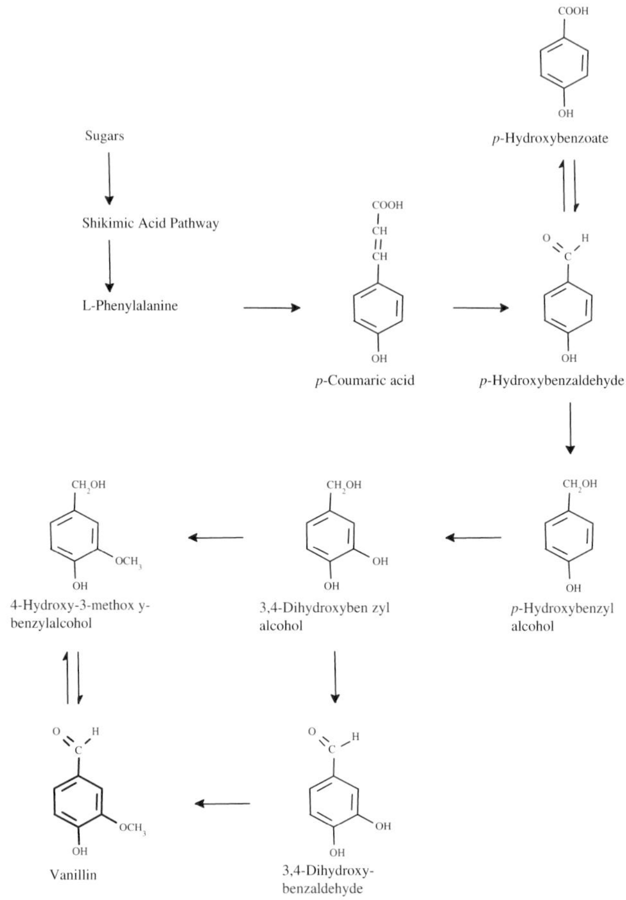

Figure 2B. Proposed Vanillin Biosynthetic Pathway in *Vanilla planifolia* (Havkin-Frenkel et al., 1996)

ring to yield vanillin. This is termed the 'benzoate pathway'. There also exists the possibility that an early intermediate in the shikimic acid pathway gives rise directly to the benzoate pool, bypassing the production of phenylpropanoids and their degradation to benzoate pathway intermediates.

2.1. Phenylalanine Ammonia-lyase (PAL) and Tyrosine Ammonia-lyase (TAL) in Vanillin Biosynthesis

L-phenylalanine ammonia-lyase (PAL, EC 4.3.15) is a key regulatory enzyme of phenylpropanoid metabolism, by catalyzing the first reaction leading to phenolic compounds and the cell wall structural polymer lignin (Jones 1984; Dixon et al., 1983). This enzyme is one of the most extensively studied in plant secondary metabolism (Bolwell et al., 1985, 1986; Achnine et al., 2002). In grasses, PAL and accompanying tyrosine ammonia-lyase (TAL) enzymatic activity, which catalyze this reaction, is the first committed step of the phenylpropanoid pathway. Rosler et al. (1977) showed that in maize PAL had a higher K_M than TAL, although the activity (*kcat*) of the former was higher. However, comparison of *kcat* over K_M ratio revealed similar values for both enzymes (0.018 and 0.023, respectively), suggesting that both PAL and TAL may have similar efficacy for the production of phenylpropanoids. Because *Vanilla planifolia* is a monocot as are the grasses, it is a reasonable assumption that these enzymes are also important in phenylpropanoid production in *Vanilla planifolia*. We were able to demonstrate, accordingly, that activity of both PAL and TAL can be found in vanilla cell culture and in the green vanilla pod, although PAL activity, measured by the accumulation of cinnamic acid, was 10-fold higher than that of TAL, measured by the accumulation of *p*-coumaric acid, as outlined before (Bolwell et al., 1985, 1986).

Feeding *Vanilla planifolia* cell culture with radioactive ^{14}C- phenylalanine or ^{14}C- tyrosine, resulted in the uptake from the reaction media of these amino acids and their subsequent conversion to proposed intermediates, in the

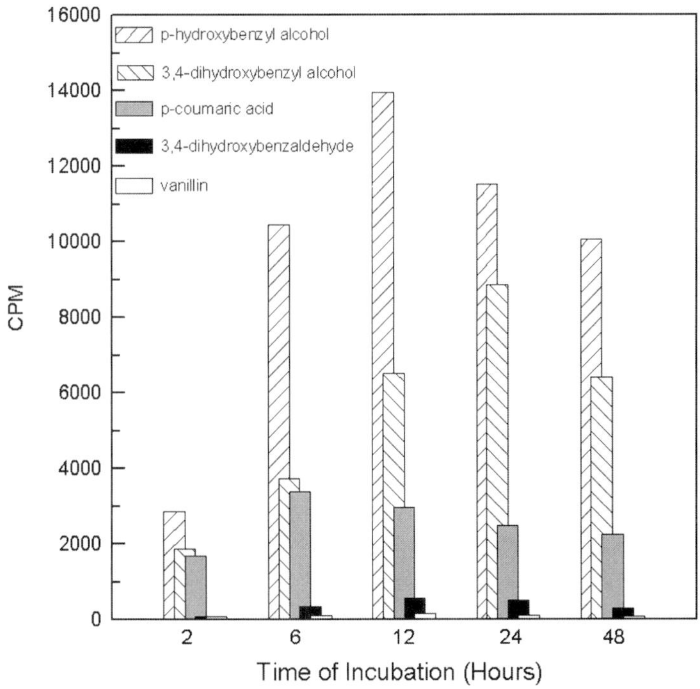

Figure 3A. Time-course of production of proposed intermediates by *Vanilla planifolia* cell culture, fed with ^{14}C phenylalanine. Activity, measured in counts per minute (CPM) represent relative amounts. Intermediates were separated and analyzed by HPLC and activity determined with a scintillation counter, as previously described (Havkin-Frenkel et al., 1996)

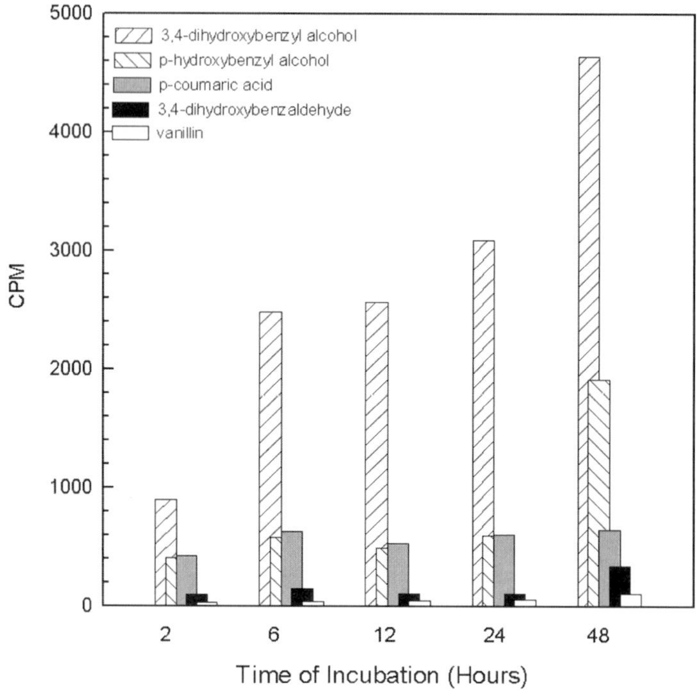

Figure 3B. Time-course of production of proposed intermediates by *Vanilla planifolia* tissue culture, fed with ^{14}C tyrosine. Activity, measured in counts per minute (CPM) represent relative amounts. Intermediates were separated and analyzed by HPLC and activity determined with a scintillation counter, as previously described (Havkin-Frenkel et al., 1996)

vanillin biosynthetic pathway (Havkin-Frenkel and Pedersen, 2000). The results shown in (Figure 3A, B,) indicate labeling of metabolites, including *trans*-cinnamic acid, *p*-coumaric acid, *p*-hydroxybenzaldehyde, *p*-hydroxybenzyl alcohol, 3,4-dihydroxybenzaldehyde, 3,4-dihydroxybenzyl alcohol that are possible intermediates in the proposed vanillin biosynthetic pathway. Similar results were obtained by on-the-vine feeding of green vanilla beans with ^{14}C phenylalanine and ^{14}C tyrosine (Fig. 4A, B). However, in both systems applied phenylalanine yielded more vanillin and proposed intermediates (approximately 10 fold higher) than applied tyrosine (Havkin-Frenkel et al., 2007).

2.2. Chain Shortening Enzyme

The formation in plants of benzoic acid derivatives (C_6-C_1) from hydroxycinnamic acids (C_6-C_3), derived from phenylalanine and tyrosine via the phenylpropanoid pathway of secondary metabolism, is well documented (Funk and Brodelius,1990a,b,1992; Yazaki et al., 1991; Schnitzel et al., 1992; Yalpani et al.,1993). A large array of metabolites represent benzoic acid derivatives in plants, such as the salicylic acid plant defense signal molecule (Verberene et al., 1999), the napthoquinone pigment shikonin (Löscher and Heide, 1994), cocaine (Bjorklund and Leete, 1992), xanthones with anti-HIV activity (El-Basyouni et al., 1964) and the anticancer drug taxol (Chu et al., 1994). However, the nature of the biochemical steps and reactions leading to shortening of the side chain of hydroxycinnamic acids to yield substituted benzoic acids is not well understood.

In plants, there may be at least five different routes to benzoic acids. One, involving the vanillin biosynthesis pathway in *V. planifolia* (Zenk, 1965) and shikonin biosynthesis in *Lithospermum erythrorhizon* (Löscher and Heide, 1994), entailing conversion of hydroxycinnamic acids to their coenzyme A esters. In this process, chain-shortening occurs in a manner similar to NAD-dependent β-oxidation of fatty acids, leading to benzoic acid. The operation of this

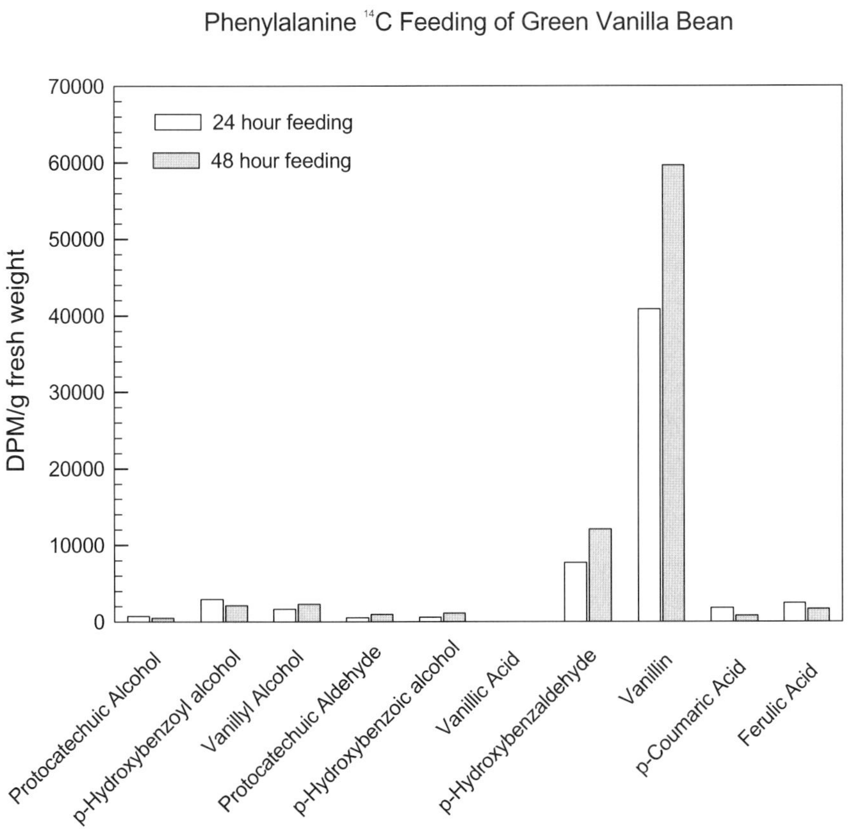

Figure 4A. Metabolites produced in green vanilla beans, measured 24 and 48 hours after feeding with ^{14}C Phenylalanine. Metabolites were separated and analyzed by HPLC and activity determined with a scintillation counter, as previously described (Havkin-Frenkel et al., 1996)

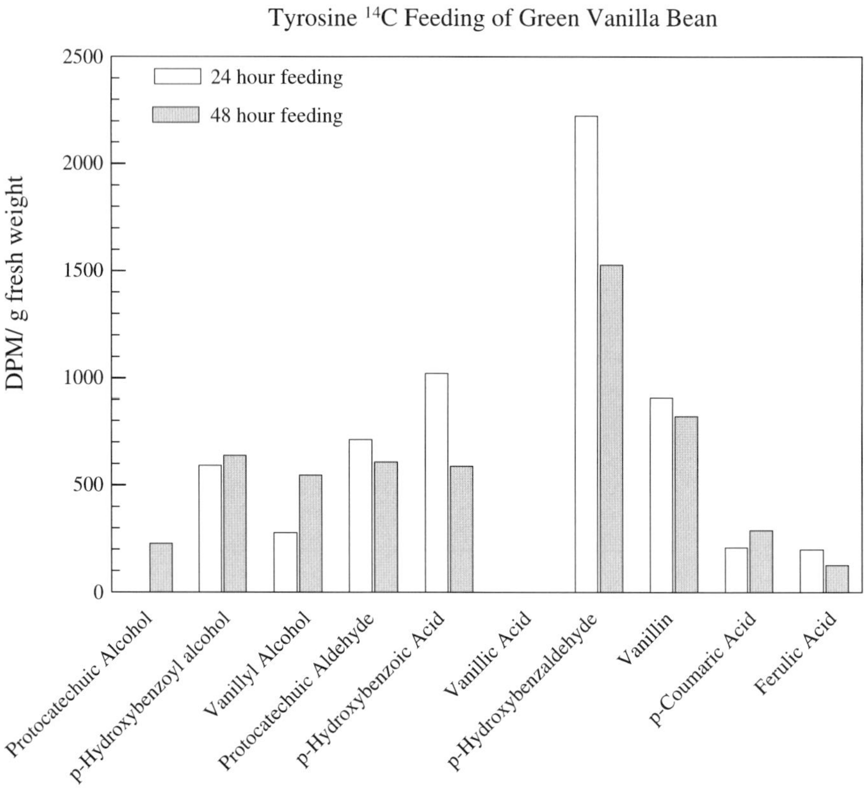

Figure 4B. Metabolites produced in green vanilla beans, measured after 24 and 48 hours after feeding with ^{14}C tyrosine. Metabolites were separated and analyzed by HPLC and activity determined with a scintillation counter, as previously described (Havkin-Frenkel et al., 1996). The intermediates shown in figures 4A and 4B are represented in the order of elution sequence from the HPLC (Havkin-Frenkel et al., 1996; Podstolski et al., 2002)

pathway is supported by radiolabeling studies (El-Basyouni et al., 1964) and verification of enzymatic activities in crude cell extracts (Löscher and Heide, 1994), although the enzymes themselves have not been purified. A second route involves a non-oxidative pathway for the conversion of 4-coumaric acid to 4-hydroxybenzaldehyde, with no co-factor requirement. This type of reaction is outlined in (Fig. 5). This reaction has been proposed to occur during formation of 4-hydroxybenzoic acid in potato tubers (French et al., 1976) and elicited carrot cell cultures (Schnitzler et al., 1992) and in an alternative pathway to shikonin in *L. erythrorhizon* (Yazki et al., 1991). In these cases, 4-hydroxybenzoic acid is formed from 4-hydroxybenzaldehyde by a separate NAD-dependent dehydrogenase. The chain-shortening enzyme itself was reported to be very unstable (French et al., 1976). A different pathway to vanillin in cell cultures of *V. plani-*

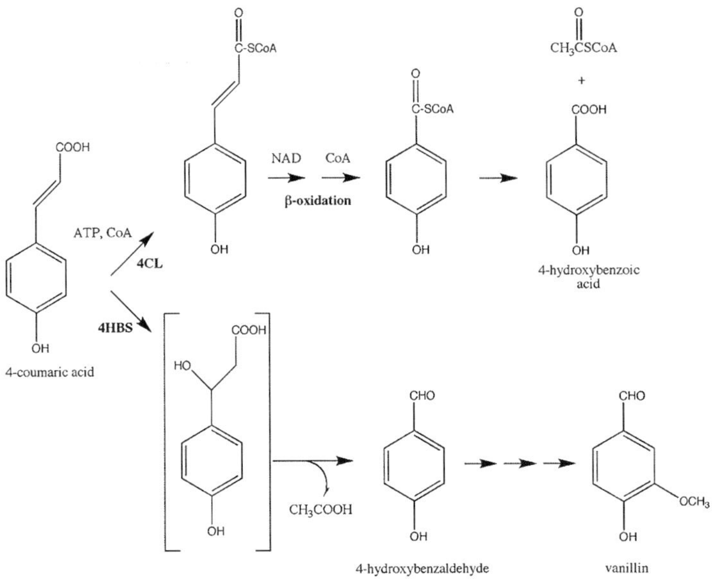

Figure 5. Oxidative and non-oxidative pathways for the chain-shortening of cinnamic acid derivatives in plants. The upper pathway shows the oxidative pathway to 4-hydroxybenzoic acid via the coenzyme A ester. The lower pathway shows the non-oxidative pathway to 4-hydroxybenzaldehyde via an unstable 4-hydroxyphenyl-β-hydroxypropionic acid intermediate (**5**). 4CL, 4-coumarate:coenzyme A ligase; TE, thioesterase; 4HBS, 4-hydroxybenzaldehyde synthase (from Podstolski et al., 2002)

folia was proposed involving methylation of the 4-position of hydroxycinnamic acids, followed by glucosylation, side-chain shortening and 4-demethylation (Funk and Brodelius,1990a). The benzoic acid precursors of xanthones in *Centaurium erythraea* appear to originate directly from the shikimate pathway rather than via cinnamic acid (El-Mawla et al., 2001). In the bacterium *Pseudomonas fluorescens*, vanillin is produced non-oxidatively from feruloyl CoA by an enzyme of the enoyl-SCoA hydratase/isomerase family (Mitra et al., 1999). Expression of the *Pseudomonas* enzyme in tobacco resulted in accumulation of glucosides of 4-hydroxy benzoic acid, vanillic acid, 4-hydroxy benzyl alcohol and vanillyl alcohol, but not vanillin (Mayer et al., 2001). Finally, recent studies have indicated that some benzoic acid derivatives, such a salicylic acid (2-hydroxybenzoic acid) and 2,3-dihydroxybenzoic acid, may be synthesized in plants from the shikimate pathway via isochorismic acid, a pathway that is also found in bacteria and which does not involve phenylalanine as an intermediate (Wildermuth et al., 2001).

We found that cell cultures of *V. planifolia* accumulate *p*-coumaric acid, 4-hydroxybenzaldehyde, 4-hydroxybenzyl alcohol, 3,4-dihydroxybenzaldehyde, 4-hydroxy-3-methoxybenzyl alcohol, as well as vanillin (Havkin-Frenkel et al., 1996). The pattern and levels of these metabolites are consistent with the suggestion

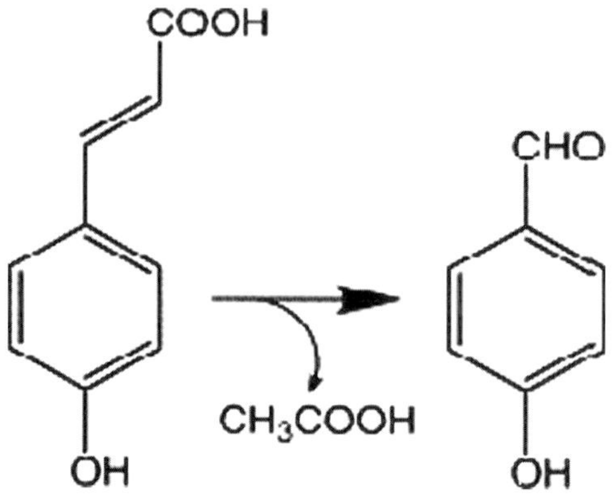

Figure 6. A schematic view of the non-oxidative chain shortening enzyme reaction, showing conversion of *p*-coumaric acid to *p*-hydroxybenzaldehyde. The process is accompanied with the production of acetic acid

that 4-hydroxybenzaldehyde is an intermediate in vanillin biosynthesis. Additional results show that there is an accumulation of acetic acid in developing green vanilla beans, a byproduct of a non-oxidative pathway for conversion of *p*-coumaric acid to 4-hydroxybenzaldehyde, also in keeping with the non-oxidative mode of chain shortening as outlined in Fig. 6.

Acetic acid is a major constituent in green vanilla beans and is found even in cured vanilla beans (Adedji et al., 1993). Our studies show, in addition, that acetic acid can be generated by the enzymatic reaction of the *p*-coumaric acid chain shortening in *in vitro* conditions (Podstolski et al., 2002). On the other hand, the acetyl CoA in the respiratory pathway may not account for acetic acid accumulation in significant quantities.

Figure 7A, showing the activity of 4HBS in developing green vanilla beans, indicates an upsurge in the enzyme activity prior to the accumulation of *p*-hydroxybenzaldehyde (Fig. 7B). Feeding experiments with radiolabeled phenylalanine or tyrosine, showing accumulation of mostly 4-hydroxybenzaldehyde, are consistent with the action of a non-oxidative chain shortening enzyme, which yields aldehyde reaction products whereas a CoA enzyme leads to the production of acids.

In conclusion, application of phenylalanine and tyrosine, the parent compounds for phenylpropanoid formation, resulted in the accumulation of vanillin, as well as putative intermediates in the vanillin biosynthetic pathway. Phenylpropanoid compounds (C_6-C_3) undergo chain shortening to C_6-C_1 metabolites in the benzoate pathway. A concomitant accumulation of acetic acid (by enzyme activity *in vitro* and in bean tissue), suggests a non-oxidative mode of chain shortening. We propose that non-oxidative chain shortening of *p*-coumaric acid, leading to the formation of *p*-hydroxybenzaldehyde, is a key reaction to the biosynthesis of vanillin, as shown in both vanilla cell culture and in green vanilla bean.

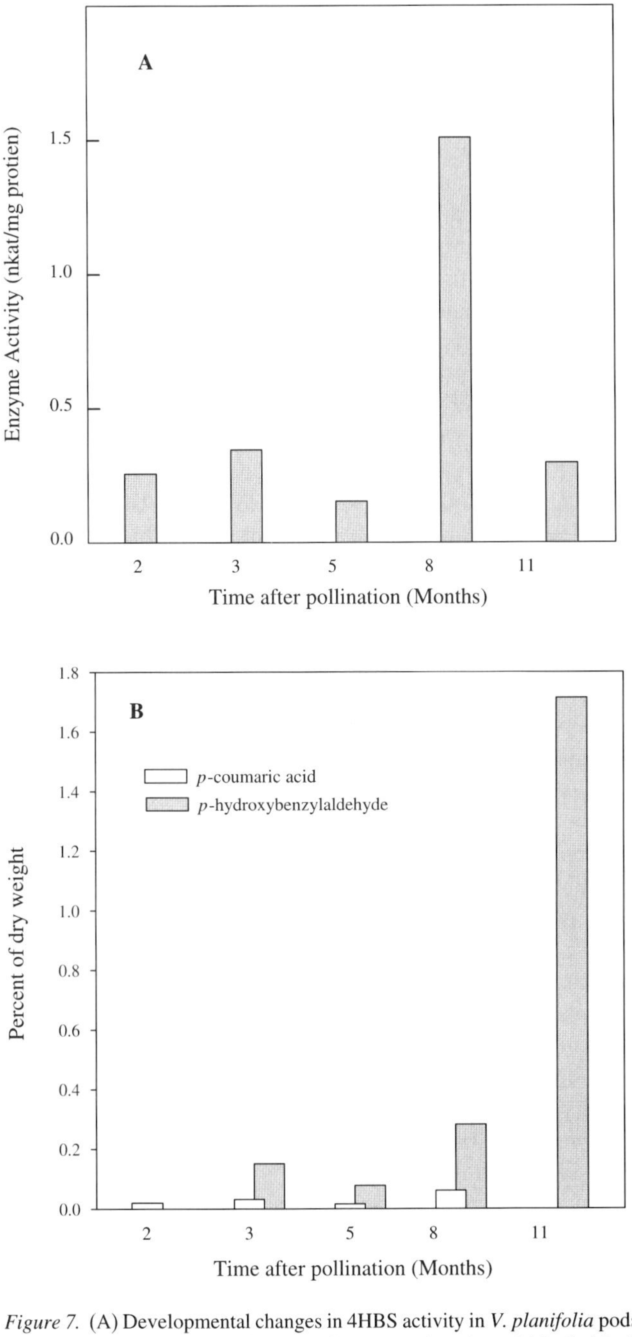

Figure 7. (A) Developmental changes in 4HBS activity in *V. planifolia* pods. (B) Developmental changes in levels of 4-coumaric acid (open bars) and 4-hydroxybenzaldehyde (shaded bars) in *V. planifolia* pods (from Podstolski et al., 2002)

2.3. Transformation of *Vanilla planifolia* with *Agrobacterium tumefaciens*

Vanilla planifolia presently in cultivation is clonally propagated either from cuttings or by tissue culture micropropagetion. It is desirable to introduce new genetic traits to cultivated vanilla, mainly resistance to *Fusarium* but also other attributes. This task may be accomplished by breeding and also by transformation of vanilla tissues. Cloned genes are transferred to tissue culture and transformed cells are grown into whole plant. We used particle bombardment as a method for gene transfer but with limited success, because injured vanilla tissue culture is slow growing, and in addition, is rich in phenolics and their release may be toxic to cells. We tested an alternative approach, using *Agrobacterium tumefaciens* as a vector for genetic material. We use A. *tumefaciens,* AGL1, containing the plasmid pMJ805, a virulent strain that is very effective in the transformation of other monocots, including cereals (Tingav et al., 1997). We used several different types of tissues, including vanilla embryo culture, callus and tissue culture-derived protocorm-like bodies, as well as sterile green house-grown vanilla plantlets (Havkin-Frenkel et al., 2003 PCT/US03/06397).

We developed a new transformation protocol, which combines particle bombardment and use of *Agrobacterium.* The procedure consists of first a bombardment with gold and next inoculation with *Agrobacterium.*

During this procedure the culture was affected by the combination of bombardment, antibiotics, and bialophos, making rapid growth difficult to achieve. Other methods of selection were devised in order to achieve recovery. In our protocol we separate the step of selection with bialophos from the co-cultivation with *Agrobacterium.* We first kill the *Agrobacterium*, allow the tissue to recover and next use favorable growth conditions followed by selection biolaphos. The above procedure is very labor intensive but will enable us to devise a method of introducing any foreign gene into vanilla tissue culture. For all types of genetic transformation success rates below 1% are very common. In vanilla transformation, as outlined above, the rate is 0.01%–0.05% (based on numbers of tissues pieces that had been used), because most of the plant material died in the process from cutting, bombardment, antibiotics or selection markers. We base these data on PCR of cultures surviving bialophos selection for a few months. We did not develop plants from the transgenic culture.

At present there has been limited success in the transformation of orchid species (Nan et al., 1997). The ability to transform vanilla plants will create an opportunity to regulate vanillin biosynthesis in *Vanilla planifolia* and to transfer genes for the regulation of the flowering cycle, and disease and stress tolerance.

2.4. Site of Vanillin Production

In an early observation on vanilla pod structure Swamy (1947) posited that vanillin is secreted “in tissues around the seeds”. Jones and Vicente (1949) also pointed out that most of the vanillin and other compounds involved in vanilla flavors were found in the middle part of the bean around the seeds. In a comprehensive study on vanilla fruit (pod) development, we found that vanillin is specifically present in the non-photosynthetic white parenchyma cells of the endocarp in the pod interior (Joel et al., 2003). Separation of these "white" inner fruit portions from the outer "green" fruit exocarp, revealed that the former contains 95% of the total vanillin found in vanilla pod (Fig. 8).

Furthermore, the white parenchymatic tissue contained proposed intermediates of vanillin biosynthesis, including vanillic acid, 4-hydroxybenzaldehyde and 4-hydroxy benzoic acid (Fig. 9), although in the glycosylated form. The extracted glucosides were hydrolyzed to the corresponding aglycones with almond β-glucosidase and identified with HPLC coupled to LC-MS, as previously described (Podstolski et al., 2002). We also used catechin-HCl, which binds to various phenolic compounds including vanillin, as a mean for localizing accumulated vanillin in the developing fruit and found that both the placenta and the adjacent endocarp parenchymatic cells were red-stained (Fig. 10A), indicating the presence of vanillin and intermediates in vanillin biosynthetic pathway in these tissues. Catechin-HCl also stained the densely packed secreted matrix that accumulates in the fruit cavity (Fig. 10B), clearly showing a descending staining gradient, from endocarp in the fruit cavity outwards. The use of this method also revealed that vanillin accumulation begins after 3 to 4 months of fruit development. However, no staining was observed in the longitudinal strips of brilliant whitish secretory tissue located in the gaps between the placentas along the central fruit cavity (Figs. 10A, B). Because vanillin is sparsely water-soluble, particularly in acidic plant vacuoles (Frenkel and Havkin-Frenkel, 2006), glycosylation of vanillin to gluco-vanillin is a likely mechanism for increasing the hydrophilicity of the compound, thus aiding in the trafficking and storage of the compound in aqueous extra-cellular regions.

Additional studies, concerned accumulation of flavor precursors and intermediates in the vanillin biosynthetic pathway during pod development. The results (Fig. 8) revealed that accumulation of these compounds was directly related to the size of inner specialized cells. Moreover, vanillin, *p*-hydroxybenzaldehyde, coumaric acid and other compounds believed to be intermediates in the vanillin biosynthetic pathway (Fig. 2B) appeared to accumulate in a manner resembling the growth of inner specialized cells. Because catechin-HCl stained the tissue region surrounding the specialized cells, but not the cells themselves, we believe that these cells synthesize flavor compounds and, apparently, secrete these products to the vicinal tissue region.

We believe that the special cells in the pod interior are dedicated to vanillin biosynthesis. We have therefore used these cells to construct a cDNA library for identification of putative genes in the vanillin biosynthetic pathway.

2.5. *O*-Methyltransferase in Vanillin Biosynthesis

The final enzymatic step in the biosynthesis of the flavor compound vanillin (4-hydroxy-3-methoxybenzaldehyde) is believed to be methylation of 3,4-dihydroxybenzaldehyde (Pak et al., 2004). Plant *O*-methytransferases (OMTs) that

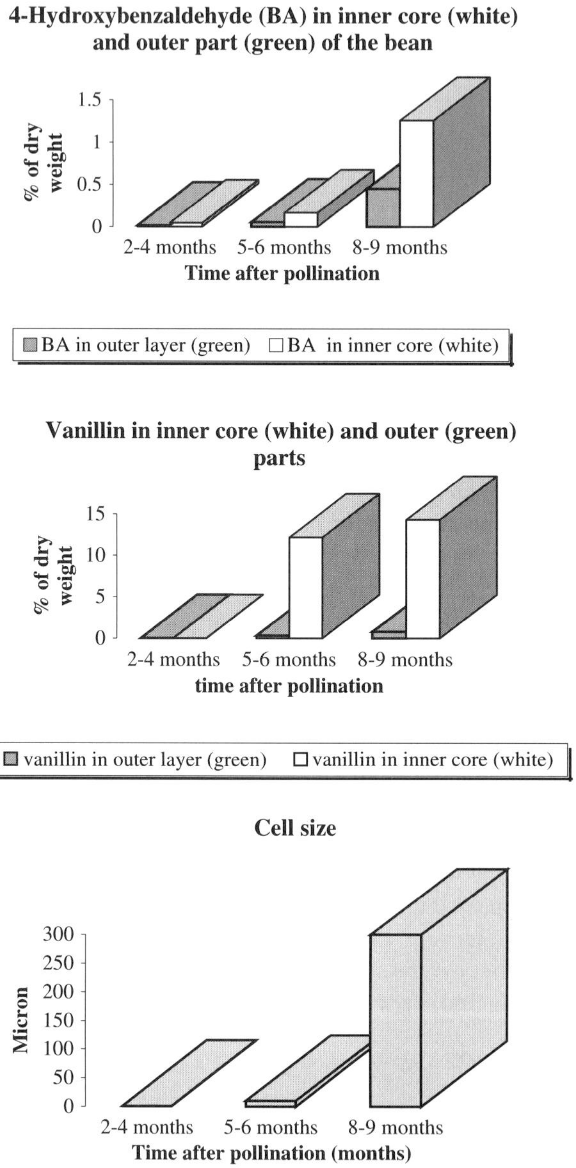

Figure 8. Distribution of 4-hydroxybenzaldehye (top) and vanillin (middle panel) in green and white tissue of vanilla pod. Also shown is cell size in relation to fruit development (bottom). From Joel et al., 2003

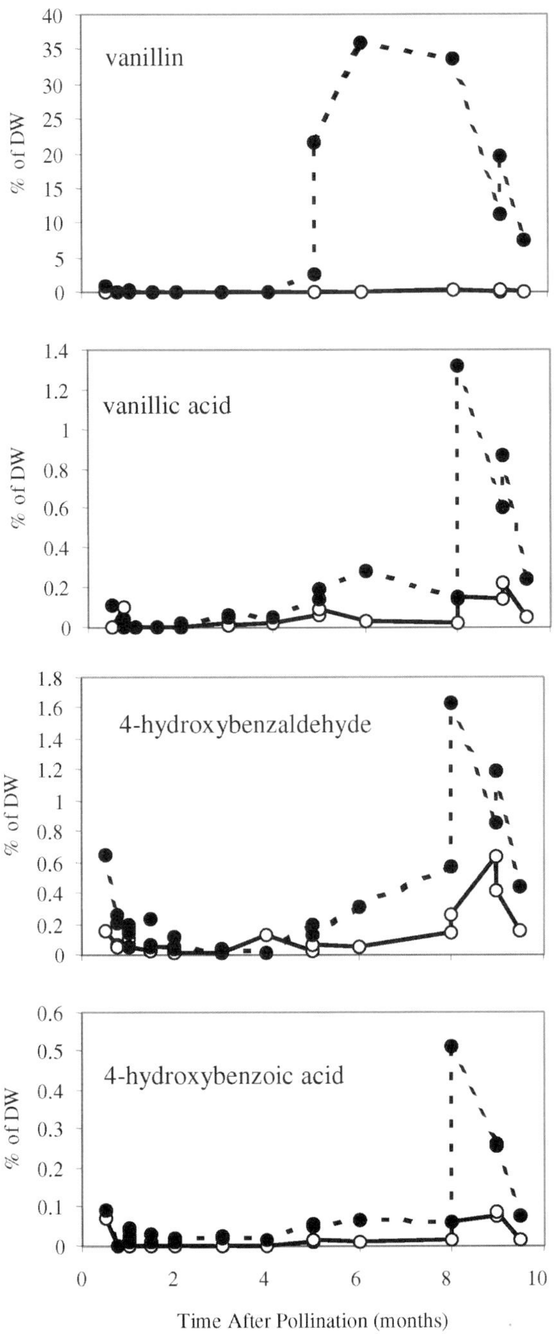
vanillin
vanillic acid
4-hydroxybenzaldehyde
4-hydroxybenzoic acid
% of DW
Time After Pollination (months)

use S-adenosylmethionine (SAM) as the methyl donor are involved in the synthesis of a diverse range of secondary products (Ibrahim et al., 1998; Schroder et al., 2002). Methylation of the 3-OH of caffeic acid and related phenylpropanoid compounds has been widely studied due to its presumed involvement in the synthesis of lignin. Caffeic acid *O*-methyltransferase (COMT) enzymes have been characterized from numerous species (Ibrahim et al., 1998). Even though caffeic acid is not the preferred substrate, these enzymes are still often referred to, for historical reasons, as COMTs. Some COMTs have been reported to have activity against 3,4- dihydroxy-benzaldehyde. In tobacco, two distinct COMTs with different substrate specificities have been characterized: tobacco class I COMT showed activity against both caffeic acid and 3,4-dihydroxybenzaldehyde with similar efficiencies, whereas class II COMT was active against 3,4-dihyroxybenzaldehyde but not caffeic acid (Maury et al., 1999). COMTs from basil (Gang et al., 2002) and strawberry (Wein et al., 2002) were found to have activity with 3,4-dihydroxybenzaldehyde at 69.4% and 140%, respectively, of their relative activity with caffeic acid. Zubieta et al. (2002) determined the crystal structure of the enzyme from alfalfa, revealing a spacious active site, which is consistent with the broad range of substrates acted upon by the enzyme. These results on the broad substrate utilization by COMTs raised the question of whether methylation of 3,4-dihydroxybenzaldehyde, in *V. planifolia*, is mediated by an enzyme specific for this substrate or whether it can occur from a COMT-like enzyme with a broad substrate range. COMT activity is expected to be present in all plant species.

We have isolated and functionally characterized a cDNA that encodes a multi-functional methyltransferase from *Vanilla planifolia* tissue cultures (Pak et al., 2004) that can catalyze the conversion of 3,4-dihydroxybenzaldehyde to vanillin (4-hydroxy-3-methoxybenzaldehyde), although 3,4-dihydroxybenzaldehyde is not the preferred substrate. The higher catalytic efficiency of the purified recombinant enzyme with the substrates caffeoyl aldehyde and 5-OH-coniferaldehyde, and its tissue distribution, suggested this methyltransferase may primarily function in lignin biosynthesis. However, since the enzyme does have 3,4-dihydroxybenzaldehyde-*O*-methyltransferase activity, it may be useful in engineering strategies for the synthesis of natural vanillin from alternate sources.

Another *O*-methyltransferase, which exclusively catalyzes the methylation of 3,4-dihydroxybenzaldehyde to vanillin in tissue extract, was found in the specialized hair cells in the inner core of the vanilla pod, but not in the green tissue regions. This tissue preparation did not exhibit activity toward caffeoyl aldehyde and 5-OH-coniferaldehyde and other typical lignin substrates, suggesting that this

Figure 9. Time course change in the content of various metabolites in the green outer tissue (solid lines) and the inner white tissue (dashed lines) of vanilla bean during pod development on the vine. Beans were harvested green at various stages of development. The various metabolites, present as glucosides, were hydrolyzed and the resulting aglycons determined as described previously (Podstolski et al., 2002). From Havkin-Frenkel et al., 2005

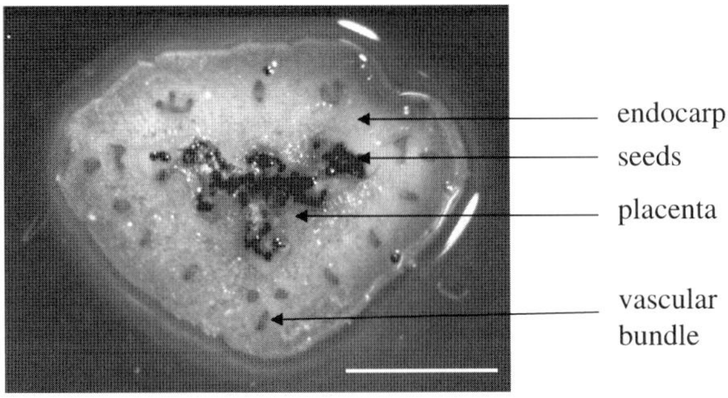

Figure 10A. Cross section of a mature green vanilla fruit after catechin-HCl staining. Vanillin stained red in the placenta and neighboring endocarp cells, and also in the secreted matrix that surrounds the seeds in the fruit cavity. The vascular bundles, rich in lignin are also stained. Scale bar = 5 mm

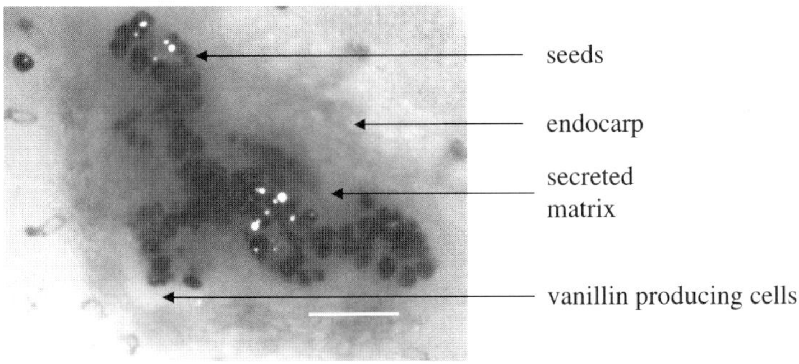

Figure 10B. The central portion of the fruit, showing a dense catechin- HCl staining of the secreted matrix in the fruit cavity around the seeds, and a weaker staining in endocarp cells adjacent to the cavity. Scale bar = 2 mm

enzyme is distinct from COMT. We are in the process of cloning and characterizing the gene(s) for this enzyme. Experimental support for the involvement of this enzyme in vanillin formation would identify the final step in the vanillin biosynthetic pathway, prior to the final stage of vanillin glycosylation and storage.

2.6. Glycosyl Transferases

Glycosylation of vanillin in the vanilla pod is the last step in vanillin biosynthesis and storage in tissues. Glycosylation appears extremely efficient in *Vanilla planifolia* (Yuana et al., 2002), accounting, perhaps, for the efficacy of this species to accumulate high levels of vanillin.

Most secondary metabolites found in plants, are O, N or S-linked to one or more sugars. Aromatic compounds, such as phenylpropanoids, quinones and flavonoids, as well as terpenoids, including monoterpenes, sesquiterpenes, diterpenes, triterpenes, or nitrogen compounds like nonprotein amino acids, cyanogenic compounds, glucosinolates, betalains and alkaloids may be glycosylated (Niggli and Pfander, 1999). When the aglycone moieties in glycosylated metabolites are released by hydrolytic action of appropriate glycosidases, hydrolytic products such as cyanogenic compounds or glucosinolates may function in defense against predators (Conn, 1993; Halkier, 1999; Beckman, 2000), attraction of insects or animals (Reuveni et al., 1999), or phytohormonal regulation of growth and development (Vevodova et al., 2001; Veach et al., 2003).

Glycosylation enhances the hydrophilicity of acceptor molecules, including secondary and other metabolites, and alters their chemical properties and subcellular location. Glycosylation often influences the bioactivity of cellular metabolites or xenophobic compounds (Bowles et al., 2005). Examples of change in chemical activity include de-toxification of biotic toxins, including pathogen-derived toxins (Poppenberger et al., 2003), or external xenobiotic compounds (Loutre et al., 2003) as well as alteration in color properties of plant pigments (Fukuchi-Mizutani et al., 2003).

An important aspect of glycosylation involves regulation of flavor perception (Kita et al., 2000) and organoleptic properties of flavor compounds from plants (DuBois and Stephenson, 1985). Accordingly, glycosylation of vanillin to glucovanillin alters the flavor perception of the compound, by analogy to other instances where glycosylation alters the organoleptic properties of natural compounds (DuBois and Stephenson, 1985). Glucovanillin hydrolysis, during the curing process of green vanilla beans is required to release vanillin in the free form and to assert the flavor value of the compound. Because vanillin is sparsely water soluble and, moreover, tends to volatilize from aqueous solutions (Frenkel and Havkin-Frenkel, 2006) an increase in hydrophilicity, stemming from glycosylation of biosynthetically formed vanillin to glucovanillin, serves to increase the water solubility of glucovanillin and, thus, efficacy to accumulate in high levels in vanilla bean tissue. The process also helps to arrest the propensity of the compound to volatilize. This view may account, perhaps, for the accumulation of intermediates in the vanillin biosynthetic pathway in the glycosylated state. From this perspective, glycosyl transferases might be important candidates for metabolic engineering applications. Background studies indicate a vast array of glycosyl transferases with substrate specificity (Bowles et al., 2005). Vanilla tissues might be then engineered to arrest the activity of glycosyl transferases active on intermediates in the vanillin biosynthetic pathway and, thereby, drive the pathway to accumulate mostly glycosylated vanillin, as final product.

3. CONCLUSIONS

Considerable information has been generated on metabolic and molecular aspects of flavor, mostly vanillin formation in vanilla beans. We have begun to address important issues in vanilla production and processing with the use of molecular and other approaches. We believe that at present the molecular approach, as a method for providing genetic markers rather than transformation, could be coupled to traditional breeding and selection for achieving desirable traits in *Vanilla planifolia*.

REFERENCES

Achnine L, Rasmussen S, Blancaflor E, Dixon RA (2002) Metabolic channeling at the entry point into the phenylpropanoid pathway:physical associations between L-phenylalanine ammonia-lyase and cinnamate 4-hydroxylase. In *Polyphenols Communications 2002*, Volume 1 (El Hadrami, I., ed.). Marrakech: Imprimerie Papeterie el Watanya, pp. 7–8.

Adedji J, Hartman TG, Ho CT (1993) Flavor characterization of different varieties of Vanilla beans. *Perfumer & Flavorist* 18:25–33.

Beckman CH (2000) Phenolic-storing cells: keys to programmed cell death and periderm formation in wilt disease resistance and in general defense responses in plants. *Physiological and Molecular Plant Pathology,* 57:101–110.

Bjorklund JA, Leete E (1992) Biosynthesis of the benzoyl moiety of cocaine from cinnamic acid via (R)-(+)-3-hydroxy-3-phenylpropanoic acid. *Phytochemistry* 31: 3883–3887.

Bolwell GP, Bell JN, Cramer CL, Schuch W, Lamb CJ, Dixon RA (1985)L-Phenylalanine ammonia-lyase from *Phaseolus vulgaris*. Characterization and differential induction of multiple forms from elicitor-treated cell suspension cultures. *Eur. J. Biochem.* 149:411–419.

Bolwell GP, Cramer CL, Lamb CJ, Schuch W, Dixon RA (1986)L-Phenylalanine ammonia-lyase from *Phaseolus vulgaris*. Modulation of the levels of active enzyme by *trans*-cinnamic acid. *Planta* 169: 97–107.

Bowles D, Isayenkova J, Lim E-K, Poppenberger B (2005) Glycosyltransferases: managers of small molecules. *Current Opinion in Plant Biology* 8:254–263.

Chu A, Furlan M, Davin LB, Zajicek J, Towers GHN, Soucy-Breau CM, Rettig SJ, Croteau R, Lewis NG (1994) Phenylbutanoid and taxane-like metabolites from needles of *Taxus brevifolia. Phytochemistry* 36: 975–985.

Conn EE (1993)-Glycosidases in plants: substrate specificity. In:Glycosidases: biochemistry and molecular biology ed. by A. ESEN Washington: American Chemical Society, p. 15–26. (ACS Symposium Series 533)

Dixon RA, Dey PM, Lamb CJ (1983) Phytoalexins: molecular biology and enzymology; Adv. *Enzymol Relat. Areas Mol. Biol* 55:1–135.

DuBois GE, Stephenson RA (1985) Diterpenoid sweeteners. Synthesis and sensory evaluation of stevioside analogues with improved organoleptic properties. *J. Med. Chem.* 28:93–98.

El-Basyouni SZ, Chen D, Ibrahim RK, Neish AC, Towers GHN (1964) The biosynthesis of hydroxybenzoic acids in higher plants. *Phytochemistry* 3: 485–492.

Fitzgerald DJ, Stratford M, Gasson MJ, Ueckert J, Bos A, Narbad A (2004) Mode of antimicrobial action of vanillin against *Escherihia coli*, *Lactobacillus plantarum* and *Listeria innnocua*. *J. Appl. Microbiol.* 94:104–113.

Fitzgerald DJ, Stratford M, Gassson MJ, Narbad A (2005) Structure-function analysis of the vanillin molecule and its antifungal properties. *J. Agric. Food Chem.* 53:1769–1775.

French CJ, Vance CP, Towers GHN (1976) Conversion of *p*-coumaric acid to *p*- hydroxybenzoic acid by cell free extracts of potato tubers and *Polyporus hispidus*. *Phytochemistry* 15:564–566.

Frenkel C, Havkin-Frenkel D (2006) Physics and chemistry of vanillin. *Perfumer & Flavorist* 31:28–36

Fukuchi-Mizutani M, Okuhara H, Fukui Y, Nakao M, Katsumoto Y,Yonekura-Sakakibara K, Kusumi T, Hase T, Tanaka Y (2003)Biochemical and molecular characterization of a novel UDPglucose:anthocyanin 3-O- glucosyltransferase, a key enzyme for blue anthocyanin biosynthesis, from Gentian. Plant Physiol.132:1652–1663

Funk C, Brodelius PE (1990a) Phenylpropanoid metabolism in suspension cultures of *Vanilla planifolia*. Andr. II. Effects of precursor feeding and metabolic inhibitors. *Plant Physiol.* 94: 95–101.

Funk C, Brodelius PE (1990b) Phenylpropanoid metabolism in suspension cultures of *Vanilla planifolia* Andr. III. Conversion of 4-methoxycinnamic acids into 4-hydroxybenzoicacids. *Plant Physiol* 94:102–108.

Funk C, Brodelius PE (1992) Phenylpropanoid metabolism in suspension cultures of *Vanilla planifolia* Andr. IV. Induction of vanillic acid formation. *Plant Physiol.* 99: 256–262.

Gang DR, Lavid N, Zubieta C, Chen F, Beuerle T, Lewinsohn E, Noel JP, Pichersky E (2002) Characterization of phenylpropene *O*-methyltransferases from sweet basil: facile change of substrate specificity and convergent evolution within a plant *O*-methyltransferase family. *Plant Cell* 14:505–519.

Gasson MJ, Kitamura Y, McLauchlan WR, Narbad A, Parr A.J, Lindsay E,Parsons H, Payne J, Rhodes MJC, Walton NJ (1998) Metabolism of ferulic acid to vanillin. *J. Biol. Chem.* 273: 4163–4170.

Guarino PA, Brown SM (1985) Liquid chromatographic determination of vanillin and related flavor compounds in vanilla extract:cooperative study. *J. Assoc. Off. Anal. Chem.* 68: 1198–1201.

Halkier B.A. (1999) Glucosinolates. *In*: Naturally occuring glycosides, ed. by R. Ikan Chichester: John Wiley and Sons,. p. 193–224.

Hartman TG, Karmas K, Chen J, Shevade A, Deagro M, Hwang H (1992)Determination of vanillin, other phenolic compounds, and flavors in vanilla bean. Ho, C.T.; Leland, C.Y.; Huang, M.T.; Eds.; Phenolic Compounds in Food and their Effect on Health 506; American Chemical Society, Washington, DC, 1992, p. 66–70.

Hartman TG (2003) Composition of vanilla beans from different geographical regions. *Vanilla* 2003: First international congress on the future of vanilla and vanillin, Princeton, NJ, USA

Havkin-Frenkel D, Pak F, Gropper S, Bount J (2007) Feeding studies of tissue culture and green beans of *Vanilla planifolia* by applied ^{14}C phenylalanine and tyrosine. In preparation.

Havkin-Frenkel D, French J, Pak F, Frenkel C (2005) Inside Vanilla,*Vanilla planifolia*, botany, curing options and future market prospects. *Perfumer and Flavorist* 30 3: 36–55

Havkin-Frenkel D, Podstolski A, Knorr D (1996) Effect of light on vanillin precursors formation by in vitro cultures of *Vanilla planifolia. Plant Cell Tissue Organ Cult* 46:169–170

Havkin-Frenkel D, Pedersen H (2000) Enzymology and flux control of vanillin biosynthetic pathway in *Vanilla planifolia* tissue culture. Regulation of Phytochemicals by Molecular Techniques, Phytochemical Society of North America, Beltsville, MD U.S.A.

Havkin-Frenkel D, Podstolski A, Dixon RA (2003) Vanillin biosynthetic pathway enzyme from *Vanilla planifolia*. U.S. Pat. Appl. Publ. 49 pp, Cont.-in-part of U.S. Ser. No. 462,576

Ibrahim RK, Bruneau A, Bantignies B (1998) Plant *O*-methyltransferases: molecular analysis, common signature and classification. *Plant Mol. Biol.* 36:1–10

Joel DM, French JC, Graf N, Kourteva G, Dixon RA, Havkin-Frenkel D (2003) A hairy tissue produces vanillin. *Israel Journal of Plant Sciences* 51:157–159

Jones DH (1984) Phenylalanine ammonia-lyase: regulation of its induction, and its role in plant development. *Phytochemistry* 23:1349–1359

Jones MA, Vicente CC (1949) Criteria for testing vanilla in relation to killing and curing methods. *J. Agr. Res.* 78:425–474.

Kita M, Hirata Y, Moriguchi T, Endo-Inagaki T, Matsumoto R, Hasegawa S, Suhayada CG, Omura M (2000) Molecular cloning and characterization of a novel gene encoding limonoid. UDP-glucosyltransferase in citrus. *FEBS Lett.* 469:173–178.

Kometani T, Tanimoto H, Nishimura T, Okasa S (1993) Glucosylation of vanillin by cultured plant cells. Biosci. Biotech. *Biochem.* 57:1290–1293.

Löscher R, Heide L (1994) Biosynthesis of *p*-hydroxybenzoate from *p*-coumarate and *p*-coumaroyl-coenzyme A in cell-free extracts of *Lithospermum erythrorhizon* cell cultures. *Plant Physiol.* 106: 271–279.

Lopez-Malo A, Alzamora SM, Argaiz A (1995) Effect of natural vanillin on germination and radical growth on moulds in fruit – based agar systems. *Food Microbiology* 12: 213–219.

Loutre C, Dixon DP, Brazier M, Slater M, Cole DJ, Edwards R (2003) Isolation of a glucosyltransferase from *Arabidopsis thaliana* active in the metabolism of the persistent pollutant 3,4-dichloroaniline. *Plant J.* 34:485–493.

Maury S, Geoffroy P, Legrand M (1999) Tobacco *O*-methyltransferases involved in phenylpropanoid metabolism. The different caffeoyl-coenzyme A/5-hydroxyferuloyl-coenzyme A 3/5-Omethltransferse and caffeic acid/5-hydroxyferulic acid 3/5-Omethyltransferase classes have distinct substrate specificities and expression patterns. *Plant Physiol.* 121:215–223.

Mayer MJ, Narbad A, Parr AJ, Parker ML, Walton NJ, Mellon FA, Michael AJ (2001) Rerouting the plant phenylpropanoid pathway by expression of a novel bacterial enoyl-CoA hydratase/lyase enzyme function. *Plant Cell* 13: 1669–1682

Mitra A, Kitamura Y, Gasson MJ, Narbad A, Parr AJ, Payne J, Rhodes MJ, Sewter C, Walton NJ (1999) 4- hydroxycinnamoyl-CoA hydratase/lyase (HCHL)-An enzyme of phenylpropanoid chain cleavage from *Pseudomonas*. *Arch. Biochem. Biophy.* 365:10–16.

Nan GL, Tang CS, Kuehnle AR, Kado CI (1997) *Dendrobium* orchids contain an inducer of *Agrobacterium* virulence genes.Physiological & Molecular Plant Pathology 51:391–399

Niggli UA, Pfander H (1999) Carotenoid glycosides and glycosyl esters *In*: Naturally occurring glycosides. ed. by R. Ikan Chichester: John Wiley and Sons. p. 193–224.

Pak FE, Gropper S, Dai WD, Havkin-Frenkel D, Belanger FC. (2002)Characterization of a multifunctional methyltransferase from the orchid *Vanilla planifolia*. Plant Cell Rep 22:959–66.

Podstolski A, Havkin-Frenkel D, Malinowski J, Blount JW, Kourteva G, Dixon RA (2002) Unusual 4-hydroxybenzaldehydesynthase activity from tissue cultures of the vanilla orchid *Vanilla planifolia*. *Phytochemistry* 61:611–620

Poppenberger B, Berthiller F, Lucyshyn D, Sieberer T, Schuhmacher R,Krska R, Kuchler K, Glossl J, Luschnig C, Adam G (2003)Detoxification of the *Fusarium mycotoxin* deoxynivalenol by a UDP-glucosyltransferase from *Arabidopsis thaliana*. *J. Biol. Chem.* 278:47905–47914.

Ranadive AS (2003) Vanilla Cultivation, Vanilla, first International Congress, Princeton, NJ. Allured Publications Inc.

Ranadive AS (2006) Inside look: chemistry and biochemistry of vanilla flavorist. *Perfumer & Flavorist* 31: 38–42.

Reuveni M, Sagi Z, Evnor D, Hetzroni A (1999), Beta-Glucosidase activity is involved in scent production in narcissus flowers. *Plant Science* 147: 19–24

Rosler J, Krekel F, Amrhein N, Schmid J (1997), Maize phenylalanine ammonia-lyase has tyrosine ammonia- lyase activity. Plant Physiol.113:178–179.

Schnitzler JP, Madlung J, Rose A, Seitz HU **(1992)** Biosynthesis of p-hydroxybenzoic acid in elicitor-treated carrot cell cultures. *Planta* 188:594–600.

Schroder G, Wehinger E, Schroder J (2002) Predicting the substrates of cloned plant *O*-methyltransferases. Phytochemistry 59:1–8.

Swamy BGL (1947) On the life history of *Vanilla planifolia*.Botanical Gazette 108:449–456.

Tingay S, McElroy D, Kalla R, Fieg, Wang M, Thronton S, Brettell R (1997) *Agrobacterium tumefaciens*-mediated barley transformation. *Plant J.* 11:1369–1376.

Veach YK, Martin RC, Mok DWS, Malbeck J, Vankova R, Mok MC (2003) *O*-Glucosylation of cis-zeatin in maize. Characterization of genes, enzymes, and endogenous cytokinins. Plant Physiol.131:1374–1380.

Verberne MC, Muljono RAB, Verpoorte R (1999) In: Hooykaas, P J.J, Hall, MA & Libbenga KR, Biochemistry and Molecular Biology of Plant Hormones. (p. 295–312). Sara Burgerhartstraat 25/ PO Box 211/1000 AE Amsterdam/Netherlands:Elsevier Science Publ B V.

Vevodova J, Marek J, Zouhar J, Brzobohaty B, Su X.D (2001) Purification, crystallization and preliminary X-ray analysis of a maize cytokinin glucoside specific β-glucosidase. Acta Crystallographica D57:140–142.

Wein M, Lavid N, Lunkenbein S, Lewinsohn E, Schwab W, Kaldenhoff R (2002) Isolation, cloning and expression of a multifunctional *O*-methyltransferase capable of forming 2,5-dimethyl-4-methoxy-3(2H)-furanone, one of the key aroma compounds in strawberry fruits. *Plant J.* 32: 755–765.

Wildermuth MC, Dewdneyj J, Wutt G, Ausubel FM (2001) Isochorismate synthase is required to synthesize salicylic acid for plant defense. *Nature* 414:562–565.

Yalpani N, Léon J, Lawton MA, Raskin I (1993) Pathway of salicylic acid biosynthesis in healthy and virus-inoculated tobacco. *Plant Physiol.* 103: 315–321.

Yazaki K, Heide L, Tabata M (1991) Formation of *p*-hydroxybenzoic acid from *p*-coumaric acid by cell free extract of *Lithospermum erythrorhizon* cell cultures. *Phytochemistry* 30:2233–2236.

Yuana, Dignum MJW, Verpoorte R (2002) Glucosylation of exogenous vanillin by plant cell cultures. *Plant Cell, Tissue and Organ Culture* 69:177–182.

Zenk MH (1965) Biosythese von Vanillin in *Vanilla Planifolia* Andr. Z. *Pflanzen Physiol.* 53:404–414.

Zubieta C, Kota P, Ferrer J-L, Dixon RA, Noel JP (2002) Structural basis for the modulation of lignin monomer methylation by caffeic acid/5-hydroxyferulic acid 3/5-O-methyltransferase. *Plant Cell* 14:1265–1277

CHAPTER 8

PATHWAY ENGINEERING OF THE PLANT VITAMIN C METABOLIC NETWORK

ARGELIA LORENCE[1] AND CRAIG L. NESSLER[2,3]
[1] *Arkansas Biosciences Institute, Arkansas State University, P.O. Box 639, State University, AR 72467*
[2] *Virginia Agricultural Experiment Station, Virginia Tech, 104 Hutcheson Hall, Blacksburg, VA 24061*
[3] *To whom correspondence should be addressed: Craig Nessler, Virginia Agricultural Experiment Station, Virginia Tech, 104 Hutcheson Hall, Blacksburg, VA 24061, cnessler@vt.edu*

Abstract: Vitamin C (ascorbic acid, AsA) is an important primary metabolite of plants that functions as an antioxidant, an enzyme cofactor, and a cell-signalling modulator in a wide array of crucial physiological processes, including biosynthesis of the cell wall, secondary metabolites and phytohormones, stress resistance, photoprotection, cell division, senescence, and growth. Humans and related primates have lost the ability to synthesize AsA and therefore must obtain it in the diet – primarily from plants. Despite its importance, our understanding of plant vitamin C biosynthesis remains incomplete. Several routes leading to AsA formation have been proposed: from D-glucose via D-mannose and L-galactose; from *myo*-inositol; from galacturonic acid, and from L-gulose. It is unclear whether these are independent pathways or whether they interlink, possibly via enzymes with nonspecific activity. Several enzymes in the vitamin C network have yet to be characterized, either biochemically or genetically, and the relative contribution of each branch to total AsA in plant tissues, and the mechanisms behind AsA homeostasis are largely unknown. Mutant analysis and transgenic studies in *Arabidopsis thaliana* and other model systems have provided important insight into the regulation, activities, integration, and evolution of individual enzymes and are already providing a knowledge base for breeding and transgenic approaches to modify the level of vitamin C in agricultural crops

Keywords: vitamin C, ascorbic acid, metabolic engineering, transgenic plants

R. Verpoorte et al. (eds.), Applications of Plant Metabolic Engineering, 197–217.
© 2007 *Springer.*

1. INTRODUCTION

Vitamin C (L-ascorbic acid, AsA) is the trivial name for the six-carbon sugar derivative L-threo-hex-2-enono-1,4-lactone. Ascorbate is synthesized from D-glucose in the liver or the kidney of most animals, except humans and non-human primates. These species do not have L-gulono-1,4-lactone oxidase (Figure 1), an enzyme essential for its synthesis (Nishikimi et al., 1994). Consequently, when humans do not ingest vitamin C in their diets, a deficiency state occurs with a wide spectrum of clinical manifestations. The clinical expression of vitamin C deficiency, scurvy, is a lethal condition unless appropriately treated. Thus in order to survive humans must ingest vitamin C, mainly from fresh fruits and produce.

Together with flavonoids, polyphenolics, and lipophilic compounds such as α-tocopherol (vitamin E), vitamin C contributes to the overall intake of antioxidants in the human diet. Increasing awareness, particularly in the developed world, of the impact of a diet rich in fresh fruits and vegetables on cancer, cardiovascular, and other diseases of longevity (lifestyle diseases) is driving a consumer-led demand for increased food choice in relation to its nutritional value, of which food antioxidant capacity is a major component. This has been one of the driving forces of the rapid expansion of the worldwide markets of dietary supplements and functional foods, with estimated total sales of $63.3 billion and $71.9, respectively, in 2004 (Hancock and Viola, 2005; Bagchi, 2006).

In plants, vitamin C has roles in a wide range of processes, including antioxidant defense, photosynthesis, cell division, growth regulation, and senescence (Smirnoff, 1996; Noctor and Foyer, 1998; Smirnoff and Wheeler, 2000; Davey et al., 2000; Smirnoff et al., 2001; Conklin and Barth, 2004; Puppo et al., 2004; Pavet et al., 2005). Additionally, ascorbate is a cofactor for enzymes, and affects the expression of genes involved in defense and hormone signaling pathways (Pastori et al., 2003; De Tullio and Arrigoni, 2004).

Recent scientific advances in understanding vitamin C biosynthesis in plants are providing unparalleled opportunities for the development of crops with enhanced vitamin C levels. These advances are summarized here.

2. CHEMISTRY OF ASCORBIC ACID AND ITS ROLE IN HUMAN HEALTH AND PLANT METABOLISM

2.1. Chemistry

Structurally ascorbic acid is related to the C_6 sugars (Figure 2), and contains an enediol group on carbons 2 and 3. Delocalization of the π-electrons over the C2-C3 conjugated enediol system stabilizes the molecule and causes the hydrogen of the C3 hydroxyl to become highly acidic, and to dissociate with a pKa of 4.13. Therefore at physiological pH, AsA exists as a monovalent anion (L-ascorbate). AsA is stable when dry, but solutions readily oxidize. The first oxidation product of AsA is the radical monodehydroascorbate (MDHA). *In vivo* MDHA is reduced back to AsA by the activity of the NADP-dependent enzyme, mono-dehydroascorbate reductase (MDHAR, EC 1.6.5.4), or by electron transfer reactions. If allowed to

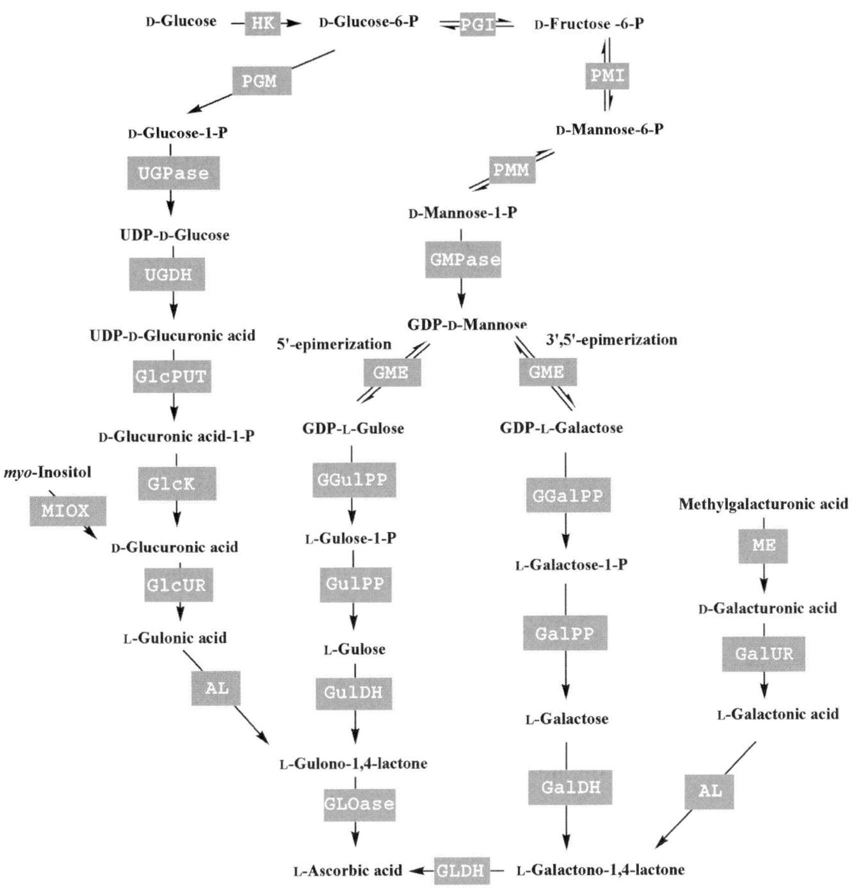

Figure 1. Proposed biosynthetic pathways of ascorbic acid (AsA) in animals and plants. In animals, vitamin C biosynthesis proceeds via D-glucuronic, and L-gulonic acids. In plants four AsA routes have been proposed. The Smirnoff-Wheeler pathway uses D-mannose and L-galactose as key intermediates. It was recently demonstrated that in addition to formation of GDP-L-galactose, GDP- D-mannose-3',5'-epimerase (GME, EC 5.1.3.18) can produce GDP-L-gulose. This observation led to proposals for al alternative route (L-gulose pathway) in which L-gulose and L-gulono-1,4-lactone are key intermediates. An alternative biogenesis for L-gulono-1,4-lactone (*myo*-inositol pathway) was recently proposed following the isolation of an Arabidopsis gene encoding *myo*-inositol oxygenase. A forth pathway operating in plants in which D-galacturonic acid is the main intermediate is also shown. Enzymes catalyzing the reactions are: HK hexokinase (EC 2.7.1.1), PGI phosphogluco isomerase (EC 5.3.1.9), PMI phosphomanno isomerase (EC. 5.3.1.8), PMM phosphomanno mutase (EC 5.4.2.8), GMPase GDP-D-mannose pyrophosphorylase (EC 2.7.7.22), GGalPP GDP-L-galactose pyrophosphatase, GalPP L-galactose-1-phosphate phosphatase, GalDH L-galactose dehydrogenase, GLDH L-galactono-1,4-lactone dehydrogenase (EC 1.3.2.3), GGulPP GDP-L-gulose-pyrophosphatase, GulPP L-gulose-1-phosphate phosphatase, GulDH L-gulose dehydrogenase, GLOase L-gulono-1,4-lactone oxidase (EC1.1.3.8), MIOX *myo*-inositol oxygenase (EC 1.13.99.1), GlcUR D-glucuronic acid reductase (EC 1.1.1.19), AL aldonolactonase (EC 3.1.1.17), ME methylesterase (EC 3.1.1.11), GalUR D-galacturonic acid reductase, PGM phosphogluco mutase (EC 5.4.2.2), UGPase UDP-glucose pyrophosphorylase (EC 2.7.7.9), UGDH UDP-glucose dehydrogenase (EC 1.1.1.22), GlcPUT glucuronate-1-phosphate uridyltransferase, GlcK glururonokinase (EC 2.7.1.43). Where EC numbers are omitted, they have not been assigned

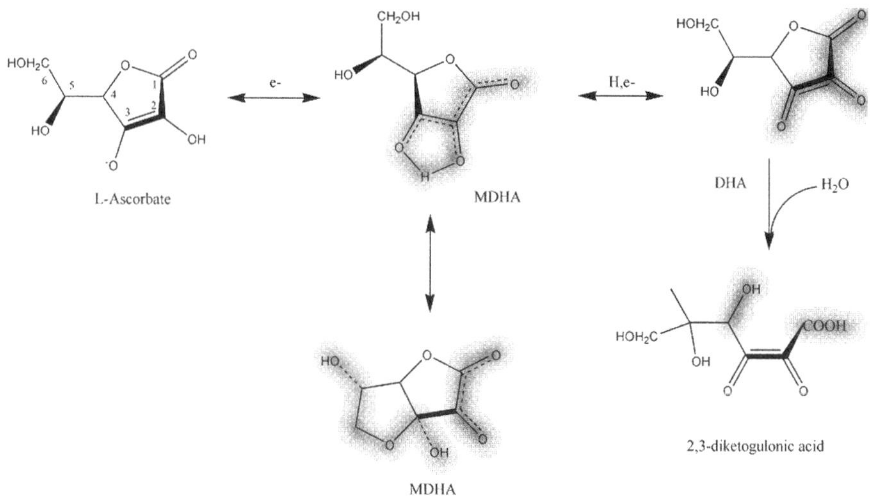

Figure 2. Oxidation of L-ascorbate

persist, two molecules of MDHA will also spontaneously disproportionate to AsA and dehydroascorbate (DHA). DHA itself is unstable and undergoes irreversible hydrolytic ring cleavage to 2,3-diketogulonic acid. DHA can be reduced to AsA in a reaction catalyzed by the enzyme DHA reductase (DHAR, EC 1.8.5.1) (reviewed in Davey et al., 2000).

2.2. Importance of AsA in Human Health

Humans, non-human primates, guinea pigs, bats, teleost fish, and some birds are unable to synthesize vitamin C and depend on dietary intake for survival (Nandi et al., 1997). In humans, deprivation of this nutrient eventually results in scurvy. Characteristics symptoms of this deficiency disease are: fatigue, bent or coiled body hair, loosening of teeth, hemorrhages, pain, and if left untreated, death. Most of these symptoms are directly attributable to a defect of the hydroxylation step during collagen synthesis. In humans, primates, and guinea pigs there is clear evidence indicating that the inability to synthesize vitamin C arose as a result of multiple mutations in the gene encoding L-gulono-1,4-lactone oxidase (GLOase; EC 1.1.3.8), the terminal enzyme of the "animal biosynthetic pathway" (Figure 1).

The current recommended dietary allowance (RDA) of vitamin C in the USA is 75 mg for women and 90 mg for men (Levine et al., 2001). Several populations have higher requirements of this vitamin. These include patients with periodontal disease, smokers (RDA 110 mg/day for women, and 125 mg/day for men), pregnant and lactating women (RDA 85 and 120 mg/day, respectively), and the elderly. High intake of the vitamin is generally well tolerated; however, a tolerable upper level was recently set at 2 g based on gastrointestinal upset that sometimes accompanies excessive dosages. In excess (>4 g/day), high levels of vitamin C and, significantly, one of its metabolites, oxalate, are found in urine and may increase the risk of renal-oxalate stone formation (Bsoul and Terezhalmy, 2004).

In addition to its antioxidant properties, vitamin C is needed for collagen synthesis, and may decrease the incidence of several diseases, such as dementia (Masaki et al., 2000), cancer (McDermott, 2000; Levi et al., 2001; Lee et al., 2003), stroke (Yokayama et al., 2000), heart disease (Rinne et al., 2000), atherosclerosis (Napoli et al., 2004), and Charcot-Marie-Tooth disease, a hereditary peripheral neuropathy (Passage et al., 2004). There is undergoing research exploring the usefulness of AsA as a potential memory-enhancer for the elderly (Parle and Dhinga, 2003), as an agent to accelerate healing/regeneration and wound repair (Catani et al., 2005), and as a pro-drug to deliver hydrogen peroxide and selectively kill cancer cells when administrated intravenously (Chen et al., 2005).

2.3. Importance of AsA in Plant Physiology

Ascorbate is the most abundant water soluble antioxidant in plant cells. It is found in most subcellular compartments, including the apoplast, and has an average cellular concentration of 2–25 mM (Horemans et al., 2000; Smirnoff, 2000a). No plant mutant completely devoid of AsA has ever been described, clearly indicating that plants unable to synthesize it are not viable. In general, three main types of biological activities can be defined for AsA in plant metabolism: its function as an enzyme cofactor, as a radical scavenger, and as a donor/acceptor in electron transport either at the plasma membrane or in the chloroplasts (Davey et al., 2000 and references therein). Additionally, AsA is a substrate for oxalate and tartrate biosynthesis, at least in certain plant species (reviewed in Loewus, 1999).

Ascorbate is cofactor or mono- and dioxygenases, which contain iron or copper at the active site. The function of AsA is to maintain the transition metal ion centers of these enzymes in a reduced form. A number of AsA-dependent dioxygenases have been described in plants. Among those are: 1-amino-cyclopropane carboxylate oxidase involved in the synthesis of ethylene, gibberellin 3-β-hydroxylase involved in the synthesis of gibberellins, and 9-*cis*-epoxy-carotein dioxygenase involved in the metabolism of absicic acid (Davey et al., 2000; Arrigoni and De Tullio, 2002; De Tullio and Arrigoni, 2004).

In both plant and animals AsA interacts enzymatically and non-enzymatically with damaging oxygen radicals and their derivatives, so-called reactive oxygen species (ROS). These detoxification reactions are an integral part of the housekeeping duties required for the aerobic existence of eukaryotic cells, and the high intracellular concentrations of AsA (up to 300 mM in the chloroplast stroma, Smirnoff, 2000a), are an indication of the importance of these functions in eukaryotic cells. In plants, the ability of AsA to interact with ROS (i.e. superoxide, singlet oxygen, hydrogen peroxide, and hydroxyl radical) implicates AsA in the modulation of processes such as lignification, cell division, cell elongation, the hypersensitive response, tolerance to stresses, and senescence (Smirnoff and

Wheeler, 2000; Conklin and Barth, 2004; Pavet et al., 2005). Unlike other low-molecular weight antioxidants (i.e. α-tocopherol, uric acid, carotenoids, flavonoids, etc), ascorbate is able to terminate radical chain reactions by disproportionation to non-toxic, non-radical products, i.e. DHA and 2,3-diketogulonic acid. Further, since AsA is only mildly electronegative, it can donate electrons to a wide range of substrates. One of the most important features of the non-enzymatic antioxidant activity of AsA is its involvement in the regeneration of vitamin E.

Ascorbate fulfills key functions in photosynthesis: it is co-substrate for violaxanthin de-epoxidase, an enzyme involved in the biosynthesis of zeaxanthin, a protoprotectan; it is a substrate for ascorbate peroxidase in the detoxification of H_2O_2, and it is an electron acceptor (as MDHA) for reduced ferredoxin in the photosynthetic electron transport chain, via the Mehler peroxidase reaction (reviewed in Smirnoff, 1996; Smirnoff and Wheeler, 2000). Under water-stress conditions, AsA becomes indispensable for the correct functioning of the antioxidant defenses in chloroplasts. In particular, its interplay with lipophilic antioxidants is essential for the protection of thylakoid membrane lipids from oxidation *in vivo* (Munné-Bosch and Alegre, 2002).

3. THE VITAMIN C METABOLIC NETWORK

3.1. The D-Mannose/L-Galactose Pathway

The *de novo* synthesis of AsA in animals was established over four decades ago, and utilizes intermediates of the D-glucuronic acid pathway (Figure 1). *In vivo*, the hexose skeleton is derived from D-glucose, and biosynthesis takes places either in the liver or the kidney. Early radiotracer studies using D-glucose labeled in either carbons C_1, C_2 or C_6 showed that the C_1 carbonyl group of AsA is mainly formed by oxidation of the C_6 carbon of D-glucose. Thus, in animals AsA follows an "inversion"-type conversion (Smirnoff et al., 2001 and references therein).

It became clear from radioisotopic labeling studies in the 1950s that in plants AsA biosynthesis does not proceed as it does in mammals, as the vast majority (~80%) of D-glucose is incorporated into AsA without inversion of the glucose skeleton. Prior to1998, the biosynthetic pathway of AsA in plants was not known. However, the characterization of the enzyme L-galactose dehydrogenase (Wheeler et al., 1998) and the identification of the gene encoding GDP-D-mannose pyrophosphorylase (Conklin et al., 1999), responsible for low AsA in the *vtc1-1* mutant (Conklin et al., 1996), enabled Smirnoff and Wheeler to propose a D-mannose/L-galactose (Man/Gal) pathway for AsA *de novo* biosynthesis (Figure 1, Wheeler et al., 1998).

In the Smirnoff-Wheeler pathway, D-mannose is converted to GDP-D-mannose (GDP-Man) and then epimerized to GDP-L-galactose. The galactose moiety is hydrolyzed via a two step process and L-galactose is reduced to its lactone and then to AsA. The early steps of this route are shared between a number of routes including cell wall polysaccharide formation and protein glycosilation (Smirnoff et al., 2001). No evidence is available directly correlating phosphomannose isomerase

(EC 5.3.1.8.) or phosphomannose mutase (EC 5.4.2.8.), with AsA biosynthesis and the activities of these enzymes are generally low in plant tissues (Smirnoff et al., 2001). GDP-Man is formed from D-mannose 1 phosphate and GTP, with a concomitant release of pyrophosphate PPi, in a reversible reaction catalyzed by a GDP-Man pyrophosphorylase (mannose 1-phosphate guanylyltransferase; EC 2.7.7.13). The Arabidopsis genome contains at least three closely related genes encoding GDP-Man pyrophosphorylase (At2g39770, At3g55590, and At4g30570; TAIR, 2006). Conklin and colleagues demonstrated that the enzyme encoded by At2g39770 is involved in the biosynthesis of AsA by showing that ozone-sensitive *vtc1* Arabidopsis mutants with a point mutation in this gene have reduced levels of both vitamin C (~70% reduction) and GDP-Man pyrophosphorylase activity (Conklin et al., 1999). Studies on the *cyt1* lines bearing other point mutations in the same gene demonstrated severe, pleiotropic effects of such mutations (Lukowitz et al., 2001). The *cyt1* mutants are deficient in cellulose, the biosynthesis of which does not require GDP-Man as a substrate, and show impaired protein glycosylation and lower content of mannose and fucose in the cell walls. The protein encoded by At2g39770, is therefore, involved in fundamental processes of protein N-glycosylation, cell-wall formation, and the synthesis of other GDP-L-fucose and GDP-L-galactose. Similarly, down-regulation of GDP-Man pyrophosphorylase activity in potato (*Solanum tuberosum*) via antisense expression of a homologous gene resulted in vitamin C deficiency, lower mannose content of cell-walls in leaves, and rapid senescence of potato plants (Keller et al., 1999).

The second step on the *de novo* synthesis of vitamin C, as originally proposed (Wheeler et al., 1998), is carried out by the GDP-Man 3',5'-epimerase (EC 5.1.3.18; At5g28840; Wolucka et al., 2001; Wolucka and Van Montagu, 2003) that catalyses a reversible conversion of GDP-D-mannose into GDP-L-galactose (Barber and Hebda, 1982). In the Man/Gal pathway, GDP-L-galactose would release L-galactose-1-phosphate by a GDP-L-galactose pyrophosphatase, enzymatic activity currently under study (DiMatteo et al., 2003; Hancock and Viola, 2005). L-Galactose-1-phosphate could undergo a deposphorylation due to the action of a recently identified phosphatase (At3g02870) that cleaves also *myo*-inositol 1 phosphate (Laing et al., 2004). Free L-galactose can then be oxidized at the C_1-position by a cytosolic L-galactose dehydrogenase (GalDH; At4g33670; Gatzek et al., 2002) to L-galactono-1,4-lactone. The later is oxidized to L-ascorbic acid by the highly specific mitochondrial inner membrane-associated L-galactono-1,4-lactone dehydrogenase (GLDH; EC 1.3.2.3; At3g47930; Oba et al., 1994; Østergaard et al., 1997; Imai et al., 1998; Siendones et al., 1999).

3.2. The L-Gulose Pathway

Biochemical studies on the cytosolic GDP-mannose 3',5'-epimerase of Arabidopsis led to the discovery of an unsuspected and novel 5'-epimerase activity responsible for the synthesis of GDP-L-gulose (Wolucka and Van Montagu, 2003). In contrast to GDP-L-galactose which serves as a donor of L-galactosyl residues for the biosynthesis of polysaccharides and glycoproteins, the presence of GDP-L-gulose is puzzling because L-gulose-containing glyco-conjugates have never been found in higher plants. The authors proposed that L-gulose serves as substrate of AsA synthesis. In this pathway L-gulose freed from GDP-L-gulose undergoes an oxidation to L-gulono-1,4-lactone by the action of the L-galactose dehydrogenase or a similar enzyme. L-gulono-1,4-lactone is not a substrate for the highly specific L-galactono-1,4-lactone dehydrogenase, and must be converted to AsA by the L-gulono-1,4-lactone oxidase (GLOase) activity detectable in the cytosol and the mitochondria (Wolucka and Van Montagu, 2003). The Arabidopsis genome contains several genes that are homologous to the rat GLOase (At1g32300, At2g46740, At2g46750, At2g46760, At5g11540, At5g56470, and At5g56490; TAIR, 2006), but the gene products have not been characterized yet.

Consistent with the key role of L-gulono-1,4-lactone in AsA biosynthesis, transgenic tobacco (*Nicotiana tobacum* var. *Xanthi*) and lettuce (*Lactuca sativa*) plants expressing a rat cDNA encoding GLOase had 4–7 fold increase in AsA levels (Jain and Nessler, 2000). Furthermore, expression of *GLOase* in wild-type (wt) Arabidopsis plants showed up to a 2-fold increase in AsA leaf content compared to controls, and all five *vtc* mutant lines expressing *GLOase* had a rescued AsA leaf content equal or higher (up to 3-fold) than wt as well as a restored phenotype (Radzio et al., 2003). These data and the current knowledge about the identity of the gene(s) mutated in the *vtc* lines indicates that an alternative pathway is present in plants, which can bypass the deficiency of GDP-mannose production of the *vtc1-1* mutant and possibly circumvent other steps in the Man/Gal pathway to synthesize AsA. Possible routes that may serve as sources of intermediates for GLOase in wt and *vtc* Arabidopsis lines include: the Man/Gal pathway, an undiscovered "animal pathway," and recently described branch pathways (i.e. the galacturonic acid and *myo*-inositol routes, see sections 3.3 and 3.4). In favor of the idea of an animal pathway in plants, recent metabolic profiling analysis has shown the presence of both galactonic and gulonic acids in Arabidopsis (Wagner et al., 2003).

3.3. The D-Galacturonic Acid Pathway

Feeding experiments using precursors have shown that the methyl ester of D-galacturonic acid causes a significant increase in the AsA content of cress seedlings and *Arabidopsis* culture cells (Isherwood et al., 1954; Davey et al., 1999). It is also known that D-galacturonic acid 1-^{14}C is metabolized to L-ascorbic acid-6-^{14}C by an inversion pathway in detached ripening strawberry (*Fragaria x ananassa* Duch.) fruit (Loewus and Kelly, 1961). A "uronic acid pathway" that accommodates both findings, the role played by the methyl ester of D-galacturonic acid as precursor and the occurrence of an inversion pathway in some plant tissues was previously proposed (Smirnoff et al., 2001). Molecular evidence of this pathway came with the cloning and characterization of a D-galacturonic acid reductase from strawberry fruit (Agius et al., 2003; Valpuesta and Botella, 2004). In this route pectin-derived D-galacturonic acid (GalUA) is reduced to L-galactonic acid, which in turn is spontaneously converted to L-galactono-1,4-lactone. This compound is the substrate of the terminal enzyme of the Man/Gal pathway (Figure 1). There are two homologous of the strawberry GalUA reductase in the Arabidopsis genome: At1g59950 and At1g59960 (TAIR, 2006).Whether the GalUA pathway operates exclusively on ripening fruits or if it is active in other organs/tissues remains unknown. However, it is important to note that AsA content has been reported to increase with ripening in other crops such as pepper (*Capsicum annuum*; Martinez et al., 2005). The current scanty evidence indicates that some protists (i.e. Chrysophyta, Bacillariophyta, and Euglenophyta) use the GalUA pathway to make AsA (Smirnoff et al., 2001).

3.4. The *Myo*-Inositol Pathway

Our group has obtained biochemical and molecular data indicating that *myo*-inositol (MI) can also serve as precursor for AsA biosynthesis in Arabidopsis (Figure 1; Lorence et al., 2004a). This MI route involves at least four enzymatic steps: 1) the oxidation of MI to glucuronic acid (GlcUA), 2) the reduction of GlcUA to L-gulonic acid, 3) the conversion of L-gulonic acid to L-gulono-1,4-lactone, and 4) the oxidation of this lactone to AsA. These reactions are catalyzed by *myo*-inositol oxygenase, GlcUA reductase, gluconolactonase, and L-gulono-1,4-lactone oxidase, respectively.

A bioinformatics approach was used to search for homologues of the pig *myo*-inositol oxygenase (MIOX, EC 1.13.99.1) in the Arabidopsis genome. Four homologs were found and named according to the chromosome where they are located as: *MIOX1* (At1g14520), *MIOX2* (At2g19800), *MIOX4* (At4g26260) and *MIOX5* (At5g56640). The cDNA encoding MIOX4 was cloned, expressed in bacteria, and demonstrated to have high specific activity. Constitutive expression of *MIOX4* in homozygous transgenic Arabidopsis lines resulted in up to 3-fold enhancement of AsA leaf content (Lorence et al., 2004a). Northern-blots (Lorence et al., 2004a) and RT-PCR analysis (unpublished results) have shown that *MIOX4* is highly expressed in flowers, and to a lesser extent in leaves.

Analysis of the protein structure showed that MIOX4/MIOX5 and MIOX1/MIOX2 are closely related. We have cloned all ORFs and confirmed the enzymatic activity of a truncated MIOX2, containing the catalytic domain (called "MIOX domain") that is common to all members, in bacterially expressed recombinant protein. Transgenic Arabidopsis lines expressing the "MIOX domain" have up to 3-fold increased AsA levels in leaves. RT-PCR experiments performed with RNA extracted from 5-week old plants showed that *MIOX1* is predominantly expressed in cauline leaves and siliques, and *MIOX5* in flowers, while *MIOX2* is highly expressed in all tissues. Assays performed with homozygous T-DNA knockout lines of *MIOX1*, 2, 4 and 5 under different light conditions revealed that *MIOX1* is the member of the family that contributes the most to the AsA content of leaf tissue in Arabidopsis. Silencing of all *MIOX* genes caused an arrest of growth in the cotyledonary stage of the T1 generation plants transformed with an *MIOX*

Domain:RNAi construct indicating the importance of the *MIOX* gene family not only for AsA biosynthesis, but also to normal growth and development (Lorence et al., 2004b). This result may be explained in part with recent evidence indicating the role of MIOX proteins in the synthesis of cell-wall precursors (Kanter et al., 2005).

MIOX is an enzyme containing non-heme iron that catalyses a four electron oxidation with the transfer of only one atom of oxygen into the product D-glucuronic acid (GlcUA). There are only two additional enzymatic steps necessary for the conversion of GlcUA to AsA. The first of those reactions is the oxidation of GlcUA to L-gulonic acid. This conversion is catalyzed by a GlcUA reductase (EC 1.1.1.19), also called gulonate dehydrogenase or L-hexonate dehydrogenase. GlcUA reductase belongs to the family of aldo/keto reductases and has been extensively studied in mammals; however, there are no reports of its enzymatic activity in plants. Our database search revealed no plant enzymes annotated as GlcUA reductase (or any of its synonyms). We expanded our analysis and found that there are close to 40 putative members of the aldo/keto reductase family in the Arabidopsis genome. Screening of T-DNA knockout lines using HPLC and spectrophotometric-based assays of some of the members of the aldo/keto reductase family allowed us to identify the SALK line 119576, which had a substantial reduction (~50%) of AsA leaf content compared to wild type plants. This line has a T-DNA inserted in ORF At2g37770. We have cloned this ORF from leaf cDNA. Experiments are in progress to test the enzymatic activity of the recombinant protein in bacteria and to examine AsA biosynthesis in transgenic Arabidopsis plants over-expressing this gene (Lorence et al., 2004c).

4. REGULATION OF THE ASA METABOLIC NETWORK

There is little information available concerning the regulation of AsA pool size in plants, which is not surprising considering the relatively recent elucidation of the biosynthetic network. In Arabidopsis foliar tissue, AsA content is developmentally controlled, being highest in young leaves and declining with age (Conklin et al., 1996; Tamaoki et al., 2003). The molecular or biochemical basis for this change is not known. Environmental conditions that increase endogenous ROS, such as ozone, drought and high temperatures, also raise AsA levels (Noctor and Foyer, 1998; Smirnoff and Wheeler, 2000; Panchuk et al., 2002; Sircelj et al., 2005). High light intensity increased AsA content in rosette leaves (Mishra et al., 1993; Grace and Logan, 1996; Smiroff and Pallanca, 1996; Gatzek et al., 2002; Tabata et al., 2002; Tamaoki et al., 2003) and this may be related to greater activity of the chloroplast ascorbate-glutathione cycle necessary for protection of the photosystem II from photoinhibition (Noctor and Foyer, 1998). The mechanism involved may be some combination of reduced catabolism of dehydroascorbate (DHA) and increased synthesis of AsA (Pallanca and Smirnoff, 2000; Smirnoff et al., 2001), although turnover studies have not yet been conducted to resolve the role of these two metabolic components in the light response. Over-expression of a DHA reductase which converts DHA to AsA, increased leaf AsA content by 2 to 4-fold in maize and tobacco, demonstrating that reduced degradation of DHA can raise the total pool substantially (Chen et al., 2003). In contrast to high light, prolonged darkness lowered AsA content, a process that could be reversed by glucose addition (Smirnoff and Pallanca, 1996). Biotic stresses such as infection with tobacco mosaic virus (Milo and Santilli, 1967), nematodes (Arrigoni et al., 1979) and nitrogen-fixating bacteria (Dalton et al., 1998) have been reported to increase AsA levels as well.

Feeding experiments have afforded some insight into possible sites of enzymatic control of biosynthesis in the mannose pathway. In non-photosynthetic tissue, exogenous AsA increased total content 5-8 fold, but reduced ^{14}C glucose incorporation into AsA by 59 to 83%, indicating feedback inhibition of synthesis by high pool size (Pallanca and Smirnoff, 2000). The terminal enzyme in the Smirnoff-Wheeler pathway, L-galactono-1,4-lactone dehydrogenase (GLDH), is not subjected to feedback modulation, since addition of L-galactono-1,4-lactone did not lessen AsA biosynthesis (Baig et al., 1970). However, light has been found to modulate the expression of the Arabidopsis *GLDH* gene as well as the protein level (Tamaoki et al., 2003). Addition of L-galactose resulted in a large increase in AsA demonstrating that supply of this substrate is limiting to synthesis and that L-galactose dehydrogenase, the penultimate enzyme, is not affected by product concentration (Wheeler et al., 1998). In contrast, the L-galactose dehydrogenase from spinach (*Spinacia oleraceae*) was found to be inhibited by the final product, suggesting feedback regulation of AsA at least in some plant species (Mieda et al., 2004). Feeding of D-mannose did not alter AsA pool size indicating that control lies between D-mannose and L-galactose. Antisense expression of GDP-D-mannose pyrophosphorylase in potato reduced AsA levels (Keller et al., 1999), suggesting that some regulation may be operating at this enzymatic level. Antisense suppression of both L-galactose dehydrogenase (Gatzek et al., 2002) and L-galactono-1,4-lactone dehydrogenase (Tabata et al., 2001) lowered AsA content only 30% and overexpression of L-galactose dehydrogenase in tobacco did not change AsA levels (Gatzek et al., 2002), indicating that these enzymes are not involved in regulation.

Both GDP-D-mannose and GDP-L-galactose feed into cell wall biosynthesis and protein glycosylation as well as AsA synthesis. The committed steps to the AsA pathway are the conversion of GDP-L-galactose to L-galactose-1-phosphate (L-Gal-1-P) and the hydrolysis of the latter compound to L-galactose. A highly specific L-Gal-1-P phosphatase from young kiwifruit (*Actinia deliciosa*) berries was recently characterized (Laing et al., 2004). Its homolog in Arabidopsis is the protein encoded by At3g02870. Interestingly, the kiwifruit enzyme was able to use not only L-Gal-1-P as a substrate, but also *myo*-inositol-1-P. Based on this result and on a phylogenetic analysis of all *myo*-inositol 1 phosphatases, the authors propose that the kiwifruit L-Gal-1-P phosphatase arose from a gene duplication and specialization from a *myo*-inositol-1-P-specific ancestor (Laing et al., 2004).

The branch mannose pathway, converting GDP-D-mannose to GDP-L-gulose, diverts mannose to AsA and appears to operate under stress conditions (Wolucka and Van Montagu, 2003). GDP-D-mannose-3',5'-epimerase (GME) has the capability to catalyze two distinct epimerizations, producing either GDP-L-galactose or GDP-L-gulose. The native enzyme, or a bacterial recombinant form, co-purified with a HSP70 heat-shock protein that may alter protein configuration and favor 5'-epimerase (GDP-L-gulose) over 3',5'-epimerase (GDP-L-galactose) activity (Wolucka and Van Montagu, 2003). In heat-stressed Arabidopsis leaves, AsA concentration increased (Panchuk et al., 2002), but no characterization of HSP70 protein content was made. GME activity is partially inhibited by AsA, L-galactono-1,4-lactone, NADH/NADPH and GDP-L-fucose and stimulated by NAD^+/$NADP^+$ (Wolucka and Van Montagu, 2003). It appears that this enzyme has the capability to respond to cell redox status, downstream product(s) concentration and stress conditions to regulate the flux of D-mannose to cell wall/glycoprotein or AsA synthesis. Structural analysis coupled to site-directed mutagenesis on the Arabidopsis GME enzyme suggest that C145 and K217 as the acid/base pair responsible for both epimerization reactions (Major et al., 2005).

The *GME* gene from rice was recently cloned (GenBank Accession No. ABI193582) and the enzyme characterized (Watanabe et al., 2006). The reaction products from GDP-D-mannose, as produced by the rice GME were also GDP-L-galactose and GDP-L-gulose. The reaction of the rice GME was inhibited by GDP, and strongly accelerated by NAD^+. The enhancement produced by NAD^+ is most likely due to a NAD binding domain present in the rice gene, but absent in the corresponding gene from Arabidopsis (Watanabe et al., 2006).

By using ^{14}C-mannose radiolabeling combined with HPLC and transcript profiling analysis, Wolucka and collaborators (2005) have shown that methyl jasmonate treatment increases the *de novo* synthesis of AsA in Arabidopsis and tobacco Bright Yellow-2 (BY-2) suspension cells. In BY-2 cells, this stimulation coincides with enhanced transcription of the genes encoding GDP-mannose 3',5'-epimerase (At5g28840) and a putative L-gulono-1,4-lactone oxidase (At2g46750). A similar effect of jasmonates as enhancers of the transcription of genes involved in the regeneration of AsA has been reported (Nishikawa et al., 2003).

A comprehensive analysis of jasmonate-regulated metabolic pathways in Arabidopsis using cDNA macroarrays was recently carried out (Sasaki-Seminoto et al., 2005). The results showed that jasmonates (JA) activate the coordinated expression of genes involved in nine metabolic pathways belonging to two functionally related groups: 1) AsA and glutathione pathways, which are important in defense responses to oxidative stress, and 2) biosynthesis of indole glucosinolate, a defense compound occurring in the Brassicaceae family. According to microarray data and Northern blot analysis, the transcripts involved in AsA synthesis/recycling that are upregulated by JA are: dehydroascorbate reductase (DHAR; At1g19570), monodehydroascorbate reductase (MDHA; At3g09940), GDP-mannose pyrophosphorylase (GMPase; *vtc1*; At2g39770), the gene mutated in the vitamin C deficient mutant *vtc2*-1 (*vtc2*, At4g26850) and a *vtc2* homologue (At5g55120). The authors confirmed that JA induced the accumulation of AsA, glutathione and cysteine, and increased the activity of DHAR. Ozone (O_3) exposure caused an induction of several genes involved in antioxidant metabolism in wt Arabidopsis. However, in JA-deficient 12-oxophytodienoate reductase (*opr3*) mutants, the induction of antioxidant genes was abolished. The *opr3* mutants were also more sensitive to O_3 exposure. These results suggest that coordinate activation of metabolic pathways mediated by JA provide resistance to environmental stresses.

It is evident from the previous paragraphs that our understanding of the regulation of the vitamin C metabolic network is still on his infancy. Most of the data accumulated up to date relates to the regulation of the Man/Gal and L-gulose pathways. Modulation of the *myo*-inositol and galacturonate routes it is still largely unknown.

A recurrent pattern however it is the fact that interfering with specific genes in the network results in abnormal development: the *vtc1-1* mutant grows slowly, is small and senesces more rapidly than wild type (Veljovic-Jovanovic et al., 2001; Barth et al., 2004), BY-2 cells transformed with an antisense construct of GLDH are abnormal in both phenotype and structure (Tabata et al., 2001), a T-DNA knockout of the gene encoding GDP-D-mannose pyrophosphorylase is embryo lethal (Smirnoff, 2000b), and an RNAi construct to the conserved region of all four *myo*-inositol oxygenase genes arrests seedling development at the cotyledonary stage (Lorence et al., 2004b). To more fully understand AsA biosynthesis in order to manipulate plant stress resistance and to increase crop nutritional value for humans, is it necessary to identify and characterize regulatory components in the pathways, and to determine the flow of intermediates through each one of the branches of the network.

Studies on AsA transport are crucial to associate the different AsA pathways to specific plant organs and/or developmental stages. Several authors have reported the presence of AsA in plant phloem (Franceschi and Tarlyn, 2002; Hancock et al., 2003), and transport of AsA from source leaves to root tips, shoots, flowers and tubers (Franceschi and Tarlyn, 2002; Tedone et al., 2004). In addition to long-distance transport, AsA translocation through different cell membranes is relevant to understand the interplay between the biosynthetic routes. However, the current knowledge in this area is very limited.

5. MANIPULATING VITAMIN C LEVELS IN PLANTS

The positive correlation between the activities of L-galactose dehydrogenase (GalDH) and L-galactono-1,4-lactone dehydrogenase (GLDH), the two terminal enzymes of the Man/Gal pathway and AsA levels in plant tissues found by several groups (Oba et al., 1994; Tabata et al., 2001; Tabata et al., 2002; Tamaoki et al., 2003) led to efforts targeting these enzymes for manipulation of vitamin C biosynthesis in plants. However, as shown in Table 1, most attempts to enhance AsA levels in plants via manipulation of GalDH and GLDH have produced little or no success. Interestingly, although BY-2 tobacco cells expressing the tobacco GLDH showed only a small increase in AsA levels, they were able to live longer in culture and became more tolerant to paraquat (methyl viologen) when compared with untransformed cells (Tokunaga et al., 2005).

Table 1. Strategies to enhance vitamin C levels in plants

Enzyme	Abbreviation	Gene source	Plant(s) transformed	AsA fold increase	Reference
Man/Gal pathway					
L-galactose dehydrogenase	GalDH	Arabidopsis	Tobacco	None	Gatzek et al., 2002
L-galactono-1,4-lactone dehydrogenase	GLDH	Cauliflower	Tobacco	0.3	Smirnoff et al., 2001
		Tobacco	BY-2 tobacco cells	2	Tokunaga et al., 2005
***myo*-Inositol pathway**					
myo-Inositol oxigenase	MIOX4	Arabidopsis	Arabidopsis	2–3	Lorence et al., 2004a
	MIOX4	Arabidopsis	Tobacco and lettuce	2–4	Lorence and Nessler, unpublished results
	MIOX Domain	Arabidopsis	Arabidopsis	2–3	Lorence et al., 2004b
L-gulono-1,4-lactone oxidase	GLOase	Rat	Lettuce and tobacco	4–7	Jain and Nessler, 2000
			Arabidopsis wild type and all *vtc* mutants	2–3	Radzio et al., 2003
L-galacturonic acid pathway					
L-galacturonic acid reductase	GalUR	Strawberry	Arabidopsis	2–3	Agius et al., 2004
AsA recycling					
Dehydroascorbate reductase	DHAR	Wheat	Tobacco and maize	2–4	Chen et al., 2003
		Rice	Arabidopsis	0.25	Ushimaru et al., 2005
Other genes not in the network					
Malate dehydrogenase	mMDH (antisense)	Tomato	Tomato	2–3	Nunes-Nesi et al., 2005

The first successful strategy to elevate the vitamin C content of plants was published by our group in 2000 (Jain and Nessler). A 4 to 7-fold increase in the AsA level of leaf tissue was obtained in three varieties of lettuce and one cultivar of tobacco by expressing the terminal enzyme of the "animal" pathway, L-gulono-1,4-lactone oxidase. The promiscuity of GLOase (reviewed in Smirnoff, et al., 2001) did not allow us to conclude whether the enzyme was using L-galactono-1,4-lactone, the precursor for AsA in the Man/Gal pathway, or the plants were producing the L-gulono-1,4-lactone. Evidence indicating the operation in plants of a pathway that resembles the animal pathway was obtained when we expressed GLOase on the *vtc1-1* mutant where the flow of intermediates of the Man/Gal route is diminished (Radzio et al., 2003). Another successful strategy our group has developed to increase vitamin C levels of different plant models has been the cloning and over-expression of members of the *myo*-inositol oxygenase gene family (Table 1, Lorence et al., 2004a and 2004b). These results have allowed us to propose a new pathway leading to vitamin C formation starting with *myo*-inositol as a precursor (Lorence et al., 2004a).

Agius et al. (2003) provided molecular evidence of a D-galacturonic pathway to AsA by expressing a strawberry L-galacturonic acid reductase in Arabidopsis. On the other hand, Chen et al. (2003) observed up to 4-fold increase in leaf AsA content in tobacco and maize as a result of over-expressing a wheat DHA reductase, and this accompanied by an increase in the AsA redox ratio (AsA:DHA) from 1.5 to 4.0. In both tobacco and maize, the level of glutathione was doubled, suggesting coordination between these two antioxidants. In a recent report, expression of a rice DHA reductase have failed at enhancing vitamin C levels in Arabidopsis however, the transgenics showed enhanced tolerance to salt stress (Ushimaru et al., 2005).

It is interesting to note that in the case of Arabidopsis, different strategies seems to coincide in a 2- to 3-fold increase of the leaf vitamin C content (Table 1), possibly reflecting the action of a feedback or some other regulatory mechanism that keeps a constant pool of AsA. This type of regulatory mechanism has been suggested previously in a turnover study performed with pea seedlings (Pallanca and Smirnoff, 2000). A degradation pathway for AsA via 4-O-oxalyl-L-threonate has been recently proposed (Green and Fry, 2005). The possible involvement of AsA catabolism as a regulatory mechanism of the AsA pool in transgenic plants with enhanced levels of this antioxidant has not been explored up to date.

Surprisingly transgenic tomato (*Lycopersicum esculentum* = *Solanum lycopersisum*) plants expressing a fragment of the mitochondrial malate dehydrogenase gene in antisense orientation and exhibiting reduced activity of this enzyme have increased levels of AsA and enhanced photosynthetic activity and aerial growth under atmospheric conditions (Nunes-Nesi et al., 2005). Several lines of evidence highlight the role of AsA in photosynthesis. AsA acts in the Mehler peroxidase reaction with ascorbate peroxidase to regulate the redox state of photosynthetic electron carriers and as a cofactor for violaxanthin de-epoxidase, an enzyme involved in xanthopyll cycle-mediated photoprotection (Smirnoff and Wheeler, 2000; Danna et al., 2003). The AsA-redox state has also recently been shown to control guard cell signaling and stomatal movement (Chen and Gallie, 2004) as well as the expression levels of both nuclear and chloroplastic components of the photosynthetic apparatus (Smirnoff, 2000b; Kiddle et al., 2003; Pastori et al., 2003). An additional function of AsA is that of cofactor of prolyl hydroxylase, an enzyme that post-translationally hydroxylates proline residues in cell-wall glycoproteins required for cell division and expansion (Smirnoff and Wheeler, 2000). Thus, when these functions are considered together with the phenotype of the malate dehydrogenase antisense plants, it is apparent that AsA represents a strong link between the major pathways of energy in leaf tissue. In support of the role of AsA for better growth of plants, vitamin C deficient (*vtc*) mutants transformed with the rat *GLOase* cDNA showed restored levels of AsA as well as better growth compared to untransformed controls and wt Arabidopsis plants (Radzio et al., 2003).

6. WHERE TO FROM HERE?

Our understanding of plant AsA biochemistry has advanced considerably in recent years. Several genes involved in AsA biosynthesis and recycling have been cloned, and transgenic plants have been generated that contain enhanced levels of this vitamin. It remains to be seen whether the improvements observed in model systems such as Arabidopsis and tobacco can be translated into viable strategies for enhancing vitamin C content of nutritionally important crops. It is expected that vitamin C-enhanced crops may have added benefits such an extended shelf life, and increased stress tolerance to soil salinity, drought, heavy metals, and pollution.

The narrow genetic basis of many crops combined with restrictions on the commercial use of genetically modified plants, has led to a surge of interest in exploring natural biodiversity as a source of novel alleles to improve the quality and nutritional value of crops (Fernie et al., 2006).Quantitative trait loci (QTL) and other mapping techniques could be a valuable tool to identify new alleles to improve vitamin C content in crops. In a recent study a high AsA QTL was identified by studying a tomato introgression line population that combines single chromosomal segments introgressed from the wild, green fruited species *Lycopersicon pennellii* in the background of the domesticated tomato, *Lycopersicon esculentum* (Rousseaux et al., 2005). Fourteen genes associated with AsA biosynthesis and recycling were recently mapped in tomato (Zou et al., 2006). This knowledge could provide the future basis for the development of new crops with enhanced vitamin C content.

ACKNOWLEDGEMENTS

Work at CN Laboratory was supported by the Interagency Metabolic Engineering Program (National Science Foundation – Metabolic Biochemistry and Integrative Plant Biology, IBN118612) and USDA/CREES 2002-3S321-11600). AL thanks the Arkansas Biosciences Institute at Arkansas State University for start-up funding for her laboratory. We thank Scott Simeon for help with art work.

REFERENCES

Agius, F, Gonzalez-Lamonthe, R, Caballero, JL, Muñoz-Blanco, J, Botella, M.A, and Valpuesta, M, 2003, Engineering increased vitamin C levels in plants by over-expression of a D-galacturonic acid reductase. *Nat Biotechnol* 21: 177–181.

Arrigoni, O, and De Tullio, MC, 2002, Ascorbic acid: much more than just an antioxidant, *Biochim Biophys Acta* **1569**: 1–9.

Arrigoni, O, Zacheo, G, Arrigoni-Liso, R, Bleve-Zacheo, T, and Lamberte, F, 1979, Relationship between ascorbic acid and resistance in tomato plants to *Meloidogyne incognita*, *Phytopathol* **69**: 579–581.

Bagchi, D, 2006, Nutraceuticals and functional foods regulations in the United States and around the world, *Toxicology* **221**: 1–3.

Baig, MM, Kelly, S, and Loewus, F, 1970, L-Ascorbic acid biosynthesis in higher plants from L-gulono-1,4-lactone and L-galactono-1,4-lactone, *Plant Physiol* **46**: 277–280.

Barth, C, Moeder, W, Klessig, DF, and Conklin, PL, 2004, The timing of senescence and response to pathogens is altered in the ascorbate-deficient Arabidopsis mutant *vitamin c-1*, *Plant Physiol* **134**: 1784–1792.

Barber, GA, and Hebda, PA, 1982, GDP-D-mannose: GDP-L-galactose epimerase from *Chlorella pyrenoidosa*, *Methods Enzymol* **83**: 522–525.

Bsoul, SA, and Terezhalmy, GT, 2004, Vitamin C in health and disease, *J Contemp Dent Pract* **2**: 1–13.

Chen, Z, and Gallie, DR, 2004, The ascorbic acid redox state controls guard cell signaling and stomatal movement, *Plant Cell* **16**: 1143-1162.

Chen, Z, and Gallie, DR, 2005, Increasing tolerance to ozone by elevating foliar ascorbic acid confers greater protection against ozone than increasing avoidance, *Plant Physiol* **138**: 1673–1689.

Chen, Z, Young, TE, Linj, J, Chang, SC, and Gallie, DR, 2003, Increasing vitamin C content of plants through enhanced ascorbate recycling, *Proc Natl Acad Sci USA* **100**: 3525–3530.

Chen, Q, Espey, MG, Krishna, MCy, Mitchell, JB, Corpe, CP, Buettner, GR, Shacter, E, and Levine, M, 2005, Pharmacologic ascorbic acid concentrations selectively kill cancer cells: action as a pro-drug to deliver hydrogen peroxide to tissues, *Proc Natl Acad Sci USA* **102**: 13604–13609.

Conklin, PL, and Barth, C, 2004, Ascorbic acid, a familiar small molecule intertwined in the response of plants to ozone, pathogens, and the onset of senescence, *Plant Cell Environ* **27**: 959–970.

Conklin, PL, Williams, EH, Last, RL, 1996, Environmental stress sensitivity of an ascorbic acid-deficient Arabidopsis mutant, *Proc Natl Acad Sci USA* **3**: 9970–9974.

Conklin, PL, Norris, SR, Wheeler, GL, Williams, EH, Smirnoff, N, and Last, RL, 1999, Genetic evidence for the role of GDP-mannose in plant ascorbic acid (vitamin C) biosynthesis, *Proc Natl Acad Sci USA* **96**: 4198–4203.

Dalton, DA, Joyner, SL, Becana, M, Iturbe-Ormaetxe I, and Chatfield, JM, 1998, Antioxidant defenses in the peripheral cell layers of legume root nodules, *Plant Physiol* **116**: 37–43.

Danna, CH, Bartoli, CG, Sacco, Fy, Ingala, LR, Santa-Maria, GE, Guiamet, JJ, and Ugalde, RA, 2003, Thylakoid bound ascorbate peroxidase mutant exhibits impaired electron transport and photosynthetic activity, *Plant Physiol* **132**: 2116–2125

Davey, W, Gilot, Cy, Persiau, G.y, Østergaard, J, Han, Y, and Van Montagu, M, 1999, Ascorbate biosynthesis in Arabidopsis cell suspension culture, *Plant Physiol* **121**: 535–543.

Davey, MW, Van Montagu, M, Inzé, D.y, Sanmartin, M.y, Kanellis, A, Smirnoff, N, Benzie, IJJ, Strain, JJ, Favell, D, and Fletcher, J, 2000, Plant L-ascorbic acid: chemistry, function, metabolism, bioavailability and effects of processing, *J Sci Food Agric* **80**: 825–860.

De Tullio, MC, and Arrigoni, O, 2004, Hopes, disillusions and more hopes from vitamin C, *Cell Mol Life Sci* **61**: 209–219.

DiMatteo, A, Hancock, RD, Ross, HA, Frusciante, L, and Viola, R, 2003, Characterization of *Chlorella pyrenoidosa* L-ascorbic acid accumulating mutants: Identification of an enhanced biosynthetic enzyme activity and cloning of the putative gene from *Arabidopsis thaliana*, *Comp Biochem Physiol* A **134**: S155.

Fernie, AR, Tadmor, Y, and Zamir, D, 2006, Natural genetic variation for improving crop quality, *Curr Opin Plant Biol* **9**: 196–202.

Franceschi, VR, and Tarlyn, NM, 2002, L-Ascorbic acid is accumulated in source leaf phloem and transported to sink tissues in plants, *Plant Physiol* **130**: 649–656.

Gatzek, S, Wheeler, GL, and Smirnoff, N, 2002, Antisense suppression of L-galactose dehydrogenase in *Arabidopsis thaliana* provides evidence for its role in ascorbate synthesis and reveals light modulated L-galactose synthesis, *Plant J* **30**: 541–553.

Grace, SC, and Logan, BA, 1996, Acclimation of foliar antioxidant systems to growth irradiance in 3 broad-leaved evergreen species, *Plant Physiol* **112**: 163–1640.

Green, MA, and Fry, SC, 2005, Vitamin C degradation in plant cells via enzymatic hydrolysis of 4-O-oxalyl-L-threonate, *Nature* **433**: 83–87.

Hancock, RD, and Viola, R, 2005, Improving the nutritional value of crops through enhancement of L-ascorbic acid (vitamin C) content: rationale and biotechnological opportunities, *J Agric Food Chem* **53**: 5248–5257.

Hancock, RD, McRae, D, Haupt, S, and Viola, R, 2003, Synthesis of L-ascorbic acid in the phloem, *BMC Plant Biol* **3**: 7.

Horemans, N, Foyer, CH, Potters, G, and Asard, H., 2000, Ascorbate function and associated transport systems in plants, *Plant Physiol Biochem* **38**: 531–540.

Imai, T, Karita, S, Shiratori, G, Hattori, M, Nunome, T, Oba, K, and Hirai, M, 1998, L-galactono-gamma-lactone dehydrogenase from sweet potato: purification and cDNA sequence analysis, *Plant Cell Physiol* **39**: 1350–1358.

Isherwood, FA, Chen, YT, and Mapson, LW, 1954, Synthesis of L-ascorbic acid in plants and animals, *Biochem J* **56**: 1–15.

Jain, AK, and Nessler, CL, 2000, Metabolic engineering of an alternative pathway for ascorbic acid biosynthesis in plants, *Mol Breed* **6**: 73–78.

Kanter, U, Usadel, B, Guerineau, F, Li, Yy, Pauly, M, and Tenhaken, R, 2005, The inositol oxygenase gene family of Arabidopsis is involved in the biosynthesis of nucleotide sugar precursors for cell-wall matrix polysaccharides, *Planta* **221**: 243–254.

Keller, R, Renz, FSy, and Kossmann, Jy, 1999, Antisense inhibition of the GDP-mannose pyrophosphorylase reduces the ascorbate content in transgenic plants leading to developmental changes during senescence, *Plant J* **19**: 131–141.

Kiddle, G, Pastori, G My, Bernard, Sy, Pignocchi, C, Antoniw, J, Verrier, PyJy, and Foyer, CH, 2003, Effects on leaf ascorbate content on defense and photosynthesis gene expression in *Arabidopsis thaliana*, *Antiox Redox Signal* **5**: 23–32.

Laing, WA, Bulley, Sy, Wright, My, Cooney, Jyy, Jensen, Dy, Barraclough, D, and MacRae E, 2004, A highly specific L-galactose-1-phosphate phosphatase on the path to ascorbate biosynthesis, *Proc.y Natl Acad Sci USA* **101**: 16976–16981.

Lee, KW, Lee, HJ, Surh, YJ, and Lee, CY, 2003, Vitamin C and cancer chemoprevention: reappraisal, *Am J Clin Nutr* **78**: 1074–1078.

Levi, F, Pasche, C, Lucchini, F, and La Vecchia, C, 2001, Dietary intake of selected micronutrients and breast-cancer risk, *Int J Cancer* **91**: 260–263.

Levine, M, Wang, Y, Padayatty, SJ, and Morrow, J, 2001, A new recommended dietary allowance of vitamin C for healthy young women, *Proc Natl Acad Sci USA* **98**: 9842–9846.

Loewus, FA, 1999, Biosynthesis and metabolism of ascorbic acid in plants and of analogs of ascorbic acid in fungi, *Phytochem* **52**: 193–210.

Loewus, FA, and Kelly, S, 1961, The metabolism of D-galacturonic acid and its methyl ester in the detached ripening strawberry, *Arch Biochem Biophys* **95**: 483–493.

Lorence, A, Chevone, BI, Mendes, P, and Nessler, CL, 2004a, *myo*-Inositol oxygenase offers a possible entry point into plant ascorbate biosynthesis, *Plant Physiol* **134**: 1200–1205.

Lorence, A, Robinson, J, Chevone, BI, Mendes, P, and Nessler, CL, 2004b, Contribution of the *myo*-inositol oxygenase (*miox*) gene family of *Arabidopsis thaliana* to ascorbate biosynthesis, Abstract submitted to the "15 International Conference on Arabidopsis Research", July 11–14, Berlin, Germany.

Lorence, A, Rogers, AM, Mendes, P, Zhang, W, Chevone, BI, and Nessler CL, 2004c, Identification and characterization of a putative glucuronic acid reductase in *Arabidopsis thaliana*, Abstract submitted to the "15 International Conference on Arabidopsis Research", July 11–14, Berlin, Germany.

Lukowitz, W, Nickle, TC, Meinke, DW, Last, RL, Conklin, PL, and Somerville, C R, 2001, *Arabidopsis cyt1* mutants are deficient in a mannose-1-phosphate guanylyltransferase and point to a requirement of N-linked glycosylation for cellulose biosynthesis, *Proc Natl Acad Sci USA* **98**: 2262–2267.

Major, L, Wolucka, BA, and Naismith, JH, 2005, Structure and function of GDP-mannose-3",5"-epimerase: an enzyme which performs three chemical reactions at the same active site, *J Am Chem Soc* **127**: 18309–18320.

Martinez, S, Lopez M, Gonzalez-Raurich, M, and Alvarez AB, 2005, The effects of ripening stage and processing systems on vitamin C content in sweet peppers (*Capsicum annuum* L), *Int J Food Sci Nutr* **56**: 45–51.

Masaki, KH, Losonczy, KG, Izmirlian, G, Foley, DJ,, Ross, GW, Petrovich, H, Havlik, R, and White LR, 2000, Association of vitamin E and C supplement use with cognitive function and dementia in elderly men, *Neurol* **54**: 1265–1272.

McDermott, JH, 2000, Antioxidant nutrients: current dietary recommendations and research update, *J Am Pharmac Assoc* **40**: 785–799.

Mieda, T, Yabuta, Y, Rapolu, M, Motoki, T, Takeda, T, Yoshimura, K, Ishikawa, T, and Shigeoka, S, 2004, Feedback inhibition of spinach L-galactose dehydrogenase by L-ascorbate, *Plant Cell Physiol* **45**: 1271–1279.

Milo, Jr, GR, and Santilli, V.y, 1967, Changes in the ascorbate concentration of Pinto bean leaves accompanying the formation of TMV-induced lesions, *Virology* **31**: 197–206.

Mishra, NyP, Mishra, RK, and Singhal, GS, 1993, Changes in the activities of antioxidant enzymes during exposure of intact wheat leaves to strong visible light at different temperatures in the presence of protein synthesis inhibitors, *Plant Physiol* **102**: 903–910.

Munné-Bosch, S, and Alegre, L, 2002, Interplay between ascorbic acid and lipophilic antioxidant defenses in chloroplasts of water-stressed *Arabidopsis* plants, *FEBS Lett* **524**: 145–148.

Nandi, A, Mukhopadhyay, CK, Ghosh, MK, Chattopadhyay, DJ, and Chatterjee, IB, 1997, Evolutionary significance of vitamin C biosynthesis in terrestrial vertebrates, *Free Rad Biol Med* **22**: 1047–1054.

Napoli, C, Williams-Ignarro, S, de Nigris, F, Lerman, LO, Rossi, L, Guarino, C, Mansueto, G, Di Tuoro, F, Pignalosa, O, De Rosa, G, Sica, V, and Ignarro, LJ, 2004, Long-term combined beneficial effects of physical training and metabolic treatment on atherosclerosis in hypercholesterolemic mice, *Proc Natl Acad Sci USA* **101**: 8797–8802.

Nishikawa, F, Kato, M, Hyodo, H, Ikoma, Y, Sugiura, M, and Yano, M, 2003, Ascorbate metabolism in harvested broccoli, *J Exp Bot* **54**: 2439–2448.

Nishikimi, M, Fukuyama, R, Minoshima, S, Shimizu, N, and Yagi, K, 1994, Cloning and chromosomal mapping of the human nonfunctional gene for L-gulono-gamma-lactone oxidase, the enzyme for L-ascorbic acid biosynthesis missing in man, *J Biol Chem* **269**: 13685–13688.

Noctor, G, and Foyer, CH, 1998, Ascorbate and glutathione: keeping active oxygen under control, *Ann Rev Plant Physiol Plant Mol Biol* **49**: 249–279.

Nunez-Nesi, A, Carrari, F, Lytovchenko, A, Smith, AMO, Loureiro, ME, Ratcliffe, RG, Sweetlove, LJ, and Fernie AR, 2005, Enhanced photosynthetic performance and growth as a consequence of decreasing mitochondrial malate dehydrogenase activity in transgenic tomato plants, *Plant Physiol* **137**: 611–622.

Oba, K, Fukui, M, Imai, Y, Iriyama, S, and Nogami, K, 1994, L-galactono-γ-lactone dehydrogenase: partial characterization, induction of activity and role in the synthesis of ascorbic acid in wounded white potato tuber tissue, *Plant Cell Physiol* **35**: 473–478.

Østergaard, J, Persiau, G, Davey, MW, Bauw, G, and Van Montagu, M, 1997, Isolation of a cDNA coding for L-galactono-γ-lactone dehydrogenase, an enzyme involved in the biosynthesis of ascorbic acid in plants, *J Biol Chem* **48**: 30009–30016.

Pallanca, JE, and Smirnoff, N, 2000, The control of ascorbic acid synthesis and turnover in pea seedlings, *J Exp Bot* **51**: 669–674

Panchuk, II, Volkov, RA, and Schoffl, F, 2002, Heat stress- and heat chock transcription factor-dependent expression and activity of ascorbate peroxidase in Arabidopsis, *Plant Physiol* **129**: 838–853.

Parle, M, and Dhingra D, 2003, Ascorbic acid: a promising memory-enhancer in mice, *J Pharmacol Sci* **93**: 129–135.

Passage, E, Norreel, JC, Noack-Fraissignes, P, Sanguedolce, V, Pizant, J, Thirion, X, Robaglia-Schlupp, A, Pellissier, JF, and Fontes, M, 2004, Ascorbic acid treatment corrects the phenotype of a mouse model of Charcot-Marie-Tooth disease, *Nature Med* **10**: 396–401.

Pastori, GM, Kiddle, G, Antoniw, J, Bernard, S, Veljovic-Janovic, S, Verrier, PJ, Graham, N, and Foyer, CH, 2003, Leaf vitamin C content modulates plant defense transcripts and regulate genes that control development through hormone signaling, *Plant Cell* **15**: 939–951.

Pavet, V, Olmos, E, Kiddle, G, Mowla, S, Kumar, S, Antoniw, J, Alvarez, ME, and Foyer CH, 2005, Ascorbic acid deficiency activates cell death and disease resistance responses in Arabidopsis, *Plant Physiol* **139**: 1291–1303.

Puppo, A, Groten, K, Bastian, F, Carzaniga, R, Soussi, M, Lucas, MM, de Felipe, MR, Harrison, J, Vanacker, H, and Foyer, CH, 2004, Legume nodule senescence: roles for redox and hormone signaling in the orchestration of the natural aging process, *New Phytologist* **165**: 683–701.

Radzio, JA, Lorence, A, Chevone, BI, and Nessler, CL, 2003, L-Gulono-1,4-lactone oxidase expression rescues vitamin C-deficient Arabidopsis (*vtc*) mutants, *Plant Mol Biol* **53**: 837–844.

Rinne, T, Mutschler, E, Wimmer-Greinecker, G, Moritz, A, and Olbrich, HG, 2000, Vitamins C and E protect isolated cardiomyocytes against oxidative damage, *Int J Cardiol* **75**: 275-281.

Rousseaux, MC, Jones, CM, Adams, D, Chetelat, R, Bennett, A, and Powell, A, 2005, QTL analysis of fruit antioxidants in tomato using *Lycopersicon pennellii* introgression lines, *Theor Appl Genet* **111**: 1396-1408.

Sasaki-Sekimoto, Y, Taki, N, Obayashi, T, Aono, M, Matsumoto, F, Sakurai, N, Suzuki, Hirai, MY, Noji, M, Saito, K, Masuda, T, Takamiya, K, Shibata, D, and Ohta, H, 2005, Coordinated activation of metabolic pathways for antioxidants and defense compounds by jasmonates and their roles in stress tolerance in Arabidopsis, *Plant J* **44**: 653–668.

Siendones, E, González-Reyes, JA, Santos-Ocaña, C, Navas, P, and Córboba, F, 1999, Biosynthesis of ascorbic acid in kidney bean, L-galactono-γ-lactone dehydrogenase is an intrinsic protein located at the mitochondrial inner membrane, *Plant Physiol* **120**: 907–912.

Sircelj, H, Tausz, M, Grill, D, and Batic, F, 2005, Biochemical responses in leaves of two apple tree cultivars subjected to progressing drought, *J Plant Physiol* **162**: 1308–1318.

Smirnoff, N, 1996, The function and metabolism of ascorbic acid in plants, *Ann Bot* **78**: 661–669.

Smirnoff, N, 2000a, Ascorbic acid: metabolism and functions of a multifaceted molecule, *Curr Opin Plant Biol* **3**: 229–235.

Smirnoff, N, 2000b, Ascorbate biosynthesis and function in photoprotection, *Philos Trans R Soc Lond B Biol Sci* **355**: 1455–1464.

Smirnoff, N, 2001, L-Ascorbic acid biosynthesis, *Vitam Horm* **61**: 241–266.

Smirnoff, N, and Pallanca, JE, 1996, Ascorbate metabolism in relation to oxidative stress, *Biochem Soc Trans* **24**: 472–478.

Smirnoff, N, and Wheeler, GL, 2000, Ascorbic acid in plants: biosynthesis and function, *Crit Rev Plant Sci* **19**: 267–290.

Smirnoff, N, Conklin, PL, and Loewus, FA, 2001, Biosynthesis of ascorbic acid in plants: a renaissance, *Annu Rev Plant Physiol Plant Mol Biol* **52**: 437–467.

TAIR, 2006, http://www arabidopsis org

Tabata, K, Oba, K, Suzuki, K, and Esaka, M, 2001, Generation and properties of ascorbic acid-deficient transgenic tobacco cells expressing antisense RNA for L-galactono-1,4-lactone dehydrogenase, *Plant J* **27**: 139–148.

Tabata, K, Takaoka, T, and Esaka, M, 2002, Gene expression of ascorbic acid-related enzymes in tobacco, *Phytochem* **61**: 631–635.

Tamaoki, M, Mukai, F, Asai, N, Nakajima, N, Kubo, A, Aono, M, and Saji, H, 2003, Light-controlled expression of a gene encoding L-galactono-γ-lactone dehydrogenase which affects ascorbate pool size in *Arabidopsis thaliana*, *Plant Sci* **164**: 1111–1117.

Tedone, L, Hancock, RD, Alberino, S, Haupt, S, and Viola, R, 2004, Long-distance transport of L-ascorbic acid in potato, *BMC Plant Biol* **4**: 16.

Tokunaga, T, Miyahara, K, Tabata, K, and Esaka, M, 2005, Generation and properties of ascorbic acid-overproducing transgenic tobacco cells expressing sense RNA for L-galactono-1,4-lactone dehydrogenase, *Planta* **220**: 854–863.

Ushimaru T, Nakagawa, T, Fujioka, Y, Daicho, K, Naito, M, Yamauchi, Y, Nonaka, H, Amako, K, Yamawaki, K, and Murata, N, 2006, Transgenic Arabidopsis plants expressing the rice dehydroascorbate reductase gene are resistant to salt stress, *J Plant Physiol* available online Nov, 2005.

Valpuesta, V, and Botella, MA, 2004, Biosynthesis of L-ascorbic acid in plants: new pathways for an old antioxidant, *Trends Plant Sci* **9**: 573–577.

Veljovic-Jovanovic, SD, Pignocchi, C, Noctor, G, and Foyer, CH, 2001, Low ascorbic acid in the *vtc-1* mutant in Arabidopsis is associated with decreased growth and intracellular redistribution of the antioxidant system, *Plant Physiol* **127**: 426–435.

Wagner, C, Sefkow, M, and Kopka, J, 2003, Construction and application of a mass spectral and retention time index database generated from plant GC/EI-TOF-MS metabolite profiles, *Phytochem* **62**: 887–900.

Watanabe, K, Suzuki, K, and Kitamura, S, 2006, Characterization of a GDP-D-mannose 3"-5"-epimerase from rice, *Phytochem* **67**: 338–346.

Wheeler, GL, Jones, MA, and Smirnoff, N, 1998, The biosynthetic pathway of vitamin C in higher plants, *Nature* **393**: 365-369.

Wolucka, BA, Persiau, G, Van Doorsselaere, J, Davey, MW, Demol, H, Vandekerckhove, J, Van Montagu, M, Zabeau, M, and Boerjan, W, 2001, Partial purification and identification of GDP-mannose 3",5"-epimerase of *Arabidopsis thaliana*, a key enzyme of the plant vitamin C pathway, *Proc Natl Acad Sci USA* **98**: 14843–14848.

Wolucka, BA, and Van Montagu, M, 2003, GDP-Mannose-3",5"-epimerase forms GDP-L-gulose, a putative intermediate for the *de novo* biosynthesis of vitamin C in plants, *J Biol Chem* **278**: 47483–47490.

Wolucka, BA, Goossens, A, and Inzé, D, 2005, Methyl jasmonate stimulates the *de novo* biosynthesis of vitamin C in plant cell suspensions, *J Exp Bot* **56**: 2527–2538.

Yokoyama, T, Date, C, Kokubo, Y, Yoshiike, N, Matsumura, Y, and Tanaka, H, 2000, Serum vitamin C concentration was inversely associated with subsequent 20-year incidence of stroke in a Japanese rural community: the Shibata study, *Stroke* **31**: 2287–2294.

Zou, L, Li, H, Ouyang, B, Zhang, J, and Ye, A, 2006, Cloning and mapping of genes involved in tomato ascorbic acid biosynthesis and metabolism, *Plant Sci*, **170**: 120–127.

CHAPTER 9

METABOLIC ENGINEERING OF TERPENOID BIOSYNTHESIS IN PLANTS

JOOST LÜCKER[1], HARRO J. BOUWMEESTER[2,4] AND ASAPH AHARONI[3]

[1] *University of British Columbia, Faculty of Land and Food Systems, Wine Research Centre, 216-2205 East Mall, Vancouver, B.C., V6T 1Z4, Canada*

[2] *Plant Research International, P.O. Box 16, 6700 AA Wageningen, The Netherlands*

[3] *Weizmann Institute of Science, P.O. Box 26, Rehovot 76100, Israel*

[4] *To whom correspondence should be addressed. Harro J. Bouwmeester, Plant Research International, P.O. Box 16, 6700 AA Wageningen, The Netherlands; e-mail: harro.bouwmeester@wur.nl*

Abstract: Metabolic engineering of terpenoids in plants is a fascinating research topic from two main perspectives. On the one hand, the various biological activities of these compounds make their engineering a new tool for improving a considerable number of traits in crops. These include for example enhanced disease resistance, weed control by producing allelopathic compounds, better pest management, production of medicinal compounds, increased value of ornamentals and fruit and improved pollination. On the other hand, the same plants altered in the profile of terpenoids and their precursor pools make a most important contribution to fundamental studies on terpenoid biosynthesis and its regulation. In this review we describe the recent results with terpenoid engineering, showing that engineering of these compounds and their derivatives in plant cells is feasible, although with some requirements and limitations. For example, in terpenoid engineering experiments crucial factors are the subcellular localization of both the precursor pool and the introduced enzymes, the activity of endogenous plant enzymes which modify the introduced terpenoid skeleton, the costs of engineering in terms of effects on other pathways sharing the same precursor pool and the phytotoxicity of the introduced terpenoids. Finally, we will show that transgenic plants altered in their terpenoid profile exert novel biological activities on their environment, for example influencing insect behavior

Abbreviations: DMAPP, dimethylallyl diphosphate; DXP, deoxyxylulose phosphate; DXR, deoxyxylulose reductoisomerase; FPP, farnesyl diphosphate; GPP, geranyl diphosphate; GGPP, geranylgeranyl diphosphate; HDR, hydroxymethylbutenyl diphosphate reductase; HMGR, 3-hydroxymethyl-3-methylglutaryl–CoA reductase; IPP, isopentenyl diphosphate; MEP, 2C-methyl-D-erythritol 4-phosphate; TPSs, terpene synthases

R. Verpoorte et al. (eds.), Applications of Plant Metabolic Engineering, 219–236.
© 2007 *Springer*.

1. INTRODUCTION

Terpenoids (or isoprenoids, or terpenes) are probably the most widespread group of natural products. They are highly variable in structure although they are all derived from one backbone molecule, the five-carbon isoprenoid unit isopentenyldiphosphate (IPP) (Fig. 1) (Gershenzon and Croteau, 1993; Sacchettini and Poulter, 1997). Terpenoids are involved in central plant processes including photosynthesis (e.g. chlorophyll side chains or carotenoids), electron transport (ubiquinone, menaquinone and plastoquinone), cell membrane architecture (sterols) and regulation of cellular development (gibberellins, abscisic acid, brassinosteroids) (Gershenzon and Croteau, 1993). In addition, they often exhibit more specialised functions in defense (phytoalexins, insect feeding and oviposition deterrents and phytotoxins) and reproduction (as attractants of pollinators and seed dispersing animals) (Fig. 1) (Pichersky and Gershenzon, 2002). Monoterpenoids and sesquiterpenoids are emitted from flowers of many plant species (Knudsen et al., 1993), where they are involved in the attraction of species-specific pollinators (Raguso and Pichersky, 1999; Pichersky and Gershenzon, 2002). Both classes of terpenoids, but also the less volatile diterpenoids, act in repelling herbivores, as reported in studies with aphids (Hori, 1998; Wang et al., 2001; Aharoni et al., 2003; Wang et al., 2004). They can also be emitted as semiochemicals after herbivore attack that attract the natural enemies of the herbivores (Fig. 1) (Pare and Tumlinson, 1997; Weissbecker et al., 1997; Bouwmeester et al., 1999; Arimura et al., 2000; Birkett et al., 2000; Baldwin, 2001; Kessler and Baldwin, 2001; van Poecke et al., 2001; Kappers et al., 2005).

In addition to their ecological functions, monoterpenoids, sequiterpenoids and diterpenoids are of high commercial importance as they are commonly used as flavor and fragrance compounds, neutraceuticals, pharmaceuticals and as industrial raw materials (Fig. 1) (Verlet, 1993; Chappell, 2004). The commercial and ecological importance of terpenoids makes their metabolic engineering most valuable (Galili et al., 2002). With respect to plant breeding, terpenoid engineering could lead to improvement of many traits. These include disease and pest resistance, weed control, improved fragrance of ornamentals and pollination of seed crops, enhanced aroma of fruits and vegetables, and production of pharmaceuticals. From a more fundamental and scientific point of view these transgenic plants can potentially advance our understanding of the regulation of terpenoid biosynthesis and the ecological interactions they are involved in. In this review we summarize the recent progress in terpenoid metabolic engineering in plants and discuss some future opportunities in this exciting field.

2. TERPENOID BIOSYNTHESIS

There are two separate pathways in plants that lead to the formation of IPP and DMAPP, the basic building blocks of the huge variety of terpenoids (Fig. 2). The mevalonate pathway is active in the cytosol and the 2*C*-methyl-D-erythritol

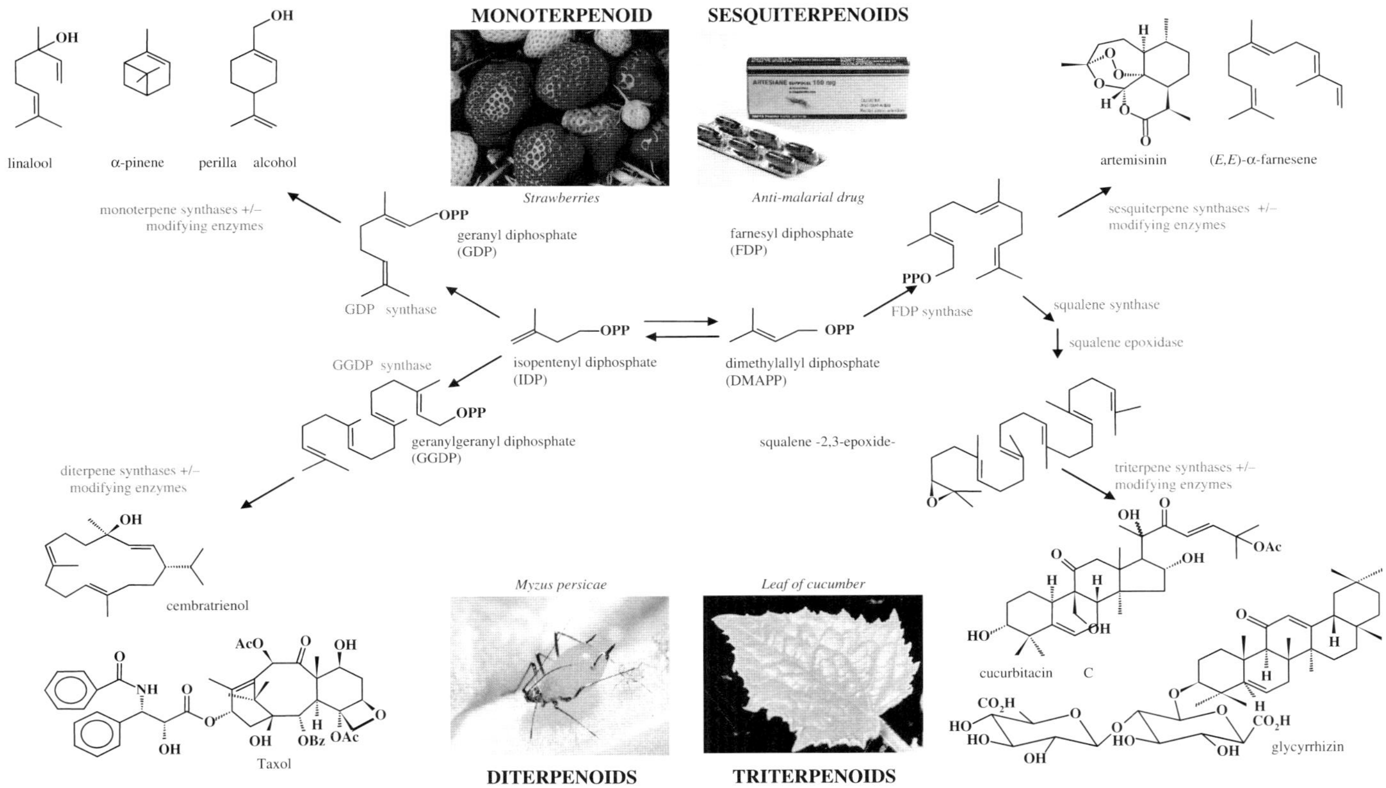

Figure 1. Schematic overview of the biosynthesis of the monoterpenoids, sesquiterpenoids, diterpenoids and triterpenoids. For some representatives of these classes structures and pictures related to their biological importance are shown. Enzymatic steps are indicated in red. Redrawn from Aharoni et al. (2005), with permission from Elsevier

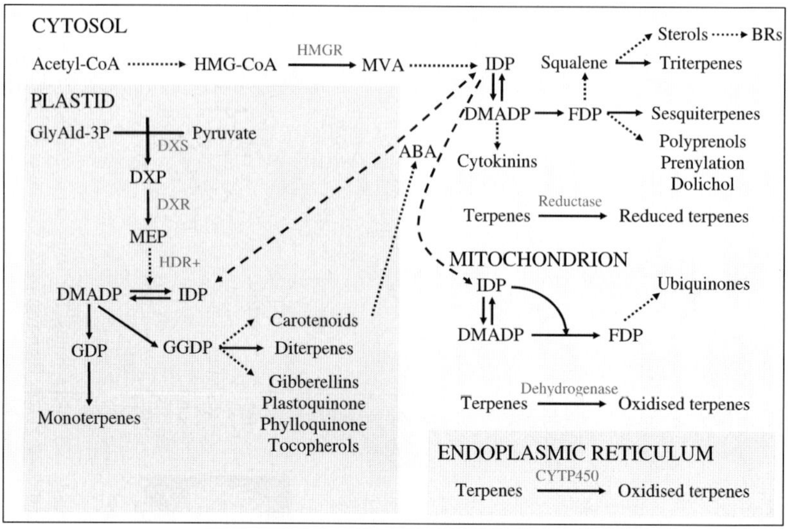

Figure 2. The isoprenoid pathway in plants. Solid, broken and dashed arrows represent single and multiple enzymatic steps and transport, respectively. Several enzymes are depicted in red: HMGR, HMG-CoA reductase; DXS, DXP synthase; DXR, DXP reductoisomerase; HDR, hydroxymethylbutenyl diphosphate reductase (marked with "+" because several enzymes are required to generate DMADP and IDP from MEP) and CYTP450 to indicate cytochrome P450 hydroxylases. HMG-CoA, hydroxymethylglutaryl CoA; MVA, mevalonic acid; IDP, isopentenyl diphosphate; DMADP, dimethylallyl diphosphate; FDP, farnesyl diphosphate; GlyAld-3P, glyceraldehyde 3-phosphate; DXP, deoxyxylulose 5-phosphate; MEP, methylerythritol 4-phosphate; GDP, geranyl diphosphate; GGDP, geranylgeranyl diphosphate; ABA, abscisic acid; BRs, Brassinosteroids. "Terpenes" include terpenes from all classes and originating from the various organelles. Redrawn from Aharoni et al. (2005), with permission from Elsevier

4-phosphate (MEP) pathway is active in the plastids (Mahmoud and Croteau, 2002; Rodriguez-Concepcion and Boronat, 2002). In both compartments these basic building blocks are used by prenyltransferases in electrophylic condensation reactions that result in the formation of various larger prenyl diphosphates (Gershenzon and Croteau, 1993). The head to tail condensation of one molecule of IPP with one molecule of DMAPP yields the C_{10} compound geranyl diphosphate (GPP), which is the immediate precursor of the monoterpenes (Fig. 2). The condensation of two IPP units with one DMAPP unit generates farnesyl diphosphate (FPP), the precursor of sesquiterpenes. The condensation of two units of FPP results in the formation of squalene, the precursor for triterpenes and sterols (Fig. 2). The condensation of three units of IPP with DMAPP provides geranyl geranyl diphosphate (GGPP), the precursor of diterpenes and the C_{40} tetraterpenes that consist of two units of GGPP. Longer prenyl (allylic) diphosphates are similarly produced by subsequent additions of IPP, as for natural rubber (Poulter and Rilling, 1981). The reactions catalyzed by these prenyltransferases are believed to be multistep and sequential in which intermediate elongation products are not released from the enzyme surface in appreciable amounts. Plant prenyltransferase proteins appear to be similar in size, share common primary structural elements including an aspartate-rich motif involved in substrate binding and require only a divalent metal ion for catalysis (Ohnuma et al., 1993; Chen et al., 1994).

3. FORMATION OF TERPENOID DIVERSITY FROM THEIR PRENYL DIPHOSPHATE PRECURSORS

The prenyl diphosphate precursors are substrates for terpene synthases that generate the enormous diversity of carbon skeletons characteristic for the terpenoids including acyclic, cyclic, those consisting of multiple ring systems and others (Fig. 2). The terpene synthases, albeit isolated from a wide variety of plant species, are quite similar in properties, and operate through electrophilic reaction mechanisms just like the prenyltransferases (Poulter and Rilling, 1981). After the formation of the primary terpene skeletons a variety of other enzymes can modify these molecules (Fig. 2). An important class is formed by the cytochrome P450 monooxygenases, that can initiate a whole chain of secondary modifications, mostly consisting of redox reactions and conjugations (Mihaliak et al., 1993). As a result, one parent hydrocarbon structure can lead to the formation of multiple structurally related derivatives bearing an identical oxygenation pattern initially established by the hydroxylase (Karp and Croteau, 1988; Little and Croteau, 1999). The terpenoids are also further modified by dehydrogenases, reductases, glycosyl- and acyl- transferases and other reaction types, which often result in drastic changes in their olfactory characteristics and volatility, stability or storability (Dudareva et al., 2004).

4. COMPARTMENTALIZATION OF TERPENOID BIOSYNTHESIS

The mevalonate pathway in the cytosol is generally considered to be responsible for precursor supply for the formation of sesquiterpenes and triterpenes (including sterols), and for the supply of precursors for protein prenylation and ubiquinone and heme-A production in the mitochondria (Fig. 2). In plastids, the MEP pathway is providing precursors for the production of isoprene, monoterpenes, diterpenes (e.g. GAs) and tetraterpenes (e.g. carotenoids). However, many experimental results with labeled precursor flux studies and terpenoid pathway engineered plants strongly indicate that there is exchange of precursors and intermediates between the cell compartments (Fig. 2) (Adam and Zapp, 1998; McCaskill and Croteau, 1998; Bick and Lange, 2003; Hemmerlin et al., 2003; Laule et al., 2003; Schuhr et al., 2003; Dudareva et al., 2004; Lücker et al., 2004a; Dudareva et al., 2005).

In the past decade, a number of research groups have used metabolic engineering approaches to alter terpenoid formation in a range of plant species. These studies exposed unanticipated caveats determining the success of pathway engineering. Crucial factors for successful engineering were for example precursor availability, enzyme localization, pH of the targeted cell compartment and the presence of endogenous, modifying enzymes that metabolize the newly introduced compounds into unexpected products (Lewinsohn et al., 2001; Lücker et al., 2001; Lavy et al., 2002; Aharoni et al., 2003; Lücker et al., 2004a; Lücker et al., 2004b). The center of attention in engineering terpenoids in plants has been the volatile mono- and sesquiterpenoids and carotenoids. Engineering of carotenoid metabolism was described in numerous reviews and therefore will only be mentioned briefly in this report (for a recent review on carotenoid engineering in plants see (Sandmann et al., 2006).

5. METABOLIC ENGINEERING OF TERPENOID PRECURSORS

One approach to engineer levels of terpenoids in plants is to alter the availability of precursors. Engineering the terpenoid precursor pool might be insufficient for increasing levels of target terpenoids and simultaneous engineering of downstream genes might be required. This could be achieved by altering expression of multiple genes in the pathway or by the use of transcription factors that activate/repress multiple steps in a pathway (Broun, 2004; Capell and Christou, 2004). The only report on a terpenoid gene regulator was the characterization of *GaWRKY1*, a cotton transcription factor that regulates the sesquiterpene synthase gene cadinene synthase (Xu et al., 2004). In this study the authors showed that overexpression of the *GaWRKY1* gene in Arabidopsis could activate the target gene promoter but they did not evaluate the levels of sesquiterpenoids in *GaWRKY1* overexpressing plants. With regard to structural genes, a number of engineering approaches aimed to affect the precursor producing enzymes (Aharoni et al., 2005). In the MEP pathway, the deoxyxylulose phosphate (DXP) synthase enzyme catalyses the first step towards IPP, and seems to be a major control point in the pathway (Fig. 2) (Wolfertz et al., 2004). Overexpression and downregulation of DXP synthase in Arabidopsis altered the levels of various isoprenoids, including chlorophylls, tocopherols, carotenoids, ABA and GAs (Estevez et al., 2001). Overexpression of the gene encoding deoxyxylulose reductoisomerase (DXR) catalyzing the next reaction downstream of DXP synthase in the MEP pathway (Fig. 2) resulted in a 50% increase of total yield of monoterpenes produced by peppermint (Mahmoud and Croteau, 2001). Apart from DXP synthase and DXR also hydroxymethylbutenyl diphosphate reductase (HDR), the enzyme encoding the final step in the MEP pathway (resulting in the formation of both IPP and DMAPP) (Fig. 2), was shown to be a rate limiting step in tomato and Arabidopsis (Botella-Pavia et al., 2004). By crossing transgenic plants overexpressing taxadiene synthase with plants that either overexpressed DXP synthase or HDR, taxadiene levels were increased by 6.5 and 13 times respectively, as compared with plants overexpressing the taxadiene synthase gene alone (Besumbes et al., 2004).

The rate limiting step in the cytosolic mevalonate pathway has been suggested to be the 3-hydroxymethyl-3-methylglutaryl-CoA reductase (HMGR) (Fig. 2) and the consequences of over expression of the corresponding gene were studied in several metabolic engineering experiments (Chappell et al., 1995; Schaller et al.,

1995; Harker et al., 2003). Total sterol levels were increased three to ten times by constitutive overexpression of HMGR in tobacco, but other terpenoids including sesquiterpenes were not altered (Chappell et al., 1995). When FPP synthase was overexpressed in Arabidopsis, sterol levels remained unaltered, but the levels of sesquiterpenes were not analyzed (Masferrer et al., 2002). In another plant species, *Artemisia annua*, overexpression of FPP synthase resulted in an increase of the antimalarial compound artemisinin (Fig. 1) (Chen et al., 2000; Han et al., 2006). The maximum level of artemisinin that could be reached in transgenic plants was about 1% of leaf dry weight. The authors speculated that formation of artemisinin is limited by yet unknown regulatory mechanisms or that a higher concentration could be phytotoxic to the plant cells (Han et al., 2006).

6. METABOLIC ENGINEERING OF MONOTERPENE PRODUCTION

When engineering monoterpene production, alteration of several steps in the MEP pathway, of GPP synthase or of the monoterpene synthases could be considered as possible points for pathway regulation (Haudenschild and Croteau, 1998; Mahmoud and Croteau, 2002). Monoterpene pathway engineering was initially attempted in essential oil containing plants (Krasnyanski et al., 1999), as these plants have specialized compartments for monoterpene production and are capable of producing large amounts of monoterpenes (Fahn, 1979; Gershenzon et al., 1989; Gershenzon et al., 2000). An attempt to modify the composition or quality of the essential oil was carried out by overexpression of a (–)-limonene synthase in peppermint. Introduction of this enzyme resulted in small qualitative and quantitative variations in the composition of the essential oil (Krasnyanski et al., 1999). In a similar experiment, a (–)-limonene synthase was introduced into transgenic peppermint and cornmint plants and similarly small changes were achieved in the monoterpene profiles (Diemer et al., 2001). In peppermint, Croteau and coworkers used metabolic engineering to change the composition of the essential oil in order to reduce the production of undesired metabolites (Mahmoud and Croteau, 2001). Menthofuran was downregulated by the introduction of an antisense construct of the gene encoding the cytochrome P450 enzyme menthofuran synthase (Bertea et al., 2001; Mahmoud and Croteau, 2001). The level of menthofuran could be reduced by 50% compared to the amount present in the wild type plant. In different experiments by the same researchers, the (–)-limonene hydroxylase enzyme was overexpressed in peppermint (Mahmoud et al., 2004). This resulted in cosupression of the P450-hydroxylase mRNA transcript in some transgenic lines which led to high accumulation of limonene in these plants (up to 80% of the essential oil), while the total amount of essential oil remained the same (Mahmoud et al., 2004).

Several reports on the engineering of terpenoids in tobacco have established this plant species as a suitable model for metabolic engineering of monoterpenes. Transgenic tobacco plants expressing the *Perilla frutescens* limonene synthase produced limonene in the leaves (Ohara et al., 2003). Production of limonene by transgenic plants overexpressing a plastid targeted limonene synthase was much higher than in plants engineered with the same protein but localized to the cytosol. In contrast, transgenic plants with the limonene synthase protein targeted to the endoplasmic reticulum (ER) did not produce limonene (Ohara et al., 2003). In another experiment, three lemon monoterpene synthases were introduced into tobacco, all under the control of the constitutive 35S CaMV promoter (Lücker et al., 2004a). This was the first example in which three foreign genes competing for the same substrate were introduced into a single plant (Capell and Christou, 2004). Transgenic tobacco plants harboring the three genes emitted significant amounts of three new terpenoids products, namely, γ-terpinene, (+)-limonene and (−)-β-pinene (Lücker et al., 2002; Lücker et al., 2004a). In addition to the main products, a large number of minor compounds were detected including: α-thujene, (+/−)-α-pinene, (+)-β-pinene, myrcene, (+/−)-sabinene, α-terpinene, (−)-limonene, *p*-cymene and terpinolene (Lücker et al., 2002; Lücker et al., 2004a). Interestingly, different transgenic lines emitted different blends of these monoterpenes (Lücker et al., 2004a). Heterologues expression in *E. coli* and enzyme assays of the recombinant proteins showed that all major and minor compounds detected *in planta* were products of the three newly introduced enzymes (Lücker et al., 2002).

Most studies showed that the monoterpenoids introduced by metabolic engineering were further modified by endogenous enzymes of the transformed plant species. For example, such modification took place when the linalool synthase (*LIS*) from *Clarkia breweri* (Dudareva et al., 1996) was introduced into petunia (Lücker et al., 2001), tomato (Lewinsohn et al., 2001), and carnation plants (Lavy et al., 2002). In tomato, linalool production was restricted to the fruit by using the fruit-specific *E8* promoter and this resulted in high levels of linalool and 8-hydroxy-linalool, probably as a consequence of the presence of an endogenous cytochrome P450 enzyme that can hydroxylate linalool (Lewinsohn et al., 2001). Carnation normally does not emit monoterpenes, but after transformation with the *C. breweri* linalool synthase, transgenic carnation plants emitted linalool from flowers and leaves, albeit at low levels. It turned out that most of the linalool formed in the transgenic plants was oxidized to linalool-oxides (Lavy et al., 2002). Introduction of *LIS* into Petunia, another plant species with no detectable monoterpene production, resulted in the production of high levels of linalool. However, the linalool was not detectable in the headspace as it was directly converted by an endogenous glycosyltransferase to non-volatile (*S*)-linalyl-β-D-glucopyranoside (Lücker et al., 2001). Tobacco plants transformed with the *LIS* gene (Lücker, 2002), also resulted in the accumulation of glycosylated forms of linalool. In contrast to Petunia, glycosylation of linalool was incomplete in tobacco since substantial amounts of free linalool were detected in the headspace (Lücker, 2002). After introduction of the strawberry linalool/nerolidol synthase gene into *Arabidopsis*, part of the produced linalool was converted to a series of oxidised and/or glycosylated metabolites (Aharoni et al., 2003). Nevertheless, a substantial amount of free linalool was produced and emitted by *Arabidopsis* leaves and flowers.

The results described above obtained by engineering several plants species with two different linalool synthase genes demonstrate that monoterpene alcohols, newly produced in tissues that normally do not accumulate such compounds are often partially converted, by oxidation and glycosylation reactions to less reactive or less phytotoxic products. High concentrations of monoterpenes such as linalool are known to be detrimental to biological tissues (Vaughn and Spencer, 1991; Weidenhamer et al., 1993; Izumi et al., 1999). The high levels of linalool derivatives produced in the different engineered plants indicate that there is a fair amount of GPP available for generating heterologous monoterpenes in the plastids (Lücker et al., 2001; Aharoni et al., 2003).

Introduction of two consecutive steps in the monoterpene biosynthetic pathway is an even more challenging metabolic engineering approach. Verhoeven and co-workers (Lücker et al., 2004b) retransformed a tobacco line already expressing the three lemon monoterpene synthase genes discussed above with a gene encoding a cytochrome P450 limonene hydroxylase from *Mentha spicata*. This resulted in the detection of a new major product in the headspace of the flowers and the leaves which was identified as (+)-*trans*-isopiperitenol (Lücker et al., 2004b). In addition to (+)-*trans*-isopiperitenol, a number of side products could be detected such as isopiperitenone and *p*-cymene (Lücker et al., 2004b). Isopiperitenone was most likely formed from isopiperitenol by an endogenous tobacco alcohol dehydrogenase. Indeed, Suga and Hirata (1990) showed that monoterpene alcohols, such as carveol, can be converted to the corresponding ketone, by cultured suspension cells of tobacco, indicating the presence of terpenol dehydrogenase activity. The detection of *p*-cymene can be explained by assuming an endogenous dehydratase activity (Lücker et al., 2004b). Two other products, tentatively identified as 1,3,8-*p*-menthatriene and 1,5,8-*p*-menthatriene could be putative intermediates in such dehydratase-initiated conversion of isopiperitenol to *p*-cymene. The engineered cytochrome P450 was most likely localised in the ER where it normally functions as part of a complex with an endogenous NADPH-cytochrome P450 reductase. As the monoterpene synthases catalyze the production of monoterpenes in the plastids, this implies that a transport mechanism between the plastids and the cytosol, possibly via lipophilic oil bodies, also operates on monoterpenes introduced through metabolic engineering. Such a transport mechanism was suggested to be present in specialized essential oil producing organs but is apparently also present in plant tissues, not explicitly adapted to the production of large amounts of monoterpenes (Bosabalidis, 1996; Bouwmeester et al., 1998; Little and Croteau, 1999).

7. METABOLIC ENGINEERING OF SESQUITERPENE BIOSYNTHESIS

Altering the production of sesquiterpenes by metabolic engineering is a more complicated task as compared to modifying monoterpene levels. This is likely a result of differences in precursor availability (Chappell, 2002). Sesquiterpene biosynthesis occurs in the cytosol and depends on the supply of FPP produced by FPP-synthase. FPP is normally produced from IPP and DMAPP that originate from the mevalonate pathway, which is tightly regulated and limited by HMGR activity. However, it was recently demonstrated that FDP can also be produced from IPP and DMAPP derived from the plastidic MEP pathway (Bick and Lange, 2003; Dudareva et al., 2005). Recently, Schnee et al. (2006) demonstrated that in certain cases significant production of sesquiterpenes in the cytosol is achievable. The authors overexpressed a maize sesquiterpene synthase (*TPS10*) that formed several sesquiterpene hydrocarbons in transgenic Arabidopsis. The transgenic plants produced a volatile profile identical to the one released by herbivore-damaged maize. The total quantity of sesquiterpenes emitted from rosette leaves of transgenics Arabidopsis was comparable to that emitted by herbivore-damaged maize plants.

When a fungal trichodiene synthase, a sesquiterpene synthase, was transformed to tobacco plants, only traces of trichodiene were formed, although enzyme activity could be easily detected (Hohn and Ohlrogge, 1991). Transformation of a chicory germacrene A synthase to Arabidopsis and an *Artemisia annua* amorpha-4,11-diene synthase to tobacco resulted in the production of only trace amounts of the expected products (Wallaart et al., 2001; Aharoni et al., 2003). Interestingly, transgenic Arabidopsis producing a strawberry dual function linalool/nerolidol synthase (*Fa*NES1) in plastids, generated a small amount of the sesquiterpene nerolidol in addition to the expected monoterpene linalool product. This indicates that there is probably already some FPP available in the plastids that is accessible to an introduced sesquiterpene synthase (Aharoni et al., 2003). A promising new approach could be the enhancement of FPP availability in the plastids by the introduction of an engineered, plastidic, FPP synthase. Concomitant introduction of an engineered, plastidic sesquiterpene synthase may then well result in strongly improved sesquiterpene production (Aharoni et al., 2005). Analogously, we showed that it is also possible to produce sesquiterpenes in plant mitochondria by targeting the *Fa*NES1 gene to this cell compartment (Kappers et al., 2005). These transgenic plants produced 3(*S*)-*E*-nerolidol and no linalool, confirming that FPP, but not GPP, is present in mitochondria. Mitochondria are the site for production of the isoprenoid ubiquinone in plants and one of the Arabidopsis FPP synthases has a mitochondrial targeting signal. Endogenous Arabidopsis enzymes converted part of the nerolidol to the C_{11} homoterpene 4,8-dimethyl-1,3(*E*),7-nonatriene [*E*-DMNT] (Kappers et al., 2005).

8. METABOLIC ENGINEERING OF DITERPENOIDS AND TRITERPENOIDS

Diterpenoids that are produced in the plastids via the MEP pathway have so far only been used once in a *de novo* metabolic engineering approach. Taxadiene synthase, catalyzing the first step in taxol biosynthesis, was introduced to Arabidopsis. Using an inducible promoter, taxadiene levels reached up to 0.6 μg/g dryweight in leaves, which was 10–60 times higher than levels obtained with (cytosolic) sesquiterpene synthases in tobacco (Hohn and Ohlrogge, 1991; Wallaart et al., 2001) but 1000 times lower than what could be achieved with monoterpenes in *Arabidopsis* (Aharoni et al., 2003). Changes in an existing diterpenoid biosynthetic pathway were achieved by George Wagner and coworkers (Wang et al., 2001). They suppressed the activity of cembratriene-ol hydroxylase catalyzing the biosynthesis of cembratriene-diol, a diterpene constituting 60% of trichome exudate weight in the tobacco cultivar investigated (Fig. 1). As a result, the concentration of cembratriene-diols decreased whereas that of its precursor cembratriene-ol increased.

With regard to triterpenoids, there are no examples of metabolic engineering yet, with the exception of the enhancement of sterol formation by increasing precursor availability through HMGR overexpression as described above. It is unclear whether no attempts were made to modify triterpene metabolism or that these attempts were unsuccessful. Indeed, the fact that triterpenes are also cytosolic products and derived from FPP, would predict that engineering – just as for sesquiterpenoids – is not straightforward.

9. FUNCTIONAL IMPLICATIONS OF METABOLIC ENGINEERING

9.1. Ecological Consequences of Engineering Terpenoids

The importance of terpenoids in the interaction of plants with other organisms implies that their modification by plant metabolic engineering will have major affects on their response to environmental cues. Petunia plants expressing the *C. breweri* linalool synthase showed a delayed and less severe natural infection by mildew than the non-transformed plants under standard greenhouse conditions (Lücker, 2002). Fruit of tomato plants transformed with the same gene were much more resistant to post-harvest pathogens than the non-transgenic controls (Lücker, 2002). Also effects of transgenic, volatile producing plants on insects have been reported. The transgenic tobacco plants transformed with the lemon monoterpene synthases, described above, were much less visited by herbivorous insects e.g. white flies but more by fruit flies compared to wildtype tobacco plants in the same greenhouse compartment (Lücker, 2002). Fruit flies have been reported to be attracted to limonene (Jacobson, 1982). In choice assays, Arabidopsis plants transformed with the strawberry linalool/nerolidol synthase, emitting higher linalool levels than the control plants, significantly repelled *Myzus persicae* (Aharoni et al., 2003). These observations were recently extended with even more convincing results: transgenic chrysanthemum (*Chrysanthemum x grandiflorum*) producing linalool repelled western flower thrips (*Frankliniella occidentalis*) (Jongsma, 2004). Increased levels of the diterpenoid cembratienol in trichome exudates of the transgenic tobacco plants (see section on Metabolic Engineering of Diterpenoids and Triterpenoids) resulted in higher resistance to aphids, also in field tests (Wang et al., 2001; Wang et al., 2004).

An even more complicated interaction between a number of organisms occurs in multi-trophic interactions (Dicke and van Loon, 2000). Here, upon herbivory, plants start *de novo* biosynthesis of volatiles (mostly terpenoids) that attract natural enemies (predators and parasitoids) of the herbivores that are foraging on the plant (Dicke and van Loon, 2000). This is a widespread phenomenon occurring in many different plant species such as cotton, potato, cucumber, *Nicotiana attenuata*, lima-bean and Arabidopsis (Weissbecker et al., 1997; Bouwmeester et al., 1999; Baldwin, 2001; van Poecke et al., 2001; Bouwmeester et al., 2003). Among the volatiles that are emitted by plants as a response to herbivory are monoterpenes such as linalool and *β*-ocimene. Experiments with transgenic potato plants, emitting linalool, showed that a change in volatile emission can affect the tritrophic interaction. Predatory mites were significantly more attracted to uninfested, transgenic, linalool producing potato plants than to uninfested wild type plants (Bouwmeester et al., 2003). The Arabidopsis plants expressing *FaNES1* with mitochondrial targeting signal and emitting (3*S*)-*E*-nerolidol and (*E*)-DMNT, described above, were attractive to carnivorous predatory mites (*Phytoseiulus persimilis*), the natural enemies of spider mites (Kappers et al., 2005). Transgenic Arabidopsis plants engineered by (Schnee et al., 2006) for the production of sesquiterpenes normally emitted by maize, attracted females of the parasitoid *Cotesia marginiventris* that located their lepidopteran hosts (the parasitoids were first exposed to the volatiles in association with their hosts).

9.2. Physiological Effects of Engineering Terpenoids

The modification of the flux in isoprenoid biosynthesis in order to generate high levels of heterologous terpenes by metabolic engineering potentially risks production of other isoprenoid derivatives and normal plant growth and development. A decrease in the precursor flux to the sesquiterpenoid pathway in transgenic monoterpene producing tobacco might be an explanation for a lower emission of β-caryophyllene from flowers of these plants (Lücker et al., 2004a). Apparently, the introduced monoterpene synthases competed for the plastidic pool of IPP, hence lowering the flux of IPP to the cytosol, and thus to sesquiterpene biosynthesis. Changes in the flux to higher terpenoids such as the carotenoids, gibberellins or abscisic acid could have dramatic physiological effects in transgenic plants. Indeed, overexpression of a phytoene synthase in tomato resulted in a dwarf phenotype that was explained by a severe inhibition of gibberellin biosynthesis by competition for the common substrate GGPP (Fray et al., 1995). Also transgenic Arabidopsis lines with a high expression of the strawberry linalool/nerolidol synthase showed growth retardation compared with control plants (Aharoni et al., 2003). Tobacco plants that were homozygous for a single lemon monoterpene synthase showed a reduction in size of about 20% compared with control plants, suggesting decreased production of gibberellins (Lücker et al., 2006). The expression of *TXS* in Arabidopsis caused retarded growth and induced a pale-green phenotype (Besumbes et al., 2004). In other cases, transgenic plants with altered terpenoid profiles were obtained without any detectable effect on the plant phenotype (Lücker et al., 2001; Ohara et al.,

2003; Kappers et al., 2005). Seeds of transgenic tobacco plants expressing a lemon monoterpene synthase displayed reduced dormancy after storage compared with control seeds (Lücker, 2002). This is likely due to the production of lower levels of abscisic acid, as a result of competition for plastidic isoprenoid precursors by the introduced monoterpene synthases. Maternal abscisic acid is responsible for the onset of dormancy in developing seeds (Bewley, 1997).

The growth reduction in the transgenic Arabidopsis and tobacco plants (Lücker, 2002; Aharoni et al., 2003) may also have been caused by a phytotoxic effect of the monoterpenes produced at high levels in cells and/or organs not adapted to deal with such high levels of terpenoids (Vaughn and Spencer, 1991, 1996; Romagni et al., 2000; Aharoni et al., 2003). Indeed, when we used the chrysanthemum Rubisco small subunit promoter [a promoter that imparts much higher expression levels than the 35S CaMV promoter; www.impactvector.com; (Outchkourov et al., 2003)] to drive expression of monoterpene synthases in potato, an extreme range of phenotypes was obtained, including severe bleaching of leaves (Aharoni et al., 2006). The negative effects of terpenoid engineering on plant growth could be overcome by better spatial and temporal control of transgene expression (Aharoni et al., 2005).

9.3. Commercial Aspects

The success in engineering a functional two-step and two-compartment pathway for monoterpene biosynthesis (Lücker et al., 2004b) and the progress in the engineering of sesquiterpene and diterpene production (Besumbes et al., 2004; Kappers et al., 2005; Schnee et al., 2006) creates possibilities of commercial interest as they theoretically enable the production of virtually any desired terpenoid molecule in plants. When this approach is used for ornamental plants, they could be genetically engineered to produce fragrant monoterpenes and monoterpene derivatives according to consumers demand. The metabolic engineering of three monoterpene synthases in tobacco, for example, imparted a fruity smell to flowers and leaves (Lücker, 2002; Lücker et al., 2006). A first attempt at sensory analysis by a human panel of these plants resulted in the detection of a significant difference between transgenic and the control plants (El Tamer et al., 2003). However, in a series of sensory attributes no significant differences could be discriminated, probably as a result of insufficient panel training (El Tamer et al., 2003). Most striking was the menthol-like fragrance of engineered tobacco plants emitting various monoterpenes including (+)-*trans*-isopiperitenol (Lücker et al., 2006).

A large number of terpenoids including mono- and sesquiterpenes are highly bioactive molecules which play a significant role in the interaction of plants with other organisms. Plants producing these compounds may display improved plant disease and pest resistance and might therefore require less pesticides. Indeed, in a series of studies on transgenic plants producing novel terpenoids, changes in the resistance to pests and diseases were observed (Wang et al., 2001; Lücker, 2002; Aharoni et al., 2003; Wang et al., 2004).

10. OUTLOOK

Now that the proof of concept has been given, metabolic engineering of terpenoids can be applied for commercial purposes, for example to produce flower specific attractants to improve pollination of seed crops or to produce volatiles attractive to humans in commercial cut flowers. The scientific challenges will be to further optimize the engineering of sesquiterpene formation and the fine-tuning of metabolic engineering by using specific promoters to target gene expression to a desirable location and time. Once specific transcription factors have been identified that could impart major changes to flux in certain branches of the isoprenoid pathway, they will also be applied to engineer terpenoid production. In addition, the instrument of metabolic engineering will create great opportunities to study the ecological importance of terpenoids in plant-plant interaction and interaction of plants with other organisms and will therefore increase our understanding of the regulation of terpenoid biosynthesis.

REFERENCES

Adam, KP, and Zapp, J, 1998, Biosynthesis of the isoprene units of chamomile sesquiterpenes, *Phytochemistry* **48**(6)**:**953–959.

Aharoni, A, Giri, AP, Deuerlein, S, Griepink, F, de Kogel, W-J, Verstappen, FWA, Verhoeven, HA, Jongsma, MA, Schwab, W, and Bouwmeester, HJ, 2003, Terpenoid metabolism in wild-type and transgenic *Arabidopsis* plants, *Plant Cell* **15**(12)**:**2866–2884.

Aharoni, A, Jongsma, MA, and Bouwmeester, HJ, 2005, Volatile science? Metabolic engineering of terpenoids in plants, *Trends Plant Sci* **10**(12)**:**594–602.

Aharoni, A, Jongsma, MA, Kim, TY, Ri, MB, Giri, AP, Verstappen, FWA, Schwab, W, and Bouwmeester, HJ, 2006, Metabolic engineering of terpenoid biosynthesis in plants, *Phytochemistry Rev***:**In Press.

Baldwin, IT, 2001, An ecologically motivated analysis of plant-herbivore interactions in native tobacco, *Plant Physiol* **127**(4)**:**1449–1458.

Bertea, CM, Schalk, M, Karp, F, Maffei, M, and Croteau, R, 2001, Demonstration that menthofuran synthase of mint (*Mentha*) is a cytochrome P450 monooxygenase: Cloning, functional expression, and characterization of the responsible gene, *Arch Biochem Biophys* **390**(2)**:**279–286.

Besumbes, O, Sauret-Gueto, S, Phillips, MA, Imperial, S, Rodriguez-Concepcion, M, and Boronat, A, 2004, Metabolic engineering of isoprenoid biosynthesis in Arabidopsis for the production of taxadiene, the first committed precursor of Taxol, *Biotechnol Bioeng* **88**(2)**:**168–175.

Bewley, JD, 1997, Seed germination and dormancy, *Plant Cell* **9**(7)**:**1055–1066.

Bick, JA, and Lange, BM, 2003, Metabolic cross talk between cytosolic and plastidial pathways of isoprenoid biosynthesis: unidirectional transport of intermediates across the chloroplast envelope membrane, *Arch Biochem Biophys* **415**(2)**:**146–154.

Bosabalidis, AM, 1996, Ontogenesis, ultrastructure and morphometry of the petiole oil ducts of celery (*Apium graveolens* L), *Flavour Frag J* **11**(5)**:**269–274.

Botella-Pavia, P, Besumbes, O, Phillips, MA, Carretero-Paulet, L, Boronat, A, and Rodriguez-Concepcion, M, 2004, Regulation of carotenoid biosynthesis in plants: evidence for a key role of hydroxymethylbutenyl diphosphate reductase in controlling the supply of plastidial isoprenoid precursors, *Plant J* **40**(2)**:**188–199.

Bouwmeester, HJ, Gershenzon, J, Konings, MCJM, and Croteau, R, 1998, Biosynthesis of the monoterpenes limonene and carvone in the fruit of carawayIDemonstration of enzyme activities and their changes with development, *Plant Physiol* **117**(3)**:**901–912.

Bouwmeester, HJ, Verstappen, FWA, Posthumus, MA, and Dicke, M, 1999, Spider mite-induced (3*S*)-(E)-nerolidol synthase activity in cucumber and lima beanThe first dedicated step in acyclic C11-homoterpene biosynthesis, *Plant Physiol* **121**(1)**:**173–180.

Bouwmeester, HJ, Kappers, IF, Verstappen, FW, Aharoni, A, Luckerhoff, LLP, Lücker, J, Jongsma, MA, and Dicke, M, 2003, Exploring multi-trophic plant-herbivore interactions for new crop protection methods, in *The International Congress Crop Science and Technology*, British Crop Protection Council, Alton, UK, Glasgow, 1123–1134.

Broun, P, 2004, Transcription factors as tools for metabolic engineering in plants, *Curr Opin Plant Biol* **7**(2)**:**202–209.

Capell, T, and Christou, P, 2004, Progress in plant metabolic engineering, *Curr Opin Biotech* **15** (2)**:**148–154.

Chappell, J, Wolf, F, Proulx, J, Cuellar, R, and Saunders, C, 1995, Is the reaction catalyzed by 3-hydroxy-3-methylglutaryl Coenzyme-A reductase a rate-limiting step for isoprenoid biosynthesis in plants, *Plant Physiol* **109**(4)**:**1337–1343.

Chappell, J, 2002, The genetics and molecular genetics of terpene and sterol origami, *Curr Opin Plant Biol* **5**(2)**:**151–157.

Chappell, J, 2004, Valencene synthase – a biochemical magician and harbinger of transgenic aromas, *Trends Plant Sci* **9**(6)**:**266–269.

Chen, DH , Ye, HC, and Li, GF, 2000, Expression of a chimeric farnesyl diphosphate synthase gene in *Artemisia annua* Ltransgenic plants via Agrobacterium tumefaciens-mediated transformation, *Plant Sci* **155**(2)**:**179–185.

Dicke, M, and van Loon, JJA, 2000, Multitrophic effects of herbivore-induced plant volatiles in an evolutionary context, *Entomol Exp Appl* **97**(3)**:**237–249.

Diemer, F, Caissard Jean, C, Moja, S, Chalchat Jean, C, and Jullien, F, 2001, Altered monoterpene composition in transgenic mint following the introduction of 4S-limonene synthase, *Plant Physiol Bioch* **39**(7–8)**:**603–614.

Dudareva, N, Cseke, L, Blanc, VM, and Pichersky, E, 1996, Evolution of floral scent in *Clarkia*: novel patterns of S-linalool synthase gene expression in the *Cbreweri* flower, *Plant Cell* **8**(7)**:**1137–1148.

Dudareva, N, Pichersky, E, and Gershenzon, J, 2004, Biochemistry of Plant Volatiles, *Plant Physiol* **135**(4)**:**1893–1902.

Dudareva, N, Andersson, S, Orlova, I, Gatto, N, Reichelt, M, Rhodes, D, Boland, W, and Gershenzon, J, 2005, The nonmevalonate pathway supports both monoterpene and sesquiterpene formation in snapdragon flowers, *P Natl Acad Sci USA* **102**(3)**:**933–938.

El Tamer, MK, Smeets, M, Holthuysen, N, Lücker, J, Tang, A, Roozen, J, Bouwmeester, HJ, and Voragen, AGJ, 2003, The influence of monoterpene synthase transformation on the odour of tobacco, *J Biotechnol* **106**(1)**:**15–21.

Estevez, JM, Cantero, A, Reindl, A, Reichler, S, and Leon, P, 2001, 1-deoxy-D-xylulose-5-phosphate synthase, a limiting enzyme for plastidic isoprenoid biosynthesis in plants, *J Biol Chem* **276**(25)**:**22901–22909.

Fahn, A, 1979, *Secretory Tissues in Plants*, Academic Press, London, pp 302.

Fray, RG, Wallace, A, Fraser, PD, Valero, D, Hedden, P, Bramley, PM, and Grierson, D, 1995, Constitutive expression of a fruit phytoene synthase gene in transgenic tomatoes causes dwarfism by redirecting metabolites from the gibberellin pathway, *Plant J* **8**(5)**:**693–701.

Galili, G, Galili, S, Lewinsohn, E, and Tadmor, Y, 2002, Genetic, molecular, and genomic approaches to improve the value of plant foods and feeds, *Crit Rev Plant Sci* **21**(3)**:**167–204.

Gershenzon, J, Maffei, M, and Croteau, R, 1989, Biochemical and histochemical localization of monoterpene biosynthesis in the glandular trichomes of spearmint (*Mentha spicata*), *Plant Physiol* **89**(4)**:**1351–1357.

Gershenzon, J, and Croteau, RB, 1993, Terpenoid biosynthesis: the basic pathway and formation of monoterpenes, sesquiterpenes and diterpenes, in: *Lipid metabolism in plants*, Moore, TS, ed, CRC Press, Boca Raton, pp 340–388.

Gershenzon, J, McConkey, ME, and Croteau, RB, 2000, Regulation of monoterpene accumulation in leaves of peppermint, *Plant Physiol* **122**(1)**:**205–214.

Han, JL, Liu, BY, Ye, HC, Wang, H, Li, ZQ, and Li, GF, 2006, Effects of overexpression of the endogenous farnesyl diphosphate synthase on the artemisinin content in *Artemisia annua* L, *Journal of Integrative Plant Biology* **48**(4)**:**482–487.

Harker, M, Hellyer, A, Clayton, JC, Duvoix, A, Lanot, A, and Safford, R, 2003, Co-ordinate regulation of sterol biosynthesis enzyme activity during accumulation of sterols in developing rape and tobacco seed, *Planta* **216**(4)**:**707–715.

Haudenschild, C, and Croteau, R, 1998, Molecular engineering of monoterpene production, *Genet Eng* **20:**267–280.

Hemmerlin, A, Hoeffler, JF, Meyer, O, Tritsch, D, Kagan, IA, Grosdemange-Billiard, C, Rohmer, M, and Bach, TJ, 2003, Cross-talk between the cytosolic mevalonate and the plastidial methylerythritol phosphate pathways in Tobacco Bright Yellow-2 cells, *J Biol Chem* **278**(29)**:**26666–26676.

Hohn, TM, and Ohlrogge, JB, 1991, Expression of a fungal sesquiterpene cyclase gene in transgenic tobacco, *Plant Physiol* **97**(1)**:**460–462.

Hori, M, 1998, Repellency of rosemary oil against *Myzus persicae* in a laboratory and in a screenhouse, *J Chem Ecol* **24**(9)**:**1425–1432.

Izumi, S, Takashima, O, and Hirata, T, 1999, Geraniol is a potent inducer of apoptosis-like cell death in the cultured shoot primordia of *Matricaria chamomilla*, *Biochem Biophys Res Commun* **259**(3)**:** 519–522.

Jacobson, M, 1982, Plants, insects, and man – their interrelationships, *Econ Bot* **36**(3)**:**346–354.

Jongsma, MA, 2004, Novel genes for control and deterrence of sucking insect pests, in *ISB News Report*, November, 1–4 http://wwwisbvtedu/news/2004/Nov04pdf.

Kappers, IF, Aharoni, A, van Herpen, T, Luckerhoff, LLP, Dicke, M, and Bouwmeester, HJ, 2005, Genetic engineering of terpenoid metabolism attracts bodyguards to Arabidopsis, *Science* **309**(5743)**:**2070–2072.

Krasnyanski, S, May, RA, Loskutov, A, Ball, TM, and Sink, KC, 1999, Transformation of the limonene synthase gene into peppermint (*Mentha* x *piperita* L) and preliminary studies on the essential oil profiles of single transgenic plants, *Theor Appl Genet* **99**(3–4)**:**676–682.

Laule, O, Furholz, A, Chang, HS, Zhu, T, Wang, X, Heifetz, PB, Gruissem, W, and Lange, BM, 2003, Crosstalk between cytosolic and plastidial pathways of isoprenoid biosynthesis in *Arabidopsis thaliana*, *P Natl Acad Sci USA* **100**(11)**:**6866–6871.

Lavy, M, Zuker, A, Lewinsohn, E, Larkov, O, Ravid, U, Vainstein, A, and Weiss, D, 2002, Linalool and linalool oxide production in transgenic carnation flowers expressing the *Clarkia breweri* linalool synthase gene, *Mol Breeding* **9**(2)**:**103–111.

Lewinsohn, E, Schalechet, F, Wilkinson, J, Matsui, K, Tadmor, Y, Nam, KH, Amar, O, Lastochkin, E, Larkov, O, Ravid, U, Hiatt, W, Gepstein, S, and Pichersky, E, 2001, Enhanced levels of the aroma and flavor compound *S*-linalool by metabolic engineering of the terpenoid pathway in tomato fruits, *Plant Physiol* **127**(3)**:**1256–1265.

Little, DB, and Croteau, RB, 1999, Biochemistry of essential oil terpenes – A thirty year overview, in: *Flavor Chemistry: 30 Years of Progress*, Teranishi, R, and Wick, EL, ed, Kluwer Academic/ Plenum Publishers, New York, pp 239–253.

Lücker, J, Bouwmeester, HJ, Schwab, W, Blaas, J, van der Plas, LHW, and Verhoeven, HA, 2001, Expression of *Clarkia S*-linalool synthase in transgenic petunia plants results in the accumulation of *S*-linalyl-β-D-glucopyranoside, *Plant J* **27**(4)**:**315–324.

Lücker, J, 2002, *Metabolic Engineering of Monoterpene Biosynthesis in Plants*, PhD Thesis, Wageningen University, pp 168.

Lücker, J, El Tamer, MK, Schwab, W, Verstappen, FWA, van der Plas, LHW, Bouwmeester, HJ, and Verhoeven, HA, 2002, Monoterpene biosynthesis in lemon (*Citrus limon*): cDNA isolation and functional analysis of four monoterpene synthases, *Eur J Biochem* **269**(13)**:**3160–3171.

Lücker, J, Schwab, W, van Hautum, B, Blaas, J, van der Plas, LHW, Bouwmeester, HJ, and Verhoeven, HA, 2004a, Increased and altered fragrance of tobacco plants after metabolic engineering using three monoterpene synthases from lemon, *Plant Physiol* **134**(1)**:**510–519.

Lücker, J, Schwab, W, Franssen, MCR, van der Plas, LHW, Bouwmeester, HJ, and Verhoeven, HA, 2004b, Metabolic engineering of monoterpene biosynthesis: Two-step production of (+)-*trans*-isopiperitenol by tobacco, *Plant J* **39**(1)**:**135–145.

Lücker, J, Verhoeven, HA, van der Plas, LHW, and Bouwmeester, HJ, 2006, Molecular engineering of floral scent, in: *Biology of Floral Scent*, Dudareva, N, and Pichersky, E, ed, CRC Press, Boca Raton, pp 321–337.

Mahmoud, SS, and Croteau, R, 2001, Metabolic engineering of essential oil yield and composition in mint by altering expression of deoxyxylulose phosphate reductoisomerase and menthofuran synthase, *P Natl Acad Sci USA* **98**(15)**:**8915–8920.

Mahmoud, SS, and Croteau, RB, 2002, Strategies for transgenic manipulation of monoterpene biosynthesis in plants, *Trends Plant Sci* **7**(8)**:**366–373.

Mahmoud, SS, Williams, M, and Croteau, R, 2004, Cosuppression of limonene-3-hydroxylase in peppermint promotes accumulation of limonene in the essential oil, *Phytochemistry* **65**(5)**:**547–554.

Masferrer, A, Arro, M, Manzano, D, Schaller, H, Fernandez-Busquets, X, Moncalean, P, Fernandez, B, Cunillera, N, Boronat, A, and Ferrer, A, 2002, Overexpression of *Arabidopsis thaliana* farnesyl diphosphate synthase (FPS1S) in transgenic *Arabidopsis* induces a cell death/senescence-like response and reduced cytokinin levels, *Plant J* **30**(2)**:**123–132.

McCaskill, D, and Croteau, R, 1998, Some caveats for bioengineering terpenoid metabolism in plants, *Trends Biotechnol* **16**(8)**:**349–355.

Ohara, K, Ujihara, T, Endo, T, Sato, F, and Yazaki, K, 2003, Limonene production in tobacco with *Perilla* limonene synthase cDNA, *J Exp Bot* **54**(393)**:**2635–2642.

Outchkourov, NS, Peters, J, de Jong, J, Rademakers, W, and Jongsma, MA, 2003, The promoter-terminator of chrysanthemum rbcS1 directs very high expression levels in plants, *Planta* **216**(6)**:**1003–1012.

Pichersky, E, and Gershenzon, J, 2002, The formation and function of plant volatiles: perfumes for pollinator attraction and defense, *CurrOpinPlant Biol* **5**(3)**:**237–243.

Rodriguez-Concepcion, M, and Boronat, A, 2002, Elucidation of the methylerythritol phosphate pathway for isoprenoid biosynthesis in bacteria and plastidsA metabolic milestone achieved through genomics, *Plant Physiol* **130**(3)**:**1079–1089.

Romagni, JG, Allen, SN, and Dayan, FE, 2000, Allelopathic effects of volatile cineoles on two weedy plant species, *J Chem Ecol* **26**(1)**:**303–313.

Sacchettini, JC, and Poulter, CD, 1997, Biochemistry – Creating isoprenoid diversity, *Science* **277**(5333)**:**1788–1789.

Sandmann, G, Romer, S, and Fraser, PD, 2006, Understanding carotenoid metabolism as a necessity for genetic engineering of crop plants, *Metabol Eng* **8**(4)**:**291–302.

Schaller, H, Grausem, B, Benveniste, P, Chye, ML, Tan, CT, Song, CT, and Chua, NH, 1995, Expression of the *Hevea brasiliensis* (HBK) MullArg 3-hydroxy-3-methylglutaryl-Coenzyme A reductase 1 in tobacco results in sterol overproduction, *Plant Physiol* **109**(3)**:**761–770.

Schnee, C, Kollner, TG, Held, M, Turlings, TCJ, Gershenzon, J, and Degenhardt, J, 2006, The products of a single maize sesquiterpene synthase form a volatile defense signal that attracts natural enemies of maize herbivores, *P Natl Acad Sci USA* **103**(4)**:**1129–1134.

Schuhr, CA, Radykewicz, T, Sagner, S, Latzel, C, Zenk, MH, Arigoni, D, Bacher, A, Rohdich, F, and Eisenreich, W, 2003, Quantitative assessment of crosstalk between the two isoprenoid biosynthesis pathways in plants by NMR spectroscopy, *Phytochemistry Rev* **2**(1–2)**:**3–16.

Suga, T, and Hirata, T, 1990, Biotransformation of exogenous substrates by plant cell cultures, *Phytochemistry* **29**(8)**:**2393–2406.

van Poecke, RMP, Posthumus, MA, and Dicke, M, 2001, Herbivore-induced volatile production by *Arabidopsis thaliana* leads to attraction of the parasitoid *Cotesia rubecula*: Chemical, behavioral, and gene-expression analysis, *J Chem Ecol* **27**(10)**:**1911–1928.

Vaughn, SF, and Spencer, GF, 1991, Volatile monoterpenes inhibit potato tuber sprouting, *Am Potato J* **68**(12)**:**821–831.

Vaughn, SF, and Spencer, GF, 1996, Synthesis and herbicidal activity of modified monoterpenes structurally similar to cimmethylin, *Weed Sci* **44**(1)**:**7–11.

Verlet, N, 1993, Commercial aspects, in: *Volatile Oil Crops: their Biology, Biochemistry, and Production*, Hay, RKM, and Waterman, PG, ed, Longman Scientific & Technical, Essex, UK, pp 137–174.

Wallaart, TE, Bouwmeester, HJ, Hille, J, Poppinga, L, and Maijers, NCA, 2001, Amorpha-4,11-diene synthase: cloning and functional expression of a key enzyme in the biosynthetic pathway of the novel antimalarial drug artemisinin, *Planta* **212**(3)**:**460–465.

Wang, E, Hall, JT, and Wagner, GJ, 2004, Transgenic *Nicotiana tabacum* L with enhanced trichome exudate cembratrieneols has reduced aphid infestation in the field, *Mol Breeding* **13**(1)**:**49–57.

Wang, EM, Wang, R, De Parasis, J, Loughrin, JH, Gan, SS, and Wagner, GJ, 2001, Suppression of a P450 hydroxylase gene in plant trichome glands enhances natural-product-based aphid resistance, *Nat Biotechnol* **19**(4)**:**371–374.

Weidenhamer, JD, Macias, FA, Fischer, NH, and Williamson, GB, 1993, Just how insoluble are monoterpenes? *J Chem Ecol* **19**(8)**:**1799–1807.

Weissbecker, B, Schutz, S, Klein, A, and Hummel, HE, 1997, Analysis of volatiles emitted by potato plants by means of a Colorado beetle electroantennographic detector, *Talanta* **44**(12)**:**2217–2224.

Wolfertz, M, Sharkey, TD, Boland, W, and Kuhnemann, F, 2004, Rapid regulation of the methylerythritol 4-phosphate pathway during isoprene synthesis, *Plant Physiol* **135**(4)**:**1939–1945.

CHAPTER 10

METABOLIC ENGINEERING OF SEED OIL BIOSYNTHETIC PATHWAYS FOR HUMAN HEALTH

HOWARD G. DAMUDE AND ANTHONY J. KINNEY
DuPont Experimental Station
Wilmington, DE 19880 USA

Abstract: Multiple studies have shown that inclusion of omega-3 long chain polyunsaturated fatty acids in the diet can have multiple health benefits including positive effects on cardiovascular and mental health. Although marine oils can be a rich source of omega-3 long chain polyunsaturated fatty acids, processed fish oil is undesirable as a food ingredient because of the associated objectionable flavors and contaminants that are difficult and cost-prohibitive to remove. Oilseed plants engineered to produce omega-3 LC PUFAs offer a safe, sustainable and cost-effective alternative to fish oils as a source of LC PUFA for food ingredients. Given the potential benefit, much recent effort from both academic and industrial research labs has been directed towards producing a land-based, oilseed-derived source of LC-PUFAs and the approaches, recent advances and future prospects will be discussed here

Keywords: Oils, fats, fatty acids, trans-fats, long chain polyunsaturated, omega-3, soybean

Abbreviations: LCPUFA – Long Chain Polyunsaturated Fatty Acids; DHA – docosahexaenoic acid; EPA – eicosapentaenoic acid; ARA – arachidonic acid; COX-2 – cyclooxygenase-2; LA – linoleic acid; ALA – alpha-linolenic acid; GLA – gamma-linolenic acid; STA – stearidonic acid; DGLA – dihomo-gamma-linolenic acid; ETA – eicosatetraenoic acid; EDA – eicosadienoic acid; ERA – eicosatrienoic acid; DPA – docosapentaenoic acid; CoA – coenzyme A; PC – phosphatidylcholine; LPAAT – lysophosphatidic acid acyltransferase; SCA – sciadonic acid; JUN – juniperonic acid

1. INTRODUCTION

Oils and fats are an important source of energy for the human body and form a vital component of many cell constituents. The main sources of fat in the human diet are vegetable oils, mostly soy, canola (oilseed rape), palm, peanut and sunflower. Many of these oils are very rich in omega-6 fatty acids and excess consumption of omega-6

R. Verpoorte et al. (eds.), Applications of Plant Metabolic Engineering, 237–247.
© 2007 *Springer*.

fatty acids leads to the depletion of omega-3 fatty acids in human body tissues, with numerous negative health consequences (Lands, 2005). Vegetable oils also contribute to the sensory characteristics of numerous processed food products, often in an hydrogenated or partially hydrogenated form. Oils are hydrogenated to reduce oxidation and increase solid content through the formation of *trans* fatty acids. Hydrogenation improves shelf-life, maintains the flavor and provides the expected mouth-feel and consistency of oil-containing foods (Stauffer, 1996). However, the negative consequences on human health of the consumption of *trans* unsaturated fatty acids has become better understood in recent years (Korver and Katan, 2006).

Initial attempts at the genetic manipulation of the fatty acid profile of oilseeds focused on redirecting fatty acid biosynthesis in the developing seed, either by blocking specific steps, such as fatty acid desaturation (Kinney and Knowlton, 1998), or introducing enzyme activities that redirected fatty acid synthesis to new end products, such as oils with medium chain fatty acids (Del Vecchio, 1996). The idea was to improve the oxidative stability of the oil without the need for hydrogenation and/or provide a solid fat functionality without the need for *trans* fatty acids. These genetic manipulations involved the introduction of one or two transgenes and were technically very successful. A number of these novel oils from transgenic oilseeds have been, or will be, commercialized (Del Vecchio, 1996; Kinney and Knowlton, 1998) and the topic has been extensively reviewed in recent years (Voelker and Kinney, 2001; Coughlan and Kinney, 2002).

Advances in gene discovery, with the advent of genomic technologies, coupled with advances in gene expression know-how, have led to the possibility of more complex manipulations of plant cell lipid metabolism, which will be discussed here.

2. LONG CHAIN POLYUNSATURATED FATTY ACIDS AND HUMAN NUTRITION

One of the most exciting areas for improving the health properties of plant seed oils is in the production of long chain polyunsaturated fatty acids (LCPUFAs). LCPUFAs containing at least 20 carbons are important in human physiology as components of cell membrane phospholipids. For instance, docosahexaenoic acid [DHA, 22:6(4,7,10,13,16,19)] is an important component of mammalian retinal and brain membranes and has been shown to play a role in the cognitive development of infants and the mental health of adults (Willatts and Forsyth, 2000; Irbarren et al., 2004; Stoll et al., 2001). In addition, numerous studies have shown cardiovascular health benefits arising from the consumption of LCPUFAs such as DHA and eicosapentaenoic acid [EPA, 20:5(5,8,11,14,17)] (Dyerberg and Bang, 1982; Simopoulos, 2006; Napier et al., 2004a; Napier et al., 2004b). LCPUFAs are also precursors to the eicosanoid family of metabolites (prostaglandins, leukotrienes, thromboxanes) (Funk, 2001; Hwang, 2000) which regulate key metabolic functions in the human body such as the inflammatory response, induction of blood clotting and regulation of blood pressure (Yaqoob, 2003). Eicosanoids derived from omega-6 LCPUFAs, such as arachidonic acid [ARA, 20:4(5,8,11,14)], are generally pro-inflammatory while those derived from omega-3 LCPUFAs, such as eicosapentaenoic acid, are anti-inflammatory (Calder, 2003; Hwang, 2000; Simopoulos, 2006). The first step in eicosanoid biosynthesis from LCPUFAs is catalyzed by the cyclooxygenase-2 (COX-2) enzyme (Hwang, 2000), against which a host of non-steroidal, anti-inflammatory pharmaceuticals have been designed (COX-2 inhibitors). The COX-2 enzyme can utilize either ARA or EPA and therefore, the anti-inflammatory action of EPA can be attributed to both competitive inhibition between EPA and ARA for COX-2 as well as the anti-inflammatory action of the omega-3-derived eicosanoids themselves (Calder, 2003). It is interesting to note that while EPA and DHA are beneficial to cardiovascular health, cardiovascular side effects have been attributed to the use of COX-2-selective inhibitors (Mukherjee et al., 2001) and some of these have been removed from sale. Clearly, an optimal balance of omega-3 and omega-6 LCPUFAs must be achieved to maintain a healthy state.

3. SOURCES OF LCPUFAS IN THE HUMAN DIET

EPA and ARA can be synthesized in the human body from the essential dietary fatty acids linoleic acid [LA, 18:2(9,12)] or alpha-linolenic acid [ALA, 18:3(9,12,15)], respectively, or can be obtained directly from the diet. Unfortunately, in modern western societies, the dietary ratio of omega-6 to omega-3 fatty acids has shifted heavily toward omega-6 fatty acids and, by some estimates, is as much as 30-fold too high (Sargent, 1997; Simopoulos, 1999). This shift is due to an overall decrease in the consumption of fish, which contain high levels of omega-3 LCPUFAs, coupled with a large increase in the consumption of omega-6-containing foods such as plant-based oils or meat and poultry fed with plant grains. Consumption of foods rich in omega-3 LCPUFAs helps to correct this imbalance by shifting the omega-6 to omega-3 fatty acid ratio to more optimal levels and leads to the observed health benefits.

Unfortunately, an increased demand for fish and fish oil high in omega-3 LCPUFAs, mainly by the aquaculture industry, is putting an even greater stress on an already overexploited resource (Pauly et al., 2000). In addition, the cost associated with removing objectionable odors and flavors, as well as contaminants such as mercury and PCBs, prohibit the use of fish oil in common foods (Jacobs et al., 2004; Hites et al., 2004). Thus, inexpensive alternative oils high in omega-3 LCPUFAs are desirable. Current alternatives to fish oil as a source of omega-3 LCPUFAs include microalgae produced by fermentation (eg. *Crypthecodinium* for DHA) but the high costs associated with the capital intensive process prohibit the practical inclusion of these oils in most foods. One alternative is to engineer the biosynthetic pathway for omega-3 LCPUFA-production into an economically sustainable host such as an oilseed crop. A commercially-relevant product would be substantially similar to fish oil having both, high levels of EPA and DHA, with negligible levels of ARA or other omega-6 pathway intermediates. Most cold water marine fish oils have very low levels of omega-6 fatty acids (2–5%), the majority of which is present as LA, an omega-3 to omega-6 ratio ranging from

5 to 11, and EPA and DHA (combined) ranging from 10–25% (Sargent, 1997). Given the potential positive health, environmental and economic benefits associated with an oilseed-based LCPUFA oil, engineering oilseeds is a very attractive but concomitantly technically-challenging proposition.

4. LCPUFA BIOSYNTHETIC PATHWAYS

Cold water marine fish and fish oils are the main source of EPA and DHA in our diet and although many of these fish are capable of synthesizing LCPUFAs from the essential fatty acids LA and ALA, the typical high content of EPA and DHA found in fish bodies can only be attained through the consumption of LCPUFAs in their diet (Bell et al., 2003). LCPUFAs in the food chain originate in marine micro-organisms or phytoplankton (eg. diatoms, golden-brown algae, green algae, blue-green algae, dinoflagellates, etc) where they are biosynthesized *de novo* from sunlight. LCPUFAs work their way up the food chain from phytoplankton to small fish such as Menhaden and finally into the larger carnivorous fish (tuna, salmon, etc).

Two classes of biochemical pathways have been identified that lead to LCPUFA biosynthesis in micro-organisms; the anaerobic polyketide synthase pathways (Metz et al., 2001) and the aerobic fatty acid desaturation/elongation pathways (Sayanova and Napier, 2004).

In the polyketide synthase pathway, large, multidomain enzymes coded for by a few ORFs systematically elongate and desaturate smaller chain fatty acids until the final fatty acid (EPA or DHA) is produced. In general, the fatty acid produced is specific to the polyketide synthase from which it was derived (Metz et al., 2001).

In the aerobic fatty acid desaturation/elongation pathways, two types of converging pathways to EPA have been characterized and these are shown in Figure 1. In both pathway types, the micro-organisms first synthesize LA and ALA from metabolic precursors. In the first pathway type (delta-6 pathway), LA and ALA are first desaturated to gamma-linolenic acid [GLA, 18:3(6,9,12)] and stearidonic acid [STA, 18:4(6,9,12,15)], respectively, by a delta-6 fatty acid desaturase. These fatty acids are then elongated by two carbons *via* the microsomal fatty acid elongation complex (Sayanova and Napier, 2004), where elongation is initiated by a substrate specificity-controlling beta-ketoacyl synthase enzyme (delta-6 elongase), to give dihomo-gamma-linolenic acid [DGLA, 20:3(8,11,14)] and eicosatetraenoic acid [ETA, 20:4(8,11,14,17)]. The final step is desaturation to ARA and EPA by a delta-5 desaturase.

In the second type of pathway (delta-9 pathway), LA and ALA are first elongated by a delta-9-specific elongase to eicosadienoic acid [EDA, 20:2(11,14)] and eicosatrienoic acid [ERA, 20:3(11,14,17)], followed by delta-8 desaturation to DGLA and ETA, respectively. A final desaturation occurs via a delta-5 desaturase as in the first pathway to generate ARA and EPA. Independent of the aerobic pathway utilized, some organisms have the added capability of converting omega-6

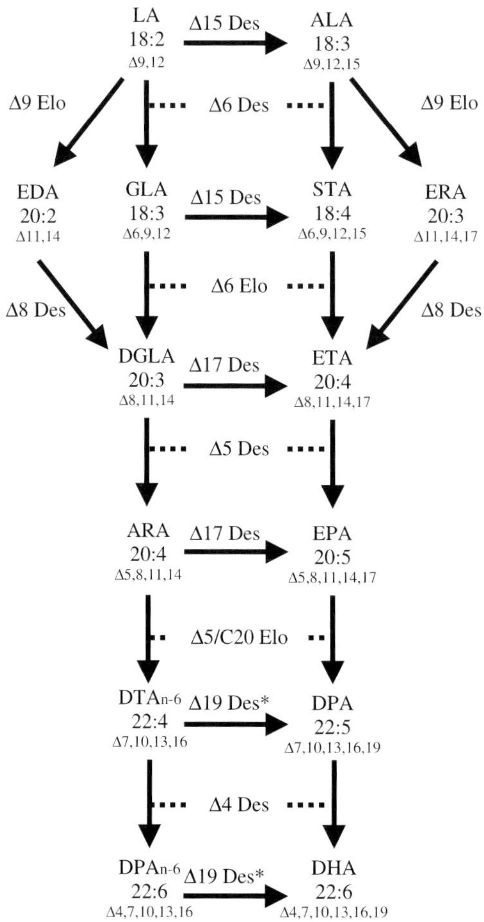

Figure 1. Microbial aerobic LC-PUFA biosynthetic pathways. The delta-6 pathway is shown with a delta-6 desaturse (Δ6 Des) and delta-6 elongase (Δ6 Elo) and the delta-9 pathway is shown with a delta-9 elongase (Δ9 elo) and a delta-8 desaturase (Δ8 Des) Both pathways converge and utilize the same delta-5 desaturase (Δ5 Des). Pathway genes are shown to utilize both omega-3 and omega-6 fatty acid substrates and in the case of Δ6 Des, Δ6 Elo and Δ5 Des, this is indicated with a dotted line. Conversion of 18 carbon, 20 carbon or 22 carbon omega-6 fatty acids to omega-3 fatty acids is indicated by the delta-15 desaturase (Δ15 Des), delta-17 desaturase (Δ17 Des) and a hypothetical delta-19 (Δ19 Des*), respectively. Elongation of ARA or EPA is indicated by the Δ5/C20 elongase (Δ5/C20 Elo) and further delta-4 desaturation by the delta-4 desaturase (Δ4 Des)

fatty acids to omega-3 fatty acids by the action of an omega-3 fatty acid desaturase (Oura and Kajiwara, 2004; Pereira et al., 2004a; Sakuradani et al., 2005; Damude et al., 2006). This can occur at either the 18-carbon (delta-15 desaturase) or 20-carbon (delta-17 desaturase) stage.

In microbes, further conversion of EPA to DHA typically occurs by delta-5/C20 elongation of EPA to docosapentaenoic acid [DPA, 20:5(7,10,13,16,19)] followed by delta-4 desaturation to DHA (Sayanova and Napier, 2004). The possibility exists for a parallel pathway acting on ARA in a similar way with some type of subsequent omega-3 desaturation (Δ19 desaturase) to DHA although this has not been reported as of yet.

In some cases, one organism may contain and/or express both classes of pathways. For instance, a complete polyketide synthase has been cloned and characterized from Thraustochytids as have elongases and desaturases (Metz et al., 2004; Qiu et al., 2001).

5. EXPRESSION OF LCPUFA PATHWAYS IN PLANTS

In recent years, a number of research groups have demonstrated the reconstitution of both types of aerobic LCPUFA pathways in plants. As of yet, expression of the anaerobic polyketide sythase pathway in plants has yet to be demonstrated, although results are expected to be published soon (J Metz, personal communication).

Among the first published descriptions of expression of an LCPUFA pathway in plants was a report describing the reconstitution of a complete delta-9 elongase pathway expressed constitutively in the model plant *Arabidopsis* (Qi et al., 2004). Individual pathway genes (delta-9 elongase, delta-8 desaturase, delta-5 desaturase) were each linked to the Ca35S promoter and in this way, EPA contents as high as 3.0% were produced in *Arabidopsis* leaves, with an ARA content of 6.6%. The overall omega-3 to omega-6 ratio of fatty acids (2.2:1) was slightly lower than that for wild-type *Arabidopsis* leaves (3.5:1), but was in the range commonly found in fish oils (Sargent, 1997). Pathway intermediates and by-products not commonly found in fish oils were also high (10.4%). Total 20-carbon fatty acid was 22.5% representing 36% elongation of 18-carbon substrates and indicating that fatty acid elongation was not particularly limiting in *Arabidopsis* when using the delta-9 elongase pathway. Although very significant as a “proof-of-concept” experiment, expression of an LCPUFA pathway in leaves would not be an economically-practical production platform.

In a separate, but parallel, effort expressing the delta-6 pathway using seed-specific promoters in both linseed and tobacco, Abbadi et al. (2004) obtained a substantially lower content of EPA and ARA (less than 2%) and a high relative abundance of pathway intermediates, mainly GLA and STA. In both cases, elongation of 18 carbon PUFA substrates was limiting with overall elongation at around only 10%. Poor elongation was attributed to the low pool of the delta-6 acyl-CoA (GLA-CoA, STA-CoA) which are the required substrates for the delta-6 elongase. Seemingly, acyl exchange between phospholipid, the substrate on which the desaturases act to form the delta-6 fatty acids, and acyl-CoA, the substrate on which the elongases act, pools was inefficient in these organisms. Linseed was chosen as a production platform because seeds naturally contain high levels of ALA, the omega-3 precursor of EPA. The overall omega-3 to omega-6 ratio of fatty acids in the LC-PUFA-expressing linseeds was good at 1.8, due mainly to high levels of ALA and relatively high levels of STA. The omega-3 to omega-6 ratio of transgenic linseed was likely lower than that of wild-type linseed (around 4:1) due to a combination of a preference of the delta-6 desaturase used for omega-6 substrates (LA) over omega-3 substrates (ALA) and availability of ALA in the sn-2 position of the phosphatidylcholine (PC) pool.

A third report, describing commercially-significant concentrations of LCPUFAs in a plant seed, was a patent application by Kinney et al. (2004). In this study a delta-6 desaturase pathway was expressed in the agronomically-important oilseed crop soybean, under control of strong, seed-specific promoters. In addition to the minimal set of delta-6 pathway-genes (delta-6 desaturase, delta-6 elongase, delta-5 desaturase), the *Arabidopsis* Fad 3 gene (Yadav et al., 1993) and the *Saprolegnia diclina* delta-17 desaturase (Pereira et al., 2004a) were also included in order to increase the omega-3 to omega-6 ratio from the approximately 0.2:1 commonly found in soybean oil to something more similar to that for fish oil (Sargent, 1997). After extensive characterization of multiple seed specific promoters and individual LCPUFA biosynthetic genes, and optimization of promoter-gene cassette combinations and orientations, in soy embryos, soybeans with a content of EPA as high as 19.5% were produced with virtually none of the omega-6 intermediate ARA. The low content of ARA was directly attributed to the use of the *S. diclina* delta-17 desaturase. Additionally, the DHA precursor, DPA (as high as 4.0%), was found in high EPA events resulting from the action of the *Mortierella alpina* elongase used. Interestingly, the *M. alpina* elongase had virtually no EPA-elongating activity when expressed in yeast (Parker-Barnes et al., 2000). In the best transgenic soybean lines, the omega-3 to omega-6 ratio was around 1:1, an approximately 5-fold improvement over that for wild-type soybean, but still lower than that commonly found in fish oil. Also, pathway intermediates and by-products were as high as 35%. Fatty acid elongation was not limiting in soybean as the total 20-carbon fatty acid was as high as 40.2% representing 56% elongation of 18-carbon substrates. Interestingly, a separate very recent study by Chen et al. (2006) expressing a similar set of LCPUFA biosynthetic genes led to very low levels of ARA (2.1%) in soy embryos. Thus, as shown in Kinney et al. (2004), the careful combination of genes with individual promoters, gene cassette orientation and screening of numerous events is crucial to obtaining the correct balance and a significant abundance of the desired LCPUFAs.

In a separate report from the same group (Damude and Yadav, 2005) the ratio of omega-3 to omega-6 fatty acids was further improved in soybean embryos by using a delta-15 desaturase from *Fusarium moniliforme* (Damude et al., 2006) in place of the *Arabidopsis* Fad3 gene. In the best event, the overall omega-3 fatty acid content was 57%. The *Fusarium* delta-15 desaturase was shown to be very active in soybean and when expressed alone, lead to ALA contents as high as 72%. It also had broad substrate specificity on omega-6 fatty acids (ALA > GLA > DGLA > ARA; Damude et al. 2006) which further increases its usefulness in an LCPUFA pathway.

6. TOWARDS A TRUE FISH-OIL REPLACEMENT

The patent application of Kinney et al. (2004) also describes the first ever demonstration of DHA in an oilseed plant in which a relative abundance of DHA of up to 3.3% was demonstrated. This DHA was produced in soy somatic embryos which are equivalent to the zygotic embryos of seeds (Kinney, 1996). In order to make the DHA, Kinney et al. (2004) added a delta-5 elongase from *Pavlova sp.* (Pereira et al., 2004b) and a delta-4 desaturase from *Schizochytrium aggregatum* (Mukerji et al., 2002), under control of seed-specific promoters, along with the EPA biosynthetic pathway described above. The delta-4 desaturase used was highly active in plants with, in some cases, close to 100% conversion of DPA to DHA.

More recently, other attempts at expressing a delta-6 pathway in oil seeds to produce both EPA and DHA have been described. In one instance, an acyl-CoA-accepting, dual delta-6/delta-5 desaturase, along with a delta-6 elongase, a delta-5 elongase and a delta-4 desaturase were expressed in *Arabidopsis* seeds (Robert et al., 2005). Reported EPA contents in this study ranged from 2.4% to 3.2% with, in some cases, an additional 0.2% as DHA. The overall omega-3 to omega-6 ratio of fatty acids was around 0.5:1 reflecting the omega-3 to omega-6 ration of the wild-type *Arabidopsis* seeds (0.5:1). These low EPA contents can be attributed to the poor delta-6 desaturase activity of the zebrafish delta-6/delta-5 desaturase when expressed in *Arabidopsis* since acyl-CoA pools of LA and ALA are not limiting (our unpublished results). Subsequently, expression of a delta-6 pathway in *Brassica juncea* seed (delta-6 desaturase, two delta-6 elongases, delta-5 desaturase), along with a delta-12 desaturase, a delta-17 desaturase, a delta-5 elongase, a delta-4 desaturase and a lysophosphatidic acid acyltransferase (LPAAT), was described (Wu et al., 2005). In this case, much higher EPA contents were obtained (as high as 15.0%) and, in addition, DHA was produced to a relative as high as 1.5%. The overall omega-3 to omega-6 ratio of fatty acids was low (on average around 0.3-0.6:1) but was still approximately 3-fold higher than that found in wild-type *Brassica* seed, probably because of the type of omega-3 desaturase used. This same omega-3 desaturase was also found to significantly increase the final EPA content by direct conversion of ARA to EPA. In this *Brassica* study, 18-carbon fatty acid elongation did not appear to be limiting, since increasing the substrate, by addition of a delta-12 desaturase, increased elongation and addition of an extra copy of delta-6 elongase only moderately boosted elongation. Although it was difficult to separate out the importance of the LPAAT in elongation of EPA, there appeared to be no effect on delta-6 elongation by the addition of this enzyme to the pathway.

7. FUTURE PROSPECTS

Given the recent success of both "proof of concept" and more agronomically-relevant research directed towards the production of LCPUFAs, such as EPA and DHA, in plants, the realization of a plant-based fish oil substitute appears to be on the horizon. Even so, key challenges remain in obtaining a plant-based oil that is substantially similar to other commercially available fish oils. The first target will be to produce a DHA content in plant oils close to that of certain marine oils, with DHA comprising at least 10% of the total fatty acids. Although the highest abundance of DHA reported so far is 3.3% the current technology will allow for this to be increased around 3-fold while maintaining an EPA abundance in the 10-15% range and thus provide an effective substitue for marine oils. The second part of the commercialization process will be to meet the first goal while keeping omega-6 intermediates (GLA, DGLA) and pathway by-products such as sciadonic acid [SCA, 20:3(5,11,14)] and juniperonic acid [JUN, 20:4(5,11,14,17)] to a minimum. We have shown that by achieving the correct balance of gene expression and metabolic flux through the pathway it will also be possible to meet this second goal in commercial oilseed plants. Of course, once a seed oil having the desired target fatty acid composition has been achieved, there remains the key regulatory and agronomic challenges that face all new transgenic crop plants.

REFERENCES

Abbadi A, Domergue F, Bauer J et al. (2004) Biosynthesis of very-long-chain polyunsaturated fatty acids in transgenic oilseeds:constraints on their accumulation. *Plant Cell* 16:2734–2748.

Bell JG, McGhee F, Campbell PJ et al. (2003) Rapeseed oil as an alternative to marine fish oil in diets of post-smolt Atlantic salmon (Salmo salar): changes in flesh fatty acid composition and effectiveness of subsequent fish oil "wash out". *Aquaculture* 218:515–528.

Calder PC (2003) n-3 polyunsaturated fatty acids and inflammation:From molecular biology to the clinic. *Lipids* 38:343–352.

Chen R, Matsui K, Ogaw M et al. (2006) Expression of Δ6,Δ5 desaturase and GLELO elongase genes from *Mortierella alpina* for production of arachidonic acid in soybean [*Glycine max* (L.) Merrill] seeds. *Plant Sci* 170:399–406.

Coughlan SJ, Kinney AJ (2002) Transgenic plants as sources of modified oils. *In*: Oksman-Caldentey KM, Barz WH (eds) Plant Biotechnology and Transgenic Plants. Marcel Dekker, New York.

Damude HG, Zhang H, Farrall L et al. (2004) Identification of bifunctional Δ12/ω3 fatty acid desaturases for improving the ratio of ω3 to ω6 fatty acids in microbes and plants. *PNAS* 103:9446–9451.

Damude H, Yadav NS (2005) Cloning and sequences of fungal Δ15 desaturases suitable for production of polyunsaturated fatty acids in oilseed plants for food or industrial uses. PCT Int. Appl.WO2005047479.

Del Vecchio AJ (1996) High laurate canola. *Inform* 7:230–243.

Dyerberg J, Bang HO (1982) A hypothesis on the development of acute myocardial infarction in Greenlanders. Scand J Clin Lab Inves Suppl 161:7–13.

Funk C (2001) Prostaglandins and Leukotriences: Advances in Eicosanoid Biology. *Science* 294: 1871–1875.

Hites RA, Foran JA, Carpenter DO et al. (2004) Global assessment of organic contaminants in farmed salmon. *Science* 303:226–229.

Hwang D (2000) Fatty acids and immune responses-a new perspective in searching for clues to mechanism. *Ann. Rev. Nutrit* 20:431–456.

Iribarren C, Markovitz JH, Jacobs DR Jr, et al. (2004) Dietary intake of n-3, n-6 fatty acids and fish: relationship with hostility in young adults-the CARDIA study. *Eur J Clin Nutr* 58:24–31.

Jacobs MN, Covaci A, Gheorghe A et al. (2004) Time Trend Investigation of PCBs, PBDEs, and Organochlorine Pesticides in Selected n-3 Polyunsaturated Fatty Acid Rich Dietary Fish Oil and Vegetable Oil Supplements; Nutritional Relevance for Human Essential n-3 Fatty Acid Requirements. *JJ Agric Food Chem* 52:1780–1788.

Kinney AJ (1996) Development of genetically engineered soybean oils for food applications. *J. Food Lipids* 3:273–292.

Kinney AJ, Knowlton S (1998) Designer oils: the high oleic soybean. In: Roller S, Harlander S (eds) Genetic Modification in the Food Industry. Blackie, London.

Kinney AJ, Cahoon EB, Damude HG et al. (2004) Production of very long chain polyunsaturated fatty acids in oilseed plants. PCT Int. Appl.WO2004071467.

Korver O, Katan MB (2006) The elimination of trans fats from spreads: how science helped to turn an industry around. *Nutr Rev* 64:275–279.

Lands WEM (2005) Dietary fat and health: the evidence and the politics of prevention: careful use of dietary fats can improve life and prevent disease. *Ann N Y Acad Sci* 1055:179–192.

Metz JG, Roessler P, Facciotti D et al. (2001) Production of polyunsaturated fatty acids by polyketide synthases in both prokaryotes and eukaryotes. *Science* 293:290–293.

Metz JG, Weaver CA, Barclay WR et al. (2004) Polyunsaturated fatty acid polyketide synthase genes and enzyme systems from Thraustochytrium and Schizochytrium and their use for preparation of bioactive molecules. PCT Int. Appl. WO2004087879.

Mukerji P, Huang Y-S, Das T et al. (2002) Protein and cDNA sequences of D4-desaturases isolated from fungi and therapeutical uses thereof. PCT Int. Appl. WO2002090493.

Mukherjee D, Nissen SE, Topol EJ (2001) Risk of cardiovascular events associated with selective COX-2 inhibitors. *JAMA* 286:954–959.

Napier JA, Beaudoin F, Michaelson LV et al. (2004) The production of long chain polyunsaturated fatty acids in transgenic plants by reverse engineering. *Biochimie* 86:785–792.

Napier JA, Sayanova O, Qi B et al. (2004) Progress toward the production of long-chain polyunsaturated fatty acids in transgenic plants. *Lipids* 39:1067–1075.

Oura T, Kajiwara S (2004) Saccharomyces kluyveri FAD3 encodes anω3 fatty acid desaturase. *Microbiology* 150:1983–1990.

Parker-Barnes JM, Das T, Bobik E et al. (2000) Identification and characterization of an enzyme involved in the elongation of n-6 and n-3 polyunsaturated fatty acids. *PNAS* 97:8284–8289.

Pauly D, Christensen V, Guenette S, et al. (2000) Towards sustainability in world fisheries. *Nature* 418:689–695.

Pereira SL, Huang YS, Bobik EG et al. (2004) A novel omega3-fatty acid desaturase involved in the biosynthesis of eicosapentaenoic acid. *Biochem J* 378:665–671.

Pereira SL, Leonard AE, HuangY-S et al. (2004) Identification of two novel microalgal enzymes involved in the conversion of the ω3-fatty acid, eicosapentaenoic acid, into docosahexaenoic acid. *Biochem J* 384:357–366.

Qi B, Fraser T, Mugford S et al. (2004). Production of very long chain polyunsaturated omega-3 and omega-6 fatty acids in plants. *Nat Biotechnol* 22:739–45.

Qiu X, Hong H, Mackenzie SL (2001) Identification of a Δ4 fatty acid desaturase from Thraustochytrium sp. involved in the biosynthesis of docosahexaenoic acid by heterologous expression in *Saccharomyxes cerevisiae* and *Brassica juncea*. *J Biol Chem* 276:31561–31566.

Robert S, Singh SP, Zhou X-R et al. (2005) Metabolic engineering of Arabidopsis to produce nutritionally important DHA in seed oil. *Funct Plant Biol* 32:473–479.

Sakuradani E, Abe T, Iguchi K et al. (2005) A novel fungal ω3-desaturase with wide substrate specificity from arachidonic acid-producing Mortierella alpina 1S-4. *Appl Microbiol Biotechnol* 66:648–654.

Sargent JR (1997) Fish oils and human diet. *Br J Nutr* 78:S5–S13.

Sayanova O, Napier JA (2004) Eicosapentaenoic acid: biosynthetic routes and the potential for synthesis in transgenic plants. *Phytochemistry* 65:147–158.

Simopoulos AP (1999) Essential fatty acids in health and chronic disease. *Am J Clin Nutr* 70:560S–569S.

Simopoulos AP (2006) Evolutionary aspects of diet, the Omega-6/Omega-3 ratio, and gene expression. In: Meskin MS, Bidlack WR, Randolph, RK (eds) Phytochemicals CRC Press LLC, Boca Raton, FL.

Stauffer CE (1996) Fats & Oils. Eagan Press, St. Paul, MN.

Stoll AL, Damico KE, Daly BP et al. (2001) Methodological considerations in clinical studies of omega 3 fatty acids in major depression and bipolar disorder. *World Rev Nutr Diet* 88:58–67.

Voelker T, Kinney AJ (2001) Variations in the biosynthesis of seed-storage lipids. *Annu Rev Plant Physiol Mol Biol* 52:335–361.

Willatts P, Forsyth JS (2000) The role of long-chain polyunsaturated fatty acids in infant cognitive development. *Prostaglandins Leukot Essent Fatty Acids* 63:95–100.

Wu G, Truksa M, Datla N et al. (2005) Stepwise engineering to produce high yields of very long-chain polyunsaturated fatty acids in plants. *Nat Biotechnol* 23:1013–1017.

Yadav NS, Wierzbicki A, Aegerter M et al. (1993) Cloning of higher plant ω-3 fatty acid desaturases. *Plant Physiol* 103:467–476.

Yaqoob P (2003) Fatty acids and the immune system: from basic science to clinical applications. *Proc Nutr Soc* 63:89–104.

CHAPTER 11

METABOLIC ENGINEERING IN SUGARCANE: ASSISTING THE TRANSITION TO A BIO-BASED ECONOMY

ROBERT G. BIRCH
Botany Department, School of Integrative Biology, The University of Queensland, Brisbane 4072 Australia

Abstract: Sugarcane is a promising crop for sustainable production of biomaterials and biofuel feedstocks, as well as sugars for food. It is a highly efficient biomass producer under tropical and sub-tropical conditions. Tissue culture and genetic transformation systems are well established and applicable to diverse commercial genotypes. It has inbuilt genetic containment features. Many cultivars are sterile under the usual commercial growing conditions. Cultivars are vegetatively propagated and they do not persist without human cultivation. Early steps in extraction remove all proteins and nucleic acids from the major food product, sucrose. A key decision in development of sugarcane for commercial production of a novel material is whether to co-produce the material with sucrose, or as an alternative to sucrose. Logical materials for production in such a large-scale industrial crop are those required in large quantities, or those for which sugarcane has a particular advantage in terms of precursor metabolite pools. This chapter summarises progress, potentials and limitations in metabolic engineering of sugarcane for enhanced yield of sucrose, higher value sugars and sugar derivatives, sucrose-derived polymers, bioplastics, aromatic compounds and waxes

1. INTRODUCTION

Sugarcane is a highly efficient biomass producer. It is widely cultivated in tropical and subtropical regions for large-scale food (sugar) and industrial (fuel ethanol, electricity co-generation) uses, with a long tradition of smaller-scale by-products including animal feeds, building materials, industrial waxes and speciality chemicals. With a well established genetic transformation system, it is one of the most appealing target crops for metabolic engineering aimed at high volume products; particularly as global thinking moves towards renewable biomaterials and biofuels as replacements for non-renewable petrochemicals.

R. Verpoorte et al. (eds.), Applications of Plant Metabolic Engineering, 249–281.
© 2007 *Springer.*

Testing of genetically modified sugarcane lines in the field has followed a pattern similar to several other crops. It commenced with reporter genes to establish the stability of transgene expression and any other effects of the gene transfer process (Hansom *et al.* 1999), then progressed to production traits including resistance to herbicides (Enriquez-Obregon *et al.* 1998; Falco *et al.* 2000; GalloMeagher and Irvine 1996; Leibbrandt and Snyman 2003), viruses (Gilbert *et al.* 2005; OGTR 1997), bacteria (Zhang *et al.* 1999), fungi (Enriquez *et al.* 2001) and insect pests (Arencibia *et al.* 1999; Braga *et al.* 2003; Setamou *et al.* 2002), then to sugar yield and quality traits (Botha and Groenewald 2001; Vickers *et al.* 2005), and pharmaceuticals (Wang *et al.* 2005) and most recently to novel sugars with potential benefits to consumers as well as producers (OGTR 2004). Additional traits are being tested in containment (Lakshmanan *et al.* 2005). Although there remain important hurdles to efficient engineering of commercially useful lines, the positive results for several of the added traits under field conditions indicate that the basic methodologies are sufficiently developed to explore metabolic engineering of output traits in sugarcane.

This chapter presents a rationale for the types of products most likely to be obtained through metabolic engineering in sugarcane, and summarises the potentials and initial achievements for several classes of these products. It also considers briefly the main generic hurdles currently facing sugarcane metabolic engineers.

2. ADVANTAGES AND LIMITATIONS OF SUGARCANE AS A TARGET CROP FOR METABOLIC ENGINEERING

2.1. Advantages

Sugarcane has several attractions as a target crop for metabolic engineering to enhance the yield of the current economic product (sucrose) or to achieve the industrial production of additional products.

- It is a highly efficient biomass producer under tropical conditions with an adequate water supply, yielding up to 80 tonnes dry weight per hectare per annum of readily harvested above-ground biomass. Of this biomass, about 80% is in millable cane stalks, distributed roughly equally between sucrose (cane sugar) and lignocellulose (fibre) with lesser amounts of other biomaterials including waxes and metabolic intermediates of potential industrial significance (Inman-Bamber *et al.* 2002; Moore *et al.* 1997).
- The sugar can be extracted fairly simply by crushing the harvested stalks (culms), and the dried fibre can be burned to generate enough steam and electricity to power not only the raw sugar extraction and crystallisation processes but (in an efficient facility) also provide a substantial surplus of renewable energy for downstream processes such as ethanol fermentation and distillation, or feed into the local electricity grid.

- The sugarcane juice stream can be processed through a series of crystallisations to maximise the yield of raw sugar for food uses, and the residual molasses used for diverse purposes including a fermentation feedstock, animal feed supplement or fertilizer. Alternatively, fewer crystallisation cycles can be used for an efficient recovery of high purity raw sugar, and the residual sugars in the remaining syrup (massecuite) used directly as a feedstock for high-volume fermentation products such as fuel ethanol.
- Pilot methods have been developed to fractionate the cane into other valuable components including surface waxes and fibrous rind before crushing the sucrose-storage parenchyma.
- Potential value in the 20% of sugarcane biomass within leaves and young stalk material, which are deliberately removed from the sugar-rich mature stalks at harvest, has barely been exploited except as a mulch (Allen *et al.* 1997; Chen and Chou 1993).

• Sugarcane is a vegetatively propagated crop, in which planted stem segments (called setts or billets) yield a 'plant crop' followed by several consecutive 'ratoon' crops grown from underground buds after each harvest (James 2004). In relation to metabolic engineering, this means that a useful primary transformant can be applied directly as an improved cultivar through the routine clonal propagation methods in the industry (Birch 1996).

• Sugarcane originated in South-East Asia (particularly the Indonesia – New Guinea region), where related species in the *Saccharum* complex could conceivably produce viable hybrid progeny. However elsewhere in its commercial range there are many locations with no species capable of unassisted hybridisation with sugarcane, and many cultivars are either male sterile or completely sterile. In these locations there is a high level of inbuilt containment against any unintended transgene movement by sexual reproduction (James 2004).
  - The climatic conditions affecting pollen fertility are well studied, and cool nights (below 18°C) away from tropical regions result in sugarcane pollen sterility (Berding 1981; Moore and Nuss 1987).

• The first stages of sucrose extraction and purification to raw sugar involve liming, heating and crystallisation steps that eliminate all nucleic acids and proteins from the resulting food product (sugar) and also from the massecuite-molasses by-product stream (Chen and Chou 1993; Klein *et al.* 1998). Therefore the large-scale food products from genetically modified sugarcane (sugar and fermentation products such as potable ethanol) will contain neither genes nor proteins expressed from endogenous genes or from any introduced gene.

• There is typically no risk associated with consumption of a gene or protein (all humans eat and digest millions of genes and protein molecules every day). Nevertheless, the absence of transgenes and protein products may be relevant in the short term to consumer perceptions and labelling requirements in some countries.

- In some regions sugarcane is consumed more or less directly as a food by occasional chewing on stalks, drinking freshly expressed juice, or use of stem pieces as uncooked skewers in certain recipes. These uses are typically absent or dispensable in industrial-scale sugarcane production.
- Burning of the fibre component for energy generation also destroys all nucleic acids and proteins.
- Any modifications that lead to the synthesis of other classes of compounds will need to be analysed separately for elimination during the sugar-extraction stages.
- Depending on the value of the target compounds, there is potential to grow some modified sugarcanes exclusively for non-food uses, with any sugar being used for industrial purposes such as fuel ethanol production.

- There is a highly efficient gene transfer and plant regeneration system for sugarcane. Although cultivars vary in the efficiency of the associated tissue-culture stages, any modern commercial sugarcane variety can apparently be transformed at a sufficient efficiency for metabolic engineering (Bower and Birch 1992; Bower *et al.* 1996).
  - Apart from the widely used microprojectile-mediated transformation system, progress has also been made with *Agrobacterium*-mediated sugarcane transformation, although so far this seems less reproducible and less readily applicable across diverse genotypes (Arencibia *et al.* 1998; Carmona *et al.* 2005).

In sum these advantages encourage the conclusion that sugarcane has a great potential for development through metabolic engineering into a secure biofactory, to help underpin the transition from a non-sustainable petrochemical-based global economy to a renewable carbohydrate-based economy.

2.2. Limitations

Although sugarcane is one of the most promising crops for sustainable production of biomaterials through metabolic engineering, there are some limitations.

- Sugarcane is not well adapted to cooler climatic regions or areas with low rainfall or irrigation water availability. Over much of the current cultivation area, 40-60 tonnes dry weight per hectare per annum of harvestable above-ground biomass is a more realistic expectation of sustainable yield.
- It is a complex polyploid in which some simple genetic tools used to elucidate biological mechanisms in diploid species are not readily applicable (Grivet and Arruda 2002).
- The remarkable property of sucrose accumulation is appealing for ease of extraction and potentially for ease of *in planta* metabolic conversion as well as downstream fermentation. However, there are limits inherent in an osmotically active storage compound that intersects many primary metabolic pathways within

sugarcane, and which is rapidly metabolised by diverse microbes once exposed in solution through the harvest or crushing of cane.

- Compared to polymeric storage compounds such as starches and oils there is also less structural diversity (indeed no structural diversity in the disaccharide sucrose) and therefore less opportunity for improvement by metabolic engineering to alter structural properties such as branching properties, chain length and degree of saturation.
- And of course, despite its remarkable capacity to convert sunlight and atmospheric CO_2 into biomaterials and bioenergy, it is unlikely that sugarcane or any other crop can be produced on a scale sufficient for the entire human population to consume at the level now reached in wealthy countries.
  - Quantum improvements in conservation, and a sense of responsibility to fellow inhabitants of our planet and our future generations, will be vital if a globally sustainable society is to be based on biomaterials and bioenergy.

3. RATIONALE FOR SELECTION OF TARGET PRODUCTS

3.1. Co-product, By-product or Alternative Product?

In sugarcane, a key decision is whether a target molecule for enhancement through metabolic engineering is intended for co-production with sucrose for the food sector, or only in cane grown for non-food industrial uses. Currently the economics strongly favour harvesting a substantial proportion of the sucrose for food uses. Even if a novel compound or an altered level of an endogenous compound caused no change in the composition of raw sugar or the purity of refined sugar, there might be sensitivities based on consumer perceptions about food from a plant variety also grown to produce industrial or pharmaceutical chemicals. Under this scenario, target compounds would need at least to replace the value differential between sugar for food and sugar as an industrial fermentation feedstock, and add sufficient value to justify the costs of separate production and processing streams for non-food cultivars.

3.2. Scale

To obtain the advantage of sugarcane as an efficient biomass producer, target compounds to be produced instead of sucrose will typically be required in large quantity and by corollary at low prices. As a benchmark, world demand for sucrose is in the order of 100 million tonnes per annum at a price below AU$1 per kilogram, and it is produced economically at a yield of about 20 tonnes of sucrose per hectare per annum. Although there is some market premium for renewable biomaterials there is unlikely to be any large-scale demand in the foreseeable future for an industrial molecule as a direct petrochemical replacement at a wholesale price much above AU$1 per kilogram. In other words, an industrial precursor bio-molecule produced instead of sucrose at a similar price would need to have properties that substantially reduced downstream production costs, or enhanced the value of the resulting products, in order to be competitive with established petrochemical feedstocks. The most likely example might be a bioplastic precursor combined with biofuel production (from non-food cultivars).

The situation is less challenging for target compounds that can be produced as co-products or by-products with sucrose for the food sector. To justify the extraction costs, these compounds are likely to be required in at least thousands of tonnes per annum at a substantial price premium (up to 10-fold) over sucrose. To be competitive with other sources, there will probably need to be an inherent biological advantage for production in sugarcane, such as the existence of an abundant precursor pool, an existing biosynthetic pathway that can be modified to advantage by slight changes, or a high energy requirement for processing that can be satisfied using sugarcane bagasse. Possibilities include a combination of sugars with perceived health benefits to consumers or compounds derived from the wax and aromatic biosynthetic pathways in sugarcane.

There may be niche markets for smaller quantities of higher-priced compounds that can be extracted from sugarcane, such as stem-surface wax components or leaf-derived antioxidants. However, unless there is a protected competitive advantage it seems inevitable that once developed such a market would be fiercely contested based on price (by competing sugarcane industries and potentially by extraction from non-sugarcane sources). A scenario can be imagined in which a substantial number of such compounds is ultimately developed, with different producers specialising in the extraction of different compounds from their biomass supply. Under such a scenario, there could be a positive impact on the economic viability of the global sugarcane industry.

It is even more difficult to make a case for the production of very low-volume and high-value pharmaceutical compounds in sugarcane. In many cases the foreseeable global demand could be met from a few hectares of greenhouse-grown plants of a non-food species, so why complicate matters by production in a high-volume food crop? However, there may be an opportunity for production in engineered sugarcane of lower-value proteins or other bio-molecules required in larger quantities, where the pricing does not justify the capital investment in cell culture facilities or the biosynthetic capacity does not exist in readily cultured cells.

3.3. Biochemical Feasibility

Although it is most obvious for its biosynthesis and storage of a large quantity of sucrose, sugarcane is also remarkable for substantial carbon flux through metabolic pathways generating cellulose fibres, lignin components, suberin and surface waxes. These pathways provide precursor pools for engineered synthesis of several classes of high value products. However, diversion of pathway intermediates through the action of introduced gene products, without severe disruption of plant development, is likely to be even more challenging than conversion of a storage pool. Feasibility depends on the detail of the endogenous biosynthetic pathways and their regulation, the kinetic properties of the intended modifying enzymes, the co-factors for the intended conversion, the ability to achieve appropriate developmental expression and compartmentation patterns, and the stability of the modifying enzyme and the conversion product within the cellular environment.

3.4. Commercial Issues

In the long term, competitive efficiency becomes a key determinant for profitable production of any commodity and is influenced by natural advantages such as climate and proximity to markets. In the case of biotechnology-derived products, there may be an intermediate period during which protected intellectual property provides some advantage (to owners) or barriers (to those temporarily excluded from use during the protected period).

Combining these considerations leads to a logical checklist that can help to compare and evaluate opportunities for profitable metabolic engineering in a crop like sugarcane (Box 1).

From such an exercise, a few classes of traits emerge as promising in the foreseeable future. (Figure 1).

4. TRAITS

4.1. Enhanced Yield of Sucrose

Enhanced sucrose content (sweetness) has undoubtedly been a priority goal in sugarcane selection by humans long before and ever since the advent of scientific plant breeding. This trait has a high commercial value and theoretically there is much room for improvement from both physiological and genetic perspectives (Jackson 2005; Moore *et al.* 1997). However, very little gain has been made in sucrose content (as a fraction of biomass or cane fresh weight) in elite sugarcane cultivars during recent decades of conventional plant breeding; in contrast with gains in total biomass (Jackson 2005).

The sugarcane culm is a challenging experimental system for the application of physiological or gene expression studies to elucidate the molecular and biochemical controls over carbon partitioning and sucrose accumulation (Moore 1995). The progress made is a credit to the determination and ingenuity of those involved (see reviews: Lakshmanan *et al.* 2005; Moore 2005; Rae *et al.* 2005; Rohwer and Botha 2001), but even with the added power of genomic approaches (Casu *et al.* 2005; Vettore *et al.* 2003; Watt *et al.* 2005) it remains insufficient to identify selectable components for conventional breeding.

Box 1. Considerations for evaluation of target compounds in sugarcane metabolic engineering

Target Compound	**Sucrose** (benchmark)	(candidate)
Product Class	Sugar	
Required Production Scale		
• Current world demand (tonnes / year)	130,000,000	
• Projected world demand (2020)	rising	
• Yield (tonnes / ha)	20	
• Area (ha)	6,500,000	
Value		
• Price (AU$ / kg wholesale)	0.5	
• Price stability	+/−30%	
• Margin over production costs	Low	
• Indirect benefits for industry or consumers *(season length, sustainability, health, product quality, price etc)*	nil	
Production Method		
• Co-production with sucrose in food cultivar	+	
• Co-production with sucrose in non-food cultivar		
• Likely effect on sucrose yield	nil	
• Alternative to sucrose		
Technical Feasibility		
• Current status	commercial	
• New biological process needs / likely constraints *(discover genes, tailor expression, improve enzyme stability, enhance cofactor availability, overcome product toxicity etc)*	nil	
• Number of transgenes required in sugarcane	nil	
• New industrial process needs / likely constraints *(extraction method, additives, product handling, safety, waste disposal etc)*	nil	
• Anticipated capital costs for new industrial facilities	nil	
• Research costs and timeline	nil	
• Background IP audited?	+	
Potential to Capture Value		
• Competitors	many	
• FTO with essential background IP?	+	
• Competitive advantage (protected?)	–	
• Timing to market	now	
• Other market risks	–	
Environmental Impact	nil	
Industry Partner(s)	established	

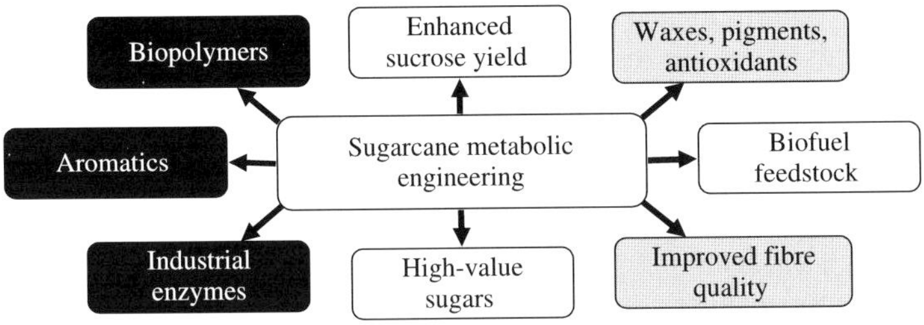

Figure 1. Examples of products that might be produced by sugarcane metabolic engineering. Compounds shown in dark boxes at left are most likely to be produced in cultivars grown for non-food purposes. Compounds shown in grey boxes at right are most likely to be by-products with sugar for food or fermentation use. Compounds in white boxes are potential major products expected to be compatible with cultivars grown for food

Interestingly, sucrose content is now typically higher in commercial sugar beet than in sugarcane (van der Poel *et al.* 1998), reinforcing the conclusion that there is physiological headroom for improvement in cane. On the other hand, cane is ahead in yield of both sugar and total biomass per hectare per annum under optimal commercial conditions. A substantial structural component seems essential for a massive above-ground crop. As renewable energy policies enhance the commercial value of the lignocellulose component in cane, there may be more interest in increasing the sucrose + fibre sum as distinct from the sucrose : fibre ratio.

The advent of gene cloning, transformation and transgene expression systems for sugarcane opened the way for naïve metabolic engineering experiments, altering the expression levels of likely key genes in sucrose metabolism (Botha *et al.* 2001; Ma *et al.* 2000; Vickers *et al.* 2005). With one possible exception (Botha and Groenewald 2001), such alterations have not enhanced sugar content in plants of elite sugarcane cultivars. Simple manipulations of genes in primary metabolism have been similarly unsuccessful in enhancing highly-selected yield traits in other economic plants (Capell and Christou 2004).

Because sucrose is the major transported photosynthate in plants, sucrose metabolism is central to plant biology. In most plants, storage materials are polymers such as starches, oils and proteins which can be removed from the active metabolic pool and accumulated at high concentration without serious osmotic consequences. Sucrose storage must involve a wonderful coordination of sucrose biosynthesis, transport, compartmentation and partitioning between storage and consumption to drive all other metabolism within sink tissues. Our understanding of the regulation of these processes is rudimentary (Koch 2004), but interesting target genes in sugarcane have been suggested (Grof and Campbell 2001). The use of kinetic properties of key enzymes of sucrose metabolism in sugarcane to model predicted consequences of altered expression levels is an exciting current approach being led by the SASRI team in South Africa (Kossmann *et al.* 2006; Rohwer and Botha 2001).

Initial results have not eliminated the concern that there is too much buffering in the control mechanisms and in the alternative routes around primary metabolic pathways in plants for improvement of highly-selected traits in elite genotypes by alteration of one or a few endogenous enzymes or metabolites. It remains to be seen whether the systems biology approach to integrate large data sets from high-throughput analysis of transcripts and metabolites overcomes or simply reinforces our perception of this practical limitation (Moore 2005; Sweetlove and Fernie 2005). Nevertheless, transgenic manipulation of key genes is a powerful complement to physiological and genomic investigation of underlying mechanisms. Clearer understanding of the mechanisms may indicate effective combinations of tailored enzyme activities, or entirely new approaches to practical enhancement of sucrose yield in commercial sugarcane.

Recently, experiments testing the capacity of sugarcane for accumulation of high-value sugars have also yielded results that challenge some current assumptions about the limits to sugar accumulation and indicate new possibilities to achieve higher sucrose yields. Introduction of a sucrose isomerase gene tailored for vacuolar compartmentation resulted in sugarcane lines with up to doubled total sugar concentrations in juice under containment glasshouse conditions. The lines with enhanced sugar accumulation also showed increased photosynthesis, sucrose transport and sink strength. This remarkable step above the former ceiling in stored sugar concentration provides a new perspective into plant source-sink relationships, and has substantial potential for enhanced food and biofuel production (Birch and Wu 2004).

4.2. High-Value Sugars

Sweetness is one of the fundamental taste sensations in humans, and enjoyment of sweet taste has been innate characteristic for tens of thousands of years, possibly arising much earlier as an indicator of safety and nutritional value in foods. Sugarcane has been cultivated since prehistoric times and sucrose has been extracted by humans for more than 1000 years for use as a food and preservative. Until recent decades it was the only widely-used refined sweetener (van der Poel *et al.* 1998). Subsequently interest has grown in other naturally occurring and synthetic sugars and sugar substitutes, particularly in wealthy countries where changed lifestyle has created a demand for low-calorie foods, but more universally in the case of nutritive sugars with wider health benefits such as lower glycaemic response and acariogenic properties (Fujii and Komoto 1991; Lina *et al.* 2002). Currently these sugars are expensive to produce by fermentation or chemical reaction. If they can be produced directly in a crop like sugarcane, their benefits are likely to be affordable by many more people while still providing an enhanced return to producers based on benefits to consumers.

The obvious advantage of sugarcane as a platform for the production of high-value sugars is an abundant supply of stored sucrose, or potentially a high flux through biosynthetic intermediates in sucrose metabolism, as substrates for conversion into the desired products.

4.3. Limitations are:

1. Sucrose is a direct substrate for relatively few useful conversions, for example the production of sucrose isomers such as isomaltulose.
2. Most of the sucrose is stored in a highly acidic and proteolytic vacuolar compartment that is (i) hostile to most introduced enzymes and (ii) metabolically isolated by the bounding tonoplast from cofactors such as ATP and NADH required for some conversions.
3. Substantial diversion of carbohydrate flux into a non-metabolized product pool in growing tissues generally interferes with normal plant development. Without appropriate osmotic control and wall synthesis for cell expansion, the plant can not construct the structural reservoir for the desired storage product. This necessitates developmental control over the expression pattern and probably also the subcellular compartmentation of the transgene product, to achieve economic yields of the novel product.

Methods have been described for the isolation of sugarcane promoters conferring potentially useful expression patterns such as mature-stem specificity (Birch and Potier 2001). Signal sequences have been established for delivery of proteins into the sugar-storage vacuoles (Gnanasambandam and Birch 2004), and also into the plastids (Gnanasambandam *et al.* 2006).

Production of isomaltulose (IM) in sugarcane was chosen as a pilot study in sugarcane metabolic engineering. Some bacteria are able to sequester sucrose (α-D-glucopyranosyl-1,2-D-fructofuranose) by conversion into isomaltulose (α-D-glucopyranosyl-1,6-D-fructofuranose). This sucrose isomer is resistant to invertases and is not metabolised by many microbes including the predominant oral microflora, conferring an advantage in many foods as an acariogenic sweetener. Isomaltulose is digested by humans with the same primary products (glucose, fructose) and ultimate caloric value as sucrose. However, because the first step involves an intestinal disaccharidase rather than salivary invertase, isomaltulose it is digested more slowly, with advantages of lower fluctuation in blood glucose and insulin concentrations. Consequently, isomaltulose has a growing market as a stable, slowly digested, acariogenic, non-hygroscopic sugar (Lina *et al.* 2002; Takazoe 1989). Its accessible carbonyl group also makes an attractive renewable starting material for the manufacture of biomaterials as eventual petrochemical replacements (Lichtenthaler and Peters 2004), but use is currently limited due to the high cost of IM production through fermentation (Schiweck *et al.* 1991).

IM biosynthesis occurs via a sucrose isomerase (SI) with no cofactor or substrate activation requirements (Wu and Birch 2005), indicating feasibility for engineered production in plants. IM is apparently not metabolised or transported in plants, but exogenous application triggers some plant sugar-sensing mechanisms and changes gene expression profiles differently from sucrose (Loreti *et al.* 2000; Sinha *et al.* 2002). In growing plant tissues, efficient conversion of sucrose into the non-metabolised isomer is lethal or disruptive (Börnke *et al.* 2002a). Tuber-specific expression of an apoplasm-targeted SI allowed partial conversion of the low soluble sucrose levels in potato tubers to IM ($\sim$15 μmol $g_{FW}{}^{-1}$) without affecting plant appearance, but with a substantial decrease in total nonstructural carbohydrate content (Börnke *et al.* 2002b; Hajirezaei *et al.* 2003).

Vacuolar targeting of a highly efficient SI allowed high IM yields (up to 440 μmol g_{FW}^{-1}) in sugarcane stems. Remarkably, the IM could be accumulated without commensurable reduction in sucrose, resulting in twice the total sugar concentration in juice from selected transgenic lines relative to their elite parent cultivar (Wu and Birch 2006). IM accumulated gradually during development, probably because of (i) continuous delivery of constitutively expressed SI into vacuoles where it is rapidly degraded by proteases, (ii) high catalytic efficiency allowing occasional IM production before SI inactivation and (iii) absence of plant enzymes for IM metabolism. For efficient commercial production of this valued sugar, it will be useful to achieve patterns of developmental expression, compartmentation and enzyme stability yielding high IM content further up the harvested stalk profile (Wu and Birch 2007).

Expression in sugarcane of a cytosol-targeted gene for the one-step synthesis of trehalose (the nonreducing disaccharide of glucose) from D-glucose and α-glucose-1-phosphate resulted in plants with 26–38 μmol g_{FW}^{-1} of trehalose in leaves. Sugar composition in stems was not reported, but the plants had no obvious phenotypic differences other than increased drought tolerance (Zhang *et al.* 2006). This work was aimed at drought resistance, but it can illustrate several of the challenges facing a metabolic engineer interested to produce trehalulose as a higher-value sugar on a commercial scale in sugarcane.

Because the cytosol comprises so little of the volume in sugarcane storage tissue, it is likely that trehalulose would have to be accumulated in the vacuole to achieve economic yields. On the other hand, one of the substrates for the synthesis (glucose-1-phosphate) is not expected to be available in the vacuole, and sugarcane is not expected to have an efficient transporter to move the desired product (trehalose) from the cytosol into the vacuole. Interestingly, while most plant species do not accumulate trehalose naturally, they do have the capability through a two-step pathway: synthesis of trehalose-6-phosphate (T-6-P) from UDP-glucose and glucose-6-phosphate by trehalose-6-phosphate synthase (TPS) then formation of trehalose by the action of T-6-P phosphatase (TPP). Over-expression of these genes aimed at trehalose accumulation has generally been harmful, probably because T-6-P is an important signal compound in plant metabolism (Cortina and Culianez-Macia 2005; Satoh-Nagasawa *et al.* 2006). By avoiding the intermediate signal compound, one-step synthesis may be a practical opportunity; if the other hurdles of expression level, developmental timing, precursor flux and product compartmentation (away from plant enzymes for trehalose breakdown) can be overcome.

The implications of signaling effects from synthesis and accumulation of novel sugars, and the limitations to their transport in plants are yet to be elucidated. Apoplastic unloading from the phloem may be relatively non-specific, as the trisaccharide kestose accumulated to high concentrations in potato tubers after synthesis in phloem (Zuther *et al.* 2004). Disaccharide transporters vary in their specificity, and commonly fail to recognize sucrose isomers such as isomaltulose and turanose (Sivitz *et al.* 2005). Nevertheless, plants can sense the isomers, with subtle or profound affects on metabolism and development in different species (Börnke *et al.* 2002a; Gonzali *et al.* 2005; Loreti *et al.* 2000; Sinha *et al.* 2002). Differential sensing and transport may be beneficial or adverse for desired commercial phenotypes, and will have to be considered or empirically tested case-by-case until a clearer understanding emerges.

4.4. Sugar Alcohols and Other Simple Sugar Derivatives

Though their production is currently tiny in comparison with sucrose, there is an established market at a higher price for sugar alcohols such as sorbitol and sugar-derived acids such as citric acid along with various oligosaccharides used in foods (Taniguchi 2004). There are also many potential conversions of carbohydrates including sucrose into non-food chemicals that can replace petroleum-sourced hydrocarbon derivatives (Lichtenthaler 2002). Examples include sucrose esters, ethers, urethanes, thio- and chloro-derivatives, and more remote conversion products with applications as surfactants and industrial monomers (Khan 1994; Lichtenthaler and Peters 2004; Schiweck *et al.* 1991). Those produced for non-food use are currently produced by chemical reaction rather than biosynthesis, and in many cases no biosynthetic path is known. Of course a decade ago the same was true of compounds like riboflavin, which today are produced entirely by genetically enhanced fermentation at a cost far lower than the conventional chemical synthesis (Bachmann *et al.* 2000; Wilke 1999).

Of the 24 building block chemicals considered most suitable for commercial production from sugars, 21 have potential biosyntheses through aerobic fermentation (most likely to be adaptable to direct conversion in engineered plants) of which seven are commercial (none yet in plants) (Werpy and Petersen 2004). Exploration of plant metabolic engineering for enhanced sorbitol biosynthesis can illustrate some of the challenges.

Some plants naturally accumulate sugar alcohols such as sorbitol, and there are informal reports from the BSES group in Australia that expression in sugarcane of the apple sorbitol-6-phosphate dehydrogenase gene for conversion of cytosolic glucose 6 phosphate into sorbitol resulted in sorbitol accumulation up to 12% of dry weight in leaves and 1% of dry weight in stems (Brumbley *et al.* 2006). Sugarcane normally accumulates sucrose to more than 50% of dry weight in stems without adverse effects, but sorbitol-accumulating sugarcane lines showed necrosis of leaves and reduced growth as previously observed in transgenic plants of other species modified for sorbitol or mannitol accumulation (Karakas *et al.* 1997; Sheveleva *et al.* 1998). In this case, substrate depletion in the metabolically active cytosolic compartment, product toxicity and secondary effects on signaling or metabolism are all possible limitations to be overcome.

4.5. Sucrose-Derived Polymers

Sugarcane has been transformed using gene constructs designed for apoplastic or vacuolar activity of the *Acetobacter diazotrophicus* fructosyltransferase (LsdA) which produces a high proportion of the trisaccharide 1-kestose (GF_2) plus free glucose from sucrose (Trujillo *et al.* 2000). The plants were reported to have a normal phenotype, but no information on sugar profiles has so far been published. Other fructosyltransferases producing fructans with higher chain lengths have been tested in sugarbeet and other plants (Ritsema and Smeekens 2003; Ritsema *et al.* 2004; Weyens *et al.* 2004). There are informal mentions from the CIGB team in Cuba of transgenic sugarcane plants accumulating either oligofructans or polyfructans in mature stems (Biotecnología 2006).

At first glance, it does not seem attractive to pursue the production of starch-like polymers in sugarcane in competition with the many plant species that have been selected for starch yield and qualities. However, the situation is likely to change as new microbial enzymes are discovered and characterised that use sucrose or hexoses as substrates for the synthesis of polymers with properties of industrial interest (van der Veen *et al.* 2004).

4.6. Bioplastics

Polyhydroxyalkanoates such as poly-3-hydroxybutyrate (PHB) are natural storage compounds in some bacteria. They can be amenable to use as biodegradable plastics, and have been produced industrially on a small scale by fermentation, with substantial interest in the possibility of more economical production in transgenic plants. There are informal reports from the BSES group in Australia that expression in sugarcane of genetic constructs designed for plastid-localised activity of the bacterial PhaA-PhaB-PhaC pathway resulted in PHB accumulation up to 2% of leaf dry weight without adverse effects on plant growth and sugar accumulation under glasshouse conditions (Brumbley *et al.* 2006). This is at the low end of reported PHB accumulation levels in transgenic plants. In some plant species yields above 3% have reduced growth, possibly due to depletion of the acetyl-CoA precursor pool; but up to 7.7% has been achieved without growth retardation in oilseeds which have greater flux of acetyl-CoA for oil biosynthesis. The minimum yield for commercial production will depend on the polymer characteristics and the extraction costs, and has been estimated at more than 15% of plant dry weight for the higher value PHBV copolymer (Slater *et al.* 1999).

4.7. Proteins

The main potential for economic protein production in a species such as sugarcane is likely to be those proteins required on a scale too large for cell culture facilities or even for containment-greenhouses. This might include certain industrial enzymes (Howard and Hood 2005), proteins with valuable fibrous or adhesive properties

(Scheller and Conrad 2005), or even proteins with potential large scale medical uses such as direct antibody therapy or serum replacements where the absence of potential human pathogens and toxins in a plant production system is desirable (Ma *et al.* 2005).

One attraction of sugarcane is the low concentration of other proteins in stem tissues (Mirkov *et al.* 2006), which should simplify purification by comparison with production from plant leaves with high levels of photosynthetic proteins such as Ribisco (Mirkov *et al.* 2002). On the other hand, the low protein concentration in sugarcane juice reflects the small contribution of the cytoplasmic compartment in which the transgene-encoded proteins are likely to be produced, and disruption of cellular compartmentation by crushing stems to harvest the products releases the highly active sugarcane vacuolar proteases (Gnanasambandam and Birch 2001).

Informal analyses by the TAMU group in the USA indicate that for a protein accumulated to 1% of total cellular proteins (TSP) in the sugarcane stem, total production costs could be around US$10 / gram, roughly equally divided between field production and purification costs (Mirkov *et al.* 2006). This yield has apparently been achieved for snowdrop lectin protein (Setamou *et al.* 2002), but estimated transgene product levels in sugarcane stems are typically in the range of 0.01% to 0.1% of TSP.

Expression of an ER-targeted human cytokine protein GM-CSF from the strongest available constitutive promoters for sugarcane yielded a maximum of 0.02% TSP in leaves (Wang *et al.* 2005). In field-grown stems, GM-CSF yield was around 500 ng/ml of juice expressed from mature internodes, or from 0.06% to 0.1% of TSP. Recovery of these yields required a construct encoding GM-CSF with a C-terminal HDEL signal for retention in the ER, and rapid dilution of the expressed juice into a buffer with a proteinase inhibitor cocktail. The purified protein was biologically active in a mammalian cell-proliferation assay, and there was no apparent phenotypic effect on the transformed sugarcane lines (Wang *et al.* 2005).

4.8. Aromatic Compounds

The aromatic hydroxyacid pHBA used in the manufacture of polymeric resins is currently produced industrially by chemical synthesis, and is also a natural intermediate in several plant and bacterial biosynthetic pathways. Toxicity and end-product inhibition have hampered attempts at overproduction in bacteria, and there is some promise for production in engineered plants where substrates are derived from the highly active phenylpropanoid pathway and products are sequestered in the vacuole as less toxic glucose conjugates. Tobacco engineered to express a bacterial pHBA biosynthetic enzyme (HCHL) that acts on a cytosolic precursor accumulated up to 1% DW free pHBA equivalent in leaves, but this was accompanied by severe depletion of phenylpropanoids, chlorosis and stunting (Mayer *et al.* 2001). A 10-fold higher yield was achieved without adverse effects on growth by plastid genetic engineering to produce CPL which uses chloroplastic chorismate as the pHBA precursor (Viitanen *et al.* 2004).

Sugarcane modified to constitutively produce cytosolic HCHL yielded pHBA-glucose conjugates (and vanillic acid conjugates as a side reaction) at levels similar to those in tobacco, also with depletion of phenylpropanoids but without severe effects on plant appearance (McQualter *et al.* 2005). Yield of the product appeared substrate-limited in lines with higher activity of the introduced enzyme. There is a substantial flux through the phenylpropanoid pathway in sugarcane stems for developmental lignification, which spreads from the vascular bundles throughout the storage parenchyma in maturing stems to contribute about 20% of mature stalk DW (Jacobsen *et al.* 1992). However, pHBA yields were low in the stem parenchyma and vascular bundles and reached 1% DW only in the fibre-rich rind (McQualter *et al.* 2005). The plastid engineering approach for higher product yields is less likely to be effective in non-photosynthetic tissues such as sugarcane storage parenchyma.

4.9. Modified Fibre Quality

Lignocellulosic fibre is the main component of bagasse, obtained after crushing and extraction of soluble sugars from harvested cane. Although many attempts have been made to develop alternative industrial uses such as building materials (Allen *et al.* 1997), and some continue at a local scale, most bagasse has traditionally been burned to provide all of the energy needed for milling of the cane and crystallisation of the raw sugar. With excess bagasse (about 40% of stalk DW) and low value for other uses, there has been little incentive for energy efficiency in sugarcane milling.

Increasing fossil fuel and petrochemical prices provide more incentive to capture the value of sugarcane bagasse as a fuel for cogeneration and as a source of useful biomaterials. The lignin content interferes with paper pulping, digestibility as a stock feed, and hydrolysis to yield fermentation feedstocks. Therefore one approach has been to down-regulate genes for lignin biosynthesis in order to reduce lignin content or to introduce genes aimed at altered lignin structure for greater extractability (Selman-Housein *et al.* 2000).

On the other hand, if lignin components can be separated with consistent properties they have industrial potential including the manufacture of thermoplastics and adhesives (Chen and Sarkanen 2006). The combined value of lignins as commodity chemicals and cellulose for hydrolysis and fermentation to ethanol as a transport fuel is potentially higher than the value of bagasse as a biofuel for direct combustion to generate electricity (Edye *et al.* 2005). The separation technologies are not yet economical but the large research effort in this direction in other crops may also create opportunities in sugarcane (Harrison *et al.* 2006; Sticklen 2006).

In the longer term, there is a possibility that engineering for novel fibre components combined with the high biomass production of the crop could renew interest in structural uses of sugarcane-derived fibres (Scheller and Conrad 2005).

4.10. Waxes

Although oilseeds are more obvious candidates for engineering of value-added lipids (Dyer and Mullen 2005), sugarcane may have a competitive advantage for some surface waxes of industrial value, particularly very-long-chain (C24-34) aldehydes and alcohols (Kunst and Samuels 2003).

4.10.1. Potential to redesign sugarcane for co-production of multiple biomaterials

Intensive selection for high sucrose yields in sugarcane has most likely selected against yields of other constituents. Growing demand for renewable biomaterials will increase the incentive to tailor sugarcane for optimal yields and value-added forms of multiple natural constituents (waxes, fibres, pigments etc.) in co-production with sucrose. Waxes are discussed here as an example of the considerations behind metabolic engineering to enhance yield or composition of 'forgotten' natural products in which sugarcane has potential advantage as a renewable source.

4.10.2. Existing commercial products from sugarcane waxes

Sugarcane produces epicuticular wax on the leaves and stems. The abundance of wax varies between genotypes, but in modern cultivars it typically comprises less than 0.1% of cane fresh weight and is most abundant near the nodes. The apparent composition varies with the wax extraction method and the material to which it is applied (whole stalks, raw juice, bagasse or filter cake). Several commercial products have been extracted.

(a) Refined hard sugarcane wax has a high concentration of high-molecular weight esters and diverse industrial uses similar to carnauba (palm) wax. Carnauba wax apparently exists on the palm leaf surface as a mixture of esters, but this is not the case for sugarcane wax, which comprises mostly polymeric long chain aldehydes and long chain primary alcohols. The ester form of sugarcane wax is mainly produced during commercial extraction, through rapid depolymerisation and oxidation of the aldehyde to the corresponding acid, and reaction with the alcohol to form the ester. This process is accelerated by heat or mineral acid (Kranz *et al.* 1960). There is a large potential market, but sugarcane wax has not historically been competitive in price (Hamilton 1995).

(b) Policosanol is a mixture of primary aliphatic alcohols (particularly 28:0 octacosanol, with some 30:0 triacontanol and 26:0 hexacosanol, and smaller amounts of other chain lengths from C24–34). It is used as a pharmaceutical, with diverse claimed health benefits, and a correspondingly high price and small market (Gouni-Berthold and Berthold 2002).

(c) Triacontanol is applied as a plant growth stimulator, particularly in Asia. It is applied in very low doses (0.1 to 10 mg per hectare, in the range for plant hormone effects rather than nutrients or crop protectants). Although studies over 25 years have provided convincing evidence for enhanced plant growth, triacontanol has yet to achieve mainstream agrochemical status, perhaps because the mechanisms for its diverse effects on plant physiology are unclear and many factors can contribute to a lack of consistency in results. If consistency can be established through improved formulation and application guidelines, the potential is for market size and price intermediate between (a) and (b) (Chen *et al.* 2002; Ries 1991).

4.10.3. Composition and biosynthesis of sugarcane wax

On the stem surface, sugarcane wax has been estimated to comprise long chain aldehydes (50%), long chain primary alcohols (5–27%), acids (3–8%), hydrocarbons (2–9%), and esters (up to 6%) (Kranz *et al.* 1960; Tulloch 1976).

The major aldehyde component appears to exist in a hemihydrate or higher polymeric form (Haas *et al.* 2001; Lamberton 1965) which readily converts into free aldehyde when distilled or dissolved in warm solvents, and rapidly oxidises to the corresponding acid if melted while exposed to the air. The acid, aldehyde and alcohol forms are biosynthetically related. All are saturated, straight chain molecules, with chain length distribution estimated at about C26 (14%), C28 (73%), C30 (7%), C32 (6%) (Kranz *et al.* 1960; Tulloch 1976). The predominant chain length is reported to vary between grass species, being C32 in maize, C28 in wheat, and C26 in barley, rye and oats (Bianchi 1995).

Waxes from other organisms can vary greatly from those produced in sugarcane, including the presence of additional components such as branched (methylated) or unsaturated chains, secondary alcohols terpenoids and diverse esters (Hamilton 1995; Kolattukudy 1976).

Factors affecting wax composition in sugarcane have not been documented, but by analogy with other plants including cereals they are likely to include genotype, source tissue, developmental stage, and environmental conditions such as light, temperature and water availability. For example, in sorghum and maize, long chain free alcohols are the major wax component in seedlings, but alcohols decrease markedly with a concomitant increase in free fatty acids in the wax from mature plants. Changes in predominant chain length with age have also been reported (Bianchi 1995).

The biosynthesis of long chain wax components commences with the plastid machinery for synthesis from acetyl-CoA and malonyl-CoA to C16 and C18 fatty acids (Figure 2).

These products are exported from the plastids and elongated (to C20 – C34 fatty acyl-CoA molecules) by microsomal (endoplasmic reticulum, membrane-associated) fatty acyl-CoA elongase (FAE) complexes (Somerville *et al.* 2000a). Each two-carbon cycle of chain elongation requires four enzymatic activities, of which the first (β-ketoacyl-CoA synthase, KCS) appears to control substrate specificity and the extent and rate of the elongation process (Blacklock and Jaworski 2002; Han *et al.* 2001; Hooker *et al.* 2002; Lassner *et al.* 1996; Millar and Kunst 1997). It is likely that several different KCS enzymes (also referred to as

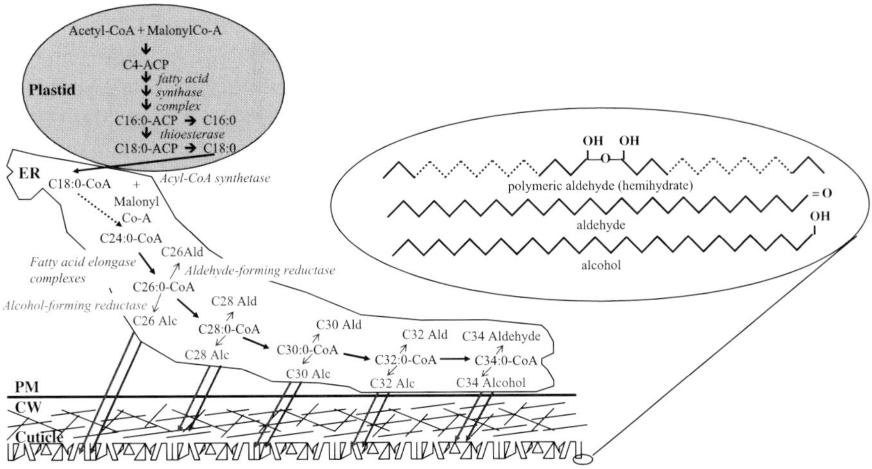

Figure 2. Biosynthetic pathway for sugarcane epicuticular wax components

elongases or condensing enzymes) operate over different chain lengths. Acyl-ACP thioesterases may also play a role in terminating the elongation process (Sandager and Stymne 2000).

Fatty alcohol synthesis involves a four-electron reduction of fatty acyl-CoA by a 58 kD membrane-associated microsomal enzyme using NADH or NADPH as a cofactor. The alcohol-generating fatty acyl-CoA reductase (FAR) probably involves an aldehyde intermediate, but the enzymes characterised to date do not release free fatty aldehyde (Vioque and Kolattukudy 1997).

Free fatty aldehydes are produced by a two-electron reduction catalysed by a separate (soluble or loosely membrane-associated) 30 kD enzyme. In peas, the aldehyde and alcohol intermediates have different size ranges and metabolic fates through hydrocarbon and ester formation. The free aldehyde appears not to be convertible by further enzymatic reduction into alcohol, and it has been speculated that there may be separate coupled systems for chain elongation, reduction and final metabolic conversions involving these separate intermediates (Vioque and Kolattukudy 1997). This divergent functional coupling has not been confirmed experimentally, and there seems less reason to expect that it might operate in sugarcane (where fatty acids, aldehydes and alcohols show similar chain length distributions, and the free aldehydes and alcohols are major wax constituents rather than precursors).

Movement of the wax components from the site of synthesis (ER) to the surface of epidermal cells is not well understood, but it may involve direct translocation or vesicular traffic from the ER to the plasma membrane, and lipid transfer proteins in the cell wall matrix (Kunst and Samuels 2003; Nawrath 2006; Somerville *et al.* 2000b).

4.10.4. *Molecular Genetics & Biotechnology of Plant Waxes*

Mutations affecting cuticular wax composition are known in several plant species, and substantial progress has been made over several decades in understanding the biochemistry of wax biosynthesis through approaches including radio-labelled precursors and the isolation of key enzymes (Bianchi 1995; Kolattukudy 1996). More recently key biosynthetic genes have begun to be cloned and characterised by techniques such as site-directed mutagenesis and over-expression or down-regulation in transgenic plants (Kunst and Samuels 2003; Nawrath 2006).

In the context of metabolic engineering for improved sugarcane wax composition and yield, key findings are:

1. Fatty acid elongase condensing enzyme (KCS) genes have been cloned from jojoba, brassica, *Lunaria* and *Arabidopsis*. They differ somewhat in both chain length specificity and elongase efficiency.

   Arabidopsis cuticular wax comprises mainly n-alkanes and ketones (unlike sugarcane), but the fatty acid/aldehyde/alcohol component does comprise mainly C30 – C26 chain lengths (like sugarcane). Of four KCS genes cloned from *Arabidopsis*, *CER6* appears to be the main contributor to stem wax and pollen coat lipid biosynthesis. The *CER6* promoter is tightly epidermis-specific. Over-expression by introduction of extra copies of the gene with its native promoter (but not the constitutive 35S promoter) doubled the wax loads in stems of some transformants. There were signs of gene silencing in some sectors and progeny (Hooker *et al.* 2002). Recently several transcriptional activator genes have been shown to affect wax accumulation and composition in *Arabidopsis*, and over-expression has conferred up to six-fold increase in cuticular wax (Aharoni *et al.* 2004; Broun *et al.* 2004).
2. A large family of 21 elongase condensing enzyme-like sequences has been identified in the *Arabidopsis* genome, and several of the cloned genes encode enzymes with different chain length specificities (Blacklock and Jaworski 2006). It will be possible to identify likely sugarcane homologs by searching the EST database against these sequences. Site-directed mutagenesis based on comparison of *Arabidopsis* and brassica KCS enzymes with slightly different product length profiles indicates that a region just after the N-terminal membrane-spanning domain plays a key role in substrate specificity (Blacklock and Jaworski 2002). Continuing research should eventually allow the prediction and tailoring of chain length specificity in KCS genes.
3. The jojoba and *Lunaria* KCS genes and *Arabidopsis FAE1* have been expressed in heterologous systems (as remote as yeast) resulting in a shift in FA profiles. *Arabidopsis* seeds contain long chain fatty acids (primarily 20:1), which are eliminated by mutations in seed-specific KCS gene *FAE1*. Expression of *FAE1* from the 35S promoter results in accumulation of 20:0, 20:1, 20:2, 20:3, 22:0, 22:1, 22:2 and 22:3 acids in other tissues, with adverse effects on plant growth. Expression from a seed-specific promoter resulted in some transgenic lines with long chain fatty acids (FA) increased from 27% to 42% of total FA, and apparent down-regulation in other lines (Lassner 1997; Millar and Kunst 1997).

4. To gain the full impact of modified KCS expression, it may also be necessary to modify expression of genes for downstream processing of FA to the final storage products. In oilseed crops this could be increased acyltransferase expression for higher triacylglyceride yield (Millar and Kunst 1997). By analogy, in sugarcane, we might require increased alcohol- or aldehyde- forming reductase.
5. An alcohol-forming fatty acid reductase (FAR) gene has been cloned from jojoba, and related genes are evident in genome databases of other plants including grasses. Eight similar genes are present in the *Arabidopsis* genome. Expression of the jojoba gene in *E. coli* and in rapeseed resulted in accumulation of fatty alcohols (or esters of the novel fatty alcohols) corresponding with the predominant fatty acids in these organisms (Metz *et al.* 2000). High-level accumulation of free fatty alcohols was not observed, and may be toxic unless the products are exported (as in the sugarcane cuticle) or converted to esters (as observed in rapeseed when the transgene was driven by an oilseed embryo-specific promoter).
6. By co-expressing foreign elongase KCS, alcohol-forming FAR and ester-forming wax synthase (WS) genes from an embryo-specific promoter, it was possible to obtain *Arabidopsis* transformants with up to 70% of seed oils converted to wax (Lardizabal *et al.* 2000).
7. Most studied mutations in wax biosynthesis do not result in an accumulation of precursors, but instead result in changed wax composition indicating increased flux of precursors through an alternative branch pathway. For example mutation in *CER4* (which is not yet cloned) reduces primary alcohol and wax ester levels, partially compensated by an increase in aldehydes (Kunst and Samuels 2003). Therefore, specific inhibition (for example by co-suppression) of alcohol- or aldehyde- forming FAR is a potential mechanism (along with over-expression of the alternative reductase) to obtain sugarcane variants with higher yields and purity of wax composed of either long-chain aldehyde or alcohol. In this context, aldehyde-forming FAR enzymes have been purified, but a corresponding gene has not yet been reported from any plant.

   It is not clear how substantial conversion to biosynthesis of one or the other major component would affect the structure or stability of wax on the sugarcane surface, or whether one of these components also functions to sequester the other in a way that avoids adverse effects on the plant. However, in different maize tissues and mutants alcohols can range from trace to 63% of waxes, and aldehydes can range from trace to 85% of waxes indicating the potential to progress to plants producing "pure" wax components (a key determinant of economic practicality for some industrial purposes).
8. The mechanisms of secretion and transport of wax components into the cuticle remain hypothetical. Lipid transfer proteins (LTP) are speculated to move the hydrophobic wax components across the predominantly hydrophilic band of the epidermal cell wall, and multiple LTP genes are known in *Arabidopsis*. At this

stage the process appears fairly non-specific, and there are no obvious transport targets for manipulation to alter wax properties or yield (Nawrath2006).

4.10.5. *Commercial Considerations & Experimental Approaches in Sugarcane*

4.10.5.1. Lessons from oilseeds Up to C18 fatty acids, the precursors for plant waxes are the same as those for lipids harvested for food and industrial use from oilseed crops. The relatively advanced work in metabolic engineering of plant oils (Murphy 2006; Thelen and Ohlrogge 2002) provides some useful lessons for work planned in waxes:

- Rational engineering for commercial outcomes requires a clear elucidation of the existing biological processes.
- The complexity of metabolic interactions generally means that it is sensible to try fine-tuning or 'domestication' of plants already synthesising useful materials before attempted engineering of unusual lipid production into a species.
- Yield and purity of the desired lipid are key determinants of economic viability for industrial use in competition with petrochemical alternatives.
- If there are substantial extraction costs, the material must have properties not readily obtained in petrochemical alternatives (or market premiums or government subsidies for renewable biomaterials in preference to non-renewable petrochemicals).

4.10.5.2. Competitive advantage Sugarcane's natural advantage is in production of very-long-chain (C24–C34) unsaturated primary aldehydes and alcohols (or corresponding acids and esters formed during certain extraction conditions). These molecules are the basis for current products:

- policosanol/octacosanol, mixed length/C28 primary alcohols used as pharmaceuticals,
- triacontanol, C30 primary alcohol used as a plant growth stimulant,
- refined hard wax of esters formed during extraction, not presently competitive in price against carnauba.

The opportunity is for increased yield and improved composition of sugarcane epicuticular waxes for industrial and pharmaceutical uses.

4.10.5.3. Factors affecting commercial strategy Before designing strategies to maximise the value of sugarcane waxes, the following questions need to be answered:

1. What is the projected market size and value for each of the current products?
2. Are there emerging markets for other components or simple derivatives of sugarcane epicuticular wax?
3. How price-sensitive are these established and emerging markets?
4. What is the potential and what are the requirements for market development (*e.g.* reliable effective formulations and application protocols for triacosanol)?

5. What is the anticipated effect of higher yield (sugarcane epicuticular wax load) on production cost? Would this be economic if accompanied by reduced sugar yield in the cane?
6. What is the extent of genotypic, developmental, and environmental variation in sugarcane epicuticular wax load and composition?
7. Is there any apparent relationship between wax load and sugar yield or any likely interference of high wax loads in current sugar industry practices?
8. What is the anticipated effect of higher purity of desired constituents in the epicuticular wax (*e.g.* a much higher alcohol:aldehyde ratio, or a much sharper chain length profile)?
9. Are there related products that would be more attractive in demand volume and pricing if they could be produced by metabolic engineering of sugarcane (*e.g.* C26–34 molecules with double bonds, methyl side chains, hydroxyl groups or other modifications at defined positions)?
10. What are the potential sources of competing products (including chemical synthesis of the same molecule)? How do they compare in product properties and prices?
11. What would be the competitive position if wax accumulation could be engineered into an oilseed crop?

4.10.5.4. Experimental approaches Depending on the outcomes of the commercial analysis of potential products, the most likely approaches to achieve increased yield and improved composition and higher value of sugarcane epicuticular waxes for industrial and pharmaceutical uses are:

1. Screening and selection for useful (yield and compositional) variants in wax production from conventional breeding (given a suitable high-throughput screen).
2. Identification of sugarcane homologs of key genes in the wax biosynthetic pathway, and elucidation of their allelic variation, expression patterns and comparative specificities and conversion efficiencies.
3. Recovery and functional confirmation of regulatory sequences for desired expression patterns (*e.g.* stem epidermal expression for wax production).
4. Over-expression and down-regulation of selected genes with a view to improved wax yield and composition.
5. Modification of sugarcane genes or introduction of selected genes from other species for desired changes in properties of wax constituents.
6. Improved extraction and purification techniques for desired components.
7. Downstream work on formulation and application of products, for market development.

5. HURDLES TO BE OVERCOME

5.1. Transgene Silencing

Silencing (or more accurately down-regulation of expression) of introduced genes is observed at some frequency with at least some genetic constructs in most tested plant species. Much is being learned about the mechanisms, which are emerging as part of a complex cellular machinery with likely roles in defence against invasive genetic elements such as viruses and transposons and in the developmental regulation of many plant genes (Brodersen and Voinnet 2006).

Silencing is a major issue in commercialising transgenic crops. The creation of lines suitable for commercialisation typically requires the production of tens or hundreds of independent transformants, followed by screens to identify a few lines with the desired commercial trait. Lines with low or unstable transgene expression or adverse changes in other commercial traits following the gene transfer process are discarded. Instability of transgene expression and consequent phenotypic instability are especially problematic, since the effects may be subtle and slow to develop. The severity may vary between species, but as a late-stage commercialisation problem the details are not typically published. Discard rates over 90% are mentioned (Finnegan and McElroy 1994), but other reports indicate no serious problem of silencing (Hawkins *et al.* 2003).

Sugarcane is unusually efficient at silencing the expression of introduced genes. This silencing occurs at both the transcriptional and post-transcriptional levels, and is possibly related to the complex polyploidy of the sugarcane genome (Birch *et al.* 2000; Birch *et al.* 1996; Hansom *et al.* 1999; Ingelbrecht *et al.* 1999). A substantial number of heterologous and sugarcane-derived promoters has now been tested in sugarcane. Many drive transgene expression in callus, but undergo efficient silencing in a very high proportion of transformants at about the time of plant regeneration. Others have been reported to drive transgene expression in plants (Lakshmanan *et al.* 2005), but to date only the maize and sugarcane Ubi promoters have driven sustained transgene expression over several generations in field-grown plants. Even Ubi constructs appear to be progressively down-regulated, typically resulting in low levels of transgene product formation (Wang *et al.* 2005). Silencing appears not to be transgene copy-number dependent (Birch *et al.* 2000; Hansom *et al.* 1999) and it has not been prevented to date by use of protective elements such as matrix attachment sequences flanking the expression cassette (Waldron *et al.* 2001; Bower and Birch, unpublished).

The development of methods to avoid or overcome this silencing is a critical requirement for practical metabolic engineering in sugarcane, other than approaches such as reduced lignification which may be accomplished by targeted down-regulation of sugarcane genes.

5.2. Expression Patterns

Most practical metabolic engineering in sugarcane will require not only sustained expression without silencing over many vegetative generations, but also tailored levels and developmental or inducible patterns of expression, appropriate to the desired effect of the transgene product in sugarcane. Despite the early emphasis on constitutive promoters (Lakshmanan *et al.* 2005), largely because at least some sustained transgene expression can be obtained using the maize Ubi-1 promoter

(Hansom *et al.* 1999), constitutive expression is unlikely to be useful for the traits discussed above. Given the efficient sugarcane transformation system, rare events where flanking sequences at a transgene integration site fortuitously confer a practically useful expression pattern can be useful in the discovery phase; but this is not practical strategy for the following stages of optimisation and commercial development of elite commercial lines.

Some of the developmental and physiological adaptations of sugarcane most relevant to metabolic engineering (e.g. accumulation of high concentrations of sucrose in the stem) are not evident in genetic model plants. It is unlikely that these model species can yield promoters with the subtle control over expression patterns necessary for efficient molecular improvement of highly adapted processes in sugarcane, so substantial effort has been made to isolate and characterise promoters associated with sugarcane genes that show commercially useful expression patterns (Birch and Potier 2001; Damaj *et al.* 2004; Mirkov *et al.* 2004; van der Merwe *et al.* 2003).

Unfortunately, the background problem of efficient transgene silencing has made it difficult to determine the portability of expression patterns from heterologous promoters into sugarcane (Potier *et al.* 2006), and the functionality with transgenes of isolated sugarcane promoters associated with useful endogenous expression patterns (Mudge *et al.* 2006).

6. PROSPECTS

It has been said that in biotechnology, there are no sensible small-scale investments (Bachmann *et al.* 2000), and given the challenges outlined above this is a fair assessment of metabolic engineering in plants. In sugarcane, the advantages of high biomass production and carbon flux through useful precursor pools indicate a bright future for renewable biomaterials production aided by metabolic engineering. Genes for enhanced yield of sucrose and for useful conversions to commodity and speciality chemicals have already been identified and shown to function in modified sugarcane lines.

Now the challenge is to achieve commercially useful production levels and stability. This requires an integration of biology and biotechnology, beyond the levels that have underpinned metabolic engineering for enhanced production of microbial biomaterials, as we understand and tailor the required developmental expression patterns, cellular compartmentation, signalling and control of source-sink relationships to get the best out of this remarkable photosynthetic biofactory.

With collaborative effort to overcome the common technical constraint of transgene silencing, it should be technically feasible to produce diverse sugarcane lines with commercially useful enhancements through metabolic engineering within the coming five to ten years.

At the same time it is essential for industry that the safety and benefits of the products are appreciated by consumers. In most countries this is being accomplished by transparent, science-based, evaluation and regulation of specific traits and crops; combined with open communication about the underlying technologies. It is heartening that several of the most promising traits in sugarcane have direct advantages for consumers and the environment as well as for industry. This combination that has been welcomed in earlier examples of plants improved using modern gene technologies (Brookes and Barfoot 2005).

REFERENCES

Aharoni A, Dixit S, Jetter R, Thoenes E, van Arkel G, Pereira A (2004) The SHINE clade of AP2 domain transcription factors activates wax biosynthesis, alters cuticle properties, and confers drought tolerance when overexpressed in *Arabidopsis. Plant Cell* **16**, 2463–2480.

Allen C, Mackay M, Aylward J, Campbell J (1997) New technologies for sugar milling and by-product modification. In 'Intensive Sugarcane Production: Meeting the Challenges Beyond 2000'. (Eds B Keating and J Wilson) pp. 267–285. (CAB International: Wallingford, UK).

Arencibia AD, Carmona ER, Cornide MT, Castiglione S, O'Relly J, Chinea A, Oramas P, Sala F (1999) Somaclonal variation in insect-resistant transgenic sugarcane (Saccharum hybrid) plants produced by cell electroporation. *Transgenic Research* **8**, 349–360.

Arencibia AD, Carmona ER, Tellez P, Chan MT, Yu SM, Trujillo LE, Oramas P (1998) An efficient protocol for sugarcane (*Saccharum* spp. L.) transformation mediated by *Agrobacterium tumefaciens. Transgenic Research* **7**, 213–222.

Bachmann R, Bastianelli E, Riese J, Schlenzka W (2000) Using plants as plants. *The McKinsey Quarterly* **2000**, 93–99.

Berding N (1981) Improved flowering and pollen fertility in sugarcane under increased night temperatures. *Crop Science* **21**, 863–867.

Bianchi G (1995) Plant waxes. In 'Waxes: Chemistry, Molecular Biology and Functions'. (Ed. RJ Hamilton) pp. 175–222. (The Oily Press: Dundee).

Biotecnología CdIGy (2006) Fructan production in industrial reactors and transgenic sugarcane. http://gndp.cigb.edu.cu/Profiles/28%20CIGBFructooligasaccharides.htm

Birch RG (1996) New gene technologies and their potential value for sugarcane. *Outlook on Agriculture* **25**, 219–226.

Birch RG, Bower R, Elliott A, Hansom S, Basnayake S, Zhang L (2000) Regulation of transgene expression: progress towards practical development in sugarcane, and implications for other plant species. In 'Plant Genetic Engineering: Towards the Third Millenium'. (Ed. AD Arencibia) pp. 118–125. (Elsevier: Amsterdam).

Birch RG, Bower R, Elliott AR, Potier BAM, Franks T, Cordeiro G (1996) Expression of foreign genes in sugarcane. In 'Proceedings of the International Society of Sugarcane Technologists XXII Congress, Cartegena, September 1995'. (Eds JH Cock and T Brekelbaum) pp. 368–373. (Tecnicana: Cali, Colombia).

Birch RG, Potier BAM (2001) Sugarcane plant promoters to express heterologous nucleic acids. International Patent Specification WO 01/18211 A1.

Birch RG, Wu L (2004) Method for increasing product yield. International Patent Specification WO 2004/099403 A1.

Blacklock BJ, Jaworski JG (2002) Studies into factors contributing to substrate specificity of membrane-bound 3-ketoacyl-CoA synthases. *European Journal of Biochemistry* **269**, 4789–4798.

Blacklock BJ, Jaworski JG (2006) Substrate specificity of *Arabidopsis* 3-ketoacyl-CoA synthases. *Biochemical and Biophysical Research Communications* **346**, 583–590.

Börnke F, Hajirezaei M, Heineke D, Melzer M, Herbers K, Sonnewald U (2002a) High-level production of the non-cariogenic sucrose isomer palatinose in transgenic tobacco plants strongly impairs development. *Planta* **214**, 356–364.

Börnke F, Hajirezaei M, Sonnewald U (2002b) Potato tubers as bioreactors for palatinose production. *Journal of Biotechnology* **96**, 119–124.

Botha FC, Groenewald JH (2001) Method for regulating or manipulating sucrose content and metabolism in sugar storing plants, e.g. increasing sucrose content, by regulating activity of pyrophosphate-dependent phosphofructokinase enzyme in plants. South Africa Patent Application ZA200101047-A 25Jul2001.

Botha FC, Sawyer BJB, Birch RG (2001) Sucrose metabolism in the culm of transgenic sugarcane with reduced soluble acid invertase activity. In 'Proceedings of the International Society of Sugarcane Technologists XXIV Congress, Brisbane, September 2001'. (Ed. DM Hogarth) pp. 588–591. (ASSCT: Mackay).

Bower R, Birch RG (1992) Transgenic sugarcane plants via microprojectile bombardment. *The Plant Journal* **2**, 409–416.

Bower R, Elliott AR, Potier BAM, Birch RG (1996) High-efficiency, microprojectile-mediated cotransformation of sugarcane, using visible or selectable markers. *Molecular Breeding* **2**, 239–249.

Braga DPV, Arrigoni EDB, Silva Filho MC, Ulian EC (2003) Expression of the Cry1Ab protein in genetically modified sugarcane for the control of *Diatraea saccharalis* (Lepidoptera: Crambidae). *Journal of New Seeds* **5**, 209–221.

Brodersen P, Voinnet O (2006) The diversity of RNA silencing pathways in plants. *Trends in Genetics* **22**, 268–280.

Brookes G, Barfoot P (2005) GM crops: the global economic and environmental impact – the first nine years 1996–2004. *AgBioForum* **8**, 187–196.

Broun P, Poindexter P, Osborne E, Jiang CZ, Riechmann JL(2004) WIN1, a transcriptional activator of epidermal wax accumulation in *Arabidopsis*. *Proceedings of the National Academy of Sciences of the United States of America* **101**, 4706–4711.

Brumbley SM, Purnell MP, Petrasovits LA, Fong Chong B, O'Neill BP, Nielsen LK (2006) Sugarcane: a versatile biofactory for accumulation of heterologous products derived from diverse precursors. In 'Tropical Crop Biotechnology Conference'. Cairns. (Ed. JM Manners) p. 39. (CSIRO).

Capell T, Christou P (2004) Progress in plant metabolic engineering. *Current Opinion in Biotechnology* **15**, 148–154.

Carmona ER, Arencibia AD, Lopez J, Simpson J, Vargas D, Sala F (2005) Analysis of genomic variability in transgenic sugarcane plants produced by *Agrobacterium tumefaciens* infection. *Plant Breeding* **124**, 33–38.

Casu RE, Manners JM, Bonnett GD, Jackson PA, McIntyre CL, Dunne R, Chapman SC, Rae AL, Grof CPL (2005) Genomics approaches for the identification of genes determining important traits in sugarcane. *Field Crops Research* **92**, 137–147.

Chen JCP, Chou C-C (1993) 'Cane Sugar Handbook: a Manual for Cane Sugar Manufacturers and their Chemists.' (Wiley: New York).

Chen XP, Yuan HY, Chen RZ, Zhu LL, Du B, Weng QM, He GC (2002) Isolation and characterization of triacontanol-regulated genes in rice (*Oryza sativa* L.): possible role of triacontanol as a plant growth stimulator. *Plant and Cell Physiology* **43**, 869–876.

Chen YR, Sarkanen S (2006) From the macromolecular behavior of lignin components to the mechanical properties of lignin-based plastics. *Cellulose Chemistry and Technology* **40**, 149–163.

Cortina C, Culianez-Macia FA (2005) Tomato abiotic stress enhanced tolerance by trehalose biosynthesis. *Plant Science* **169**, 75–82.

Damaj MB, Mirkov ET, *et al.* (2004) New isolated nucleic acid comprises jasmonate-induced protein (JAS) promoter and an exogenous nucleic acid, useful as a promoter for altering carbon metabolism in a plant cell or for driving expression of insecticidal proteins in sugarcane. PCT Patent Application WO2004062366-A2.

Dyer JM, Mullen RT(2005) Development and potential of genetically engineered oilseeds. *Seed Science Research* **15**, 255–267.

Edye LA, Doherty WOS, Blinco JA, Bullock GE (2005) The sugarcane biorefinery: energy crops and processes for the production of liquid fuels and renewable commodity chemicals. In 'Australian Society of Sugar Cane Technologists 27th Conference'. Bundaberg. (Ed. DM Hogarth) pp. 9–22.

Enriquez-Obregon GA, Vazquez-Padron RI, Prieto-Samsonov DL, De la Riva GA, Selman-Housein G (1998) Herbicide-resistant sugarcane (*Saccharum officinarum* L.) plants by *Agrobacterium*-mediated transformation. *Planta* **206**, 20–27.

Enriquez GA, Arencibia A, *et al.* (2001) Sugarcane biotechnology and genetic engineering experience at CIGB. In 'International Society of Sugar Cane Technologists. Proceedings of the XXIV Congress, Brisbane, Australia, 17–21 September 2001. Volume 2' pp. 578–579).

Falco MC, Neto AT, Ulian EC (2000) Transformation and expression of a gene for herbicide resistance in a Brazilian sugarcane. *Plant Cell Reports* **19**, 1188–1194.

Finnegan J, McElroy D (1994) Transgene inactivation: plants fight back! *Bio/Technology* **12**, 883–888.

Fujii S, Komoto M (1991) Novel carbohydrate sweeteners in Japan. *Zuckerindustrie* **116**, 197–200.

GalloMeagher M, Irvine JE (1996) Herbicide resistant transgenic sugarcane plants containing the bar gene. *Crop Science* **36**, 1367–1374.

Gilbert RA, Gallo-Meagher M, Comstock JC, Miller JD, Jain M, Abouzid A (2005) Agronomic evaluation of sugarcane lines transformed for resistance to sugarcane mosaic virus strain E. *Crop Science* **45**, 2060–2067.

Gnanasambandam A, Birch RG (2001) The challenge of protein targeting to the sugarcane vacuoles for metabolic engineering. In 'Proceedings of the International Society of Sugarcane Technologists XXIV Congress, Brisbane, September 2001'. (Ed. DM Hogarth) pp. 667–669. (ASSCT: Mackay).

Gnanasambandam A, Birch RG (2004) Efficient developmental mis-targeting by the sporamin NTPP vacuolar signal to plastids in young leaves of sugarcane and *Arabidopsis*. *Plant Cell Reports* **24**, 435–447.

Gnanasambandam A, Polkinghorne IG, Chamberlain D, Birch RG (2006) Efficient targeting of green fluorescent protein to plastids and endoplasmic reticulum in twelve (monocot and dicot) crops. *Submitted.*

Gonzali S, Novi G, Loreti E, Paolicchi F, Poggi A, Alpi A, Perata P (2005) A turanose-insensitive mutant suggests a role for *WOX5* in auxin homeostasis in *Arabidopsis thaliana*. *Plant Journal* **44**, 633–645.

Gouni-Berthold I, Berthold HK (2002) Policosanol: clinical pharmacology and therapeutic significance of a new lipid-lowering agent. *American Heart Journal* **143**, 356–365.

Grivet L, Arruda P (2002) Sugarcane genomics: depicting the complex genome of an important tropical crop. *Current Opinion in Plant Biology* **5**, 122–127.

Grof CPL, Campbell JA (2001) Sugarcane sucrose metabolism: scope for molecular manipulation. *Australian Journal of Plant Physiology* **28**, 1–12.

Haas K, Brune T, Rucker E (2001) Epicuticular wax crystalloids in rice and sugar cane leaves are reinforced by polymeric aldehydes. *Journal of Applied Botany* **75**, 178–187.

Hajirezaei MR, Börnke F, Peisker M, Takahata Y, Lerchl J, Kirakosyan A, Sonnewald U (2003) Decreased sucrose content triggers starch breakdown and respiration in stored potato tubers (*Solanum tuberosum*). *Journal of Experimental Botany* **54**, 477–488.

Hamilton RJ (1995) Commercial waxes: their composition and applications. In 'Waxes: Chemistry, Molecular Biology and Functions'. (Ed. RJ Hamilton) pp. 257–310. (The Oily Press: Dundee).

Han JX, Luhs W, Sonntag K, Zahringer U, Borchardt DS, Wolter FP, Heinz E, Frentzen M (2001) Functional characterization of beta-ketoacyl-CoA synthase genes from *Brassica napus* L. *Plant Molecular Biology* **46**, 229–239.

Hansom S, Bower R, *et al.* (1999) Regulation of transgene expression in sugarcane. In 'Proceedings of the International Society of Sugarcane Technologists XXIII Congress, New Delhi, February 1999'. (Ed. V Singh) pp. 278–290. (STAI: New Dehli).

Harrison MD, Kato M, Becker DK, Harding R, Bullock G, Dale JL (2006) Switching on enzymes to break down sugarcane waste for fermentation to bioethanol. In 'Tropical Crop Biotechnology Conference'. Cairns. (Ed. JM Manners) p. 97. (CSIRO).

Hawkins S, Leple JC, Cornu D, Jouanin L, Pilate G (2003) Stability of transgene expression in poplar: a model forest tree species. *Annals of Forest Science* **60**, 427–438.

Hooker TS, Millar AA, Kunst L (2002) Significance of the expression of the CER6 condensing enzyme for cuticular wax production in *Arabidopsis*. *Plant Physiology* **129**, 1568–1580.

Howard JA, Hood E (2005) Bioindustrial and biopharmaceutical products produced in plants. *Advances in Agronomy* **85**, 91–124.

Ingelbrecht IL, Irvine JE, Mirkov TE (1999) Posttranscriptional gene silencing in transgenic sugarcane. Dissection of homology-dependent virus resistance in a monocot that has a complex polyploid genome. *Plant Physiology* **119**, 1187–1197.

Inman-Bamber NG, Muchow RC, Robertson MJ (2002) Dry matter partitioning of sugarcane in Australia and South Africa. *Field Crops Research* **76**, 71–84.

Jackson PA (2005) Breeding for improved sugar content in sugarcane. *Field Crops Research* **92**, 277–290.

Jacobsen KR, Fisher DG, Maretzki A, Moore PH (1992) Developmental changes in the anatomy of the sugarcane stem in relation to phloem unloading and sucrose storage. *Botanica Acta* **105**, 70–80.

James G (2004) 'Sugarcane.' (Blackwell Science Ltd: Oxford).

Karakas B, OziasAkins P, Stushnoff C, Suefferheld M, Rieger M (1997) Salinity and drought tolerance of mannitol-accumulating transgenic tobacco. *Plant Cell and Environment* **20**, 609–616.

Khan R (1994) Sucrose: its potential as a raw material for food ingredients and for chemicals. In 'Sucrose: Properties and Applications'. (Eds M Mathlouthi and P Reiser) pp. 264–278. (Blackie: London).

Klein J, Altenbuchner J, Mattes R (1998) Nucleic acid and protein elimination during the sugar manufacturing process of conventional and transgenic sugar beets. *Journal of Biotechnology* **60**, 145–153.

Koch K (2004) Sucrose metabolism: regulatory mechanisms and pivotal roles in sugar sensing and plant development. *Current Opinion in Plant Biology* **7**, 235–246.

Kolattukudy PE (1976) Introduction to natural waxes. In 'Chemistry and Biochemistry of Natural Waxes'. (Ed. PE Kolattukudy) pp. 1–15. (Elsevier: Amsterdam).

Kolattukudy PE (1996) Biosynthetic pathways of cutin and waxes and their sensitivity to environmental stresses. In 'Plant Cuticles'. (Ed. G Kerstiens) pp. 83–108. (BIOS Scientific: Oxford).

Kossmann J, Groenewald J-H, Botha FC (2006) Metabolic engineering of sucrose synthesis in sugarcane. In 'Tropical Crop Biotechnology Conference'. Cairns. (Ed. JM Manners) p. 20. (CSIRO).

Kranz ZH, Lamberton JA, Murray KE, Redcliffe AH (1960) Sugar-cane wax II: an examination of the constituents of sugar-cane cuticle wax by gas chromatography. *Australian Journal of Chemistry* **13**, 498–505.

Kunst L, Samuels AL (2003) Biosynthesis and secretion of plant cuticular wax. *Progress in Lipid Research* **42**, 51–80.

Lakshmanan P, Geijskes RJ, Aitken KS, Grof CLP, Bonnett GD, Smith GR (2005) Sugarcane biotechnology: The challenges and opportunities. *In Vitro Cellular & Developmental Biology-Plant* **41**, 345–363.

Lamberton JA (1965) The long-chain aldehydes of sugar-cane wax. *Australian Journal of Chemistry* **18**, 911–913.

Lardizabal KD, Metz JG, Sakamoto T, Hutton WC, Pollard MR, Lassner MW (2000) Purification of a jojoba embryo wax synthase, cloning of its cDNA, and production of high levels of wax in seeds of transgenic *Arabidopsis*. *Plant Physiology* **122**, 645–655.

Lassner MW (1997) Transgenic oilseed crops: a transition from basic research to product development. *Lipid Technology* **9**, 5–9.

Lassner MW, Lardizabal K, Metz JG (1996) A jojoba beta-ketoacyl-CoA synthase cDNA complements the canola fatty acid elongation mutation in transgenic plants. *The Plant Cell* **8**, 281–292.

Leibbrandt NB, Snyman SJ (2003) Stability of gene expression and agronomic performance of a transgenic herbicide-resistant sugarcane line in South Africa. *Crop Science* **43**, 671–677.

Lichtenthaler FW (2002) Unsaturated O- and N-heterocycles from carbohydrate feedstocks. *Accounts of Chemical Research* **35**, 728–737.

Lichtenthaler FW, Peters S (2004) Carbohydrates as green raw materials for the chemical industry. *Comptes Rendus Chimie* **7**, 65–90.

Lina BAR, Jonker D, Kozianowski G (2002) Isomaltulose (Palatinose ®): a review of biological and toxicological studies. *Food and Chemical Toxicology* **40**, 1375–1381.

Loreti E, Alpi A, Perata P (2000) Glucose and disaccharide-sensing mechanisms modulate the expression of alpha-amylase in barley embryos. *Plant Physiology* **123**, 939–948.

Ma H, Albert H, Paull R, Moore P (2000) Metabolic engineering of invertase activities in different subcellular compartments affects sucrose accumulation in sugarcane cells. *Australian Journal of Plant Physiology* **21**, 1021–1030.

Ma JKC, Chikwarnba R, Sparrow P, Fischer R, Mahoney R, Twyman RM (2005) Plant-derived pharmaceuticals – the road forward. *Trends in Plant Science* **10**, 580–585.

Mayer MJ, Narbad A, Parr AJ, Parker ML, Walton NJ, Mellon FA, Michael AJ (2001) Rerouting the plant phenylpropanoid pathway by expression of a novel bacterial enoyl-CoA hydratase/lyase enzyme function. *Plant Cell* **13**, 1669–1682.

McQualter RB, Chong BF, Meyer K, Van Dyk DE, O'Shea MG, Walton NJ, Viitanen PV, Brumbley SM (2005) Initial evaluation of sugarcane as a production platform for p-hydroxybenzoic acid. *Plant Biotechnology Journal* **3**, 29–41.

Metz JG, Pollard MR, Anderson L, Hayes TR, Lassner MW (2000) Purification of a jojoba embryo fatty acyl-coenzyme A reductase and expression of its cDNA in high erucic acid rapeseed. *Plant Physiology* **122**, 635–644.

Millar AA, Kunst L (1997) Very-long-chain fatty acid biosynthesis is controlled through the expression and specificity of the condensing enzyme. *The Plant Journal* **12**, 121–131.

Mirkov ET, Monclin JP, Barrilleaux A, Irvine JE, Moonan F, Mirkov TE (2002) Extracting, refining recombinant proteins from transgenic plant material, comprises fractionating juice extracted from the material using multiple-stage filtering process followed by membrane type filtration and ion exchange. PCT Patent Application WO200283715-A.

Mirkov TE, Damaj MB, Gonzalez J, Molina J, White SG, Nikolov Z (2006) Sugarcane as a biofactory for the economic production of low to medium value proteins: reality or wishful thinking? In 'Tropical Crop Biotechnology Conference'. Cairns. (Ed. JM Manners) p. 21. (CSIRO).

Mirkov TE, Damaj MB, *et al.* (2004) New isolated nucleic acid comprises an O-methyl transferase (OMT) promoter and an exogenous nucleic acid, useful as a promoter for altering carbon metabolism in a plant cell or for driving expression of insecticidal proteins in sugarcane. PCT Patent Application WO2004062365-A2.

Moore PH (1995) Temporal and spatial regulation of sucrose accumulation in the sugarcane stem. *Australian Journal of Plant Physiology* **22**, 661–679.

Moore PH (2005) Integration of sucrose accumulation processes across hierarchical scales: towards developing an understanding of the gene-to-crop continuum. *Field Crops Research* **92**, 119–135.

Moore PH, Botha FC, Furbank RT, Grof CPL (1997) Potential for overcoming physio-biochemical limits to sucrose accumulation. In 'Intensive Sugarcane Production: Meeting the Challenges Beyond 2000'. (Eds BA Keating and JR Wilson) pp. 141–155. (CAB International: Wallingford, UK).

Moore PH, Nuss KJ (1987) Flowering and flower synchronization. In 'Sugarcane Improvement Through Breeding'. (Ed. DJ Heinz) pp. 273–311. (Elsevier: Amsterdam).

Mudge SR, Osabe K, Casu RE, Bonnett GD, Manners JM, Birch RG (2006) Sugarcane allelic variation, allelic expression patterns and implications for the isolation of sugarcane promoters. In 'Tropical Crop Biotechnology Conference'. Cairns. (Ed. JM Manners) p. 42. (CSIRO).

Murphy DJ (2006) Molecular breeding strategies for the modification of lipid composition. *In Vitro Cellular & Developmental Biology-Plant* **42**, 89–99.

Nawrath C (2006) Unraveling the complex network of cuticular structure and function. *Current Opinion in Plant Biology* **9**, 281–287.

OGTR (1997) PR-72: Field test of sugarcane modified for resistance to sugarcane mosaic virus. www.ogtr.gov.au/pdf/volsys/pr72.pdf.

OGTR (2004) DIR 051/2004 – Field trial of genetically modified (GM) sugarcane expressing sucrose isomerase. www.ogtr.gov.au/ir/dir051.htm

Potier BAM, Snyman SJ, Huckett BI (2006) Evaluating tissue-specific promoters from maize and sorghum for expression in sugarcane. In 'Tropical Crop Biotechnology Conference'. Cairns. (Ed. JM Manners) p. 41. (CSIRO).

Rae AL, Grof CPL, Casu RE, Bonnett GD (2005) Sucrose accumulation in the sugarcane stem: pathways and control points for transport and compartmentation. *Field Crops Research* **92**, 159–168.

Ries S (1991) Triacontanol and its 2nd messenger 9-beta-L(+)-adenosine as plant-growth substances. *Plant Physiology* **95**, 986–989.

Ritsema T, Smeekens SCM (2003) Engineering fructan metabolism in plants. *Journal of Plant Physiology* **160**, 811–820.

Ritsema T, Verhaar A, Vijn I, Smeekens S (2004) Fructosyltransferase mutants specify a function for the beta-fructosidase motif of the sucrose-binding box in specifying the fructan type synthesized. *Plant Molecular Biology* **54**, 853–863.

Rohwer JM, Botha FC (2001) Analysis of sucrose accumulation in the sugar cane culm on the basis of in vitro kinetic data. *Biochemical Journal* **358**, 437–445.

Sandager L, Stymne S (2000) Characterisation of enzymes determining fatty acid chain length in developing seeds of *Limnanthes douglasii*. *Journal of Plant Physiology* **156**, 617–622.

Satoh-Nagasawa N, Nagasawa N, Malcomber S, Sakai H, Jackson D (2006) A trehalose metabolic enzyme controls inflorescence architecture in maize. *Nature* **441**, 227–230.

Scheller J, Conrad U (2005) Plant-based material, protein and biodegradable plastic. *Current Opinion in Plant Biology* **8**, 188–196.

Schiweck H, Munir M, Rapp KM, Schneider B, Vogel M (1991) New developments in the use of sucrose as an industrial bulk chemical. In 'Carbohydrates as Organic Raw Materials'. (Ed. FW Lichtenthaler) pp. 57–94. (Wiley-VCH: Weinheim).

Selman-Housein G, Lopez MA, Ramos O, Caroma ER, Arencibia AD, Menendez E, Miranda F (2000) Towards the improvement of sugarcane bagasse as raw material for the production of paper pulp and animal feed. In 'Plant Genetic Engineering: Towards the Third Millenium'. (Ed. AD Arencibia) pp. 189–193. (Elsevier: Amsterdam).

Setamou M, Bernal JS, Legaspi JC, Mirkov TE, Legaspi BC (2002) Evaluation of lectin-expressing transgenic sugarcane against stalkborers (Lepidoptera : Pyralidae): Effects on life history parameters. *Journal of Economic Entomology* **95**, 469–477.

Sheveleva EV, Marquez S, Chmara W, Zegeer A, Jensen RG, Bohnert HJ (1998) Sorbitol-6-phosphate dehydrogenase expression in transgenic tobacco – High amounts of sorbitol lead to necrotic lesions. *Plant Physiology* **117**, 831–839.

Sinha AK, Hofmann MG, Romer U, Kockenberger W, Elling L, Roitsch T (2002) Metabolizable and non-metabolizable sugars activate different signal transduction pathways in tomato. *Plant Physiology* **128**, 1480–1489.

Sivitz AB, Reinders A, Ward JM (2005) Analysis of the transport activity of barley sucrose transporter HvSUT1. *Plant and Cell Physiology* **46**, 1666–1673.

Slater S, Mitsky TA, *et al.* (1999) Metabolic engineering of *Arabidopsis* and *Brassica* for poly(3-hydroxybutyrate co 3 hydroxyvalerate) copolymer production. *Nature Biotechnology* **17**, 1011–1016.

Somerville C, Browse J, Jaworski JG, Ohlrogge JB (2000a) Genetic engineering of lipids. In 'Biochemistry and Molecular Biology of Plants'. (Eds BB Buchanan, W Gruissem and RL Jones) pp. 518–525. (ASPP: Rockville, MD, USA).

Somerville C, Browse J, Jaworski JG, Ohlrogge JB (2000b) Lipids. In 'Biochemistry and Molecular Biology of Plants'. (Eds BB Buchanan, W Gruissem and RL Jones) pp. 456–527. (ASPP: Rockville, MD, USA).

Sticklen M (2006) Plant genetic engineering to improve biomass characteristics for biofuels. *Current Opinion in Biotechnology* **17**, 315–319.

Sweetlove LJ, Fernie AR (2005) Regulation of metabolic networks: understanding metabolic complexity in the systems biology era. *New Phytologist* **168**, 9–23.

Takazoe I (1989) Palatinose – an isomeric alternative to sucrose. In 'Progress in Sweeteners'. (Ed. TH Grenby) pp. 143–167. (Elsevier: Barking).

Taniguchi H (2004) Carbohydrate research and industry in Japan and the Japanese society of applied glycoscience. *Starch* **56**, 1–5.

Thelen JJ, Ohlrogge JB (2002) Metabolic engineering of fatty acid biosynthesis in plants. *Metabolic Engineering* **4**, 12–21.

Trujillo LE, Arrieta JG, *et al.* (2000) Strategies for fructan production in transgenic sugarcane (*Saccharum* spp. L.) and sweet potato (*Ipomoea batatas* L.) plants expressing the *Acetobacter diazotrophicus* levansucrase. In 'Plant genetic engineering: towards the third millennium: Proceedings of the International Symposium on Plant Genetic Engineering, Havana, Cuba, 6–10 December, 1999.' pp. 194–198).

Tulloch AP (1976) Chemistry of waxes of higher plants. In 'Chemistry and Biochemistry of Natural Waxes'. (Ed. PE Kolattukudy) pp. 235–287. (Elsevier: Amsterdam).

van der Merwe J, Groenewald S, Botha FC (2003) Isolation and evaluation of a developmentally regulated sugarcane promoter. *Proceedings of the South African Sugar Technologists' Association* **77**, 146–149.

van der Poel P, Schiweck H, Schwartz T (1998) 'Sugar Technology: Beet and Cane Sugar Manufacture.' (Verlag Dr. Albert Bartens KG: Berlin).

van der Veen BA, Potocki-Veronese G, Albenne C, Joucla G, Monsan P, Remaud-Simeon M (2004) Combinatorial engineering to enhance amylosucrase performance: construction, selection, and screening of variant libraries for increased activity. *FEBS Letters* **560**, 91–97.

Vettore AL, da Silva FR, *et al.* (2003) Analysis and functional annotation of an expressed sequence tag collection for tropical crop sugarcane. *Genome Research* **13**, 2725–2735.

Vickers JE, Grof CPL, Bonnett GD, Jackson PA, Morgan TE (2005) Effects of tissue culture, biolistic transformation, and introduction of PPO and SPS gene constructs on performance of sugarcane clones in the field. *Australian Journal of Agricultural Research* **56**, 57–68.

Viitanen PV, Devine AL, Khan MS, Deuel DL, Van Dyk DE, Daniell H (2004) Metabolic engineering of the chloroplast genome using the *Echerichia coli ubiC* gene reveals that chorismate is a readily abundant plant precursor for p-hydroxybenzoic acid biosynthesis. *Plant Physiology* **136**, 4048–4060.

Vioque J, Kolattukudy PE (1997) Resolution and purification of an aldehyde-generating and an alcohol-generating fatty acyl-CoA reductase from pea leaves ({IPisum sativum} L). *Archives of Biochemistry and Biophysics* **340**, 64–72.

Waldron J, Reyes MEC, Hamerli D, Birch RG, Carroll BJ (2001) Tomato DNA sequences for resisting transgene silencing in sugarcane. In 'Proceedings of the International Society of Sugarcane Technologists XXIV Congress, Brisbane, September 2001'. (Ed. DM Hogarth) pp. 665–666. (ASSCT: Mackay).

Wang ML, Goldstein C, Su W, Moore PH, Albert HH (2005) Production of biologically active GM-CSF in sugarcane: a secure biofactory. *Transgenic Research* **14**, 167–178.

Watt DA, McCormick AJ, Govender C, Carson DL, Cramer MD, Huckett BI, Botha FC (2005) Increasing the utility of genomics in unravelling sucrose accumulation. *Field Crops Research* **92**, 149–158.

Werpy T, Petersen G (2004) 'Top Value Added Chemicals from Biomass.' (U.S. Department of Energy: Oak Ridge, TN).

Weyens G, Ritsema T, *et al.* (2004) Production of tailor-made fructans in sugar beet by expression of onion fructosyltransferase genes. *Plant Biotechnology Journal* **2**, 321–327.

Wilke D (1999) Chemicals from biotechnology: molecular plant genetics will challenge the chemical and the fermentation industry. *Applied Microbiology and Biotechnology* **52**, 135–145.

Wu L, Birch RG (2005) Characterization of the highly efficient sucrose isomerase from *Pantoea dispersa* UQ68J and cloning of the sucrose isomerase gene. *Applied and Environmental Microbiology* **71**, 1581–1590.

Wu L, Birch RG (2007) Doubled sugar content in sugarcane plants modified to produce a sucrose isomer. *Plant Biotechnology Journal* **5**, 109–117.

Zhang L, Xu J, Birch RG (1999) Engineered detoxification confers resistance against a pathogenic bacterium. *Nature Biotechnology* **17**, 1021–1024.

Zhang SZ, Yang BP, Feng CL, Chen RK, Luo JP, Cai WW, Liu FH (2006) Expression of the *Grifola frondosa* trehalose synthase gene and improvement of drought-tolerance in sugarcane (*Saccharum officinarum* L.). *Journal of Integrative Plant Biology* **48**, 453–459.

Zuther E, Kwart M, Willmitzer L, Heyer AG (2004) Expression of a yeast-derived invertase in companion cells results in long-distance transport of a trisaccharide in an apoplastic loader and influences sucrose transport. *Planta* **218**, 759–766.

CHAPTER 12

SINGLE-CHAIN Fv ANTIBODY STIMULATES BIOSYNTHESIS OF SECONDARY METABOLITES IN PLANTS

WARAPORN PUTALUN, HIROYUKI TANAKA AND YUKIHIRO SHOYAMA*

Faculty of Pharmaceutical Science,
Nagasaki International University,
2825-7 Hausutenbosu-cho, Sasebo, Nagasaki 859-3298, Japan

Abstract: A recombinant antibody fragment-single chain-fragment-variable (scFv) antibody can control the concentration of solasodine glycosides in hairy root cultures of *Solanum khasianum* transformed by the anti-solamargine (As)-scFv gene. The characterization of the As-scFv protein expressed in *Escherichia coli* was identical to those of the parent monoclonal antibody (MAb). Up to 220 ng recombinant As-scFv was expressed per milligram of soluble protein in transgenic hairy root cultures of *S. khasianum*. The concentration of solasodine glycosides was 2.3-fold higher in the transgenic than in the wild-type hairy root reflecting by the soluble As-scFv level and antigen binding activities. These phenomena suggested that the scFv antibody expressed in transgenic hairy roots controlled the antigen level representing a novel plant breeding methodology that can produce secondary metabolites, recommending that the method can be called as a missile-type breeding

Abbreviations: As: Anti-solamargine, CaMV: cauliflower mosaic virus, CDR: complementarity determining region, ELISA: enzyme-linked immunosorbent assay, ER: endoplasmic reticulum, IPTG: isopropyl-β-D-thiogalactopyranoside, K_d, dissociation constant, MAb: monoclonal antibody, MALDI: matrix assisted laser desorption/ionization, PCR: polymerase chain reaction, RPAS: recombinant phage antibody system, scFv: single-chain fragment-variable, SDS PAGE: sodium dodecyl sulfate-polyacrylamide gel electrophoresis, SUC2: the N-terminal yeast s peptide, TSP: total soluble protein

* To whom correspondence should be addressed. Phone or Fax: +81-92-642-6580. E-mail: shoyma@nill.ac.jp

R. Verpoorte et al. (eds.), Applications of Plant Metabolic Engineering, 283–295.
© 2007 *Springer.*

1. INTRODUCTION

The application of singel-chain fragment-variable (scFv) antibodies engineering as a diagnostic tool in plant pathology has recently been investigated (Ziegler et al., 1995; Harper et al., 1997; Hoogenboom, 1997) since they possess the same monovalent binding specificity and affinity as their parent antibodies (Tang et al., 1995). Moreover, scFv has opened new fields in plant research, such as protection against viral attack (Tavladoki et al., 1993; Fiedler and Cornad, 1995; Fecker et al., 1996, 1997), and studies on plant growth (Artsaenko et al., 1995, Harper and Ziegler, 1999).

In our ongoing efforts to prepare MAbs for setting up analytical systems for naturally occurring bioactive compounds (Loungratana et al., 2004; Kim et al., 2004; Morinaga et al., 2005; Lu et al., 2005, Zhu et al., 2006), we prepared anti-solamargine MAb (Ishiyama et al., 1996) and a simple staining method called Eastern blotting (Tanaka et al., 1997) and rapid immunoaffinity column separation (Putalun et al., 1999). Solasodine glycosides have anti-neoplastic activity in mice (Cham et al., 1987) and anti-skin carcinoma properties in humans (Daunter and Cham, 1990; Cham et al., 1991). Furthermore, the solasodine glycosides,

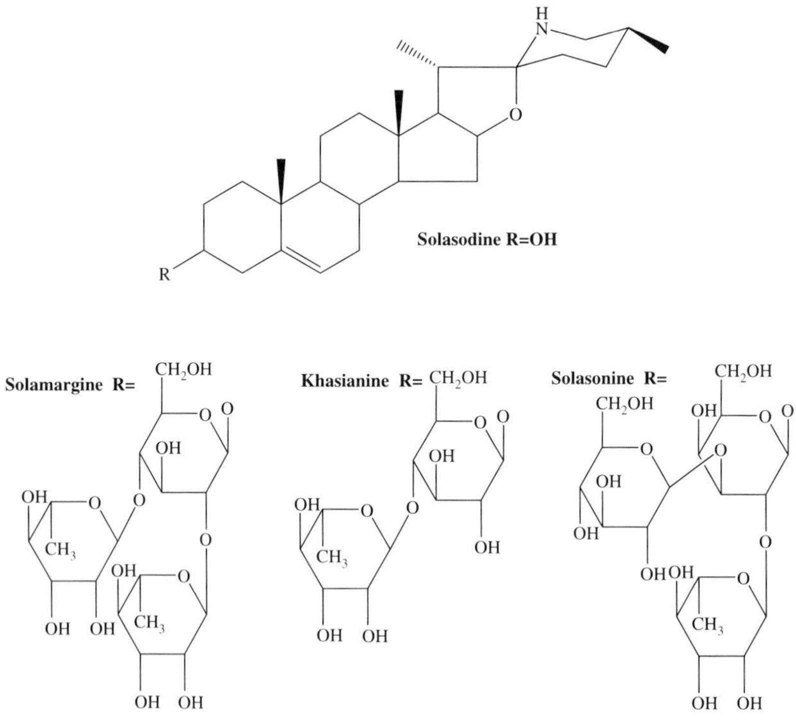

Figure 1. Structures of solasodine glycosides

solamargine and solasonine (Fig. 1), inhibit the acetylcholine esterase (Roddick, 1989). α-Solamargine is a more powerful anti-fungal agent than ketoconazole (Chataing et al., 1998) and solamargine can inhibit herpes simplex virus type I (Ikeda et al., 2000).

It is evident that the natural resources of adrenocortical and sex hormones, particularly from *Dioscorea* spp., are becoming scarce, therefore, glycoalkaloids from *Solanum* spp. as indicated the structures in Fig. 1, have increased interest as starting materials for the production of steroid hormones. We developed micropropagation systems to improve the quality of medicinal plants (Tanaka et al., 1997), and planned to genetically engineer a variety of *S. khasianum* that can yield greater quantities of solasodine glycosides by expressing an anti-solamargine (As)-scFv gene. In this chapter we show evidence that a functional As-scFv antibody produced in transgenic *S. khasianum* can increase the concentration of solasodine glycosides (Putalun et al., 2003).

2. CONSTRUCTION OF scFv

Recently the scFv antibody engineering of plants has shown to increase resistance to plant viruses (Tavladoki et al., 1993; Fiedler and Cornad, 1995; Fecker et al., 1996, 1997). In our ongoing studies to improve quality of medicinal plants we started to construct recombinant scFv (As-scFv) for small antigens like solasodine glycosides to increase secondary metabolites by the accumulation of antigen-antibody complexes in As-scFv expressing transformed plant organs. The As-scFv was constructed from cDNAs encoding the VH and VL variable regions of As-MAb (SM-BD9) (Ishiyama et al., 1996) as described below.

We expressed scFv using an *E.coli* expression vector essentially according to the protocols of the recombinant phage antibody system (RPAS) purification and expression modules. We extracted mRNA from hybridoma cells secreting the As-MAb, SM-BD9 (Ishiyama et al. 1996) and synthesized first-strand cDNA primed by random hexamers using murine reverse transcriptase. PCR was performed using VH and VL primers from the scFv module of the RPAS kit and KOD DNA polymerase. The cDNAs of the VH and VL chain were assembled by seven incubation cycles using a DNA linker fragment encoding the amino acid sequence $(Gly_4Ser)_3$. The resulting fragment was ligated into the pCANTAB5E phagemid from which high affinity scFv protein was selected. The DNA encoding scFv fused to an E-tag and an amber stop codon was amplified by PCR from the template pCANTAB5E. The gene was subcloned into the pET-28a (+) expression vector to contain His-and T7-tags at the N-terminus and an E-tag at the C-terminus.

Figure 2 shows the nucleotide and deduced amino-acid sequences of As-scFv. The hypervariable regions or complementarity-determining regions (CDRs) were identified as indicated in Fig. 2. Linker fragment is shown in italics. Nucleotide sequences encoding respective complementarity-determining regions (CDRs) are underlined. The nucleotide sequence data appear in the GenBank nucleotide

ATGGCCCAGGTGCAACTGCAGGAGTCAGGACCTGAACTGGTGAAGCCT

M A Q V Q L Q E S G P E L V K P

GGGGCTTCTGTGAAGATATCCTGCAAGACCTCTGGATACACATTCACT

G A S V K I S C K T S G Y T F T

GGAAACCTCATACACTGGGTGCAGCAGAGTCCTGGAAAGAGCCTTGAG

G N L I H W V Q Q S P G K S L E

HCDR1

TGGATTGGACATATAAATCCTAACAATGGTGGTACTGTCTACAACCAG

W I G H I N P N N G G T V Y N Q

HCDR2

AAGTTCACGGGCAAGGCCACATTGACTGTAGACAAGTTATCCAACACA

K F T G K A T L T V D K L S N T

GCCCACATGGAGTTCCGCAGCCTGACATCTGAGGATTCCGCAGTCTAT

A H M E F R S L T S E D S A V Y

TACTGTGCAAGATCGGGACGTGACTACCCTTACTGGGGCCAAGGGATC

Y C A R S G R D Y P Y W G Q G I

HCDR3

ACGGTCACCGTCTCCTCA*GGTGGAGGCGGTTCAGGCGGAGGTGGCTCT*

T V T V S S *G G G G S G G G G S*

------- **Linker** -------

*GGCGGTGGCGGATCG*GACATCGAGCTCACTCAGTCTCCAAAACTCATG

G G G G S D I E L T Q S P K L M

TCTACATTAGTAGGAGACAGGGTCAGCGTCACCTGCAAGGCCAGTCAG

S T L V G D R V S V T C K A S Q

AATGTATTTAGTCATGTAGCCTGGTATCAACAAAAACCAGGGCAATCT

N V F S H V A W Y Q Q K P G Q S

LCDR1

CCTAAAGCATTGATTTATTCGACATCCAACCGGTACAGTGGAGTCCCT

P K A L I Y S T S N R Y S G V P

LCDR2

GATCGCTTCACAGGCAGTGGATCTGGGACAGATTTCACTCTCACCATC

D R F T G S G S G T D F T L T I

AGCAATGTGCAGTCTGAAGACTTGGCAGAATATTTCTGTCAGCAATAT

S N V Q S E D L A E Y F C Q Q Y

AACACCTATCCGTACACGTTCGGCGGGGGGACCAAGCTGGAGCTGAAA

N T Y P Y T F G G G T K L E L K

LCDR3

CGG

R

Figure 2. Nucleotide and deduced amino-acid sequences of anti-solamargine single chain fragment-variable antibody (As-scFv). Linker fragment is shown in italics. Nucleotide sequences encoding respective complementarity determining regions (CDRs) are underlined

sequence database under the accession number AF332008. The sequence was compared with those compiled in the Kabat antibody sequence database (Martin, 1996). The VH region of MAb (SM-BD9) belonged to the mouse heavy chain subgroup II (A), whereas the VL region belonged to the mouse kappa light chain V group.

3. EXPRESSION OF scFv IN *E.COLI* AND ITS CHARACTERIZATION

After induction with IPTG, a large amount of recombinant protein was produced in the bacterial cytoplasm as inclusion bodies. The scFv protein refolded by rapid dilution was purified using a metal chelate affinity column resulting in a good yield of 12.5 mg/100ml culture medium. Samples were resolved by electrophoresis on 12.5% SDS-polyacrylamide gels (PAGE) and stained with Coomassie brilliant blue. Fig. 3 shows the SDS-PAGE analysis of scFv purified by metal chelate affinity column chromatography. Matrix- assisted laser desorption/ionization (MALDI)-tof mass spectrometry confirmed the purity of scFv and the amino acid sequence (Fig. 4).

The competitive MAb ELISA was identical to the scFv assay except for the anti-E-tag antibody step. With the exception of solamargine and anti-E-tag antibody addition, scFv binding inhibition by direct ELISA was determined in the same way as for competitive ELISA for solamargine, and cross-reactivity of other steroidal compounds with scFv was measured in the same manner as for solamargine (Ishiyama et al., 1996).

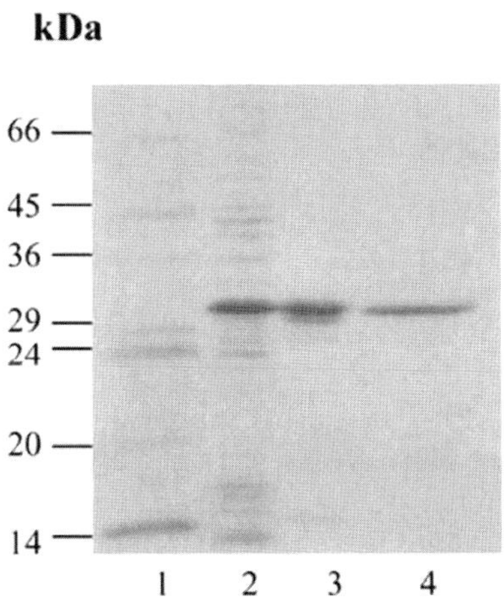

Figure 3. SDS-PAGE analysis and Western blotting of scFv purified by metal chelate affinity column chromatography. Samples were resolved by electrophoresis on 12.5% SDS-polyacrylamide gels and stained with Coomassie brilliant blue. Lanes: 1, Protein molecular weight markers (kDa), 2, whole cell extract, 3, unpurified scFv inclusion bodies, 4, purified scFv protein (3 μg)

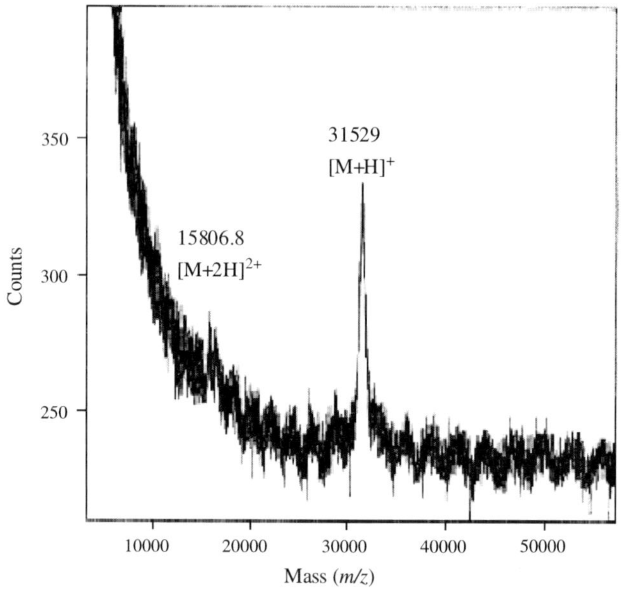

Figure 4. MALDI-tof MS spectrum of purified scFv against solamargine

Table 1 indicates the cross reactivity of As-scFv and MAb (SM-BD9) against various steroidal compounds. Cross-reactions of As-scFv fragments against solasodine glycosides with different sugar moieties such as khasianine (80.7%), solasonine (92.1%), 3-O-β-D-glucopyranosy1 solasodine (112.9%), O-α-L-rhamnosyl-(1->2)-3-O-β-D-glucopyranosyl-solasodine (98.5%), 3-O-β-D-galacopyranosyl solasodine (115.4%), O-β-D-gucopyranosy1-(1->3)-3-O-β-D-galactopyranosy solasodine (118.2%) and isoanguivine (150.4%) were almost identical to those for the MAb (Putalum et al., 2000).

4. scFv PROTEIN EXPRESSION IN PLANT AND ITS CHARACTERIZATION

Expression of the scFv gene in plants is controlled by the constitutive cauliflower mosaic virus (CaMV) 35S promoter (Odell et al., 1985) in a plant expression vector. For ER targeting and retention, a leader sequence derived from yeast and a KDEL sequence were integrated into the N-and C-termini of As-scFv, respectively, to construct a plant expression vector cassette (pAW16; Fig. 5). The As-scFv protein was directed to the ER by the N-terminal yeast signal peptide (SUC2) and retained in the ER using the C-terminal KDEL sequence (Artsaenko et a1., 1995; Schouten et a1., 1996; Spiegel et a1., 1999). Specific expression and retention of scFv protein in the ER are essential for stable accumulation in plants. Adding the KDEL sequence to the C-terminus tripled scFv protein accumulation compared with expression

Table 1. Cross-reactivities of anti-solamarigine MAb-BD9 and scFv against steroidal compounds

Compound	MAb(SM-BD9) (%)	scFv (%)
Solamarigine	100.00	100.00
Solasonine	92.10	89.81
Khasianine	80.67	82.13
3-O-β-D-glucopyranosy1 solasodine	112.90	116.98
O-α-L-rhamnosyl-(1->2)-3-O-β-D-glucopyranosyl-solasodine	98.46	97.43
3-O-β-D-galacopyranosyl solasodine	115.39	114.96
O-β-D-gucopyranosy1-(1->3)- 3-O-β-D-galactopyranosy solasodine	118.20	111.08
Isoanguivine	150.40	146.56
Solasodine	43.80	52.28
Solaverine I	4.04	4.91
Solaverine II	7.85	7.24
12-Hydroxysolamargine	2.04	2.03
12-Hydroxysolasonine	1.80	1.17
Tomatine	2.06	2.72
Tomatigine	0.27	0.36
Ergosterol	0.29	0.13
α-Chaconine	0.14	<0.12
Xylosyl-β-solamargine	<0.12	<0.12
α-Solanine	<0.12	<0.12
Solanidine	<0.12	<0.12
Dioscin	<0.12	<0.12

The cross-reactivites were determined according to Weiler's equation (Weiler et al.,1976)

without the KDEL sequence (data not shown), indicating that the C-terminal KDEL sequence acts to stabilize the As-scFv protein.

Plant expression vectors were transferred with the pRK 2013 helper plasmid into *Agrobacterium rhizogenes* by triparental mating and were used to transform *S. khasianum* using standard procedures (Pollard and Walker, 1990). Transgenic hairy roots were cultured on hormone-free half strength MS medium containing kanamycin. Kanamycin-resistant hairy roots were produced after infection with *A.rhizogenes* to yield 37 clones of transgenic hairy roots. We

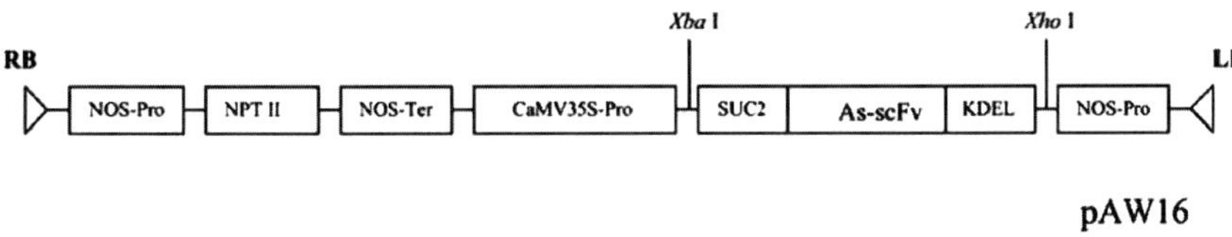

Figure 5. Schematic diagram of plant expression vector. *CaMV35SPro* Cauliflower mosaic virus 35S promoter, *NOS-Pro* nopaline synthase promoter sequence, *NOS-Ter* nopaline synthase terminator sequence, *NPT II* neomycin phosphotransferase II marker gene, *scFv* single-chain Fv gene, *SUC2* yeast signal-peptide-encoding sequence, *KDEL* endoplasmic reticulum retention signal

analyzed As-scFv protein expression and solasodine glycoside concentrations in the cloned roots. To assess the ability of plants to produce functional As-scFv protein, each transgenic hairy root line was analyzed by comparison with purified mouse Fab using an ELEA. Up to 220 ng soluble scFv was expressed per milligram total soluble protein (TSP) extracted from the hairy roots. The molecular weight of As-scFv protein produced by the transgenic hairy roots was the same as that of the protein produced in *E. coli*, whereas wild type hairy roots did not express As-scFv protein according to Western blotting (Fig. 6).

The molecular weight of the protein from hairy roots expressing As-scFv confirmed by MALDI-tof mass spectrometry. A peak representing scFv $[M+H]^+$ appeared at 31,529 *m/z*, in good agreement with the molecular weight (31,504) calculated from the complete amino acid sequence as shown in Fig. 4.

After purification on an immobilized metal chelate affinity column, we analyzed the binding kinetics of the extract from transgenic hairy roots with scFv protein, original anti-solamargine MAb and scFv expressed in *E.coli* using a competitive ELISA (Friguet et al., 1996). The average values of the dissociation constant (K_d) of the MAb (SM-BD9), scFv from *E.coli* and scFv from hairy root cultures were $4.44 \pm 0.41 \times 10^{-9}$, $2.86 \pm 0.25 \times 10^{-9}$ and $2.43 \pm 0.23 \times 10^{-9}$ M, respectively. These results suggest that the functional affinity constant of the scFv protein derived from hairy roots is similar to that of scFv expressed in *E.coli* and the original MAb. Moreover, dot blots showed that the specific binding activities of the two scFv proteins and the original MAb were the same (data not shown).

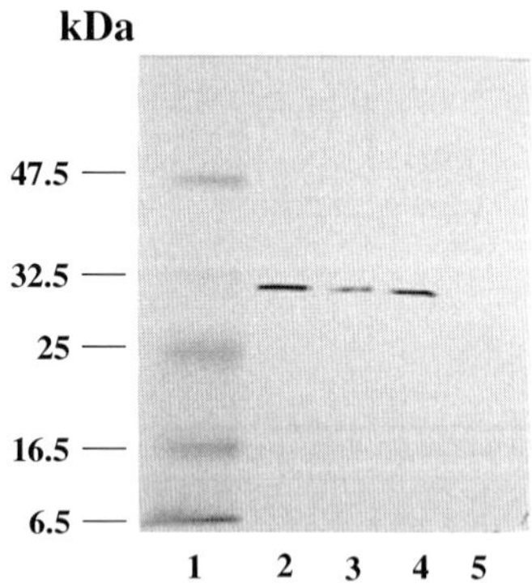

Figure 6. Western blots of soluble protein from transgenic hairy root expressing As-scFv. Lanes: 1, Protein molecular weight markers (kDa), 2, 200 ng scFv from *Escherichia coli*, 3, affinity-purified scFv from hairy root clone SR2, 4, affinity-purified scFv from hairy root clone SR19, 5, soluble protein from wild-type hairy root

5. DETERMINATION OF SOLASODINE GLYCOSIDES CONCENTRATION IN HAIRY ROOTS AND REGENERATED PLANTS

The concentration of solasodine glycosides in all clones of As-scFv transgenic hairy roots ranged from 62 to 100 μg/g dry wt, about 1.4 to 2.2-fold higher than that of the wild type hairy roots, which contain 43–45 μg/g dry wt. Of 37 cloned transgenic hairy roots, 7 produced a two-fold amount of As-scFv compared with the wild type. Figure 7 shows that the solasodine glycoside concentration increased in proportion to the soluble scFv level, therefore, the expression level of As-scFv seems to be important for the increase in the solasodine glycosides concentration. It is evident that As-scFv protein enhances their biosynthesis in transgenic hairy roots. It was estimated that the accumulation of solasodine alkaloids might come about through an antigen-antibody complex in transgenic hairy roots expressing As-scFv.

Further to confirm this result, we regenerated plantlets from hairy roots as indicated in Fig. 8. After cultivation of regenerated *S. khasianum* plants for 6 months (Fig. 9), the fruits of transgenic *S. khasianum* contained 126.95 ± 6.03 μ g/mg dry wt. solasodine glycosides, which was over twice that produced by *S. khasianum* transformed by wild *A.rhizogenes* (53.43 ± 3.90 μ g/mg dry wt). We discovered that transformation with the scFv fragment improves the yield of a

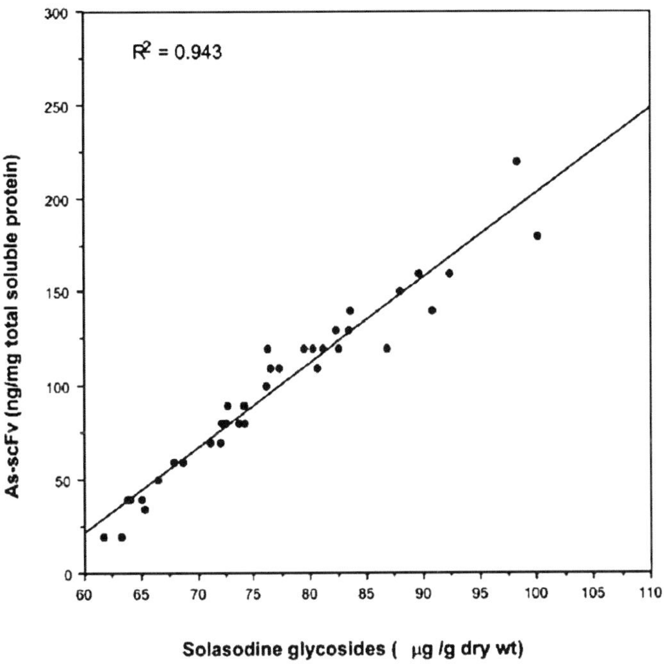

Figure 7. Comparison of solasodine glycoside concentrations from hairy root clones and scFv expression levels

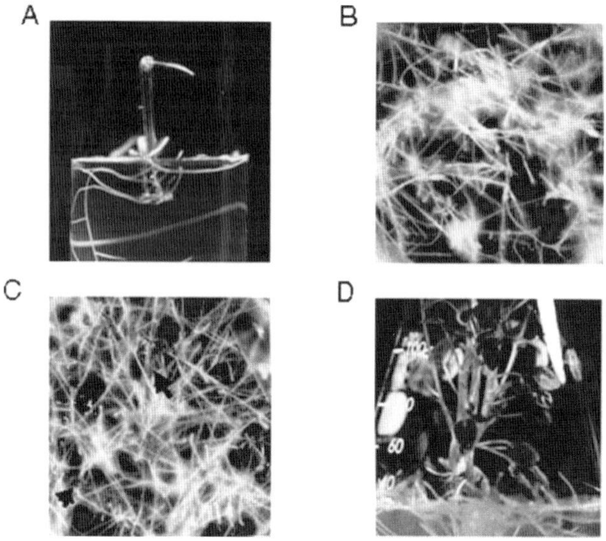

Figure 8. Development of hairy root after infected with *A. rhizogenes* (A), transgenic hairy roots (B), shoot production from transgenic hairy roots (C), regeneration of transgenic plant from hairy root of *S. khasianum*

Figure 9. Pictures of control (A) and transgenic *S. khasianum* (B). Leaves and fruits of control (C) and transgenic *S. khasianum* (D)

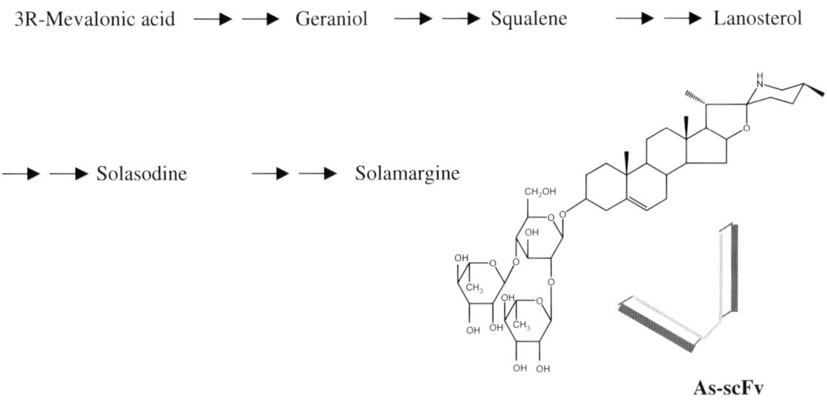

Figure 10. A presumed control mechanism of anti-solamargine scFv on biosynthetic of solasodine glycosides in transgenic *S. khasianum*

pharmacologically active compound, although Artsaenko et al. (1995) reported that the expression of anti-abscisic acid scFv creates a wilty phenotype in transgenic tobacco, suggesting that they produce higher levels of abscisic acid than wild type plants. The biosynthetic pathway of solasodine glycosides proposed by Ehmke et a1. (1995) have not yet been fully elucidated. The system established in this review may open up new plant breeding methodologies for pharmacologically active glycosides such as ginsenosides, glycyrrhizin and crocin without a requirement to understand the biosynthetic pathway as summarized in Fig. 10. This system also allows the recovery of both antigen and scFv protein, which can be used for further analytical studies. Finally we recommend that this methodology which can control biosynthesis of a targeted secondary metabolite by means of a single specific gene can be called missile-type molecular breeding.

ACKNOWLEDGEMENTS

This work was supported in part by a Grant-in-Aid from JSPS's Asian CORE Program, the Ministry of Education, Culture, Sports, Science and Technology, Japan.

REFERENCES

Artsaenko O, Peisker M, Nieden U, Fiedler U, Weiler EW, Muntz K,Conrad U (1995) Expression of a single-chain Fv antibody against abscisic acid creates a wilty phenotype in transgenic tobacco. *Plant J* 8: 745–750.

Bradford MM (1976) A rapid and sensitive method for the quantitation of microgram quantities of protein utilizing the principle of protein-dye binding. *Anal Biochem* 72:248–254.

Cham BE, Gilliver M, Wilson L (1987) Antitumour effects of glycoalkaloids isolated from *Solanum sodomaeum*. *Planta Med* 53:34–36.

Cham BE, Daunter B, Evans RA (1991) Topical treatment of malignant and premalignant skin lesions by very low concentration of a standard mixture (BEC) of solasodine glycosides. *Cancer Lett* 59: 183–192.

Chataing B, Concepclon JL, Lobaton R, Usubillaga A (1998) Inhibition of Trypanosoma cruzi growth in vitro by *Solanum* alkaloids: a comparison with ketoconazole. *Planta Med* 64:31–36.

Daunter B, Cham BE (1990) Solasodine glycosides: in vitro preferential cytotoxicity for human cancer cells. *Cancer Lett* 55:209–220.

Ehmke A, Ohmstede D, Eilert U (1995) Steroidal glycoalkaloids in cell and shoot teratoma cultures of *Solanum dulcamara*. *Plant Cell Tissue Organ Cult* 43:191–197.

Fecker LF, Kaufmann A, Commandeur U, Commandeur J, Koenig R, Burgermeister W (1996) Expression of single chain antibody fragment (scFv) specific for beet necrotic yellow vein virus coat protein or 25 kDa protein in *Escherichia coli* and *Nicotiana benthamiana*. *Plant Mol Bio* 132: 979–986.

Fecker LF, Koenig R, Obermeier C (1997) *Nicotiana benthamiana* plants expressing beet necrotic yellow vein virus (BNYVV) coat protein-specific scFv are partially protected against the establishment of the virus in the early stages of infection and its pathogenic effects in the late stages of infection. *Arch Virol* 142:1857–1863.

Fiedler U, Cornad U (1995) High level production and long-term storage of engineered antibodies in transgenic tobacco seeds. *Biotechnology* 13:1090–1093.

Friguet B, Djavadi-Ohaniance L, Goldberg ME (1996) Immunoochemical analysis of protein conformation. In: Creighton TE (ed) Protein structure, a practical approach, 2nd edn. Oxford University Press,Oxford, pp 335–342.

Harper K, Ziegler A (1999) Recombinant antibodies, applications in plant science and plant pathology. Scottish Crop Research Institute,Dundee, UK.

Harper K, Kerschlbaumer R, Ziegler A, Macintosh SM, Cowan GH, Himmler G (1997) A scFv-alkaline phosphatase fusion protein which detects potato leafroll luteovirus in plant extracts by ELISA. *J Virol Meth* 63: 237–242.

Hoogenboom HR (1997) Designing and optimizing library selection strategies for generating high-affinity antibodies. *Trends Biotechnol* 15: 62–70.

Huston JS, McCartney J, Tai MS, Mottola-Hartshorn C, Jin D, Warren F, Keck P, Oppermenn H (1993) Medicinal applications of single chain antibodies. *Int Rev Immunol* 10: 195–217.

Ikeda T, Ando J, Miyazono A, Zhu XH, Tsumagari H, Nohara T, Yokomizo K, Uyeda M (2000) Anti-herpes virus activity of *Solanum* steroidal glycosides. *Biol Pharm Bull* 23: 363–364.

Ishiyama M, Shoyama Y, Murakami H, Shinohara H (1996) Production of monoclonal antibodies development of an ELISA for solamargine. *Cytotechnology* 18: 153–158.

Kim JS, Tanaka H, Shoyama Y (2004) Immunoquantitative analysis for berberine and its related compounds using monoclonal antibodies in herbal medicines. *Analyst* 129:87–91.

Loungratana P, Tanaka H, Shoyama Y (2004) Production of monoclonal antibody against ginkgolic acids in *Ginkgo biloba* Linn. *Am J Chin Med* 32: 33–48.

Lu Z, Masaki T, Shoyama Y, Tanaka H (2006) Construction and expression of a single chain Fv fragment against pharmacologically active paeoniflorin in *Escherichia coli*, and its potential use in an enzyme-linked immunosorbent assay. *Planta Med* 72:151–155.

Martin ACR (1996) Accessing the Kabat antibody sequence database by computer. *Proteins-Structure, Function and Genetics* 25: 130–133.

Morinaga O, Tanaka H, Shoyama Y (2006) Detection and quantification of ginsenoside Re in ginseng samples by a chromatographic immunostaining method using monoclonal antibody against ginsenoside Re. *J Chromatogr B* 830: 100–104.

Odell JT, Nagy F, Chua NH (1985) Identification of DNA sequence required for activity the cauliflower mosaic virus 35S promoter. *Nature* 313: 810–812.

Pollard JW, Walker JM (1990) Methods in molecular biology 6: plant cell and tissue culture. Humana, Clifton, N.J.

Putalun W, Tanaka H, Shoyama Y (1999) Rapid separation of solasodine glycosides by an immunoaffinity column using antisolamargine monoclonal antibody. *Cytotechnology* 31: 151–156.

Putalun W, Tanaka H, Yahara S, Lhieochaiphan S, Shoyama Y (2000)Survey of solasodine type glycoalkaloids by western blotting and ELISA using anti-solamargine monoclonal antibody. *Biol Pharm Bull* 23: 72–75.

Roddick JG (1989) The acetylcholinesterase-inhibitory activity of steroidal glycoalkaloids and their aglycone. *Phytochemistry* 28:2631–2636.

Schouten A, Roosien J, Engelen FA, Jong GAM, Borst AWM, Zilverentant JF, Bosch D, Stiekema WJ, Gommers F, Schots A, Bakker J (1996) The C-terminal KDEL sequence increases the expression level of a single-chain antibody designed to be targeted to both the cytosol and the secretory pathway in transgenic tobacco. *Plant Mol Biol* 30:781–793.

Spiegel H, Schillberg S, Sack M, Holzem A, Nahring J, Monecke M, Liao YC, Fischer R (1999) Accumulation of antibody fusion proteins in the cytoplasm and ER of plant ce1ls. *Plant Sci* 149:63–71.

Tanaka H, Putalun W, Tsuzaki C, Shoyama Y (1997) A simple determination of steroidal alkaloid glycosides by thin-layer chromatography immunostaining using monoclonal antibody against solamargine. *FEBS Lett* 404: 279–282.

Tang PM, Foltz LA, Mahoney WC, Schueler PA (1995) A high affinity dioxin-binding protein displayed on M13 is functionally identical to the native protein. *J Biol Chem* 270: 7829–7835.

Tavladoki P, Benvenuto E, Trinca S, Martinis DD, Cattarleo A, Galeffi P (1993) Transgenic plants expressing a functional single-chain Fv antibody are specifically protected from virus attack. *Nature* 336: 469–472.

Weiler EW, Zenk MH (1976) Radioimmunoassay for determination of digoxin and related compounds in *Digitalis lanata*. *Phytochemistry* 15: 1537–1545.

Ziegler A, Torrance L, Macintosh SM, Cowan GH, Maya MA (1995)Cucumber mosaic cucumovirus antibodies from a synthetic phase display library. *Virology* 214: 1302–1305.

Zhu S, Shimokawa S, Shoyama Y, Tanaka H (2006) A novel analytical ELISA-based methodology for pharmacologically active saikosaponins. *Fitoterapia* 77: 100–108.

CHAPTER 13

METABOLIC ENGINEERING OF SULFUR ASSIMILATION IN PLANTS

Molecular and biochemical analysis of serine acetyltransferase and cysteine synthase

MASAAKI NOJI[1], AND KAZUKI SAITO[1,2,3]

[1] *Graduate School of Pharmaceutical Sciences, Chiba University, Yayoi-cho 1-33, Inage-ku, Chiba 263-8522, Japan*

[2] *RIKEN Plant Science Center, 1-7-22 Suehiro-cho, Tsurumi-ku, Yokohama 230-0045, Japan*

[3] *To whom correspondence should be addressed. Department of Molecular Biology and Biotechnology, Graduate School of Pharmaceutical Sciences, Chiba University, Yayoi-cho 1-33, Inage-ku, Chiba 263-8522, Japan; e-mail: ksaito@faculty.chiba-u.jp*

Abstract: Sulfur and sulfur-containing products play important roles in plant cells. In order to enhance the ability of the production of sulfur-containing compounds by the metabolic engineering of sulfur assimilation in plants, we constructed transgenic plants over-expressing cysteine synthase (CSase), or serine acetyltransferase (SATase) gene, key enzymes for cysteine biosynthesis in plants. Transgenic plants over-expressing CSase gene were highly tolerant to sulfur dioxide, sulfite, oxidative stress, and cadmium. Besides, transgenic plants over-expressing SATase gene showed the over-accumulation of sulfur-containing compounds, cysteine and glutathione (GSH). These results indicate that the over-expression of CSase or SATase gene is promising for the metabolic engineering of sulfur assimilation in plants

Keywords: sulfur assimilation; cysteine biosynthesis; cysteine synthase; serine acetyltransferase

Abbreviations: GSH, glutathione; PCs, phytochelatins; Cd, cadmium; SO_2, sulfur dioxide; OAS, *O*-acetyl-L-serine; APS, adenosine 5′-phosphosulfate; SATase, serine acetyltransferase; CSase, cysteine synthase; Rubisco, ribulose-1,5-bisphosphate carboxylase

R. Verpoorte et al. (eds.), Applications of Plant Metabolic Engineering, 297–309.
© 2007 *Springer.*

1. INTRODUCTION AND BACKGROUNDS OF STUDY

Sulfur is an essential nutrient for growth of plants. Plants assimilate inorganic sulfur into cysteine through a cysteine biosynthetic pathway (Leustek and Saito, 1999; Saito, 2000; Saito, 2004). This cysteine biosynthetic pathway is very important for the sulfur cycle in nature, because the inorganic sulfur in the environment (*e.g.*, sulfate ion in the soil and sulfur dioxide in the air) is first fixed into cysteine, an organic compound containing the sulfur atom, mainly by this sulfur assimilation pathway in plants. Cysteine is then incorporated into proteins and glutathione, or serves as the sulfur donor of methionine and sulfur-containing secondary products in plants.

Sulfur and sulfur-containing products play significant roles in various situations in plant cells. Sulfur is required as a macro essential element for the biosynthesis of primary metabolites such as amino acids, proteins and vitamins. Besides, sulfur forms the thiol groups of glutathione (GSH) and thioredoxin, which are directly responsible for the modulation of intracellular redox state. Moreover, some sulfur-containing products like phytochelatins (PCs), GSH and choline sulfate are involved in the alleviation of the damage from various stresses, *e.g.*, heavy metal, oxidative, or biotic stress, *etc.* Sulfur is also included in some important secondary metabolites (*e.g.*, chamalexin, glucosinolates and alliine), and the sulfur in these compounds contributes to their characteristic bioactivity.

Because of such usefulness of sulfur and sulfur-containing products in plant cells, it is hoped to enhance the ability of the production of sulfur-containing compounds by the metabolic engineering of sulfur assimilation, via the cysteine biosynthetic pathway in plants. One of the anticipated outcomes by the metabolic engineering of sulfur assimilation in plants is to develop a transgenic plant which is tolerant to a variety of environmental pollutions. A number of heavy metals cause serious world wide environmental and human health problems in elevated concentrations in the soil (Clemens, 2001; Hall, 2002). PCs, of the basic structure (γ-Glu-Cys)n-Gly (n = 2–11), synthesized from GSH by phytochelatin synthase in plants when exposed to heavy metals, play an important role in detoxification of toxic metals such as cadmium (Cd) and arsenic (Cobbett, 2000). The detoxification mechanism involves complex formation of PCs with the heavy metals and their storage into the vacuole. GSH is synthesized from cysteine by a two-step enzymatic reaction catalyzed by γ-glutamylcysteine synthetase and GSH synthetase, both regulated at transcriptional and translational levels (Xiang and Oliver, 1998). Furthermore, GSH biosynthesis is strongly regulated by the cysteine availability (Strohm et al., 1995). Another serious problem for the global environment is environmental pollution caused by sulfur-containing compounds, for instance, sulfur dioxide (SO_2), hydrogen sulfide, sulfite and sulfate ions. In particular, gaseous SO_2 influences human health and the global ecological system of animals and plants (Wellburn, 1994; Murray, 1997). One of the most common types of visible injury to plants caused by SO_2 is foliar damage such as chlorosis and necrosis. Moreover, when the gaseous SO_2 encounters moisture, considerable amounts of SO_2 are converted into sulfite and sulfate. These are critical components of acid rain and haze (Wellburn, 1994). Although sulfur-containing pollutants are toxic for plants at higher concentration, plants are also able to assimilate these compounds into cysteine. Thus, the detoxification mechanisms of these environmental pollutants (*i.e.* heavy metals and sulfur-containing pollutants) are strongly associated with the sulfur assimilation pathway in plants. The development of transgenic plants, which resist and detoxify heavy metals and sulfur-containing pollutants, by the metabolic engineering of sulfur assimilation pathway, seems to be promising to overcome these problems for the environment.

For the purpose of more effective metabolic engineering of sulfur assimilation in plants, it is essential to understand in detail the sulfur assimilation and cysteine metobolism at molecular level, and then to engineer the most important and forceful step in the biosynthetic pathway by a rational way. Sulfur assimilation and biosynthetic pathway of cysteine in plants involves several enzymatic reactions (Fig. 1). The cysteine biosynthesis consists of two biochemical pathways: the pathway of the activation and reduction of sulfate into sulfide, and the pathway supplying the amino acid moiety, which is derived from serine through *O*-acetyl-L-serine (OAS), and then yielding cysteine by the reaction of incorporating the sulfide moiety into β-position of alanine. Three enzymes, ATP sulfurylase, adenosine 5′-phosphosulfate (APS) reductase and sulfite reductase are concerned with the reduction of sulfate into sulfide. This reduction pathway takes place almost exclusively in chloroplasts. On the other hand, two enzymes, serine acetyltransferase (SATase) and cysteine synthase (*O*-acetylserine (thiol)-lyase) (CSase) are involved in the final step of cysteine biosynthesis. This final step of cysteine biosynthesis exists in three major compartments of plant cells, cytosol, chloroplasts, and mitochondria. The presence of SATase (Smith, 1972; Ascano and Nicholas, 1977; Brunold and Suter, 1982; Ruffet et al., 1995) and cysteine synthase (Brunold and Suter, 1989; Lunn et al., 1990; Droux et al., 1992; Rolland et al., 1992; Yamaguchi and Masada, 1995;

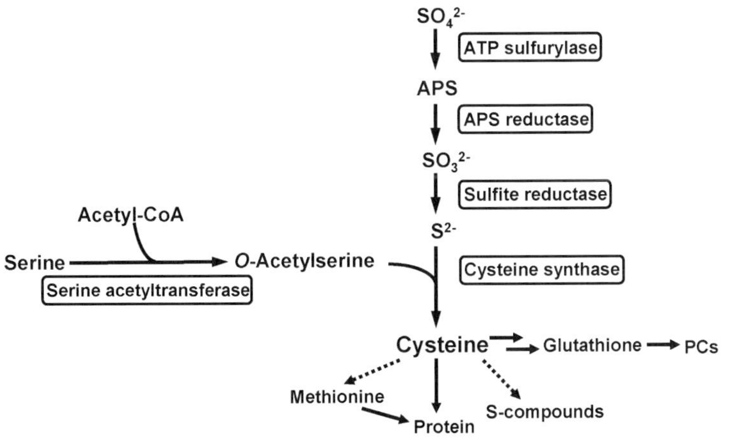

Figure 1. Sulfur assimilation in plants. APS, adenosine 5′-phoshosulfate; PCs, phytochelatins

Kuske et al., 1996) activities has been demonstrated in these three compartments from several plants.

Among several enzymatic steps of sulfur assimilation, we focused on CSase and SATase, which appear to be the key enzymes for cysteine biosynthesis in plants, for the metabolic engineering of sulfur assimilation in plants. In this chapter, we describe the molecular and biochemical analysis of CSase and SATase, and the metabolic engineering of sulfur assimilation in plants by over-expressing each enzyme's gene.

2. TRANSGENIC PLANTS OVER-EXPRESSING CSase GENE

CSase catalyzes the formation of cysteine from OAS and hydrogen sulfide, with the release of acetic acid at the final step of the cysteine biosynthetic pathway. This enzyme requires pyridoxal phosphate as a cofactor.

We constructed transgenic tobacco (*Nicotiana tabacum*) plants carrying either spinach cytosolic CSase A cDNA (Saito et al., 1992), designated 3F plants, or chimeric CSase A cDNA fused with the sequence for chloroplast-targeting transit peptide of pea ribulose-1,5-bisphosphate carboxylase (Rubisco) small subunit, designated 4F plants (Saito et al., 1994). The 3F and 4F transgenic plants showed enhanced CSase activity in the cytosol and in the chloroplasts, respectively. The leaf discs of these transgenic tobacco plants showed partial tolerance to SO_3^{2-} (Saito et al., 1994). In order to intensify the CSase activity more, we crossed 3F plants with 4F plants to generate F_1 transgenic tobacco, in which CSase activities were enhanced both in the cytosol and in the chloroplasts (Noji et al., 2001). CSase activity in the cell-free extract of F_1 was about 5-fold higher than that of non-transformant plant. In the chloroplast fraction, CSase activity of F_1 was at the same level as that of 4F, and about 6-fold higher than those of 3F and non-transformant. In the cytosol fraction, CSase activity of F_1 was at the same level as that of 3F, and about 4-fold higher than those of 4F and non-transformant. These results indicated that F_1, in which CSase activities were enhanced both in the chloroplasts and in the cytosol, had the highest activity of CSase in these transgenic tobacco plants.

The F_1 transgenic plants were highly tolerant to toxic 1 μL L^{-1} sulfur dioxide and 20 mM sulfite (Noji et al., 2001). To clarify the mechanism of resistance to SO_2 in transgenic plants, we exposed the plants to 0.1μL L^{-1} SO_2 and then measured the contents of cysteine and GSH in the leaves. During the SO_2 fumigation, cysteine and GSH contents were increased significantly in F_1 plants, but not in control plants, suggesting that the F_1 transgenic plants may fix the atmospheric SO_2 into cysteine and GSH more efficiently than the control plants. Thus, the tolerance to SO_2 in transgenic plants may be caused, at least partly, by the efficient assimilation of sulfur into cysteine and GSH. Since SO_2 is highly soluble in water, SO_2 absorbed by plants changes into SO_3^{2-} rapidly. Some part of SO_3^{2-} is used for the biosynthesis of cysteine, and the rest is oxidized into SO_4^{2-} by sulfite oxidase (Eilers et al., 2001). Produced SO_4^{2-} are stored in vacuoles, and subsequently used for the biosynthesis of cysteine (Rennenberg, 1984). During oxidation of SO_3^{2-} to SO_3^{2-} in plant cells, active oxygen species are produced as secondary toxic substances (Hänsch et al., 2006). Therefore, the toxicity of SO_2 may be due to the generation of the active oxygen species. The leaves of F_1 transgenic tobacco plants also exhibited the increased resistance to paraquat, a herbicide generating active oxygen species (Noji et al., 2001). The activities of enzymes involved in scavenging active oxygen species (superoxide dismutase, catalase, and ascorbate peroxidase), however, were not enhanced in F_1 plant. The ascorbate-GSH cycle is an enzymatic scavenging mechanism for the removal of active oxygen species (De Kok and Stulen, 1993). GSH levels respond to the availability of cysteine (Strohm et al., 1995), therefore F_1 possessing high activity of CSase was expected to have high ability to synthesize GSH. The increased GSH is most likely to be involved in the detoxification of active oxygen; accordingly F_1 showed high tolerance to active oxygen species. Taken together, these results suggest that the tolerance of F_1 plants to sulfur-containing pollutants may be caused not only by the efficient assimilation of sulfur into cysteine but also by the detoxification of active oxygen species by GSH.

Furthermore, the 3F, 4F and F_1 plants over-expressing CSase showed tolerance to Cd, while the wild-type tobacco showed severe toxicity symptoms (Fig. 2) (Kawashima et al., 2004). The wild-type plants became necrotic and ceased to grow. In contrast, transgenic plants showed attenuated symptoms and continued to grow. The F_1 transgenic plants under Cd condition showed more expanded leaves than 3F and 4F plants. This result indicated that F_1 transgenic tobacco, in which CSase activity in the chloroplasts and cytosol is enhanced, showed an improved heavy metal tolerance compared with 3F and 4F plants. The heavy metal tolerance observed in the transgenic plants over-expressing CSase suggests that the increased

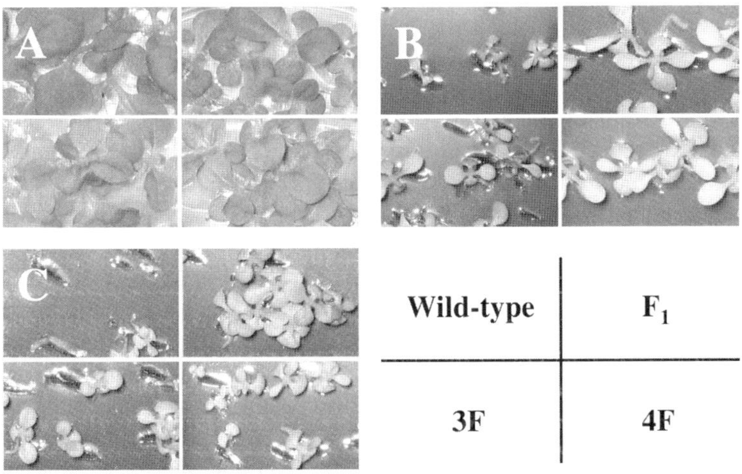

Figure 2. Resistance of transgenic plants to cadmium. Control wild-type tobacco and transgenic plants over-expressing CSase (F_1, 3F and 4F) were grown for three weeks on agar of Cd-free medium (A) or in the presence of 250 μM Cd (B), 500 μM Cd (modified from Kawashima et al., 2004)

thiol contents in these transgenics are mostly liked to be involved in efficient synthesis of PCs, which are responsible for the detoxification of Cd. Since the levels of GSH and cysteine are higher in F_1 plants than 3F and 4F plants (Noji et al., 2001), these higher levels of thiols would be beneficial to detoxify Cd. The accumulation of Cd in F_1 transgenic plants, grown for three weeks on agar medium supplemented with $CdCl_2$, was determined by ICP MS (Kawashima et al., 2004). In shoots, the total amount of Cd per g tissue and per one plant in transgenic F_1 tobacco was larger than in wild-type plants. In roots, the total content of Cd per one plant in F_1 transgenic plants was higher than in wild-type tobacco. However, Cd content per g tissue in roots was higher in wild-type plants, indicating that in F_1 plants Cd does not accumulate in roots but effectively transport Cd into shoots. Cd is possibly transported from root to shoot in a potential detoxification mechanism in the F_1 tobacco plants more effectively. The fact that Cd accumulation per plant is enhanced in F_1 transgenic plants is promising for further application of phytoremediation of heavy metals.

Although the transgenic plants over-expressing the CSase gene demonstrated excellent results about the tolerance to some environmental pollutants, cellular contents of cysteine and GSH were not dramatically increased under normal growth conditions. By feeding experiment of biosynthetic precursor of cysteine using isolated chloroplasts of transgenic plants over-expressing CSase gene, a supply of OAS was a limiting factor for the biosynthetic flux to cysteine (Saito et al., 1994). Consequently, a favorable target for a metabolic engineering of sulfur assimilation was the step involved in the supply of OAS, *i.e.* the enzymatic step catalyzed by SATase.

3. TRANSGENIC PLANTS OVER-EXPRESSING SATase GENE

SATase catalyzes the formation of OAS from L-serine and acetyl-CoA. SATase is placed at the entry step from serine metabolism to cysteine biosynthesis; therefore, it would be logical to postulate the existence of regulation of cysteine biosynthesis at this step. It is known that SATase can be multimeric, and is able to form a complex in association with CSase (Bogdanova and Hell, 1997; Ruffet et al., 1994; Droux et al., 1998; Nakamura et al., 1988; Nakamura and Tamura, 1990). In bacteria, SATase activity is inhibited by micromolar concentrations of L-cysteine (Kredich and Tomkins, 1966). OAS serves as an inducer of cysteine regulon besides as a substrate for cysteine synthesis in bacteria (Kredich, 1996), and OAS is a dominant limiting factor for cysteine formation (Saito et al., 1994). This enzymatic step catalyzed by SATase is therefore a promising target for metabolic engineering in order to enhance the ability of sulfur assimilation in plants.

In the Arabidopsis genome, five genes (*Serat1;1*, *Serat2;1*, *Serat2;2*, *Serat3;1*, *and Serat3;2*) encode SATase. All five SATase isoforms were characterized regarding different sensitivity to cysteine-feedback regulation, subcellular localization (Noji et al., 1998; Kawashima et al., 2005) (Table 1). The activity of cytosolic SATase (Serat1;1) was inhibited by L-cysteine at a physiological concentration

Table 1. Kinetic constants of recombinant Arabidopsis SATase proteins

Isoforms	Subcellular localization	*Km* [mM]		Feedback inhibition by L-Cys	Inhibition *(Ki)* by L-Cys [μM]	
		L-Ser	Acetyl-CoA		L-Ser	Acetyl-CoA
Serat1;1	cytosol	2.71	0.28	sensitive	10.8 (noncompetitive)	7.4 (competitive)
Serat2;1	chloroplasts	1.64	0.16	insensitive	–	–
Serat2;2	mitochondria	1.68	0.02	insensitive	–	–
Serat3;1	cytosol	121.4	24.5	insensitive	–	–
Serat3;2	cytosol	39.5	45.1	sensitive	17.3 (noncompetitive)	2.5 (competitive)

–, No inhibition by L-cysteine

in an allosteric manner, but the plastidic (Serat2;1) and mitochondrial (Serat2;2) forms were not subject to this feedback regulation. The cytosolic Serat1;1 regulates the OAS concentration strictly by the feedback inhibition by cysteine, since OAS is not only the key intermediate of cysteine formation but also regarded as a regulatory molecule of the whole sulfur assimilation pathway (Saito, 2000; Saito 2004). In chloroplasts however, high OAS levels have to be maintained even at high cysteine concentrations for the biosynthesis of methionine and glutathione by feedback-insensitive SATase isoforms in this compartment (Noji et al., 1998). Cytosolic localization of Serat3;1 and Serat3;2 was indicated by the analyses of the GFP fusion proteins directed by the transit peptides of these SATases. Surprisingly, Serat3;2 and Serat3;1 showed contrasting properties with respect to sensitivity of cysteine-feedback inhibition (Table 1). Both Serat3;1 and Serat3;2 have low substrate affinities (Table 1), suggesting low enzyme activities *in vivo*. From these findings, a minor role of these two isoforms in flux regulation towards cysteine synthesis may be inferred.

In order to increase the sulfur-containing products in plants by over-expressing SATase, it is advisable to abolish the cysteine-feedback regulation of SATase by artificial modification of SATase protein. We therefore determined the site required for the allosteric inhibition of watermelon SATase by cysteine. The subcellular localization of watermelon SATase is cytosol, and its activity is inhibited by cysteine (Table 2). Exhaustive analysis of mutated watermelon SATase indicated that the sensitivity of SATase to L-cysteine was determined by the co-operation of the N-terminal and C-terminal regions outside of the catalytic domain. Futhermore, amino acid residues responsible for the cysteine inhibition in the C-terminal region were identified (Inoue et al., 1999). The C-terminal region from Pro-276 to Phe-285 of watermelon SATase, where the amino acid sequences are completely identical among feedback-inhibition sensitive SATases, differs from Serat2;1 from *Arabidopsis thaliana* at 5 amino acid positions, Gly-277, Glu-278, Ser-279, His-282 and Phe-285. Thus, five site-directed SATase mutants were

Table 2. Comparison of the L-cysteine inhibition (IC50) and kinetic constants of watermelon SATase and several point mutants

Mutant	L-cysteine inhibition IC_{50} (μM)	*Km (mM) for* L-Ser	Acetyl-CoA
Watermelon SATase	2.9 ± 0.3	0.59	0.13
SATG277C	77.8 ± 5.2	0.97	0.06
SATE278L	1.7 ± 0.1	0.26	0.38
SATS279T	2.6 ± 0.1	0.42	0.16
SATH282Q	10.1 ± 0.6	1.1	0.10
SATF285Y	4.9 ± 0.1	0.63	0.09

constructed for detailed analysis of this C-terminal region in watermelon SATase. Each five residue, Gly-277, Glu-278, Ser-279, His-282, and Phe-285 in watermelon SATase was substituted with the corresponding amino acid residue found in Serat2;1 at these positions (Fig. 3). The mutant proteins (SATG277C, SATE278L, SATS279T, SATH282Q and SATF285Y) were expressed in *E. coli*, and the catalytic and regulatory properties were investigated using partially purified recombinant proteins. The sensitivity of SATG277C to L-cysteine was markedly decreased from $IC_{50} = 2.9\,\mu M$ to $IC_{50} = 77.8\,\mu M$ (Table 2). The sensitivity of the mutant, SATH282Q, was slightly decreased to $IC_{50} = 10.1\,\mu M$. However, replacement of 3 other amino acid residues, Glu-278, Ser-279, and Phe-285, did not substantially change the sensitivity to L-cysteine. Several point mutations affected the *Km* values of watermelon SATase for L-serine and acetyl-CoA (Table 2). These results confirm that Gly-277 in watermelon SATase is primarily responsible for inhibition by L-cysteine, and that His-282 is secondarily involved.

```
                264                         294
WaSATase    ••••GGKEKPSQLEDIPGESMDHTSFISEWSDYII
Serat2;1    ••••GGKENPRKHDKIPCLTMDQTSYLTEWSDYVI
                **** *      **    ** **   ***** *
                284                         314

SATG277C    ••••GGKEKPSQLEDIPCESMDHTSFISEWSDYII
SATE278L    ••••GGKEKPSQLEDIPGLSMDHTSFISEWSDYII
SATS279T    ••••GGKEKPSQLEDIPGETMDHTSFISEWSDYII
SATH282Q    ••••GGKEKPSQLEDIPGESMDQTSFISEWSDYII
SATF285Y    ••••GGKEKPSQLEDIPGESMDHTSYISEWSDYII
```

Figure 3. Point mutations in the C-terminal region in watermelon SATase. The residues, Gly-277, Glu-278, Ser-279, His-282 and Phe-285 in watermelon SATase were replaced with corresponding amino acids of Serat2;1 from Arabidopsis to give mutants, SATG277C, SATE278L, SATS279T, SATH282Q and SATF285Y, respectively (Inoue et al., 1999). *Asterisk* indicates an identical amino acid residue between Serat2;1 and WaSATase. *Boxed* region from Pro-276 to Phe-285 in WaSATase is shown as an essential region for the L-cysteine inhibition. The *double underlined* amino acids represent the point mutations in each SATase mutant. *Key.* WaSATase, watermelon SATase

To reveal the role of SATase in the regulation of cysteine synthesis and to enhance the ability of cysteine synthesis in plants, we made four kinds of transgenic *Arabidopsis*, SATW, SATM, RuSATW, and RuSATM, in which watermelon SATase gene or mutant SATase gene, SATG277C (Inoue et al., 1999), is overexpressed. In SATW and SATM, produced SATase was localized in cytosol; whereas in RuSATW and RuSATM produced SATase was localized in chloroplasts, since SATase gene was fused with the Rubisco small subunit transit peptide gene in these chimeric genes. In SATW and RuSATW, wild-type SATase produced was highly sensitive to the feedback inhibition by L-cysteine. However, mutated SATase produced in SATM and RuSATM was insensitive to this feedback regulation because of the point mutation of Gly-277 to Cys. These gene expressions were controlled by the strong enhancer element El2 and the 35S promoter. The contents of OAS and cysteine in 3-week-old transgenic *Arabidopsis* overexpressing SATG277C were increased significantly as compared to the wild-type *Arabidopsis* (Fig. 4) (Noji and Saito, 2002). The overexpression of SATG277C caused the overaccumulation of OAS and cysteine, but overexpression of wild-type SATase gene did not. In cell-free extracts of SATW and RuSATW, we detected highly enhanced SATase activity by the *in vitro* assay, however, the enzymatic activity of SATase is presumably inhibited by L-cysteine that exists in plant cells. Therefore, overaccumulation of OAS and cysteine did not remarkably occur in the transgenic plants, SATW and RuSATW, overexpressing wild-type SATase gene compared to SATM and RuSATM overexpressing mutated SATase gene. Interestingly, GSH contents were increased not only in SATM and RuSATM plants but also in SATW and RuSATW plants. These results suggest that the metabolic flux through cysteine to GSH is likely enhanced in all transgenic plants, although the steady-state concentration of cysteine is not increased in SATW and RuSATW plants (Noji and Saito, 2002). These *in vivo* results through transgenic plant experiments confirmed that feedback inhibition to SATase activity by L-cysteine is important for the regulation of the cysteine synthesis at least for cytosolic SATase-catalyzed cysteine formation. These observations strongly imply that the cancellation of feedback-regulation system of SATase is indispensable for large scale production of cysteine and GSH in transgenic plants over-expressing SATase gene.

4. CONCLUSIONS

Metabolic engineering of sulfur assimilation in plants is expected to increase the production of sulfur-containing compounds because sulfur and sulfur-containing products play important roles, for example, as nutrients and stress-mitigated compounds in plant cells. By the over-expression of CSase, the transgenic plants were predictably tolerant to sulfur-containing pollutants and Cd. The modification of amino acid residues in SATase protein, which was intended to invalidate the feedback inhibition of SATase activity by cysteine, enables transgenic plants to produce sulfur-containing compounds in large quantities. Thus, the over-expression of CSase or SATase, the key enzymes for cysteine biosynthesis, is an effective

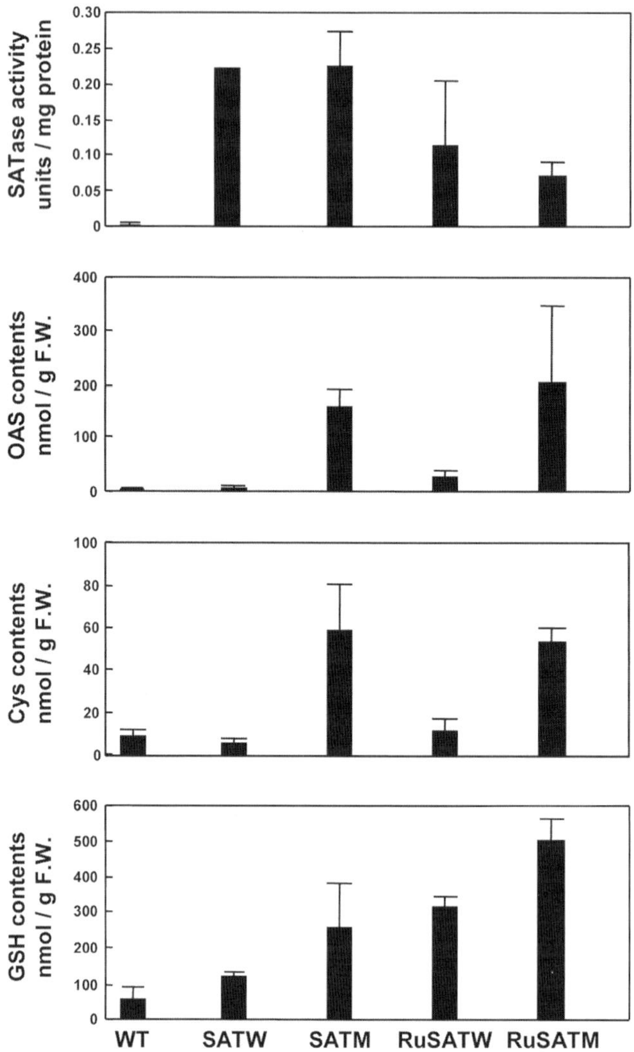

Figure 4. SATase activity and OAS, cysteine and GSH contents of transgenic Arabidopsis overexpressing SATase. SATase activity and OAS, cysteine and GSH contents were determined from thirty transgenic plants from each transgenic line (Noji and Saito, 2002). Data are the means of triplicate analyses ±S.D

approach for the engineering of sulfur assimilation in plants. Attempts to improve the sulfur assimilation in plants by the over-expression of CSase or SATase gene have been, in fact, widely carried out, and remarkable successes have been observed, for example, transgenic plants increasing the level of sulfur-containing metabolites (Harms et al., 2000; Wirtz and Hell, 2003), transgenic plants tolerant to oxidative stress (Blaszczyk et al., 1999; Youssefian et al.,2001), hydrogen sulfide gas (Youssefian et al., 1993), and Cd (Domínguez-Solís et al., 2001). Hence, the transgenic plants by metabolic engineering of sulfur assimilation seem to be of great use in phytoremediation of sulfur-containing pollutants from the atmosphere or heavy metals from soil.

REFERENCES

Ascano A, Nicholas DJD (1977) Purification and properties of *O*-acetyl-L-serine sulphydrylase from wheat leaves. *Phytochemistry* 16: 889–893.

Blaszczyk A, Brodzik R, Sirko A (1999) Increased resistance to oxidative stress in transgenic tobacco plants overexpressing bacterial serine acetyltransferase. *Plant J* 20: 237–243.

Bogdanova N, Hell R (1997) Cysteine synthesis in plants: Protein-protein interactions of serine acetyltransferase from *Arabidopsis thaliana*. *Plant J* 11: 251–262.

Brunold C, Suter M (1982) Intracellular localization of serine acetyltransferase in spinach leaves. *Planta* 155: 321–327.

Brunold C, Suter M (1989) Localization of enzymes of assimilatory sulfate reduction in pea roots. *Planta* 179: 228–234.

Clemens S (2001) Molecular mechanisms of plant metal tolerance and homeostasis. *Planta* 212: 475–486.

Cobbet CS (2000) Phytochelatin and their roles in heavy metal detoxification. *Plant Physiol* 123: 825–833.

De Kok LJ, Stulen I (1993) Role of glutathione in plants under oxidative stress. In: De Kok LJ, Stulen I, Rennenberg H, Brunold C, Rauser WE (eds) Sulfur Nutrition and Assimilation in Higher Plants. Regulatory Agricultural and Environmental Aspects. SPB Academic Publishing, The Hague, pp 125–138.

Domínguez-Solís JR, Gutiérrez-Alcalá G, Romero LC, Gotor C (2001) The cytosolic *O*-acetylserine (thiol) lyase gene is regulated by heavy metals and can function in cadmium tolerance. *J Biol Chem* 276: 9297–9302.

Droux M, Martin J, Sajus P, Douce R (1992) Purification and characterization of *O*-acetylserine (thiol) lyase from spinach chloroplasts. *Arch Biochem Biophys* 295: 379–390.

Droux M, Ruffet ML, Douce R, Job D (1998) Interactions between serine acetyltransferase and *O*-acetylserine (thiol) lyase in higher plants: structural and kinetic properties of the free and bound enzymes. *Eur J Biochem* 255: 235–245.

Eilers T, Schwarz G, Brinkmann H, Witt C, Richter T, Nieder J, Koch B, Hille R, Hänsch R, Mendel RR (2001) Identification and biochemical characterization of *Arabidopsis thaliana* sulfite oxidase: A new player in plant sulfur metabolism. *J Biol Chem* 276: 46989–46994.

Hänsch R, Lang C, Riebeseel E, Lindigkeit R, Gessler A, Rennenberg H, Mendel RR (2006) Plant sulfite oxidase as novel producer of H_2O_2: Combination of enzyme catalysis with a subsequent non-enzymatic reaction step. *J Biol Chem* 281: 6884–6888.

Hall JL (2002) Cellular mechanisms for heavy metal detoxification and tolerance. *J Exp Bot* 336: 1–11.

Harms K, von Ballmoos P, Brunold C, Höfgen R, Hesse H (2000) Expression of a bacterial serine acetyltransferase in transgenic potato plants leads to increased levels of cysteine and glutathione. *Plant J* 22: 335–343.

Inoue K, Noji M, Saito K (1999) Determination of the sites required for the allosteric inhibition of serine acetyltransferase by L-cysteine on plants. *Eur J Biochem* 266: 220–227.

Kawashima CG, Berkowitz O, Hell R, Noji M, Saito K (2005) Characterization and expression analysis of a serine acetyltransferase gene family involved in a key step of the sulfur assimilation pathway in Arabidopsis. *Plant Physiol* 137: 220–230.

Kawashima CG, Noji M, Nakamura M, Ogra Y, Suzuki KT, Saito K, (2004) Heavy metal tolerance of transgenic tobacco plants over-expressing cysteine synthase. *Biotech Lett* 26: 153–157.

Kredich NM (1996) Biosynthesis of cysteine. In: Neidhardt FC, Curtiss R III, Ingraham JL, Lin ECC, Low KB, Magasanik B, Renznikoff WS, Riley M, Schaechter M, Umbargar HE (eds) *Escherichia coli* and *Salmonella typhimurium*: Cellular and Molecular Biology, 2nd edn, Vol 1. American Society for Microbiology, Washington DC, pp. 514–527.

Kredich NM, Tomkins GM (1966) The enzymic synthesis of L-cysteine in *Escherichia coli* and *Salmonella typhimurium*. *J Biol Chem* 241: 4955–4965.

Kuske CR, Hill KK, Guzman E, Jackson PJ (1996) Subcellular location of *O*-acetylserine sulfhydrylase isoenzymes in cell cultures and plant tissues of *Datura innoxia* mill. *Plant Physiol* 112: 659–667.

Lunn JE, Droux M, Martin J, Douce, R (1990) Localization of ATP-sulfurylase and *O*-acetylserine (thiol) lyase in spinach leaves. *Plant Physiol* 94: 1345–1352.

Leustek T, Saito K (1999) Sulfate transport and assimilation in plants. *Plant Physiol* 120: 637–643.

Murray F (1997) Urban air pollution and health effects. In: Brune D, Chapman DV, Gwynne MD, Pacyna JM (eds) The Global Environment: Science, Technology and Management. Scandinavian Science Publisher, Weinheim, pp. 585–598.

Nakamura K, Hayama A, Masada M, Fukushima K, Tamura G (1988) Purification and some properties of plant serine acetyltransferase. *Plant Cell Physiol* 29: 689–693.

Nakamura K, Tamura G (1989) Five isoforms of cysteine synthase in rape leaves. *Agric Biol Chem* 53: 2537–2538.

Noji M, Inoue K, Kimura N, Gouda A, Saito K (1998) Isoform-dependent differences in feedback regulation and subcellular localization of serine-acetyltransferase involved in cysteine biosynthesis from *Arabidopsis thaliana*. *J Biol Chem* 273: 32739–32745.

Noji M, Saito K (2002) Molecular and biochemical analysis of serine acetyltransferase and cysteine synthase towards sulfur metabolic engineering in plants. *Amino Acids* 22: 231–243.

Noji M, Saito M, Nakamura M, Aono M, Saji H, Saito K (2001) Cysteine synthase overexpression in tobacco confers tolerance to sulfur-containing environmental pollutants. *Plant Physiol* 126: 973–980.

Rennenberg H (1984) The fate of excess sulphur in higher plants. *Ann Rev Plant Physiol* 35: 121–153.

Rolland N, Droux M, Douce R (1992) Subcellular distribution of *O*-acetylserine (thiol) lyase in cauliflower (*Brassica oleracea* L.) inflorescence. *Plant Physiol* 98: 927–935.

Ruffet ML, Droux M, Douce R (1994) Purification and kinetic properties of serine acetyltransferase free of *O*-acetylserine (thiol) lyase from spinach chloroplasts. *Plant Physiol* 104: 597–604.

Ruffet ML, Lebrun M, Droux M, Douce R (1995) Subcellular distribution of serine acetyltransferase from *Pisum sativum* and characterization of an *Arabidopsis thaliana* putative cytosolic isoform. *Eur J Biochem* 227: 500–509.

Saito K (2000) Regulation of sulfate transport and synthesis of sulfur-containing amino acids. Curr Opin Plant Biol 3: 188–195.

Saito K (2004) Sulfur assimilatory metabolism: The long and smelling road. *Plant Physiol* 136: 2443–2450.

Saito K, Kurosawa M, Tatsuguchi K, Takagi Y, Murakoshi I (1994) Modulation of cysteine biosynthesis in chloroplasts of transgenic tobacco overexpressing cysteine synthase [*O*-acetylserine (thiol) lyase]. *Plant Physiol* 106: 887–895.

Saito K, Miura N, Yamazaki M, Hirano H, Murakoshi I (1992) Molecular cloning and bacterial expression of cDNA encoding a plant cysteine synthase. *Proc Natl Acad Sci USA* 89: 8078–8082.

Smith IK (1972) Studies of L-cysteine biosynthetic enzymes in *Phaseolus vulgaris* L. *Plant Physiol* 50: 477–479.

Strohm M, Jouanin L, Kunert KJ, Pruvost C, Polle A, Foyer CH, Rennenberg H (1995) Regulation of glutathione synthesis in leaves of transgenic poplar overexpressing glutathione synthetase. *Plant J* 7: 141–145.

Wellburn A (1994) Air Pollution and Climate Change: The Biological Impact 2nd edn. Longman Scientific & Technical, England.

Wirtz M, Hell R (2003) Production of cysteine for bacterial and plant biotechnology: Application of cysteine feedback-insensitive isoforms of serine acetyltransferase. *Amino Acids* 24: 195–203.

Xiang C, Oliver D (1998) Glutathione metabolic genes coordinately respond to heavy metals and jasmonic acid in Arabidopsis. *Plant Cell* 10:1539–1550.

Yamaguchi T, Masada M (1995) Comparative studies on cysteine synthase isozymes from spinach leaves. *Biochem Biophys Acta* 1251: 91–98.

Youssefian S, Nakamura M, Orudgev E, Kondo N (2001) Increased cysteine biosynthesis capacity of transgenic tobacco overexpressing an *O*-acetylserine (thiol) lyase modifies plant responses to oxidative stress. *Plant Physiol* 126: 1001–1011.

Youssefian S, Nakamura M, Sano H (1993) Tobacco plants transformed with the *O*-acetylserine (thiol) lyase gene of wheat are resistant to toxic levels of hydrogen sulfide gas. *Plant J* 4: 759–769.

CHAPTER 14

APPROACHES TO QUALITY PLANT BASED MEDICINE: SIGNIFICANCE OF CHEMICAL PROFILING

PRAVEEN K. SAXENA[1], IAN B. COLE[2], AND SUSAN J. MURCH[2]

[1] *Department of Plant Agriculture, University of Guelph, Guelph, Ontario, Canada, N1G 2W1*

[2] *Chemistry, I.K. Barber School of Arts & Sciences, University of British Columbia Okanagan, Kelowna, British Columbia, Canada, V1V 1V7*

1. PLANT-BASED MEDICINE

The World Health Organization has estimated that 65–80% of the world population's primary health care needs are met through plant-based traditional medicines. In developing countries plant-based medicines have always been and remain the primary source of medical care. Increasing popularity and acceptance of plant-based medicines in the Western world (referred to as Neo-Western Herbalism) seen in recent years is considered a result of the quest for longevity and improved quality of life. The desire of individuals to take responsibility of their own health and the confidence that plants are "natural and therefore safe and effective" has further contributed to the phenomenal growth of this sector. The market for herbal remedies has risen by 4–10% per year for the past decade in North America and Europe, and is now in excess of US$20 billion. The marketplace for Natural Health Products (NHPs) is rapidly changing in response to a number of reported problems and on-going issues with inconsistent results of major clinical trials of plant-based medicines. Newly introduced or upcoming legislations in Canada by the Natural Health Products Directorate (NHPD) at Health Canada, and in the USA by the Food and Drug administration are likely to have major impact on how plant-based medicines are developed, manufactured, and marketed. Ultimately, these changes will improve the safety and the efficacy of the products through the development of new technologies and approaches to production of plant-based medicines that comply with the regulations.

R. Verpoorte et al. (eds.), Applications of Plant Metabolic Engineering, 311–330.
© 2007 *Springer.*

2. GENERAL ISSUES AND PROBLEMS WITH PLANT-BASED MEDICINE

The issues with plant-based medicine include little understanding of the unique physiology of the plant, the non-existent technologies for optimized growth, production, harvest or handling, and the adulteration and contamination of plant-based medicine with misidentified plant species, plant and soil born microbes and environmental pollutants compromising the safety and efficacy of many products (Murch et al., 2000a; Murch and Saxena, 2006). Extreme variations of medicinal metabolites in plant-based medicine pose challenges for conducting clinical research to ensure safety and efficacy. The extract of even one single plant is an inordinately complex mixture of chemical constituents, making the double-blind, placebo-controllcd trials very difficult.

There are additional problems associated with traditional field production or wild-harvest since plant species are complex biological systems with varying capabilities for absorption of nutrients and contaminants from the environment. Plant-based medicines are commonly prepared as whole tissues. Preparations may be made from unwashed leaf tissues to preserve bioactive molecules in surface glands that would be ruptured in cleaning or from roots of plants that are several years old. It is not uncommon, then, for the plant tissues and the resulting products to contain soil, bacteria, fungi, insects and environmental pollutants. One example is the recently reported contamination of root tissues of Goldenseal (*Hydrastis canadiensis*) with environmental contaminants (Liu et al., 2004). The products of the North American native medicinal plant were found to have high levels of aluminum (35 mg), lead (286 μg), cadmium (18 μg), nickel (784 μg), and other metals (Liu et al., 2004). In other examples, pharmaceuticals similar to Sildenafil have been found as contaminants in various natural products that claim to increase vigour and sexual performance and advisories have been issued in Canada and the USA. The chemical diversity of plant materials means that frequently whole plant preparations contain bioactive compounds that are not known to be present. A commercial preparation of Feverfew (*Tanacetum parthenium*) was found to contain the mammalian neurotransmitter melatonin (Murch et al., 1997). Finally, the natural practice of wild-harvest actually decimates natural populations endangering genetic diversity. It has been estimated that upto 10,000 medicinal plants are endangered worldwide (Leaman, 2001).

3. PLANT-BASED MEDICINES: CANADIAN CONTEXT

Health Canada surveys indicate that 71% of Canadians regularly take vitamins and minerals, herbal products, homeopathic medicines and the like – products that have come to be known as natural health products (NHPs) with Canadian sales exceeding $1.5 billion annually. The problems associated with NHPs prompted legislation under the newly created “Natural Health Products Directorate” whose goal is to ensure a regulatory framework for safety quality and efficacy of natural health products. These regulations are the result of a comprehensive and inclusive consultation process with Canadian consumers, academics, health care practitioners and industry stakeholders. The NHPD regulations utilize a staged approach planned to be carried out over a period of 6 years. Under these regulations, all products are to have specific labels describing medicinal and non-medicinal ingredients, their quantity, source and shelf-life, among many others requirements. Under the legislation, it will be permissible to make a medical claim on whole plant preparations and therapeutic alternatives, but the onus will be on manufacturers and producers to provide products free from contamination and with standardized medicinal activity. These products will be required to originate from transparent sources with processing in compliance with GMP standard operating protocols. A large number of products in the current market may not meet these requirements and the manufacturers must adopt new and improved production practices to ensure the overall quality of their products.

4. ST. JOHN'S WORT (*HYPERICUM PERFORATUM*): AN EXAMPLE OF COMPLEX CHEMISTRY

The chemical profiles of medicinal plants are affected by conditions of cultivation, manufacturing, marketing, and distribution. The factors that affect biochemical profiles and secondary metabolite production in plants include physiological, genetic, and environmental (photoperiod, climate, soil conditions, nutrient availability, moisture) variables. Secondary metabolite content is also subject to time of harvesting, storage, drying, extraction and processing for final packaging. In the following section we use an extensively investigated medicinal species *Hypericum perforatum* commonly known as St. John's wort, as an example to illustrate the widespread occurrence of chemical variability and the importance of chemical profiling of medicinal plants.

The medicinal properties of St. John's wort were first discovered by Greek physician Hippocrates, fifth century B.C.E. Since then, St. John's wort has been the subject for numerous studies, and clinical trials with the objective of determining its efficacy for a wide variety of ailments such as depression, cancer, inflammation and viral infections. There are at least 37 distinct common names for *Hypericum perforatum* representing at least 19 different cultures scattered across the world. The Ethnomedical information on *Hypericum perforatum* lists more than 80 reported uses as an antidepressant, emmenagogue, anthelmintic, spasmolytic, diuretic, antirheumatic, astringent, and in the healing of wounds, colds, stomach ache, headache, paralysis, tetanus, spinal convulsion, ulcers, gastrointestinal disorders, jaundice, goiter, diabetes, urinary bladder diseases, kidney diseases, inflammations, skin wounds, insect bites, herpes, insomnia, hair loss and bites of venomous animals. Additional activities of the plant and its extracts are contained within the several hundred references to scientific studies demonstrating efficacy in treatment of depression, cancer, aids and other serious ailments. In scientific studies, St. John's wort has been shown to effectively treat mild to moderate depression (Szegedi et al., 2005). Its high efficacy and tolerability is comparable to other antidepressants while lacking major side effects, making it a safe antidepressant (Bilia et al., 2002). Currently, St. John's wort products are one of the top five phytopharmaceuticals in North America (National Centre for Complementary and Alternative Medicine, 2001).

St. John's wort consists of a number of different chemical constituents, many of them with unknown actions. Thus, the issues of medicinal efficacy of whole plant preparations are complex and need further research. Hypericin and hyperforin, the two main naphthodianthrone compounds, are widely recognized for their medicinal benefits. However, isolated preparations of these two compounds do not have the same medicinal efficacy as the whole plant preparations. Common preparations of St. John's wort include tinctures, tablets, and capsules, and are prepared to standardized concentrations of with hypericin and hyperforin at 0.3% and 3–5% respectively (Blumenthal et al., 2003). While several trials of St. John's wort have shown that extracts were at least as effective as conventional prescription antidepressant medications, others showed no efficacy or limited efficacy leading to the conclusion that the evidence of efficacy is "inconsistent, confusing, and difficult to interpret" (Linde et al., 2005).

5. ST. JOHN'S WORT AND DRUG INTERACTIONS

St. John's wort is a popular anti-depressant among cancer patients due to the psychological side-effects associated with illness. This is a cause for concern as hyperforin, a bioactive constituent of St. John's wort, is also known to affect metabolism of several drugs (Wentworth et al., 2000), including immunosuppressants, protease inhibitor, and active ingredients of cancer drugs. For example Mathijssen et al. (2002) found that St. John's wort interferes with Irinotecan (a common cancer drug) metabolism, and induce cytochrome P450 CYP3A4 isoform expression. Additionally, plasma levels of metabolites of Irinotecan were significantly lower in patients taking St. John's wort at the same time (Mathijssen et al., 2002). Consumption of St. John's wort has also been shown to reduce the bioavailability of certain drugs such as indinavir (HIV protease inhibitor), cyclosporine (immunosuppressant), and digoxin (cardiac glycoside) (Perloff et al., 2001) due to the induction of P-glycoprotein (pGp) by St. John's wort. Perloff et al. (2001) have shown that administration of both St. John's wort and hypericin independently results in an increase in the levels pGp, an ATP-dependant drug transporter. Alongside the effects on drug bioavailability and metabolism, these interactions may have potentially dangerous consequences in many other forms. For example, transplant patients have a greater chance of graft rejection due to the low cyclosporine. A levels resulting from pGp activation and cytochrome P450 induction, attributed to St. John's wort supplementation (Mansky et al., 2002). Furthermore, the naphthodianthrone structure of hypericin resembles the anthraquinone ring structure found in many cancer drugs and can antagonize the action of chemotherapeutics (Peebles et al., 2001).

6. CHEMICAL CONTENT OF ST. JOHN'S WORT PREPARATIONS

Several studies have assessed the hyperforin and hypericin content in commercial St. John's wort products with the objective of testing consistency, as well as batch-to-batch reproducibility. In a survey of products, only one of the eight products tested had the required hyperforin content of at least 3%, while hypericin content generally exceeded the proposed limit. High variability was also observed between manufacturers when testing batch-to-batch reproducibility (Wurglics et al., 2001; Michelitsch et al., 2003). Draves and Walker (2003) further investigated hypericin and pseudohypericin content in St. John's wort products and found that only two of the 42 products had levels that came within 10% of their claim. Furthermore, in most cases, the hypericin content as indicated on the label was overestimated by a factor of two (Draves and Walker, 2003). The reason for such large discrepancy between actual content and label claim may be due to the use of analytical protocols used by manufacturers. The older methods recommended by the US Pharmacopea were a simple spectrophotometric assay that did not distinguish between different naphthodianthrones, and other compounds that absorbed at 562 nm resulting in overestimations of medicinal content. More recent methods have implemented a much more sensitive chromatographic method with fluorescent detection that is more accurate but results in overall lower quantifications. These studies demonstrate the need for accurate methods to determine medicinal content, the lack of consistency in chemical profile of St. John's wort tissues and the need for better protocols for preparation and storage of medicinal plant products.

7. SOURCES OF CHEMICAL COMPLEXITY

Complex genetics of secondary metabolite biosynthesis, adulterations with misidentified plant species, and plant-environment interactions affect the chemical profile of St. John's wort preparations (Murch and Saxena, 2006). St. John's wort reproduces in at least 11 different ways which is a major cause of chemical diversity in this species. (Pank et al., 2003). A recent, most comprehensive study of the genetic diversity and reproductive biology of the facultative apomitic ecotypes of St. John's wort concluded the reproductive strategies in this species to be much more dynamic and versatile than previously expected (Barcaccia et al., 2006). This study further emphasized that experimental trials performed with a few genotypes in a chosen environment can lead to results that deviate significantly from the dynamics of the same genotypes in their source populations (Barcaccia et al., 2006). The hypericin content of St. John's wort tissues is highest in the flowers and when the flower is in full bloom with levels decrease during the over blooming period (Cellárová, 2003; Seidler-Lozykowska, 2003). Growth environment including light, temperature, and nutrition, and the time of harvest can effectively alter the phytochemical profile of St. John's wort tissues (Buter et al., 1998; Briskin and Gawienowski, 2001; Seidler-Lozykowska, 2003; Murch et al., 2002; Zobayed et al., 2005a, 2000b; Couceiro et al., 2006; Murch and Saxena, 2006). Insect herbivory can also change hypericin and hyperforin content of the tissues by as much as 30-100% (Sirvent et al., 2003). Additionally, St. John's wort preparations may also be affected indirectly by adulteration with other species such as *Hypericum maculatum*, *Hypericum barbatum*, and *Hypericum montanum* (American Botanical Council, 1997). Finally, the concentrations and bioavailability of phytochemicals are affected by the conditions of processing (Gaedcke, 2003; Meier, 2003).

8. CHEMICAL PROFILING: IMPORTANCE OF METABOLOME ANALYSIS

At least 190 different compounds have been reported in the literature on St. John's wort including triterpenes, sesquiterpenes, monoterpenes, xanthones, carotenoids, quinoids, phenylpropenoids, flavonols, indole alkaloids, and compounds of unknown structure (Table 1). Flavonoids such as rutin have been reported at concentrations in excess of 1.6% (Dorossiev, 1985). Rutin, hyperoside, isoquercitrin, avicularin, quercitrin, and quercetin, form some of the constituents of a flavonoid-rich extract of St. John's wort that has strong antioxidant activity and medicinal potential (Zou et al., 2004). Recent technologies such as metabolomics are likely to identify numerous novel metabolites in medicinal plants. Conceptually the term metabolomics refers to a comprehensive analysis of the whole complement of small metabolites within a complex system such as a plant cell using analytical chemistry with dedicated statistical, bioinformatics and data mining strategies (Aharoni et al., 2002; Hall et al., 2002; Hall, 2006). The technologies of metabolomics complement genomics data arrays as a great number of genes have been sequenced but have not been assigned putative functions (Fiehn, 2001). Plant metabolomics offers a valuable technology for the assessment of metabolic profiles under changing environments and is of particular significance in medicinal plant research in view of the environmental regulation of their medicinal chemistry (Murch et al., 2004).

We have previously shown the presence of over 2000 compounds in metabolomic analysis of a widely used medicinal plant Huang-qin (*Scutellaria baicalensis*) (Murch et al., 2004). Huang-qin extracts were found to contain 781 compounds of putative medicinal importance as determined by a database search. In addition, the extracts also indicated the presence of previously unidentified amino-derivatives of baicalin and wogonin, two major medicinal components of this species. Surprisingly, Huang-qin also contained a metabolite with the same net formula as hyperforin, previously thought to be unique to Saint John's wort (Figure 1). Recently, we separated metabolites of in vitro grown St. John's wort tissues into multiple extracts based upon their polarity and acid/base chemistry and subject them to perform FTMS (Fourier Transform Ion Cyclotron Mass Spectrometry) analysis. A series of internal standards were used and FTMS analyses were then performed with multiple extracts using the methods established earlier (Murch et al., 2004; Figure 1). The metabolome analysis of St. John's wort detected the expected phytocompounds and differences in concentration of specific metabolites in the different germplasm lines

Table 1. Chemicals reported in the literature on St. John's wort (*Hypericum perforatum* L)

Compound	Chemical Class	Reference
Adehyperfolin	Alicyclic	Maisenbacher and Kovar, 1992
Alkanes(C28, C30)	Alkane	Mathis and Ourisson, 1964a
Alkanols(C24, C26, C28)	Alkanol	Mathis and Ourisson, 1964a
Amentoflavone	Flavonoid	Buter et al., 1998
β-Amyrin	Triterpene	Hooper and Chandler, 1984
Anthrone	Quinoid	Kovac and Imamovic, 2000
Apigenin	Flavone	Brantner et al., 1994
Arachidic Acid	Lipid	Stojanovic et al., 2003
Ascorbic Acid	Vitamin	Khalmatov, 1957
Astilbin	Flavonoid	Butterweck et al., 2000
Avicularin	Flavonol	Wu et al., 2001
Behenic Acid	Lipid	Stojanovic et al., 2003
Benzyl Benoate	Benzenoid	Baser et al., 2002
Biapigenin	Flavonoid	Buter et al., 1998
Butenol, 2-methyl	Alkanol	Nahrstedt and Butterweck, 1997
Butyric Acid, γ-Amino	Amino Acid	Franke et al., 1997
Cadalene	Sesquiterpene	Cakir et al., 1997
Cadiforin, Hydroperoxy	Terpenoid	Rucker et al., 1995
Δ-Cadinene	Sesquiterpene	Baser et al., 2002
γ-Cadinene	Sesquiterpene	Baser et al., 2002
α-Cadinol	Sequiterpene	Baser et al., 2002
Caffeic Acid	Phenylpropenoid	Holzl and Ostrowski, 1987
Carene	Monoterpene	Cakir et al., 1997
α-Carotene	Carotenoid	Omarova and Artamonova, 1997
β-Carotene	Carotenoid	Omarova and Artamonova, 1997
Carvacrol	Monoterpene	Erken et al., 2001
Carvacrol Methyl Ether	Monoterpene	Erken et al., 2001
Caryophyllene	Sesquiterpene	Mathis and Ourisson, 1964b
Caryophyllene Oxide	Sesquiterpene	Baser et al., 2002
β-Caryophyllene	Sesquiterpene	Nogueira et al., 1999
Cathechin (+)	Flavonoid	Kitanov, 1983
Catechin, Epi: (−)	Flavonoid	Olliver et al., 1985
Chlorogenic Acid	Phenylpropenoid	Jurgenliem and Nahrstedt, 2002
Chlorophyll A	Pigment	Omarova and Artamonova, 1997
Chlorophyll B	Pigment	Omarova and Artamonova, 1997
Choline	Alkaloid	Chaplinskaya, 1959
Chromen-4-one,5-Hydroxy-7-Methoxy-3	Chromone	De Shan et al., 2004
α-Copaene	Sesquiterpene	Baser et al., 2002
Cryptochlorogenic Acid	Phenylpropenoid	Jurgenliem and Nahrstedt, 2002
α-Cuprenene	Sesquiterpene	Weyerstahl et al., 1995
Cynaidin-3-O-α-L-Rhamnoside	Flavonoid	Jurgenliem and Nahrstedt, 2002
Cyclohexanone,3-methyl-4-6-di-(3-methyl-2-butenyl)-2-(2-methyl-1-oxopropyl)-3-(4-methyl-3-pentenyl)	Alicyclic	De Shan et al., 2001

(Continued)

Table 1. (Continued)

Compound	Chemical Class	Reference
Cyclohexanone,3-methyl-4-6-di-(3-methyl-2-butenyl)-2-(2-methyl-1-oxopropyl)-3-(4-methyl-3-pentenyl)	Alicyclic	Vajs et al., 2003
Cyclopseudohypericin	Quinoid	Giese, 1980
Cymene	Monoterpene	Erken et al., 2001
Daucosterol	Steroid	Yin et al., 2001
Decanal	Alkanal	Mathis and Ourisson, 1964b
Decane,2-methyl	Alkane	Mathis and Ourisson, 1964c
Docosane	Alkane	Stojanovic et al., 2003
Dodec-3-ene,1-(2-methyl-1-oxo-propyl)-2-12-dioxo-3-10-β-bis-(3-methyl-but-2-enyl)-6-β-(1-methyl1-hydroxy-ethyl)-11-β-methyl-11-a-(4-methyl-pen-3-enyl)-5-oxa-tricyclo-[6.3.1.0-4,8]	Triterpene	Trifunovic et al., 1998
Dodecanol	Alkanol	Baser et al., 2002
Dotriancontane	Alkanol	Stojanovic et al., 2003
Emodin Anthrone	Quinoid	Nahrstedt and Butterweck, 1997
Encosane	Alkane	Stojanovic et al., 2003
Eriodictyl	Flavanone	Nahrstedt and Butterweck, 1997
Erucic Acid	Lipid	Stojanovic et al., 2003
Farnesene	Sesquiterpene	Erken et al., 2001
Foeniculin	Flavonol	Jamsen et al., 1999
Funebrene	Sesquiterpene	Baser et al., 2002
Furan-2-one,Dihydro: 5-methyl-5-(4-8-12-trimethyl-tridecyl)	Diterpene	De Shan et al., 2004
Furoadhyperforin	Alicyclic	Vajs et al., 2003
Furohyperforin	Alicyclic	Verotta et al., 1999
Gallic Acid	Benzenoid	Jela and Tamara, 1981
Germacrene D	Sesquiterpene	Nogueira et al., 1999
Guaijaverin	Flavonol	Jurgenliem and Nahrstedt, 2002
Heneicosane	Alkane	Stojanovic et al., 2003
Hentriacontane	Alkane	Stojanovic et al., 2003
Heptacosane	Alkane	Stojanovic et al., 2003
Heptane,2-4-Dione,5-Methyl	Alkanone	Weyerstahl et al., 1995
Heptane-2-4-Dione,6-Methyl	Alkanone	Weyerstahl et al., 1995
Hexacosanol	Alkanol	Brondz et al., 1983
Hexacosane	Alkane	Stojanovic et al., 2003
Humulene	Sesquiterpene	Mathis and Ourisson, 1964b
Hypercinin,Cyclo-Pseudo	Quinoid	Wagner and Bladt, 1994
Hyperfolin	Terpenoid	Rucker et al., 1995
Hyperforin	Alicyclic	Gurevich et al., 1971
Hyperforin, oxy	Alicyclic	Bergonzi et al., 2001
Hyperforin, Pyrano[7-28-B]	Alicyclic	Vajs et al., 2003

Hyperforin-8-1-Hemiacetal,8-Hydroxy	Alicyclic	Verotta et al., 2000
Hypericin	Quinoid	Jensen and Miller, 1951
Hypericin, Proto	Quinoid	Poutaraud et al., 2001
Hypericin,Proto-Pseudo	Quinoid	Haeberlein et al., 1992
Hypericin,Pseudo	Quinoid	Jurgenliem and Nahrstedt, 2002
Hyperoside	Flavonol	Brantner et al., 1994
Imanin	Structure Unknown	Aizenman, 1969
Ishwarane	Sesquiterpene	Weyerstahl et al., 1995
Kaempferol	Flavonol	Mueller and Schaefer, 1996
Kielcorin	Xanthone	Nielsen and Arends, 1978
Lauric Acid	Lipid	Stojanovic et al., 2003
Leucocyanidin	Flavonoid	Kitanov and Blinova, 1987
Lignoceric Acid	Lipid	Stojanovic et al., 2003
Limonene	Monoterpene	Mathis and Ourisson, 1964c
Linoleic Acid	Lipid	Omarova and Artamonova, 1999
Linolenic Acid	Lipid	Omarova and Artamonova, 1999
Lutein	Carotenoid	Costes, 1967
Luteolin	Flavone	Dias et al., 1998
Luteolin,6-C-Prenyl	Flavone	Dias et al., 1998
Luteolin-3′-5-Dimethyl Ether	Flavone	Dias et al., 1998
Luteolin-3′-Glucoside	Flavone	Dias et al., 1998
Luteolin-5-Glucoside	Flavone	Dias et al., 1998
Luteoxanthin	Carotenoid	Costes, 1967
Mangiferin	Xanthone	Seabra et al., 1992
Mangiferin,Iso	Xanthone	Kitanov and Nedialkov, 1998
Melatonin	Indole Alkaloid	Murch et al., 1997
Miquelianin	Flavonol	Seabra et al., 1992
α-Muurolene	Sesquiterpene	Baser et al., 2002
Myrcene	Monoterpene	Mathis and Ourisson, 1964c
Myriaboric Acid	Triterpene	Yin et al., 2001
Myricetin	Flavonol	Chatterjee et al., 1998
Myricetin-3-O-β-D-Glucoside	Flavonol	Akhtardzhiev et al., 1984
Myristic Acid	Lipid	Costes, 1967
Neoxanthin	Carotenoid	Omarova and Artamonova, 1997
Nicotinic Acid	Alkaloid	Rolleri, 1943
Nonacosane	Alkane	Stojanovic et al., 2003
Nonadecane	Alkane	Stojanovic et al., 2003
Nonane	Alkane	Mathis and Ourisson, 1964c
Nonane,3-methyl	Alkane	Erken et al., 2001
Novoimanin	Structure Unknown	Aizenman, 1969
Octacosanol	Alkanol	Brondz et al., 1983
Octancosane	Alkane	Stojanovic et al., 2003
Octanal	Alkanal	Mathis and Ourisson, 1964b
Octane, 2-methyl	Alkane	Nogueira et al., 1999
Oleoresin	Structure Unknown	Quintanilla et al., 2000
Orientin,iso	Flavone	Jurgenliem and Nahrstedt, 2002

(Continued)

Table 1. (Continued)

Compound	Chemical Class	Reference
Oxepahyperforin	Alicyclic	Verotta et al., 2000
Palmitic Acid	Lipid	Omarova and Artamonova, 1999
Palmitoleic Acid	Lipid	Stojanovic et al., 2003
Pentacosane	Alkane	Stojanovic et al., 2003
Perflavit	Flavonoid	Biochinov et al., 1965
Phloroglucinol	Benzenoid	Grujic-Vasic et al., 1986
Phytol	Diterpene	De Shan et al., 2004
α-Pinene	Monoterpene	Baser et al., 2002
β-Pinene	Monoterpene	Mathis and Ourisson, 1964c
Pinocembrin-7-methyl ether	Flavanone	Jurgenliem and Nahrstedt, 2002
Procyanidin A-2	Flavonoid	Ploss et al., 2001
Procyanidin B-1	Flavonoid	Ploss et al., 2001
Procyanidin B-2	Flavonoid	Ploss et al., 2001
Procyanidin B-3	Flavonoid	Ploss et al., 2001
Procyanidin B-5	Flavonoid	Ploss et al., 2001
Procyanidin B-7	Flavonoid	Ploss et al., 2001
Procyanidin C-1	Flavonoid	Ploss et al., 2001
Protocatechuic Acid	Benzenoid	Jurgenliem and Nahrstedt, 2002
Protohypericin	Quinoid	Jamsen et al., 1999
Pyrocatechol	Benzenoid	Grujic-Vasic et al., 1986
Pyrogallol	Benzenoid	Grujic-Vasic et al., 1986
Pyrohyperforin	Alicyclic	Shan et al., 2000
Quercetin	Flavonol	Maksyutina and Kolesnikov, 1964
Quercetin-3-(6′-O-acetyl)-Galactoside	Flavonol	Yin et al., 2001
Quercetin-3-O-(2′-acetyl)-β-D-Galactoside	Flavonol	Jurgenliem and Nahrstedt, 2002
Quercetin-3-O-Xyloside	Flavonol	Seabra et al., 1992
Quercetin-Galacturonide	Flavonol	Jamsen et al., 1999
Quercetrin	Flavonol	Kartnig and Heydel, 1993
Quercitin,Iso	Flavonol	Akhtardzhiev et al., 1984
Quercitrin	Flavonol	Von Eggelkraut et al., 2002
Quercitrin, Iso	Flavonol	Brantner et al., 1994
Quinic Acid,3-O-Cis-Para-Coumaroyl	Phenylpropenoid	Jurgenliem and Nahrstedt, 2002
Quinic Acid,3-O-Para-Coumaroyl	Phenylpropanoid	Brolis et al., 1998
Quinic Acid,3-O-Trans-Para-Coumaroyl	Phenylpropanoid	Jurgenliem and Nahrstedt, 2002
Rescorcinol	Benzenoid	Grujic-Vasic et al., 1986
Rutin	Flavonol	Razhinskaite, 1971
Sabinene	Monoterpene	Cakir et al., 1997
Scopoletin	Courmarin	Khosa and Bhatia, 1982
α-Selenene	Sesquiterpene	Baser et al., 2002
β-Selenene	Sesquiterpene	Baser et al., 2002
α-Selinene	Sesquiterpene	Erken et al., 2001
β-Selinene	Sesquiterpene	Erken et al., 2001
β-Sitosterol	Steroid	Yin et al., 2001
Skyrin-6-O-α-Arabinofuranoside	Quinoid	Wirz et al., 2000

Skyrin-6-O-β-Glucopyranoside	Quinoid	Wirz et al., 2000
Skyrin-6-O-β-Xylopyranoside	Quinoid	Wirz et al., 2000
Taraxasterol	Triterpene	Hooper and Chandler, 1984
γ-Terpinene	Monoterpene	Cakir et al., 1997
Tetracosanol	Alkanol	Brondz et al., 1983
Tetracosane	Alkane	Stojanovic et al., 2003
Tetradecanol	Alkanol	Erken et al., 2001
Tetradecanoid Acid	Lipid	Baser et al., 2002
Thymol	Monoterpene	Erken et al., 2001
Tricontan-1-ol	Alkanol	Brondz et al., 1983
Triacontane	Alkane	Stojanovic et al., 2003
Tricosane	Alkane	Stojanovic et al., 2003
Tritriacontane	Alkane	Stojanovic et al., 2003
Trollichrome	Carotenoid	Costes, 1967
Trollizanthin	Carotenoid	Costes, 1967
Tryptamine,5-hydroxy	Iindole Alkaloid	Murch et al., 2000
Umbelliferone	Coumarin	Khosa and Bhatia, 1982
Undecane	Alkane	Mathis and Ourisson, 1964c
Ursolic Acid	Triterpene	Yin et al., 2001
Violaxanthin	Carotenoid	Costes, 1967
Xanthone,1-3-5-6-Tetrahydroxy	Xanthone	Dias et al., 2001
Xanthone,1-3-5-Trimethoxy	Xanthone	Muruganandam et al., 2000
Xanthone,1-3-6-7-Tetrahydroxy	Xanthone	Dias et al., 2001
Xanthone, 1-Hydroxy-3-5-Dimethoxy	Xanthone	Muruganandam et al., 2000
Zeaxanthin	Carotenoid	Omarova and Artamonova, 1997

(Figure 2). St. John's wort was found to contain a total of 2817 distinct compounds with mass varying from 100 to 2000 mwt. Comparison of the whole metabolome to a database of natural products found 1151 compounds that putatively match medicinal metabolites. Twenty compounds in this pool had not previously been identified in St. John's wort and all were thought to be characteristic of different species (Table 2). The presence of these and other compounds is indicative of the vast resource of as-yet-unidentified medicinal metabolites in St. John's wort. Plant based medicines often use multiple combinations of many different species. Plant species such as Huang-qin have been used in combination with as many as 20 other medicinal constituents. Unknown compounds with a range of medicinal activities in these mixtures pose serious health concerns due to potential interactions with conventional drugs. The results obtained with metabolomics provide further evidence of the chemical complexity of whole plant preparations and emphasize the need for complete analyses of individual species. Broadening our understanding of the complex chemistry of medicinal species will provide new perspectives for understanding human responses to whole plant treatments.

It is noteworthy that St. John's wort also contains high amounts of neuroregulators melatonin and serotonin (Murch et al., 1997; 2000b; 2002). There are more than 11,200 citations in PubMed for therapeutic role of melatonin including sleep disorders, depression, aging, and many different types of cancer (Reiter et al., 2001;

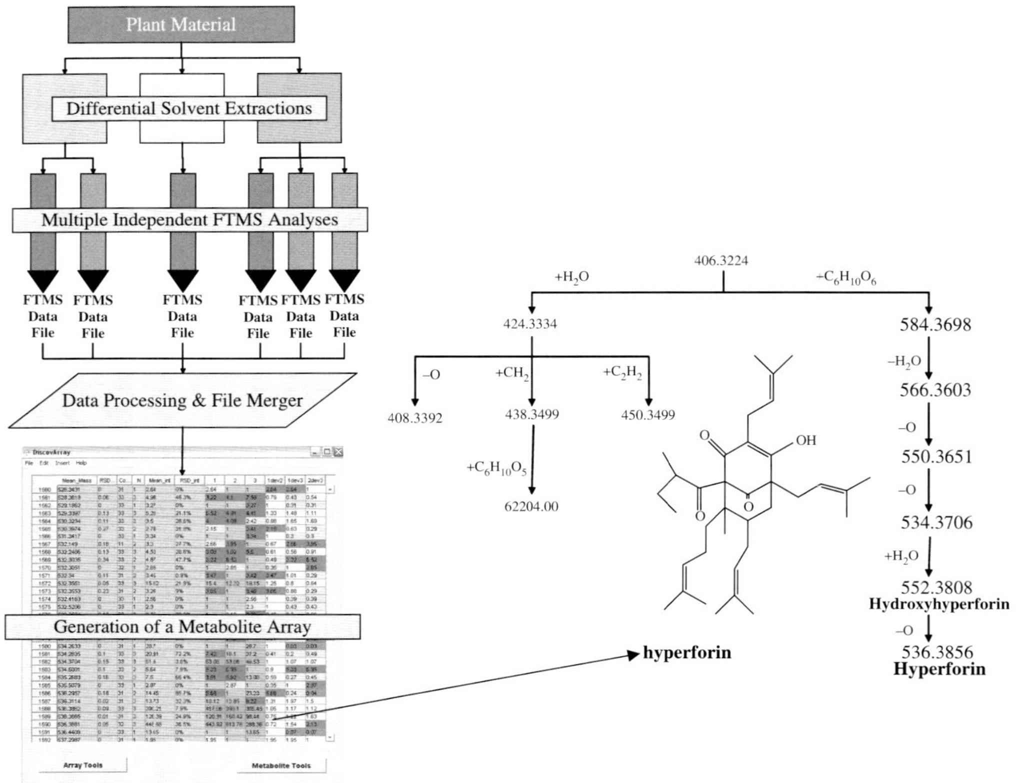

Figure 1. Schematic representation of differential extraction routines for metabolomic analyses (left panel) and indentification of the structure of hyperforin and related metabolites (right panel) in *Scutellaria baicalensis*. (modified from Murch et al. 2004). Hyperforin is considered a characteristic compound of St. John's wort

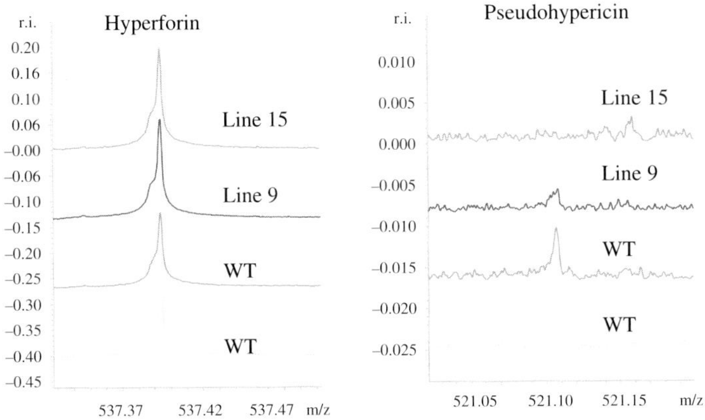

Figure 2. Metabolome analysis of different germplasm lines of St. John's wort (*Hypericum perforatum*). A. Differences in spectra of hyperforin. B. Differences in spectra of pseudohypericin

Boutine et al., 2005). Melatonin is an interesting ubiquitous indoleamine compound that has been implicated in almost all the main physiological functions of the body and mind. The presence of melatonin in St. John's wort has significant potential of affecting the fundamental processes involved in depression, insomnia and various types of cancer.

Table 2. Twenty examples of putatively identified medicinal metabolites in St. John's wort

Compound	T_Mass	Formula	Species	Botanical Family
Acrovestone	554.2880	C32H42O8	*Acronychia pedunculata*	Rutaceae
Andrachamine	257.2355	C15H31NO2	*Andrachne aspera*	Euphorbiaceae
Annodienin	588.4754	C37H64O5	*Annona jahnii*	Annonaceae
Annojahnin	590.4910	C37H66O5	*Annona jahnii*	Annonaceae
Ardisenone	496.2825	C30H40O6	*Ardisia iwahigensis*	Myrsinaceae
Cassine	297.2668	C18H35NO2	*Senna spectabilis*	Leguminosae
Conoidene	326.1518	C20H22O4	*Silene conoidea*	Caryophyllaceae
Cristatin	340.2038	C22H28O3	*Lepidagathis cristata*	Acanthaceae
Diinsinin	704.1741	C36H32O15	*Sarcophyte piriei*	Balanophoraceae
Disepalin	666.5071	C39H70O8	*Disepalum anomalum*	Annonaceae
Europine	329.1838	C16H27NO6	*Heliotropium sp.*	Boraginaceae
Geijerene	162.1409	C12H18	*Pimpinella anisum*	Rutaceae
Hypercalin	484.3189	C30H44O5	*Hypericum calycinum*	Guttiferae
Lateriflorone	578.2516	C33H38O9	*Garcinia lateriflora*	Guttiferae
Nimbothalin	366.1831	C23H26O4	*Azadirachta indica*	Meliaceaea
Pfaffic acid	440.3290	C29H44O3	*Pfaffia paniculata*	Amaranthaceae
Polycerasoidol	358.2144	C22H30O4	*Orophea enneandra*	Annonaceae
Ponticaepoxide	182.0732	C13H10O	*Artemisia annua*	Compositae
Striatol	442.3083	C28H42O4	*Ptycanthus striatus*	Asclepiadaceae

9. CONTROLLED ENVIRONMENT PRODUCTION AND CHEMICAL CONSISTENCY

In order to fully maximize the medicinal value of herbals therapeutics, it is crucial to understand the plant as a whole system with regards to physiology, botany and chemistry (Murch et al., 2000a, 2000c). Due to lesser expense and more convenience, however, trends for pharmaceutical industries have been to isolate and manufacture single constituents that may be associated with a certain beneficial effect. This philosophy, unfortunately, may not be the best applicable for plant-based medicines. Practically, the production of whole plant material with consistent, complex chemistry requires application of the modern techniques of biotechnology. In vitro plant propagation and production systems rely on the capacity of plant cells to de-differentiate and redirect their developmental pathway in suitable growth environment. The correct combination of growth regulators and environmental conditions effectively stimulate the cells from differentiated tissues to form new shoots, roots, somatic embryos, and whole plants. Technologies for in vitro plant production of St. John's wort have been reported for the production of large numbers of genetically and physiologically similar plants in a short time frame and small amount of space (Murch et al., 2000d). This genetic and physiological homogeneity can be exploited in order to produce plants that exhibit a uniform chemical profile in a highly controlled sterile environment. The advantages of in vitro production of medicinal plants for basic research and the development of commercial products have been discussed earlier (Murch et al., 2000a, 2000c). Briefly, the benefits of controlled environment production of medicinal plants include seasonally independent production, minimum contamination with microbes, heavy metals, mycotoxins and foreign plant material, and chemically consistent tissues (Murch et al., 2000c; Murch and Saxena, 2006).

The commercial application of in vitro regeneration technologies for the plant-based medicine industry requires the development of additional protocols for large-scale production. Potential approaches to large-scale production include development of aseptic bioreactors, controlled environment greenhouses and hydroponic systems. The production of cloned St. John's wort plantlets in many different systems including hydroponic greenhouse systems (Murch et al., 2002), various bioreactors (Zobayed et al., 2004), controlled environment growth chambers (Zobayed and Saxena, 2003) has been reported previously. (Zobayed et al. 2005) have grown St. John's wort in growth rooms which allow manipulation of many physical and chemical growth conditions such as light, temperature and nutrition that influence chemical profiles. While the comparison between field and greenhouse production may be complicated due to many factors including the differences in the source material of and the environmental and geographic parameters, the evidence that St. John's wort produced in controlled environment systems is of equal or better quality than field-grown tissues is growing consistently (Murch et al., 2002; Murch and Saxena, 2006; Zobayed et al., 2005; Mosaleeyanon et al., 2005). Greenhouse grown medicinal plants are relatively new to plant based medicines, but the technology is likely to play a significant role in the production of high quality, chemically-consistent plant material as shown for Saint John's wort (Murch et al., 2002) and Echinacea species (Zheng et al., 2006).

10. CONCLUDING REMARKS

Development of plant based medicines requires comprehensive understanding of plant systems including biological, chemical, genetic, and agronomic aspects. Chemical consistency at all stages of manufacturing processes such as extraction, stability, shelf-life and purity is of utmost importance to ensure medicinal efficacy and consumer safety. Unfortunately, there is a lack of solid scientific evidence with respect to the unique physiology of medicinal plants and their bioactive constituents. It is also important to recognize that medicinal plants, like any other plant or animal species, have co-evolved with other biological entities. The plant historically described as medicinal may have partially or fully lost the spectrum of metabolites responsible for the mechanism of disease prevention and cure. Environmental changes over time, unfortunately most of them devastating to the natural ecosystems, are likely to be a major cause of the adaptations and concomitant changes in the medicinal plant metabolism. The unique natural diversity of medicinal plants is being continually lost with consistent rise in demand resulting in indiscriminate harvest and commercialization requiring vegetative cloning. The rituals, respect, and restrictions associated with medicinal plant conservation and their compassionate use have been lost in translation of indigenous wisdom by the corporate minds. Many of the seemingly trivial rituals such as picking medicinal flowers on specific days or hours of the day are reflective of the intrinsic indigenous knowledge of plant chemistry. Clearly, complete chemical profiling of medicinal plant tissues is crucial to assess the safety and efficacy plant-based medicinal products.

There has been remarkable progress in our understanding of the biochemistry of medicinal plants due to the development of new analytical tools such as metabolomics, proteomics, and genomics. These newly designed biochemical and molecular techniques will facilitate discovery of new compounds in medicinal plants, and more interestingly, may identify compounds of medicinal efficacy in yet unexploited, commonly used plant species. As well, there has been tremendous progress in technologies for controlled environment production of whole plants and specific plant cell, tissues, and organs. Together, the application of such tools with controlled plant production techniques provides a framework for understanding the biology of medicinal plants and their use as medicines.

ACKNOWLEDGEMENTS

This research was supported by the Natural Sciences and Engineering Research Council of Canada.

REFERENCES

Aharoni, A, Ric de Vos, CH, Verhoeven, HA, Maliepaard, CA, Kruppa, G, Bino, R, and Goodenowe, DB 2002, Nontargeted Metabolome Analysis by Use of Fourier Transform Ion Cyclotron Mass Spectrometry. *OMICS* 6: 217–34.

American Botanical Council 1997, American Herbal Pharmacoepea and Therapeutic Compendium HerbalGram.

Aizenman, BY 1969, Antibiotic preparations from St. John's wort (*Hypericum perforatum*). *Mikrobiol ZH* 31: 128.

Akhtardzhiev, KH, Koleva, M, Kitanov, G, Ninov, S 1984, Pharmacognostic study of representatives of *Arum, Althaea* and *Hypericum* species. *Farmatsiya (SOFIA)* 34: 1–6.

Barcaccia, G, Arzenton, F, Sharbel, F, Varotto, S , Parrini, P, and Lucchin, M 2006, Genetic diversity and reproductive biology in ecotypes of the facultative apomict Hypericum perforatum L Heredity 96, 322–334.

Baser, KHC, Ozek, T, Nuriddinov, HR, and Demirci, AB 2002, Essential oils of two Hypericum species from Uzbekistan. *Chem Nat Comp* 38: 54–57.

Bergonzi, MC, Bilia, AR, Galloris, S, Guerrini, D, and Vincieri, FF 2001, Variability in the content of the constituents of Hypericum perforatum L and some commercial extracts. *Drug Dev Ind Pharm* 27: 491–497.

Bilia, AR, Gallori, S, and Vincieri, FF 2002, St John's wort and depression – Efficacy, safety and tolerability – an update. *Life Sci* 70: 3077–3096.

Biochinov, A, Drumev, D, Gakhniyan, R, and Akhtardzhiev, K 1965, Bioflavonoid from Hypericum perforatum possessing vitamin P activity. *Farmatsiya (SOFIA)* 15: 92.

Boutin, JA, Audinot, V, Ferry, G, and Delagrange, P 2005, Molecular tools to study melatonin pathways and actions. *Trends Pharmacol Sci* 26: 412–419.

Brantner, A, Kartnig, T, Quehenberger, F 1994, Comparative phytochemical investigation of Hypericum perforatum L and Hypericum maculatum Crantz. *Sci Pharm* 62: 261–276.

Briskin, DP, Gawienowski, MC 2001, Differential effects of light and nitrogen on production of hypericins and leaf glands in *Hypericum perforatum. Plant Physiol Biochem* 39: 1075–1081.

Brolis, M, Gabetta, B, Fuzzati, N, Pace, R, Panzeri, F, and Peterlongo, F 1998, HPLC-DAD-MS identification and HPLC-UV quantification of the major constituents of Hypericum perforatum Fitoterapia 69: 27–28.

Brondz, I, Greibrokk, T, Aasen, AJ 1983, N-1-Alkanols of *Hypericum perforatum J Nat Prod* 46: 940.

Buter, B, Orlacchio, C, Soldati, A, Berger, K 1998, Significance of genetic and environmental aspects in the field cultivation of *Hypericum perforatum. Planta Med* 64: 431–437.

Butterweck, V, Jurgenliemk, G, Nahrstedt, A, Winterhoff, H 2000, Flavonoids from *Hypericum perforatum* show antidepressant activity in the forced swimming test. *Planta Med* 66: 3–6.

Cellárová, E 2003, Culture and Biotechnology of Hypericum. Hypericum The genus Hypericum. E Ernst. New York, Taylor and Francis 31: 65–76.

Cakir, A, Duru, ME, Harmandar, M, Ciriminna, R, Passannanti, S, Piozzi, F 1997, Comparison of the volatile oils of *Hypericum scabrum* L and *Hypericum perforatum* L from Turkey. *Flavour* FragJ 12: 285–287.

Chaplinskaya, 1959, The determination of choline-like substances in the herb of *Hypericum perforatum. Aptechn Delo* 81: 58.

Chatterjee, SS, Noldner, M, Koch, E, Erdelmeier, C 1998, Antidepressant activity of *Hypericum perforatum* and hyperforin: The neglected possibility. *Pharmacopsychiatry Suppl* 31: 7–15.

Costes, C 1967, Carotenoid pigments in the flowers and petals of *Hypericum perforatum. Ann Physiol Veg* 9 2: 157–177.

Couceiro, MA, Afreen, F, Zobayed, SMA, and Kozai, T 2006, Variation in concentrations of major or bioactive compounds of St. John's wort: Effects of harvesting time, temperature and germplasm. *Plant Sci* 170:1 128–134.

De Shan, M, Hu, LH, Chen, ZL 2000, Pyrohyperforin, a new prenylated phlorogluinol from *Hypericum perforatum. Chin Chem Lett* 11 : 701–704.

De Shan, M, Hu, LH, Chen, ZL 2001, Three new hyperforin analogues from *Hypericum perforatum*. *J Nat Prod* 64: 127–130.

De Shan, M, An, TY, Hu, LH, Chen, ZL 2004, Diterpene derivative and chromone from *Hypericum perforatum Nat Prod Res* 18 : 15–19.

Dias, ACP, Tomas-Barberan, FA, Fernandes-Ferreir, M, Ferreres, F 1998, Unusual flavonoids produced by callus of *Hypericum perforatum*. *Phytochem* 48: 1165–1168.

Dias, ACP, Seabra, RM, Andrade, IB, Ferreres, F, Ferreira, MF 2001, Xanthone production in calli and suspended cells of *Hypericum perforatum*. *J Plant Physiol* 158 : 821–827.

Dona, M, Aica, ID, Pezzato, E, Sartor, L, Calabrese, F, Barbera, MD, Donella-Deana, A, Appendino, G, Borsarinin, A, Caniato, R, Garbisa, S 2004, Hyperforin inhibits cancer invasion and metastasis. *Can Res* 64: 6225–6232.

Dorossiev, I 1985, Determination of flavonoids in *Hypericum perforatum*. *Pharmazie* 40: 585–586.

Draves, AH, Walker, SE 2003, Analysis of the hypericin and pseudohypericin content of commercially available St John's wort preparations. *Can J Clin Pharmacol* 10:114–118.

Erken, S, Malyer, H, Demirei, F, Demirci, B, Baser, KHC 2001, Chemical investigations of one *Hypericum* species growing in turkey 1. *Chem Nat Comp* 37: 434–438.

Franke R, Schenk R, Bauermann U 1997. Variability in *Hypericum perforatum* L. breeding lines. *Acta Hort (ISHS)* 502:167–174.

Fiehn, O 2001, Combining genomics, metabolome analysis, and biochemical modeling to understand metabolic networks. *Comparative Funct Genom* 2 : 155–168.

Gaedcke, F 2003. St John's wort herb extracts: manufacturing, standardization and characterization. In: Hypericum The genus Hypericum. E Ernst. New York, Taylor and Francis 31: 49–64.

Giese, AC 1980, Hypericism (Review). *Photochem. Photobiol Rev* 5: 229–255.

Grujjic-Vasic, J, Bosnic, T, Jovanovic, M 1986. The examining of isolated tannins and their astringent effect. *Planta Med* 52: 548–554.

Gurevich, AI, Dobrynin, VN, Kolosov, MN, Popravko, SA, Ryabova, ID, Chernov, BK, Derbentseva, NA, Aizenman, BE, Garagulya, AD 1971, Hyperforin, an antibiotic from *Hypericum perforatum*. *Antibiotiki* 16 : 510.

Haeberlein, H, Tschiersch, KP, Stock, S, Hoelzl, J 1992, St John's wort (*Hypericum perforatum* L) Part I. Identification of an additional naphthodianthrone. *Pz Wiss* 5: 169–174.

Hall, RD 2006, Plant metabolomics: from holistic hope, to hype, to hot topic. *New Phytologist* 169: 453–468.

Hall, RD, Beale, M, Fiehn, O, Hardy, N, Sumner, L, Bino, R 2002, Plant metabolomics: The missing link in functional genomics strategies. *Plant Cell* 14:1437–1440.

Holzl, J, Ostrowski, E 1987, HPLC analysis of the constituents and the variety in populations. *Dtsch Apoth ZTG* 127: 1227–1230

Hooper, SN, Chandler, RF 1984, Herbal remedies of the maritime Indians: Phytosterols and triterpenes of 67 plants. *J Ethnopharmacol* 102: 181–194.

Kovac-Besovic, EE, Imamovic, J 2000, Determination of hypericin and rutin in St John's wort, Hypericum perforatum L by thin layer chromatography. *Farm Glas* 56: 37–43.

Jamsen, SJ, Kemsem, AG, Cornett, C, Bjemsdottir, I, Taylor, S, Wright, B, Wilson, ID 1999, High-performance liquid chromatography on line coupled to high field NMR and mass spectrometry for structure elucidation of constituents of *Hypericum perforatum* L *Anal Chem* 71 : 5235–5241.

Jela, GV, Tamara, B 1981, Study of plant oxyaromatic acids Arch Farm 31: 273–278.

Jensen, LB, Miller, WA 1951. Antibiotic from St John's wort as a food preservative. PATENT-US-2,550,266

Jurgenliem, G, Nahrstedt, A 2002, Phenolic compounds from *Hypericum perforatum Planta Med* 68 : 88–91.

Kartnig, T, Heydel, B 1993, Effects of visible and ultraviolet light on the production of hypericins and flavonoids in cell cultures of Hypericum perforatum Planta Med Suppl 59: A654.

Khalmatov, KK 1957, Quantitative changes in tannin substances and in ascorbic acid in plants. *Dokl Akad Nauk Uzb* 14: 35.

Khosa, RL, Bhatia, N 1982, Antifungal effect of *Hypericum perforatum* Linn. *J Sci REs Pl Med* 3: 49–50.

Kitanov, G 1983, Determination of the absolute configuration of catechins isolated from *Hypericum perforatum*. *Farmatsiya* 33: 19–22.

Kitanov, GM, Blinova, KF 1987, Modern state of the chemical study of species of the genus Hypericum. *Chem Nat Comp* 2: 151–166.

Kitanov, GM, Nedialkov, PT 1998, Mangiferin and isomangiferin in some Hypericum species. *Biochem Syst Ecol* 26: 647–653.

Koperdakova, J, Brutovska, R, Cellarova, E 2004, Reproduction pathway analysis of several *Hypericum perforatum* L somaclonal families. *Hereditas* 140: 34–41.

Kovac-Besovic, EE, Imamovic, J 2000, Determination of hypericin and rutin in StJohn's wort, *Hypericum perforatum* L by thin layer chromatography. *Farm Glas* 56: 37–43.

Linde, K, Berner, M, Egger, M, Mulrow, C 2005, St John's wort for depression – Meta-analysis of randomised controlled trials. *Brit J Psychiatry* 186: 99–107.

Liu, CZ, Murch, SJ, Jain, JC and Saxena, PK 2004, *In vitro* regeneration for germplasm conservation and elimination of heavy metal contamination of Goldenseal (*Hydrastis canadensis* L) In Vitro Cell. *Dev Biol* 40:75–79.

Maksyutina, NP, Kolesnikov, DG 1964, Extraction of hyperin and quercetin from *Hypericum perforatum*. *Med Prom* 18: 41.

Maisenbacher, P, Kovar, KA 1992, Adhyperforin: A homologue of hyperforin from *Hypericum perforatum*. Planta Med 58: 291–293.

Mansky, PJ, Straus, SE, 2002, St John's wort: more implications for cancer patients. *J Natl Cancer I* 94: 16 1187–1188.

Mathijssen, RHJ, Xie, R, McLeod, HL, Verweij, J, Palmer, P, Karlsson, MO, Sparreboom, A 2002, Pharmacokinetic and pharmacogenetic analysis of irinotecan (CPT-11) given in combination with R115777. *Eur J Can* 38: S44–S44.

Mathis, C, Ourisson, G 1964, Chemotaxonomic study of the genus Hypericum V. Identification of several non-volatile constituents of *Hypericum perforatum* L *Phytochem* 3: 379.

Mathis, C, Ourisson, G 1964, Chemotaxonomic study of the genus Hypericum. Distribution of sesquiterpenes, monoterpene alcohols and saturated aldehydes from the essential oil of Hypericum. *Phytochem* 3: 377–378.

Mathis, C, Ourisson, G 1964, Chemotaxanomic study of the genus Hypericum III. The distribution of saturated hydrocarbons and monoterpenes from the essential oil of Hypericum. *Phytochem* 3: 133–141.

Meier, B 2003. Herbal medicinal products of St John's wort Manufacturing and quality control. In: Hypericum The genus Hypericum. E Ernst. New York, Taylor and Francis 31: 106–136.

Michelitsch, A, Wurglics, M, 2003, Electrochemical oxidation of hyperforin on glassy carbon electrode and determination in herbal medicinal products. *Electroanalysis* 15:9 797–802.

Mosaleeyanon, K, Zobayed, SMA, Afreen, F, Kozai, T 2005, Relationships between net photosynthetic rate and secondary metabolite contents in St John's wort Plant Sci 169:523–531.

Murch, SJ, Simmons, CB and Saxena, PK 1997, Melatonin in feverfew and other medicinal plants. *Lancet* 350: 1598 1599.

Murch, SJ, KrishnaRaj, S and Saxena, PK, 2000a, Phytopharmaceuticals: Problems, limitations and solutions. Scientific Reviews of Alternate Medicine 4:33–38.

Murch, SJ, KrishnaRaj, S and Saxena, PK, 2000b, Tryptophan is a precursor for melatonin and serotonin biosynthesis in in vitro regenerated St John's wort (Hypericum perforatum L cv Anthos) plants. *Plant Cell Reports* 19: 698 704.

Murch, SJ, KrishnaRaj, S and Saxena, PK 2000c, Phytopharmaceuticals:Mass-production, standardization and conservation. *Scientific Reviews of Alternate Medicine* 4:39–43.

Murch, SJ, Choffe, KL, Victor, JMR, Slimmon, TY, KrishnaRaj, S and Saxena, PK 2000d, Thidiazuron-induced regeneration from hypocotyl cultures of St John's wort (*Hypericum perforatum* cv Anthos) *Plant Cell Reports* 19:576–581.

Murch, SJ and Saxena, PK, 2002, Mammalian neurohormones: Potential significance in sexual and asexual reproduction of St John's wort (*Hypericum perforatum* L)? *Naturwissenschaften* 89:555–560.

Murch, SJ, Rupasinghe, HPV and Saxena, PK 2002, An in vitro and hydroponic growing system for hypericin, pseudohypericin, and hyperforin production of St John's Wort (*Hypericum perforatum* L CV New Stem). *Planta Medica* 68:1–5.

Murch, SJ, Rupasinghe, HPV, Goodenowe, D and Saxena, PK 2004, A metabolomic analysis of medicinal diversity in Huang-qin (*Scutellaria baicalensis* Georgi) genotypes: discovery of novel compounds *Plant Cell Reports* 23: 419–425.

Murch, SJ, and Saxena, PK 2006, St John's wort (*Hypericum perforatum* L.): Challenges and strategies for production of chemically-consistent plants. *Can J Plant Sci* (in press)

Mueller, W.E, Schaefer, C 1996, St John's wort: In vitro investigation on Hypericum extract, hypericin and kaempferol as antidepressant. *Dtsch Apoth ZTG* 136: 17–22.

Muruganandam, AV, Ghosal, S, Bhattacharya, SK, 2000, The role of xanthones in the antidepressant activity of Hypericum perforatum involving dopaminergic and serotonergic GIC system. *Biogenic Amines* 15 5: 553–567.

National Centre for Complementary and Alternative Medicine (NC- 399 CAM), St John's Wort (Fact Sheet), National Institutes of Health, 400 Bethesda, 2001.

Nahrstedt, A, Butterweck, V 1997, Biologically active and other chemical constituents of the herb Hypericum perforatum L Pharmacopsychiatry 30 Supp 2: 129–134.

Nielsen, H, Arends, P 1978, Structure of the xanthonolignoid kielcorin. *Phytochem* 17 : 2040–2041.

Nogueira, T, Durante, F, Tavares, R, Curto, MJM, Capelo, J, Freitas, AC 1999, Comparative study of the aromas of Hypericum L species from Portugal using olfactroscopy. *Flavour Fragrance J* 14: 195–199.

Ollivier, B, Balnsard, C, Maillard, C, Vidal, E, Boudon, G 1985, Separation and identification of phenolic acids in parietary (*Parietaria officinalis* L.) and Saint John's wort (*Hypericum perforatum* L.) by HPLC and UV. *J Pharm Belg* 40: 173–177.

Omarova, MA, Artamonova, NA 1997, Liposoluble pigments from the herb *Hypericum perforatum Chem Nat Comp* 33: 691–692.

Omarova, MA, Artamonova, NA, 1999, Classes and fatty-acid composition of lipids from *Hypericum perforatum*. *Chem Nat Comp* 35: 684–685.

Pank, F, Matzk, F, Kastner, U, Bluthner, WD, de Garcia, EF, Meister, A, Ryschka, U, Schumann, G, 2003, Reproductive diversity and strategies for breeding in St John's wort (*Hypericum perforatum* L) *Euphytica* 134: 77–84.

Peebles, KA, Baker, RK, Kurz, EU, Schneider, BJ, Kroll, DJ 2001, Catalytic inhibition of human DNA topoisomerase II alpha by hypericin, a naphthodianthrone from St John's wort (*Hypericum perforatum*). *Biochem Pharmacol* 62: 1059–1070.

Perloff, MD, von Moltke, LL, Stormer, E, Shader, RI, Greenblatt, DJ 2001, Saint John's wort: An in vitro analysis of P-glycoprotein induction due to extended exposure. *Brit J Pharmacol* 134: 1601–1608.

Ploss, O, Petereit, F, Nahrstedt, A, 2001, Procyanidins from the herb of *Hypericum perforatum*. *Pharmazie* 56: 509–511.

Poutaraud, A, Gregorio, FD, Tin, VCF, Girardin, P, 2001, Effect of light on hypericins contents in fresh flowering top parts and in an extract of St John's wort (*Hypericum perforatum*). *Planta Med* 67: 254–259.

Quintanilla, AE, Ramirez, BA, Abernd, A, Pardo, ZJ, Diaz, AJ, Pamies, MD, Carrion-Gutierrez, MA, 2000, Oleoresin of *Hypericum perforatum* L. A method for obtaining said oleoresin and the use thereof. PATENT-PCT INT APPL-01 66, 121, 24.

Razhinskaite, DK, 1971, Active substances of St John's wort 2. The flavonoids and their quantitative dynamics Liet Tsr Mokslu Akad Darb Ser B 1: 89.

Reiter, RJ, Tan, DX, Burkhardt, S, Manchester, LC, 2001, Melatonin in plants. *Nutr Rev* 59 : 286–290.

Roscetti, G, Franzese, O, Comandini, A, Bonmassar, E 2004, Cytotoxic activity of *Hypericum perforatum* L. on K562 erythroleukemic cells: Differential effects between methanolic extract and hypericin Phyto Res 18 : 66–72.

Rolleri, F, 1943, Occurrence of nicotinic acid and nicotinamide in curative plants. *Arch Pharm* 281:118.

Rucker, G, Manns, D, Hartmann, R, Bonsels, U, 1995, Peroxides as constituents of plants Part 19. A C50-Hydroperoxide from *Hypericum perforatum*. *Arch Pharm (WEINHEIM)* 328 : 725–730.

Seabra, RM, Vasconcelos, MH, Costa, MAC, Alves, AC, 1992, Phenolic compounds from *Hypericum perforatum* and *H undulatum Fitoterapia* 63: 473–474.

Seidler-Lozykowska, K 2003, Secondary metabolites content of Hypericum sp. in different stages and plant parts. Hypericum The genus Hypericum. E Ernst New York, Taylor and Francis 31: 100–105.

Sirvent, TM, Krasnoff, SB, and, Gibson, D 2003, Induction of hypericins and hyperforins in *Hypericum perforatum* in response to damage by herbivores. *J Chem Ecol* 29:2667–2681.

Stojanovic, G, Palic, R, Tarr, CH, Reddy, CM, Marinkovic, O 2003, N-alkanes and fatty acids on Hypericum perforatum, Hypericum maculatum and Hypericum olympicum Biochem Syst Ecol 31: 223–226.

Szegedi, A, 2005, Acute treatment of moderate to severe depression with hypericum extract WS 55 70 (St John's wort): Randomised controlled double blind non-inferiority trial versus paroxetine 330: 503.

Trifunovic, S, Vajs, V, Macura, S, Juranic, N, Diarmati, Z, Jankov, R, Milosavljevic, S 1998, Oxidation products of hyperforin from *Hypericum perforatum. Phytochem* 49: 1305–1310.

Vajs, V, Vugdelija, S, Trifunovic, S, Karadzic, I, Juranic, N, Macura, S, Milosvljevic, S 2003, Further degradation product of hyperforin from *Hypericum perforatum* (St John's wort). *Fitoterapia* 74: 439–444.

Verotta, L, Appendino, G, Belloro, E, Jakupovic, J, Bombardelli, E 1999, Furohyperforin, a prenylated phloroglucinol from St John's wort (*Hypericum perforatum*). *J Nat Prod* 62: 770–772.

Verotta, L, Appendino, G, Jakupovic, J, Bombardeli, E 2000, Hyperforin analogues from St John's wort (*Hypericum perforatum*). *J Nat Prod* 63: 412–415.

Von Eggelkraut Gottanka, SG, Abed, SA, Muller, W, Schmidt, PC 2002, Quantitative analysis of the active components and the by-products of eight dry extracts of *Hypericum perforatum* L. (St John's wort). *Phytochem Anal* 13: 170–176.

Wagner, H, Bladt, S 1994, Pharmaceutical quality of Hypericum extracts. *J Geriat Psychiat Neurol* 7 S1: S65–S68.

Wentworth, JM, Agnostini, M, Love, J, Schwabe, JW, Chatterjee, VKK, 2000, St John's wort, a herbal antidepressant, activates the steroid X receptor. *Jour of Endocrinology* 166: R11.

Weyerstahl, P, Splittgerber, U, Marschall, H, Kaulvk, K, 1995, Constituents of the leaf essential oil of *Hypericum perforatum* L. from India. *Flavour Fragrance J* 10 6: 365–370.

Wirz, A, Simmen, U, Heilmann, J, Clais, I, Meier, B, Sticher, O, 2000, Bisanthraquinone glycosides of *Hypericum perforatum* with binding inhibition to CHR-1 receptors. *Phytochem* 55 : 941–947.

Wu, Y, Li, P, Zhou, SD 2001, Chemical constituents of *Hypericum perforatum. Chung Ts'Ao Yao* 32: 206.

Wurglics, M, Westerhoff, K, Kaunzinger, A, Baumeister, A, Dressman, J, and, Schubert-Zsilavecz, M, 2001, Batch-to-batch reproducibility of St John's wort preparations. *Pharmacopsychiatry* 34: S152–S156.

Yin, ZQ, Ye, WC, Zhao, SX 2001, Studies on chemical constituents of *Hypericum perforatum. Chung Ts'Ao Yao* 32: 487–488.

Zheng, Y, Dixon, M, Saxena, PK 2006, Greenhouse production of Echinacea purpurea (L) and E angustifolia using different growing media, NO3–/NH4+ ratios and watering regimes. *Can J Plant Sci* (In Press)

Zobayed, SMA, Murch, SJ, Rupasinghe, HPV, Saxena, PK 2003, Elevated carbon supply altered by hypericin and hyperforin contents of St John's wort (*Hypericum perforatum*) grown in bioreactors. *Plant Cell Tiss Org Cult* 5: 143–149.

Zobayed, SMA, Murch, SJ, Rupasinghe, HPV and Saxena, PK, 2004, In vitro production and chemical characterization of St John's wort (*Hypericum perforatum* L. CV New Stem) *Plant Sci* 166:333–340.

Zobayed, SMA, Saxena, PK 2004, Production of St John's wort plants under controlled environment for maximizing biomass and secondary metabolites. In Vitro Cell. *Dev Biol-Pant* 40: 108–114.

Zobayed, SMA, Afreen, F, Kozai, T 2005, Temperature stress can alter the photosynthetic efficiency and secondary metabolite concentrations in St John's wort. *Plant Phys Biochem* 43: 977–984.

Zou, YP, Lu, YH, Wei, DZ 2004, Antioxidant activity of a flavonoid-rich extract of *Hypericum perforatum* L. *in vitro J Ag Food Chem* 52:5032–5039.

INDEX